INTERNATIONAL SERIES O
ON PHYSIC

INTERNATIONAL SERIES OF MONOGRAPHS ON PHYSICS

144. T. R. Field: *Electromagnetic scattering from random media*
143. W. Götze: *Complex dynamics of glass-forming liquids – a mode-coupling theory*
142. V. M. Agranovich: *Excitations in organic solids*
141. W. T. Grandy: *Entropy and the time evolution of macroscopic systems*
140. M. Alcubierre: *Introduction to 3 + 1 numerical relativity*
139. A. L. Ivanov, S. G. Tikhodeev: *Problems of condensed matter physics – quantum coherence phenomena in electron–hole and coupled matter–light systems*
138. I. M. Vardavas, F. W. Taylor: *Radiation and climate*
137. A. F. Borghesani: *Ions and electrons in liquid helium*
136. C. Kiefer: *Quantum gravity*, Second edition
135. V. Fortov, I. Iakubov, A. Khrapak: *Physics of strongly coupled plasma*
134. G. Fredrickson: *The equilibrium theory of inhomogeneous polymers*
133. H. Suhl: *Relaxation processes in micromagnetics*
132. J. Terning: *Modern supersymmetry*
131. M. Mariño: *Chern–Simons theory, matrix models, and topological strings*
130. V. Gantmakher: *Electrons and disorder in solids*
129. W. Barford: *Electronic and optical properties of conjugated polymers*
128. R. E. Raab, O. L. de Lange: *Multipole theory in electromagnetism*
127. A. Larkin, A. Varlamov: *Theory of fluctuations in superconductors*
126. P. Goldbart, N. Goldenfeld, D. Sherrington: *Stealing the gold*
125. S. Atzeni, J. Meyer-ter-Vehn: *The physics of inertial fusion*
123. T. Fujimoto: *Plasma spectroscopy*
122. K. Fujikawa, H. Suzuki: *Path integrals and quantum anomalies*
121. T. Giamarchi: *Quantum physics in one dimension*
120. M. Warner, E. Terentjev: *Liquid crystal elastomers*
119. L. Jacak, P. Sitko, K. Wieczorek, A. Wojs: *Quantum Hall systems*
118. J. Wesson: *Tokamaks*, Third edition
117. G. E. Volovik: *The universe in a helium droplet*
116. L. Pitaevskii, S. Stringari: *Bose–Einstein condensation*
115. G. Dissertori, I. G. Knowles, M. Schmelling: *Quantum chromodynamics*
114. B. DeWitt: *The global approach to quantum field theory*
113. J. Zinn-Justin: *Quantum field theory and critical phenomena*, Fourth edition
112. R. M. Mazo: *Brownian motion – fluctuations, dynamics, and applications*
111. H. Nishimori: *Statistical physics of spin glasses and information processing – an introduction*
110. N. B. Kopnin: *Theory of nonequilibrium superconductivity*
109. A. Aharoni: *Introduction to the theory of ferromagnetism*, Second edition
108. R. Dobbs: *Helium three*
107. R. Wigmans: *Calorimetry*
106. J. Kübler: *Theory of itinerant electron magnetism*
105. Y. Kuramoto, Y. Kitaoka: *Dynamics of heavy electrons*
104. D. Bardin, G. Passarino: *The standard model in the making*
103. G. C. Branco, L. Lavoura, J. P. Silva: *CP violation*
102. T. C. Choy: *Effective medium theory*
101. H. Araki: *Mathematical theory of quantum fields*
100. L. M. Pismen: *Vortices in nonlinear fields*
99. L. Mestel: *Stellar magnetism*
98. K. H. Bennemann: *Nonlinear optics in metals*
96. M. Brambilla: *Kinetic theory of plasma waves*
94. S. Chikazumi: *Physics of ferromagnetism*
91. R. A. Bertlmann: *Anomalies in quantum field theory*
90. P. K. Gosh: *Ion traps*
88. S. L. Adler: *Quaternionic quantum mechanics and quantum fields*
87. P. S. Joshi: *Global aspects in gravitation and cosmology*
86. E. R. Pike, S. Sarkar: *The quantum theory of radiation*
83. P. G. de Gennes, J. Prost: *The physics of liquid crystals*
73. M. Doi, S. F. Edwards: *The theory of polymer dynamics*
69. S. Chandrasekhar: *The mathematical theory of black holes*
51. C. Møller: *The theory of relativity*
46. H. E. Stanley: *Introduction to phase transitions and critical phenomena*
32. A. Abragam: *Principles of nuclear magnetism*
27. P. A. M. Dirac: *Principles of quantum mechanics*
23. R. E. Peierls: *Quantum theory of solids*

Theory of fluctuations in superconductors

ANATOLY LARKIN

W.B. Fine Theoretical Physics Institute of Minnesota University, USA

and

Landau Institute for Theoretical Physics, Moscow, Russia

ANDREI VARLAMOV

COHERENTIA-INFM, UdR "Tor Vergata", Consiglio Nazionale di Ricerca, Rome, Italy

OXFORD
UNIVERSITY PRESS

OXFORD
UNIVERSITY PRESS

Great Clarendon Street, Oxford OX2 6DP

Oxford University Press is a department of the University of Oxford.
It furthers the University's objective of excellence in research, scholarship,
and education by publishing worldwide in

Oxford New York

Auckland Cape Town Dar es Salaam Hong Kong Karachi
Kuala Lumpur Madrid Melbourne Mexico City Nairobi
New Delhi Shanghai Taipei Toronto

With offices in

Argentina Austria Brazil Chile Czech Republic France Greece
Guatemala Hungary Italy Japan South Korea Poland Portugal
Singapore Switzerland Thailand Turkey Ukraine Vietnam

Oxford is a registered trade mark of Oxford University Press
in the UK and in certain other countries

Published in the United States
by Oxford University Press Inc., New York

First published in paperback 2009

ISBN 978-0-19-852815-9 (Hbk.)

ISBN 978-0-19-956483-5 (Pbk.)

Printed in Great Britain
on acid-free paper by
CPI Antony Rowe, Chippenham, Wiltshire

To our parents, in memoriam

PREFACE

The theory of phase transitions and the theory of superconductivity are the two summits of statistical physics which were conquered in the second part of the twentieth century. They are described in detail in many books. These two summits are connected by the mountain-range, which is called the theory of superconducting fluctuations. We hope that this book will serve as the guide along this mountain-range.

The storm of the first summit was started by Landau [1], who proposed his phenomenological description of the second-order phase transition that is widely known today. But in any business it is important to have certain corner stones, meaning the results, which leave no doubts about their correctness. Such a corner stone for the physics of phase transitions was provided by Onsager in 1944 by his exact solution of the two–dimensional ($2D$) Ising model [2].

We now live in an artificial world and it is often easier to make an experiment on a $2D$ object than on a $3D$ one. Moreover, the theory of phase transitions for the Ising model is now used not only in natural sciences, but also in other areas, such as money counting [3]. Forty years ago physics still was a natural science, which studied our $3D$ world. So, the majority of physicists at that time knew neither the Ising model nor the Onsager solution and believed in the Landau theory [1]. Landau himself was among the first who clearly understood that the Onsager solution gives an example of what is happening close to critical points in real materials, violating the orthodox Landau theory.

The physicists continued to climb the peak of phase transitions from two directions. The first direction was to utilize Onager's exact solution to find the general laws of phase transitions. The hypothesis of universality was formulated in the papers of Vaks et al. [4] and Kadanoff [5]. According to this hypothesis all physical systems are divided into classes of a different symmetry of the order parameter, and the critical behavior for all the systems belonging to the same class are, essentially, identical. Even more important was the hypothesis of scaling proposed by Pokrovsky and Patashinskii, Kadanoff, Gribov and Migdal (Jr.) and Polyakov [5–8]. According to this hypothesis all physical parameters close to the phase transition point are determined by a single correlation length which increases when the system approaches the phase transition point. The scaling hypothesis enabled Halperin and Hohenberg [9] to find singularities of kinetic coefficients.

The second trail to the peak began from the Landau theory and included systematic analysis of the fluctuation corrections to it. Lee and Yang [10] considered a weakly non-ideal Bose-gas, taking the interaction into account by means of perturbation theory. They have found that a first-order transition occurs in

Bose-gas. It became clear pretty soon that this result was an artifact of the perturbation theory, which does not work close to the phase transition point. The fluctuation corrections to Landau's theory not very close to the transition point, where they are still small, were found by Levanyuk [11].

The methods of quantum field theory allowed Khmelnitsky and Larkin to segregate the most divergent fluctuation contributions appearing from so-called parquet diagrams and sum up these contributions [12]. They studied phase transition in uni-axial ferroelectric or ferromagnet with dipole–dipole interactions, where the angular dependence of the interaction increased the effective dimensionality of the real $3D$ system up to four. This circumstance allowed them to solve the problem of accounting for the fluctuations close to the phase transition exactly and to find that the specific heat singularity has the form of the power of logarithm.

The paper by Larkin et al. [12] had two important appendices of a methodical nature. In one the effect of the symmetry of the order parameter on the singularity at the transition point of a non-physical $4D$ system was considered. For the order parameter being an n-component vector the specific heat has more complex singularity depending on the number of the order parameter components n. In the second appendix the same result as in the body of paper was obtained using the new to condensed matter theory, method of multiplicative renormalization group (RG). Being equivalent to the summation of parquet diagrams, this method was simpler and later it found applications in different fields of condensed matter theory.

At the same time Di Castro and Iona-Lasinio [13] also applied the general ideas of the RG method to the theory of phase transitions and showed that the presence of the fixed point in the RG equation at a finite coupling constant confirms the scaling hypothesis.

The next step in the study of the cross-over from logarithmic laws in the $4D$ case to power-laws in any lower dimension was performed for dimension $D = 4 - \varepsilon$ by Wilson and Fisher [14]. Their single-loop RG approach gave the critical indices in the leading order in ϵ. Next order corrections, $\sim \varepsilon^2$, were calculated by Wilson [15] and Abrahams and Tzuneto [16].

After these works the scaling hypothesis became a theory. Papers by Wilson [17] signified the passage through the summit. The significance of Wilson's theory goes far beyond the physics of phase transitions. He demonstrated that the renormalization of the action had better be done, while earlier people had renormalized the Green functions. After that, the renormalization group became a real working tool, which gave jobs to many theorists.

Another important summit of statistical physics is the theory of superconductivity. Between the discovery, by Kamerling–Onnes in 1911, of the phenomenon of superconductivity [18] and the formulation in 1957 of the microscopic theory [19] almost half a century passed. Let us remember that the creation of the theory of superfluidity in liquid helium took only four years.

The first theories of superconductivity were phenomenological. So in 1934 brothers F. and H. London wrote down the equations [20] for the electrodynamics of a superconductor which explained the Meissner–Ochsenfeld effect: complete expulsion of the magnetic flux from the bulk superconductor [21]. This discovery was especially important in superconductivity, namely, this effect demonstrated that superconductivity is a new state of matter. In 1936 Landau and Peierls created the theory of intermediate state [22]. Soon Landau formulated his phenomenological theory of the second-order phase transitions [1]. In 1950, Ginzburg and Landau (GL) successfully applied the ideas of this theory to superconductivity [23]. They did not have a clear understanding about the physical origin of the order parameter of such phase transition and just supposed that it was described by some complex charged field. This allowed them to derive the equations which described almost all properties of superconductors known at the moment. According to De Gennes [24] the GL theory is a bright example of the manifestation of the physical intuition. The greatest success of the GL theory became Abrikosov's explanation [25] of the Shubnikov's phase [26] where superconductivity and the magnetic field peacefully coexist. Abrikosov demonstrated that the magnetic field can penetrate into superconductor in the form of vortices ordering in the perfect lattice.

In order to create the theory of superconductivity it was enough to demonstrate how the gap appears in the excitations spectrum. Such a gap does not exist in the spectrum of the ideal electron gas. A perturbative accounting for the interaction also did not result in the gap opening. But such approximation was not really justified, the interelectron interaction in metals is not small. Therefore it was necessary to give up the ideas of the perturbation approach and to work out the theory of the strongly interacting electron system in a metal. At this time similar problems rose in high energy physics and they had been successfully resolved in the framework of the new Feynman diagrammatic technique. These ideas were transferred to statistical physics and soon the methods of quantum field theory were applied to problems of condensed matter theory.

At the beginning of the 1950s the isotopic effect was discovered in superconductors [27–29] and it became clear that for the theory of superconductivity not only was the electron system important, but the phonon one too. Migdal constructed the theory of strong electron-phonon interaction [30] but even in this way he did not succeed to find the gap in the excitation spectrum. This is due to the fact, that in all these efforts the important phenomenon of the Bose condensation had been lost. It was difficult to involve it in the theoretical models: electrons in metals obey the Fermi statistics and, in view of their strong Coulomb repulsion, it seemed there was no way to unify them into composed Bose particles.

In 1956, Cooper [31] found that it was enough to have a weak attraction between particles of the degenerate Fermi liquid to obtain the formation of the bound states, now called Cooper pairs. Soon after this discovery Bardeen, Cooper and Schrieffer proposed the microscopic theory (BCS theory) of superconductivity as the theory of the Cooper pairs Bose condensation [19].

Almost at the same time in Russia, Bogolyubov succeeded in solving the problem of superconductivity by the method of the approximate second quantization [32], and a little bit later Gor'kov proposed the solution of the problem in the framework of the Green functions formalism [33]. This method was very effective and it gave to Gor'kov the possibility of demonstrating that the phenomenological GL equations follow from the BCS theory in the limit $T \to T_c$ [34].

Both GL and BCS theories are formulated within the framework of the mean-field approximation (MFA). Here it is necessary to mention that usually the MFA allows one to get only the qualitative picture of a phenomenon. Fortunately in the case of superconductivity, this method works quantitatively. It was Ginzburg [35] who demonstrated that in clean bulk superconductors the fluctuation phenomena become important only in a very narrow ($\sim 10^{-12}\,K$) region in the vicinity of the transition temperature. Aslamazov and Larkin [36] demonstrated that the fluctuation region in dirty superconducting films is determined by the resistance per film unit square and could be much wider than in bulk samples. Even more importantly they demonstrated the presence of fluctuation effects beyond the critical region, and not only in thermodynamic but in kinetic characteristics of superconductors too. They have discovered the phenomenon which is called *paraconductivity* today: the decrease of the resistance of superconductor in the normal phase, still at $T > T_c$. Simultaneously this phenomenon was experimentally observed by Glover [37] and his results were found in perfect agreement with the Aslamazov–Larkin (AL) theory. Since this time the variety of fluctuation effects have been discovered. Their manifestation have been investigated also today, especially in new superconducting systems.

The characteristic feature of superconducting fluctuations is their strong dependence on temperature and magnetic fields in the vicinity of phase transition. This allows us to definitely separate the fluctuation effects from other contributions and to use them as the source of information about the microscopic parameters of a material. Accounting for fluctuation effects is necessary in the design of superconducting devices. Many ideas of the theory of superconducting fluctuations have been used in other fields of condensed matter theory, e.g. in developing of the theory of quantum fluctuations. Finally, the theory of fluctuations is itself simply very beautiful and we hope that the reader will enjoy it reading this book.

In the fluctuation theory, as in modern statistical physics on the whole, two methods have been mainly used: they are the diagrammatic technique and the method of functional (continual) integration over the order parameter. Each of them has its own advantages and disadvantages and in different parts of this book we will use the former or the latter.

The years of the fluctuation boom coincided with the greatest development of the diagrammatic methods of many body theory in condensed matter physics. These methods turned out to be extremely powerful: any physical problem, after its clear formulation and the writing down of the Hamiltonian, can be reduced to the summation of some classes of diagrams. The diagrammatic technique allows us in a unique way to describe the quantum and classical fluctuations, the thermodynamical, and transport effects. The diagrammatic technique is especially suited to problems containing a small parameter: in this case it is possible to restrict their summation to the ladder approximation only. In the theory of superconducting fluctuations one such small parameter exists: as we will show below, it is the so-called Ginzburg–Levanyuk number $Gi_{(D)}$ which is expressed as some power of the small parameters T_c/E_F and $\hbar/E_F\tau$, with τ as the electron scattering time. In the vicinity of transition, superconducting fluctuations influence different physical properties of metal and lead to the appearance of small corrections to corresponding physical characteristics in a wide range of temperatures. Due to the above mentioned smallness of $Gi_{(D)}$ these corrections can be evaluated quantitatively in the wide enough temperature region. On the other hand, their specific dependence on the nearness to the critical temperature $T - T_c$ allows us to separate them in experiments from other effects.

In the description of the effect of fluctuations on thermodynamic properties of the system the method of functional integration turned out to be simpler. The ladder approximation in the diagrammatic approach is equivalent to the Gaussian approximation in functional integration. The method of functional integration turns out to be more effective in the case of strong fluctuations, for instance, in the immediate vicinity of the phase transition. The final equations of the renormalization group carried out by means of functional integrations turn out to be equivalent to the result of the summation of the parquet diagrams series. Nevertheless, the former derivation is much simpler.

There is one other reason why we will use both methods and will carry out some results in both ways. In its explosive development the last decades physics became an "oral science." In the process of communication near a blackboard it is difficult to write and to read some cumbersome formulae. The language of diagrams is much more comprehensive: by drawing them the speaker demonstrates that this one is small, and that this one has to be taken into account for this and so on reasons clear to the experienced listener. The success of the diagrammatic technique, in some sense, is similar to the success of geometry in Ancient Greece, where the science was "oral" too.

This advantage of the diagrammatic technique becomes its disadvantage when there is no direct communication between the speaker and listener. It is difficult to learn the diagrammatic technique by reading a textbook on your own, when no one helps you to find the necessary insight on a complex graph. Maybe, because of similar reasons geometry disappeared in Middle Ages when direct communication between scientists was minimal while "written" algebra had continued to develop.

The layout of the book is the following. It consists of four parts. The first part is devoted to the description of the phenomenological method, while the second is devoted to the presentation of the tools of microscopic theory. These two parts are written in detail and may serve as a textbook. Two last parts can be considered as a hand book and guide to the numerous recent results of the theory of fluctuations.[1] They are devoted to the description of superconducting fluctuation manifestations of different physical properties in a variety of superconducting systems: conventional superconductors, layered superconducting systems, superconducting nanodrops, tunnel and Josephson structures, thin and nano- wires, superconductors in the Berezinskii–Kosterlitz–Thouless state, strongly disordered superconductors, etc. Here we have done our best to provide the necessary details but it is supposed that in the case of particular interest the reader will refer to the original articles cited in corresponding sections. The book is concluded by the concise presentation of the recent theories of HTS superconductivity in this or that way involving the ideas of fluctuations.

Finally, we want to acknowledge here all the people without whose inputs this book would not have appeared. These are all our coauthors of the original papers used in the book. We are specially grateful to our colleagues and friends B. Altshuler, G. Balestrino, V. Galitski, D. Geshkenbein, L. Glazman, D. Livanov, Yu. N. Ovchinnikov, R. S. Thompson, A. Rigamonti, L. Romano, V. Tognetti, collaboration and discussions with whom helped us in writing this work. We want to remember our colleague and friend, the late Lev Aslamazov. One of us (A.V.) would like to acknowledge the hospitality of the W. B. Fine Theoretical Physics Institute of Minnesota University, Landau Institute for Theoretical Physics and personally Tatiana Larkina. The authors acknowledge financial support of NSF Grant No.DRM-0439026 (USA), projects COFIN2004 and FIRB of the Italian Ministry of Science and Education, the hospitality of the International Center for Theoretical Physics (Trieste, Italy), Aspen Center for Physics (USA) and Kavli Institute for Theoretical Physics (Santa Barbara, USA).

Anatoly Larkin, Andrei Varlamov

Minneapolis – Moscow – Rome
2001–2004

[1] The total number of publications in the field of superconducting fluctuations is measured by tens of thousands, so we do not pretend in any degree to the completeness of the relative bibliography. Our references sooner have the service role, providing readers the omitted details. More references one can find in the Proceedings of relatively recent congresses devoted to fluctuation phenomena in superconductors [38, 39].

PREFACE TO THE PAPERBACK EDITION

Four years passed since the first publication of this book. Unfortunately, one of its authors, Anatoly Larkin, deceased in 2005.

For the paperback edition the book was revised, and the revealed misprints and mistakes were corrected. Chapter 4 (Chapter 10 in the hardback edition) was rewritten in view of the considerable progress achieved recently in investigations of the Nernst–Ettingshausen effect in high-temperature superconductors above critical temperature, as well as in conventional superconductors in the fluctuating regime. Chapter 11 was also extended, and a few new figures were added.

I would like to express my thanks to the colleagues who made comments about the first edition of the book, which I took into account when revising the text for the paperback edition.

Andrei Varlamov

Rome
November, 2008

CONTENTS

PART I

PHENOMENOLOGY OF FLUCTUATIONS: GINZBURG–LANDAU FORMALISM

1

INTRODUCTION

A major success in low temperature physics was achieved with the introduction by Landau of the notion of quasiparticles. According to his hypothesis, the properties of many body interacting systems at low temperatures are determined by the spectra of some low-energy, long-living excitations (quasiparticles). Another milestone of many-body theory is the mean field approximation (MFA), which allowed considerable progress to be made in the theory of phase transitions. Phenomena which cannot be described by the quasiparticle method or by MFA are usually called fluctuations. The Bardeen–Cooper–Schrieffer (BCS) theory of superconductivity is a successful example of the use of both MFA and the quasiparticle description. The success of the theory for traditional superconductors was determined by the fact that fluctuations give small corrections to the MFA results.

During the first half of the century, after the discovery of superconductivity, the problem of fluctuation smearing of the superconducting transition was not even considered. In bulk samples of traditional superconductors the critical temperature T_c sharply divides the superconducting and the normal phases. It is worth mentioning that such behavior of the physical characteristics of superconductors is in perfect agreement with both the GL phenomenological theory (1950) [23] and the BCS microscopic theory of superconductivity (1957) [19].

The characteristics of high temperature and organic superconductors, low-dimensional and amorphous superconducting systems studied today strongly differ from those of the traditional superconductors discussed in textbooks. The transitions turn out to be much more smeared out. The appearance of superconducting fluctuations above the critical temperature leads to precursor effects of the superconducting phase occurring while the system is still in the normal phase, sometimes far from T_c. The conductivity, the heat capacity, the diamagnetic susceptibility, the sound attenuation, etc. may increase considerably in the vicinity of the transition temperature.

The first numerical estimation of the fluctuation contribution to the heat capacity of a superconductor in the vicinity of T_c was done by Ginzburg in 1960 [35]. In that paper he showed that superconducting fluctuations increase the heat capacity even above T_c. In this way fluctuations change the temperature dependence of the specific heat in the vicinity of the critical temperature where, according to the phenomenological Landau theory of second-order phase transitions, a jump should take place. The range of temperatures where the fluctuation correction to the heat capacity of a bulk, clean, conventional superconductor is relevant was estimated by Ginzburg[2] to be

$$Gi = \frac{\delta T}{T_c} \sim \left(\frac{T_c}{E_F}\right)^4 \sim 10^{-12} \div 10^{-14}, \tag{1.1}$$

where E_F is the Fermi energy. The correction occurs in a temperature range δT many orders of magnitude smaller than that accessible in real experiments.

In the 1950s and 1960s the formulation of the microscopic theory of superconductivity, the theory of type-II superconductors, and the search for high-T_c superconductivity attracted the attention of researchers to dirty systems, superconducting films and filaments. In 1968, in papers by Aslamazov and Larkin [36], and Maki [40], and a little later in a paper by Thompson [41], the fundament of the microscopic theory of fluctuations in the normal phase of a superconductor in the vicinity of the critical temperature were formulated. This microscopic approach confirmed Ginzburg's evaluation [35] for the width of the fluctuation region in a bulk clean superconductor. Moreover, it was found that the fluctuation effects increase drastically in thin dirty superconducting films and whiskers. In the cited papers it was demonstrated that fluctuations affect not only the thermodynamical properties of a superconductor but its dynamics too. Simultaneously the fluctuation smearing of the resistive transition in bismuth amorphous films was found experimentally by Glover [37], and it was perfectly fitted by the microscopic theory.

In the BCS theory [19] only the Cooper pairs forming a Bose-condensate are considered. Fluctuation theory deals with the Cooper pairs out of the condensate. In some phenomena these fluctuation Cooper pairs behave similarly to quasiparticles but with one important difference. While for the well defined quasiparticle the energy has to be much larger than its inverse life time, for the fluctuation Cooper pairs the "binding energy" E_0 turns out to be of the same order. The Cooper pair life time τ_{GL} is determined by its decay into two free electrons. Evidently, at the transition temperature the Cooper pairs start to condense and $\tau_{\mathrm{GL}} = \infty$. Therefore it is natural to suppose from dimensional analysis that $\tau_{\mathrm{GL}} \sim \hbar / k_B(T - T_c)$. The microscopic theory confirms this hypothesis and gives the exact coefficient:

$$\tau_{\mathrm{GL}} = \frac{\pi\hbar}{8k_B(T - T_c)}. \tag{1.2}$$

Another important difference of the fluctuation Cooper pairs from quasiparticles lies in their large size $\xi(T)$. This size is determined by the distance by which the electrons forming the fluctuation Cooper pair move apart during the pair

[2]The expression for the width of the strong fluctuation region in terms of the Landau phenomenological theory of phase transitions was obtained by Levanyuk [11]. So in the modern theory of phase transitions the relative temperature width of the fluctuation region is called the Ginzburg–Levanyuk parameter $Gi_{(D)}$, where D is the effective sample dimensionality.

lifetime τ_{GL}. In the case of an impure superconductor the electron motion is diffusive with the diffusion coefficient $\mathcal{D} \sim v_F^2 \tau$ (τ is the electron scattering time[3]), and $\xi_d(T) = \sqrt{\mathcal{D}\tau_{\text{GL}}} \sim v_F \sqrt{\tau \tau_{\text{GL}}}$. In the case of a clean superconductor, where $k_B T \tau \gg \hbar$, impurity scattering no longer affects the electron correlations. In this case the time of electron ballistic motion turns out to be less than the electron–impurity scattering time τ and is determined by the uncertainty principle: $\tau_{bal} \sim \hbar / k_B T$. Then this time has to be used in this case for the determination of the effective size instead of τ: $\xi_{\text{c}}(T) \sim v_F \sqrt{\hbar \tau_{\text{GL}} / k_B T}$. In both cases the coherence length grows with the approach to the critical temperature as $\epsilon^{-1/2}$, where

$$\epsilon = \ln \frac{T}{T_{\text{c}}} \approx \frac{T - T_{\text{c}}}{T_{\text{c}}} \tag{1.3}$$

is the reduced temperature.[4] We will write down coherence length in the unique way ($\xi = \xi_{c,d}$):

$$\xi(T) = \frac{\xi}{\sqrt{\epsilon}}. \tag{1.4}$$

The microscopic theory in the case of an isotropic Fermi surface gives for the coherence length ξ the precise expression (see Appendix A).

Finally it is necessary to recognize that fluctuation Cooper pairs can really be treated as classical objects, but that these objects instead of Boltzmann particles appear as classical fields in the sense of Rayleigh–Jeans. This means that in the general Bose–Einstein distribution function only small energies $\mathcal{E}(p)$ are involved and the exponent can be expanded:

$$n(p) = \frac{1}{\exp(\mathcal{E}(p)/k_B T) - 1} = \frac{k_B T}{\mathcal{E}(p)}. \tag{1.5}$$

That is why the more appropriate tool to study fluctuation phenomena is not the Boltzmann transport equation but the GL equation for classical fields. Nevertheless, at the qualitative level the treatment of fluctuation Cooper pairs as particles with the concentration $n_s^{(D)} = \int n(p) d^D p / (2\pi\hbar)^D$ often turns out to be useful.[5]

Below it will be demonstrated, in the framework of both the phenomenological GL theory and the microscopic BCS theory, that in the vicinity of the transition

[3]Strictly speaking τ in the majority of future results should be understood as the electron transport scattering time τ_{tr}. Nevertheless, as is well known, in the case of isotropic scattering these values coincide; so for the sake of simplicity we will use hereafter the symbol τ.

[4]In the literature the reduced temperature is often denoted as τ. We will use here the notation ϵ since the symbol τ is used to represent the relaxation time.

[5]This particle density is defined in the (D)-dimensional space. This means that it determines the normal volume density of pairs in the $3D$ case, the density per square unit in the $2D$ case and the number of pairs per unit length in $1D$. The real $3D$ concentration n can be defined too: $n = n_s^{(2)}/d$, where d is the thickness of the film and $n = n_s^{(1)}/S$, where S is the wire cross-section.

$$\mathcal{E}(p) = \alpha k_B(T - T_c) + \frac{\mathbf{p}^2}{2m^*} = \frac{1}{2m^*}\left[\hbar^2/\xi^2(T) + \mathbf{p}^2\right]. \qquad (1.6)$$

Far from the transition temperature the dependence $n(p)$ turns out to be more sophisticated than (1.5), nevertheless one can always write it in the form

$$n(p) = \frac{m^* k_B T}{\hbar^2}\xi^2(T) f\left(\frac{\xi(T)p}{\hbar}\right). \qquad (1.7)$$

In classical field theory the notions of the particle distribution function $n(p)$ (proportional to $\mathcal{E}^{-1}(p)$ in our case) and Cooper pair mass m^* are poorly determined. At the same time, the characteristic value of the Cooper pair center of mass momentum can be defined and it turns out to be of the order of $p_0 \sim \hbar/\xi(T)$. So for the combination $m^*\mathcal{E}(p_0)$ one can write $m^*\mathcal{E}(p_0) \sim p_0^2 \sim \hbar^2/\xi^2(T)$. In fact, the particles's density enters into many physical values in the combination n/m^*. As a consequence of the above observation it can be expressed in terms of the coherence length:

$$\frac{n_s^{(D)}}{m^*} = \frac{k_B T}{m^*\mathcal{E}(p_0)}\left(\frac{p_0}{\hbar}\right)^D \sim \frac{k_B T}{\hbar^2}\xi^{2-D}(T). \qquad (1.8)$$

p_0^D here estimates the result of momentum integration.

For example, we can evaluate the fluctuation Cooper pair's contribution to conductivity by using the Drude formula:

$$\sigma = \frac{n_s^{(D)} e^2 \tau}{m^*} \Rightarrow \frac{k_B T}{\hbar^2} d^{D-3}\xi^{2-D}(T)(2e)^2\tau_{\mathrm{GL}}(\epsilon) \sim \epsilon^{D/2-2}. \qquad (1.9)$$

Analogously a qualitative understanding of the increase in the diamagnetic susceptibility above the critical temperature may be obtained from the well-known Langevin expression for the atomic susceptibility:[6]

$$\chi = -\frac{e^2}{c^2}\frac{n_s^{(D)}}{m^*}\left\langle R^2\right\rangle \Rightarrow -\frac{4e^2}{c^2}\frac{k_B T}{\hbar^2} d^{D-3}\xi^{4-D}(T) \sim -\epsilon^{D/2-2}. \qquad (1.10)$$

Here we used the ratio (1.8).

Besides these examples of the direct influence of fluctuations on superconducting properties, indirect manifestations by means of quantum interference in the pairing process and of renormalization of the density of one-electron states in the normal phase of a superconductor take place. These effects, being much more sophisticated, have a purely quantum nature, and in contrast to the direct Cooper pair contributions require microscopic consideration.

[6]This formula is valid for the dimensionalities $D = 2, 3$, when the fluctuation Cooper pair has the ability to "rotate" in the applied magnetic field and the average square of the rotation radius is $< R^2 >\sim \xi^2(T)$. "Size" effects, important for low-dimensional samples, will be discussed later on.

2

FLUCTUATION THERMODYNAMICS

2.1 Ginzburg–Landau theory

2.1.1 *GL functional*

Let us consider the model of metal being close to transition to the superconducting state. The complete description of its thermodynamic properties can be done through the calculation of the partition function:[7]

$$Z = \operatorname{tr}\left\{\exp\left(-\frac{\widehat{\mathcal{H}}}{T}\right)\right\}. \tag{2.1}$$

As discussed in the Introduction, in the vicinity of the superconducting transition, side by side with the fermionic electron excitations, fluctuation Cooper pairs of a bosonic nature appear in the system. As already mentioned, they can be described by means of classical bosonic complex fields $\Psi(\mathbf{r})$ which can be treated as "Cooper pair wave functions". Therefore the calculation of the trace in (2.1) can be separated into a summation over the "fast" electron degrees of freedom and a further functional integration carried out over all possible configurations of the "steady flow" Cooper pairs wave functions:

$$Z = \int \mathfrak{D}^2\Psi(\mathbf{r})\mathcal{Z}[\Psi(\mathbf{r})], \tag{2.2}$$

where

$$\mathcal{Z}[\Psi(\mathbf{r})] = \exp\left(-\frac{\mathcal{F}[\Psi(\mathbf{r})]}{T}\right) \tag{2.3}$$

is the system partition function in a fixed bosonic field $\Psi(\mathbf{r})$, already summed over the electronic degrees of freedom.

The "steady flow" of wave functions means that they are supposed to vary over a scale much larger than the interatomic distances. The classical part of the Hamiltonian, dependent on bosonic fields, may be chosen in the spirit of the Landau theory of phase transitions. However, in view of the space dependence of wave functions, Ginzburg and Landau included in it additionally the first nonvanishing term of the expansion over the gradient of the fluctuation field. Symmetry analysis shows that it should be quadratic. The weakness of the field coordinate dependence allows us to omit the high order terms of such

[7] Hereafter $\hbar = k_B = c = 1$.

an expansion. Therefore, the classical part of the Hamiltonian of a metal close to superconducting transition related to the presence of the fluctuation Cooper pairs in it (so-called GL functional) can be written as:[8]

$$\mathcal{F}[\Psi(\mathbf{r})] = F_N + \int dV \left\{ a|\Psi(\mathbf{r})|^2 + \frac{b}{2}|\Psi(\mathbf{r})|^4 + \frac{1}{4m}|\nabla\Psi(\mathbf{r})|^2 \right\}. \tag{2.4}$$

Let us discuss the coefficients of this functional. In accordance with the Landau hypothesis, the coefficient a goes to zero at the transition point T_c and depends linearly on $T - T_c$. Then $a = \alpha T_c \epsilon$; all the coefficients α, b and m are supposed to be positive and temperature independent. Concerning the magnitude of the coefficients, it is necessary to make the following comment. One of these coefficients can always be chosen arbitrarily: this option is related to the arbitrariness of the Cooper pair wave function normalization. Nevertheless, the product of two of them is fixed by dimensional analysis: $ma \sim \xi^{-2}(T)$. Another combination of the coefficients, independent of the wave function normalization and temperature, is α^2/b. One can see that it has the dimensionality of the density of states (DOS). Since these coefficients were obtained by a summation over the electronic degrees of freedom, the only reasonable candidate for this value is the one electron DOS ν (for one spin at the Fermi level). One can notice that the arbitrariness of the order parameter amplitude results in the ambiguity in the choice of the Cooper pair mass, introduced in (2.4) as $2m$. Indeed, this value enters in (2.6) as the product with the coefficient α, hence one of these parameters has to be set down.

In the phenomenological GL theory normalization of the order parameter Ψ is usually chosen in such a way that the coefficient m corresponds to the free electron mass. At that the coefficient α for D-dimensional clean superconductor is determined by the expression

$$\alpha_{(D)} = \frac{2D\pi^2}{7\zeta(3)} \frac{T_c}{E_F}. \tag{2.5}$$

Yet, the other normalization when the order parameter, denoted as $\Delta(\mathbf{r})$, coincides with the value of the gap in spectrum of one-particle excitations of a homogeneous superconductor turns out to be more convenient. As it will be shown below, in vicinity of T_c the microscopic theory allows to present the free energy of superconductor in the form of the GL expansion namely over the powers of $\Delta(\mathbf{r})$. At that turn out to be defined also the exact values of the coefficients α and b:

$$4m\alpha T_c = \xi^{-2}; \alpha^2/b = \frac{8\pi^2}{7\zeta(3)}\nu, \tag{2.6}$$

where $\zeta(x)$ is the Riemann zeta function, $\zeta(3) = 1.202$.

[8]For simplicity in this subsection the magnetic field is assumed to be zero.

Let us stress that at such choice of the order parameter normalization the GL parameter $C = 1/4m$ turns out to be dependent on the concentration of impurities.

2.1.2 *Heat capacity jump*

As the first step in the Landau theory of phase transitions Ψ is supposed to be independent of position. This assumption in the limit of sufficiently large volume V of the system allows us a calculation of the functional integral in (2.2) by the method of steepest descent. Its saddle point determines the equilibrium value of the order parameter

$$|\widetilde{\Psi}|^2 = \begin{cases} -\alpha T_c \epsilon / b, \text{ when } \epsilon < 0, \\ 0, \text{ when } \epsilon > 0. \end{cases} \tag{2.7}$$

The part of the free energy related to the transition is determined by the minimum of the functional (2.4):

$$F = (\mathcal{F}[\Psi])_{\min} = \mathcal{F}[\widetilde{\Psi}] = \begin{cases} F_N - \frac{\alpha^2 T_c^2 \epsilon^2}{2b} V, \text{ when } \epsilon < 0, \\ F_N, \text{ when } \epsilon > 0. \end{cases} \tag{2.8}$$

From the second derivative of Eq. (2.8) one can find an expression for the jump of the specific heat capacity at the phase transition point:

$$\begin{aligned} \Delta C = C_S - C_N &= \frac{T_c}{V}\left(\frac{\partial S_S}{\partial T}\right) - \frac{T_c}{V}\left(\frac{\partial S_N}{\partial T}\right) \\ &= -\frac{1}{VT_c}\left(\frac{\partial^2 F}{\partial \epsilon^2}\right) = \frac{\alpha^2}{b} T_c = \frac{8\pi^2}{7\zeta(3)} \nu T_c. \end{aligned} \tag{2.9}$$

Let us mention that the jump of the heat capacity was obtained because of the system volume was taken to infinity first, and after this the reduced temperature ϵ was set equal to zero.

2.1.3 *GL equations*

Introduced by Ginzburg and Landau the free energy functional in the form (2.4) became the cornerstone of their phenomenological theory of superconductivity [23] which was later rederived by Gor'kov [34] in the framework of the microscopic theory [19]. Due to its relative simplicity GL theory has been successfully used until now.

Let us start from the discussion of the generalization of the functional (2.4) which is necessary to make in the presence of magnetic field. First of all the functional must be gauge invariant, therefore the momentum operator $-i\nabla$ must be substituted by its gauge invariant form $-i\nabla - 2e\mathbf{A}(\mathbf{r})$.[9] Moreover, the presence

[9] The precise value of the effective charge $e^* = 2e$ could not be determined in the framework of the GL phenomenology. It was found in the Gor'kov's microscopic rederivation of their equations [34].

of a magnetic field results in the accumulation of some residual energy of the magnetic field in the volume of superconductor. Finally, the superconductor itself interacts with the external magnetic field $\mathbf{H}$. Taking into account these three observations one can write the generalization of the functional (2.4) in the form

$$\mathcal{F}[\Psi(\mathbf{r})] = F_n + \int dV \left\{ a|\Psi(\mathbf{r})|^2 + \frac{b}{2}|\Psi(\mathbf{r})|^4 + \frac{1}{4m}|\left(-i\nabla - 2e\mathbf{A}(\mathbf{r})\right)\Psi(\mathbf{r})|^2 \right. \\ \left. + \frac{[\nabla \times \mathbf{A}(\mathbf{r})]^2}{8\pi} - \frac{\nabla \times \mathbf{A}(\mathbf{r}) \cdot \mathbf{H}}{4\pi} \right\}. \tag{2.10}$$

The equations of the GL theory of superconductivity appear as the conditions of minimization of Eq. (2.10) over $\Psi^*(\mathbf{r})$ and the vector-potential $\mathbf{A}(\mathbf{r})$. They describe the distribution of the order parameter, currents and magnetic field in superconductor being placed in the external magnetic field $\mathbf{H}$ and staying in equilibrium.

In our further presentation we will be mostly interested in accounting for fluctuations, i.e. will consider the deviations from such equilibrium. This will be performed by functional integration of the proper expressions over fluctuation fields. Nevertheless for the convenience of the further reading we will present here the derivation of the GL equations.

Let us start from the calculation of the variation of the functional (2.10) under the variation of $\Psi^*(\mathbf{r})$:

$$\delta_\Psi \mathcal{F} = \int dV \left\{ a\Psi(\mathbf{r})\delta\Psi^*(\mathbf{r}) + b\Psi(\mathbf{r})|\Psi(\mathbf{r})|^2\delta\Psi^*(\mathbf{r}) \right. \\ \left. + \frac{1}{4m}\left[i\nabla\delta\Psi^*(\mathbf{r}) - 2e\mathbf{A}(\mathbf{r})\delta\Psi^*(\mathbf{r})\right]\left[-i\nabla\Psi(\mathbf{r}) - 2e\mathbf{A}(\mathbf{r})\Psi(\mathbf{r})\right] \right\} = 0. \tag{2.11}$$

In order to move $\delta\Psi^*(\mathbf{r})$ out of the parentheses one can use the identity

$$\nabla \cdot (\delta\Psi^* \cdot \mathbf{a}) = \delta\Psi^* \nabla \cdot \mathbf{a} + \mathbf{a} \cdot \nabla\left(\delta\Psi^*(\mathbf{r})\right), \tag{2.12}$$

where $\mathbf{a} = -i\nabla\boldsymbol{\Psi}(\mathbf{r}) - 2e\mathbf{A}(\mathbf{r})\Psi(\mathbf{r})$ and to present the corresponding integral in the second line of (2.11) as

$$\int dV \left(\nabla\delta\Psi^*(\mathbf{r}) \cdot \mathbf{a}\right) = -\int dV \delta\Psi^* \nabla \cdot \mathbf{a} + \int dV \nabla \cdot \left(\delta\Psi^*(\mathbf{r}) \cdot \mathbf{a}\right). \tag{2.13}$$

The last integral can be transformed according to the Gauss theorem into a surface integral and finally one gets:

$$\delta_\Psi \mathcal{F} = \int dV \left\{ a\Psi(\mathbf{r}) + b\Psi(\mathbf{r})|\Psi(\mathbf{r})|^2 + \frac{1}{4m}\left(i\nabla - 2e\mathbf{A}(\mathbf{r})\right)^2 \gneqq(\mathbf{r}) \right\} \delta\Psi^*(\mathbf{r})$$

$$+\oint_S \{-i\nabla\Psi(\mathbf{r})-2e\mathbf{A}(\mathbf{r})\Psi(\mathbf{r})\}\,\delta\Psi^*(\mathbf{r})dS$$

This expression can be set identically equal to zero for an arbitrary $\delta\Psi^*(\mathbf{r})$ only in the case when both expressions in the parentheses are equal to zero. Such a requirement gives us the first GL equation and the boundary condition for it:

$$a\Psi(\mathbf{r})+b\Psi(\mathbf{r})|\Psi(\mathbf{r})|^2+\frac{1}{4m}\left(i\nabla-2e\mathbf{A}(\mathbf{r})\right)^2\Psi(\mathbf{r})=0, \tag{2.14}$$

$$\left[-i\nabla\boldsymbol{\Psi}(\mathbf{r})-2e\mathbf{A}(\mathbf{r})\Psi(\mathbf{r})\right]\cdot\mathbf{n}=0, \tag{2.15}$$

where $\mathbf{n}$ is the unit vector normal to the superconductor surface. The minimization of the functional (2.10) over $\Psi(\mathbf{r})$ results in the equation for $\Psi^*(\mathbf{r})$ which is complex conjugate to (2.14).

The obtained equation contains two variables: the order parameter and the vector-potential. In order to obtain the complete system of the GL differential equations one has to perform the variation procedure over the last variable $\mathbf{A}(\mathbf{r})$ that results in the Maxwell type equation relating the vector-potential with the superconducting current:

$$\nabla\times\nabla\times\mathbf{A}(\mathbf{r})=4\pi\mathbf{j}_s, \tag{2.16}$$

$$\mathbf{j}_s=-\frac{ie}{2m}\left(\Psi\nabla\Psi^*-\Psi^*\nabla\Psi\right)-\frac{2e^2}{m}|\Psi|^2\mathbf{A}. \tag{2.17}$$

In the homogeneous case the order parameter $\Psi=\widetilde{\Psi}=const$ does not depend on coordinates and the system of equations (2.16)-(2.17) is reduced to:

$$4\pi\mathbf{j}_s=\lambda^{-2}\mathbf{A}, \tag{2.18}$$

$$\lambda^{-2}=\frac{8\pi e^2}{m}|\widetilde{\Psi}|^2. \tag{2.19}$$

The value $|\widetilde{\Psi}|^2$ in this case is determined by Eq. (2.7). Choosing the coefficient α in it according to (A.12) one finds that the equilibrium value of the order parameter $|\widetilde{\Psi}|^2$ coincides with the superfluid density n_s[10] which relates the supercurrent $\mathbf{j}_s$ with the vector-potential $\mathbf{A}$ in the microscopic theory of superconductivity [19]:

$$\mathbf{j}_s=-n_s\frac{(2e)^2}{2m}\mathbf{A}. \tag{2.20}$$

Let us notice that applying the operation $\nabla\times$ to both sides of (2.16) one can obtain the London equation:

[10] the superfluid density $n_s=1/2N_s$, where N_s is the density of superconducting electrons of the London theory.

$$\nabla \times \nabla \times \mathbf{B} = \lambda^{-2}\mathbf{B}. \tag{2.21}$$

The value λ is the penetration depth of magnetic field in superconductor and it is related to the superfluid density by means of (2.20).

The system of GL equations (2.14), (2.16), (2.17) takes an especially simple form for the dimensionless order parameter, normalized on its equilibrium value in homogeneous case (2.7) $\psi(r) = \Psi(\mathbf{r})/\widetilde{\Psi} = (b/|a|)^{1/2}\,\Psi(\mathbf{r})$:

$$\xi^2(T)\left(i\nabla - \frac{2\pi}{\Phi_0}\mathbf{A}\right)^2 \psi - \psi + \psi|\psi|^2 = 0, \tag{2.22}$$

$$\nabla \times \nabla \times \mathbf{A}(\mathbf{r}) = -\frac{\Phi_0}{4\pi\lambda^2}(\psi\nabla\psi^* - \psi^*\nabla\psi) - \frac{|\psi|^2}{\lambda^2}\mathbf{A}, \tag{2.23}$$

$$\left[i\nabla + \frac{2\pi}{\Phi_0}\mathbf{A}\right]\psi \cdot \mathbf{n} = 0. \tag{2.24}$$

2.2 Fluctuation contribution to heat capacity

2.2.1 *Zero dimensionality: the exact solution*

In a system of finite volume the fluctuations smear out the jump of the heat capacity. Let us demonstrate this in the example of a small superconducting sample with the characteristic size $d \ll \xi(T)$.

In this case the space independent mode $\Psi_0 = \Psi\sqrt{V}$ defines the main contribution to the free energy:

$$Z_{(0)} = \int d^2\Psi_0 \exp\left(-\frac{\mathcal{F}[\Psi_0]}{T}\right) = \pi\int d|\Psi_0|^2 \exp\left(-\frac{(\,a|\Psi_0|^2 + \frac{b}{2V}|\Psi_0|^4)}{T}\right)$$
$$= \sqrt{\frac{\pi^3 VT}{2b}}\exp(x^2)(1-\operatorname{erf}(x))\Big|_{x=a\sqrt{\frac{V}{2bT}}}. \tag{2.25}$$

By evaluating the second derivative of this exact result [42] one can find the temperature dependence of the heat capacity of the superconducting granule (see Fig. 2.1 taken from [42]). One can see that this function is analytic in temperature, therefore fluctuations remove phase transition in the $0D$ system. The smearing of the heat capacity jump takes place in the region of temperatures in the vicinity of T_{c0} where $x \sim 1$, i.e.

$$\epsilon_{cr} = Gi_{(0)} = \frac{\sqrt{7\zeta(3)}}{2\pi}\frac{1}{\sqrt{\nu T_{c0} V}} \approx 13.3\left(\frac{T_{c0}}{E_F}\right)\sqrt{\frac{\xi_0^3}{V}}.$$

Here T_{c0} and ξ_0 are the mean field critical temperature and the zero temperature coherence length (see Appendix A) of the appropriate bulk material. It is

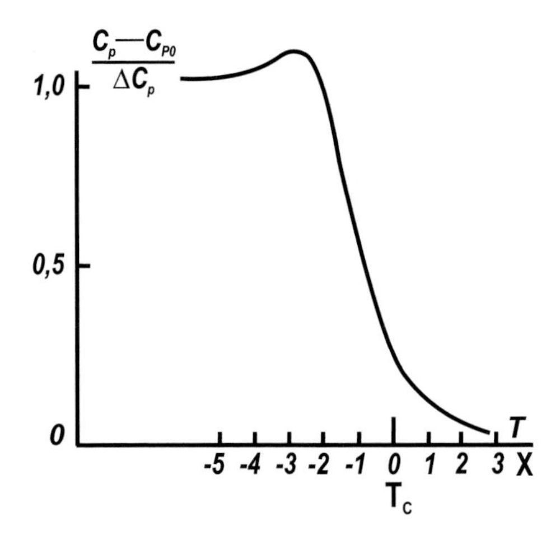

FIG. 2.1. Temperature dependence of the heat capacity of superconducting grains in the region of the critical temperature [42].

interesting that the width of this smearing does not depend on impurities concentration. From this formula one can see that the smearing of the transition is very narrow ($\epsilon_{cr} \ll 1$) when the granule volume $V \gg (\nu T_c)^{-1}$. This criterion means that the average spacing between the levels of the dimensional quantization:

$$\delta = (\nu V)^{-1} \tag{2.26}$$

still remains much less than the value of the mean field critical temperature, T_{c0}.

Far above the critical region, where $Gi_{(0)} \ll \epsilon \ll 1$, one can use the asymptotic expression for the $\mathrm{erf}(x)$ function and find

$$F_{(0)} = -T \ln Z_{(0)} = -T \ln \frac{\pi}{\alpha \epsilon}. \tag{2.27}$$

Calculation of the second derivative gives an expression for the fluctuation part of the heat capacity in this region:

$$\delta C_{(0)} = \frac{1}{V\epsilon^2}. \tag{2.28}$$

The experimental study of the heat capacity of small Sn particles in the vicinity of the transition was done in [43].

2.2.2 *Arbitrary dimensionality: case $T \geq T_c$*

Let us start with the discussion of the fluctuation contribution to the heat capacity in the normal phase of a superconductor of arbitrary dimensionality. In this case the functional integral in Eq. (2.3) with the free energy in the form (2.4) due to the presence of the depending on $\nabla \Psi$ term cannot be carried out exactly and the problem turns out to be much more difficult than in the $0D$ case discussed above.

One can estimate the fluctuation contribution to heat capacity for a specimen of an arbitrary effective dimensionality on the basis of the following observation. The volume of the specimen may be divided into regions of size $\xi(T)$, which are weakly correlated to each other. Then the whole free energy can be estimated as the free energy of one such $0D$ specimen (2.27), multiplied by their number $N_{(D)} = V\xi^{-D}(T)$:

$$F_{(D)} = -TV\xi^{-D}(T)\ln\frac{\pi}{\alpha\epsilon}. \tag{2.29}$$

This formula gives the correct temperature dependence of the free energy not too close to T_c for the specimens of the even dimensionalities. As we will demonstrate below, a more accurate treatment removes the $\ln\epsilon$ dependence from it in the case of the odd dimensions.

Let us restrict ourselves to the region of temperatures beyond the immediate vicinity of transition, where this correction is still small. In this region one can omit the fourth-order term in $\Psi(\mathbf{r})$ with respect to the quadratic one and write down the GL functional, expanding the order parameter in a Fourier series:

$$F[\Psi_{\mathbf{k}}] = F_N + \sum_{\mathbf{k}}[a + \frac{\mathbf{k}^2}{4m}]|\Psi_{\mathbf{k}}|^2 = F_N + \alpha T_c \sum_{\mathbf{k}} \left(\epsilon + \xi^2\mathbf{k}^2\right)|\Psi_{\mathbf{k}}|^2. \tag{2.30}$$

Here $\Psi_{\mathbf{k}} = \frac{1}{\sqrt{V}}\int \Psi(\mathbf{r})e^{-i\mathbf{kr}}dV$ and the summation is carried out over the wave vectors $\mathbf{k}$ (fluctuation modes). For the specimen of dimensions L_x, L_y, L_z $k_i L_i = 2\pi n_i$. The functional integral for the partition function (2.3) can be factored out to a product of Gaussian type integrals over these modes:

$$Z = \prod_{\mathbf{k}} \int d^2\Psi_{\mathbf{k}} \exp\left\{-\alpha(\epsilon + \frac{\mathbf{k}^2}{4m\alpha T_c})|\Psi_{\mathbf{k}}|^2\right\}. \tag{2.31}$$

Carrying out these integrals, one gets the fluctuation contribution to the free energy:

$$F(\epsilon > 0) = -T\ln Z = -T\sum_{\mathbf{k}} \ln\frac{\pi}{\alpha(\epsilon + \frac{\mathbf{k}^2}{4m\alpha T_c})}. \tag{2.32}$$

The appropriate correction to the heat capacity of a superconductor at temperatures above the critical temperature may thus be calculated. We are interested in the most singular term in ϵ^{-1}, therefore the differentiation over temperature can be again replaced by that over ϵ:

$$\delta C_+ = -\frac{1}{VT_c}\left(\frac{\partial^2 F}{\partial\epsilon^2}\right) = \frac{1}{V}\sum_{\mathbf{k}} \frac{1}{(\epsilon + \frac{\mathbf{k}^2}{4m\alpha T_c})^2}. \tag{2.33}$$

The result of the summation over $\mathbf{k}$ strongly depends on the linear sizes of the sample, i.e. on its effective dimensionality. As is clear from (2.33), the scale with which one has to compare these sizes is determined by the value $(4m\alpha T_c\epsilon)^{-1/2}$ which, as was already mentioned above, coincides with the effective size of the Cooper pair, $\xi(T)$. Thus, if all dimensions of the sample considerably exceed $\xi(T)$, one can integrate over $(2\pi)^{-3}L_xL_yL_zdk_xdk_ydk_z$ instead of summing over n_x, n_y, n_z. In the case of arbitrary dimensionality the fluctuation correction to the heat capacity turns out to be

$$\delta C_+ = \frac{V_D}{V}\int \frac{1}{(\epsilon + \frac{\mathbf{k}^2}{4m\alpha T_c})^2}\frac{d^D\mathbf{k}}{(2\pi)^D} = \vartheta_D\frac{V_D}{V}\frac{(4m\alpha T_c)^{\frac{D}{2}}}{\epsilon^{2-\frac{D}{2}}}, \tag{2.34}$$

where $V_D = V, S, L, 1$ for $D = 3, 2, 1, 0$. For the coefficients ϑ_D it is convenient to write an expression valid for an arbitrary dimensionality D, including fractional ones. For a space of fractional dimensionality we just mention that the momentum integration in spherical coordinates is carried out according to the rule: $\int d^D\mathbf{k}/\left(2\pi\right)^D = \mu_D\int k^{D-1}dk$, where

$$\mu_D = \frac{D}{2^D\pi^{D/2}\Gamma\left(1 + D/2\right)} \tag{2.35}$$

and $\Gamma(x)$ is a gamma-function. The coefficient ϑ_D in (2.34) can be also expressed in terms of the gamma function:

$$\vartheta_D = \frac{\Gamma(2 - D/2)}{2^D\pi^{D/2}} \tag{2.36}$$

yielding $\vartheta_1 = 1/4, \vartheta_2 = 1/4\pi$ and $\vartheta_3 = 1/8\pi$.

In the case of small particles with characteristic sizes $d \lesssim \xi(\epsilon)$ the appropriate fluctuation contribution to the free energy and the heat capacity coincides with the asymptotics of the exact results (2.27) and (2.28). From the formula given above it is easy to see that the role of fluctuations increases when the effective dimensionality of the sample or parameter $m\alpha$ decrease. As it follows from the microscopic theory such a decrease of $m\alpha$ takes place in dirty superconductors with a short electron mean free path.

2.2.3 *Arbitrary dimensionality: case $T < T_c$*

The general expressions (2.2) and (2.4) allow one to find the fluctuation contribution to heat capacity below T_c. In order to this let us restrict ourselves to the region of temperatures not very close to T_c from below, where fluctuations are sufficiently weak. The order parameter can be written as the sum of the equilibrium $\widetilde{\Psi}$ (see (2.7)) and fluctuation, $\psi(\mathbf{r})$, parts:

$$\Psi(\mathbf{r}) = \widetilde{\Psi} + \psi(\mathbf{r}). \tag{2.37}$$

Keeping in (2.4) the terms up to the second order in $\psi(\mathbf{r})$ and up to the fourth order in $\widetilde{\Psi}$, one can find

$$\mathcal{Z}[\widetilde{\Psi}] = \exp(-\frac{a\widetilde{\Psi}^2 + b/2\widetilde{\Psi}^4}{T})\prod_{\mathbf{k}}\int d\operatorname{Re}\psi_{\mathbf{k}} d\operatorname{Im}\psi_{\mathbf{k}} \times \tag{2.38}$$
$$\times \exp\left\{-\frac{1}{T}[(3b\widetilde{\Psi}^2 + a + \frac{\mathbf{k}^2}{4m})\operatorname{Re}^2\psi_{\mathbf{k}} + (b\widetilde{\Psi}^2 + a + \frac{\mathbf{k}^2}{4m})\operatorname{Im}^2\psi_{\mathbf{k}}]\right\}.$$

Let us remind, that the value $\widetilde{\Psi}$ we chose as real. Carrying out the integral over the real and imaginary parts of the order parameter one can find an expression for the fluctuation part of the free energy:

$$F = -\frac{T}{2}\sum_{\mathbf{k}}\left\{\ln\frac{\pi T_c}{3b\widetilde{\Psi}^2 + a + \frac{\mathbf{k}^2}{4m}} + \ln\frac{\pi T_c}{b\widetilde{\Psi}^2 + a + \frac{\mathbf{k}^2}{4m}}\right\}. \tag{2.39}$$

Let us discuss this result. It is valid both above and below T_c. The two terms in it correspond to the contributions of the modulus and phase fluctuations of the order parameter. Above T_c $\widetilde{\Psi} \equiv 0$ and these contributions are equal: phase and modulus fluctuations in the absence of $\widetilde{\Psi}$ represent just two equivalent degrees of freedom of the scalar, complex order parameter. Below T_c, the symmetry of the system decreases (see (2.7)). The order parameter modulus fluctuations remain of the same type as those above T_c, while the character of the phase fluctuations, in accordance with the Goldstone theorem, changes dramatically. The energy of such fluctuations tends zero when $k \to 0$ and in this limit it does not depend more on the phase ψ.

Substitution of Eq. (2.7) into Eq. (2.39) results in the disappearance of the temperature dependence of the phase fluctuation contribution and, calculating the second derivative, one sees that only the fluctuations of the order parameter modulus contribute to the heat capacity. As a result the heat capacity, calculated below T_c, turns out to be proportional to that found above:

$$\delta C_- = 2^{\frac{D}{2}-1}\delta C_+.$$

Hence, in the framework of the theory proposed we found that the heat capacity of the superconductor tends to infinity at the transition temperature. Strictly speaking, the restrictions of the above approach do not allow us to discuss seriously this divergence at the critical point itself. The calculations in principle are valid only in that region of temperatures where the fluctuation correction is small. We will discuss below the quantitative criteria for the applicability of this perturbation theory.

2.3 Fluctuation diamagnetism

2.3.1 *Qualitative preliminaries*

In this section we discuss the effect of fluctuations on the magnetization and the susceptibility of a superconductor above the transition temperature. Being the precursor effect to the Meissner diamagnetism, the fluctuation-induced magnetic susceptibility has to be a small correction with respect to the diamagnetism of a superconductor but it can be comparable to or can even exceed the value of the normal metal diamagnetic or paramagnetic susceptibility and can be easily measured experimentally. As was already mentioned in the Introduction the temperature dependence of the fluctuation induced diamagnetic susceptibility can be qualitatively analyzed on the basis of the Langevin formula, but some precautions in the case of low-dimensional samples have to be made.

As regards the $3D$ case we would like to mention here that Eq. (1.10), presented in terms of $\xi(T)$, has a wider region of applicability than the GL one. Indeed, in the GL region ($|\epsilon| \gtrsim Gi$) $\xi(\epsilon) \sim \epsilon^{-1/2}$ and such presentation is trivial. The situation changes in the immediate vicinity of the transition point where the Cooper pair interaction becomes important. The presence of fluctuation Cooper pairs themselves effects on the value of coherence length. In accordance with the scaling hypothesis (we will discuss it below) the coherence length is the only scale of length at these temperatures ($|\epsilon| \lesssim Gi$,). It still depends on temperature as a power function: $\xi(\epsilon) \sim \epsilon^{-\nu}$, but the interaction of fluctuations changes its critical exponent ν with respect to the GL value $\nu = 1/2$. The same scaling hypothesis allows us to write down for diamagnetic susceptibility the general relation

$$\chi_{(3)} \sim -e^2 T\xi(T) = -\chi_P \epsilon^{-1/2} \begin{cases} 1, \text{ when } \epsilon \gg Gi, \\ \left(\frac{\epsilon}{Gi}\right)^{1/2-\nu}, \text{ when } \epsilon \ll Gi. \end{cases} \tag{2.40}$$

Let us stress that this expression is valid also in the region of critical fluctuations, i.e. in the immediate vicinity of the transition temperature. Here, in order to define the scale of fluctuation effects, we have introduced the Pauli paramagnetic susceptibility $\chi_P = e^2 v_F / 4\pi^2$.

Moreover, the Langevin formula allows us to extend the estimation of the fluctuation diamagnetic effect to the other side beyond the GL region: to high temperatures $T \gg T_c$. The coherence length far from the transition becomes a slow function of temperature. In a clean superconductor, far from T_c, $\xi(T) \sim v_F/T$, so one can write

$$\chi_{(3c)}(T \gg T_c) \sim -e^2 T\xi(T) \sim -\chi_P \tag{2.41}$$

and see that the fluctuation diamagnetism turns out to be of the order of the Pauli paramagnetism even far from the transition. More precise microscopic calculations of $\chi_{(3)}(T \gg T_c)$ lead to the appearance of $\ln^2(T/T_c)$ in the denominator of (2.41).

In the $2D$ case, Eq. (1.10) is applicable for the estimation of $\chi_{(2)}$ in the case when the magnetic field is applied perpendicular to the plane, allowing $2D$ rotations of fluctuation Cooper pairs in it:

$$\chi_{(2c)}(T) \sim e^2 \frac{n}{m} < R^2 > \sim e^2 T \xi^2(T) \sim -\chi_P \frac{E_F}{T - T_c}. \tag{2.42}$$

This result is valid for a wide range of temperatures and exceeds the Pauli paramagnetism by factor E_F/T even far from the critical temperature (we consider the clean case here).

For a thin film ($d \ll \xi(T)$) placed perpendicular to the magnetic field the fluctuation Cooper pairs behave like effective $2D$ rotators. The formula (1.10) still can be used, though one has to take into account that the susceptibility in this case is calculated per unit square of the film. So for the realistic case (from the experimental point of view) of the dirty film, one has to just use in (2.42) the expression (A.4) for the coherence length:

$$\chi_{(2d)} \sim \frac{e^2 T}{d}\xi^2(T) \sim -\chi_P\left(\frac{l}{d}\right)\frac{T_c}{T-T_c}. \tag{2.43}$$

Let us discuss now the important case of a layered superconductor (e.g. example, a HTS). It is usually supposed that the electrons move freely in conducting planes separated by a distance s. Electron motion in the perpendicular direction has a tunneling character, with effective energy J. The related velocity and coherence length can be estimated as $v_z = \partial E(\mathbf{p})/\partial p_\perp \sim J/p_\perp \sim sJ$ and $\xi_{z,(c)} \sim sJ/T$ for the clean case. In the dirty case the anisotropy can be taken into account in the spirit of formula (A.4) yielding $\xi_{z,(d)} \sim \sqrt{\mathcal{D}_\perp/T} \sim sJ\sqrt{\tau/T}$.

We start from the case of a weak magnetic field applied perpendicular to layers. The effective area of a rotating fluctuation pair is $\xi_x(\epsilon)\xi_y(\epsilon)$. The density of Cooper pairs in the conducting layers (1.8) has to be modified for the anisotropic case. Its isotropic $3D$ value is proportional to $1/\xi(\epsilon)$, that now has to be read as $\sim 1/\sqrt{\xi_x(\epsilon)\xi_y(\epsilon)}$. The anisotropy of the electron motion leads to a concentration of fluctuation Cooper pairs in the conducting layers and hence, to an effective increase of the Cooper pairs density of $\sqrt{\xi_x(\epsilon)\xi_y(\epsilon)}/\xi_z(\epsilon)$ times with respect to its isotropic value. This increase is saturated when $\xi_z(\epsilon)$ reaches the interlayer distance s, so finally the anisotropy factor appears in the form $\sqrt{\xi_x(\epsilon)\xi_y(\epsilon)}/\max\{s, \xi_z(\epsilon)\}$ and the square root in its numerator is removed in the Langevin formula (1.10), rewritten for this case

$$\chi_{(layer,\perp)}(\epsilon, H \to 0) \sim -e^2 T\frac{\xi_x(\epsilon)\xi_y(\epsilon)}{\max\{s, \xi_z(\epsilon)\}}. \tag{2.44}$$

The existence of a crossover between the $2D$ and $3D$ temperature regimes in this formula is evident: as the temperature tends to T_c the diamagnetic susceptibility temperature dependence changes from $1/\epsilon$ to $1/\sqrt{\epsilon}$. This happens when the reduced temperature reaches its crossover value $\epsilon_{cr} = r$ ($\xi_z(\epsilon_{cr}) \sim s$). The anisotropy parameter

$$r = \frac{4\xi_z^2(0)}{s^2} = \frac{J^2}{T}\begin{cases} \frac{\pi\,\tau}{4}, & \text{when } T\tau \ll 1 \\ \frac{7\zeta(3)}{8\pi^2 T}, & \text{when } T\tau \gg 1 \end{cases} \tag{2.45}$$

plays an important role in the theory of layered superconductors.[11]

[11] We use here a definition of r following from microscopic theory (see section 8.3).

It is interesting to note that this intrinsic crossover, related to the spectrum anisotropy, has an opposite character to the geometric crossover which happens in thick enough films when $\xi(T)$ reaches d. In the latter case the characteristic $3D$ $1/\sqrt{\epsilon}$ dependence taking place far enough from T_c (where $\xi(T) \ll d$), is changed to the $2D$ $1/\epsilon$ law (see (2.43)) in the immediate vicinity of transition (where $\xi(T) \gg d$) [44]. It is worth mentioning that in a strongly anisotropic layered superconductor the fluctuation-induced susceptibility may considerably exceed the normal metal dia- and paramagnetic effects even relatively far from T_c [45, 46].

Let us consider a magnetic field applied along the layers. First, it is necessary to mention that the fluctuation diamagnetic effect disappears in the limit $J \sim \xi_z \to 0$. Indeed, for the formation of a circulating current it is necessary to tunnel at least twice, so

$$\chi_{(layer,\|)} \sim -e^2 T \frac{\xi_z^2}{\max\{s, \xi_z(T)\}} \sim -\chi_P(\frac{sJ}{v_F}) \frac{J/T}{\sqrt{\epsilon}\max\{\sqrt{\epsilon}, J/T\}}.$$

In the general case of an anisotropic superconductor, choosing the z-axis along the direction of magnetic field H, the following extrapolation of the results obtained may be written

$$\chi \sim -e^2 T \frac{\xi_x^2(\epsilon)\xi_y^2(\epsilon)}{\max\{\mathfrak{a}, \xi_x(\epsilon)\}\max\{\mathfrak{b}, \xi_y(\epsilon)\}\max\{\mathfrak{s}, \xi_z(\epsilon)\}}. \tag{2.46}$$

This general formula is useful for the analysis of the fluctuation diamagnetism of anisotropic superconductors or samples of some specific shape. It is also applicable to the case of a thin film ($d \ll \xi_z(\epsilon)$) placed perpendicular to the magnetic field: it is enough to replace $\xi_z(\epsilon)$ by d in (2.46). Yet the formula (2.46) cannot be applied to the cases of a thin film in parallel field, wires and granules. In those cases the Langevin formula (1.10) can still be used with the replacement of $\langle R^2 \rangle \to d^2$:

$$\chi_{(D)} \sim \chi_P \left(\frac{T}{v_F}\right) \xi^{2-D} d^{D-1} \sim \epsilon^{D/2-1}.$$

The magnetic field dependence of the fluctuation part of free energy in these cases is reduced only to accounting for the quadratic shift of the critical temperature versus magnetic field.[12]

Let us stress that Eq. (2.46) determines the magnetic susceptibility only for weak enough magnetic fields $H \ll \Phi_0/[\xi_x(\epsilon)\xi_y(\epsilon)] = H_{c2}(\epsilon) \ll H_{c2}(0)$.

[12] Let us mention that for $3D$ systems or in the case of a film placed in a perpendicular magnetic field the critical temperature depends on H linearly, while the magnetic field dependent part of the free energy for $H \ll H_{c2}^*(-\epsilon)$ (the line $H_{c2}^*(-\epsilon)$ is mirror-symmetric to the $H_{c2}(\epsilon)$ with respect to y-axis passing through $T = T_c$) is proportional to H^2.

2.3.2 Zero-dimensional diamagnetic susceptibility

For quantitative analysis of the fluctuation diamagnetism we start from the GL functional for the free energy written down in the presence of the magnetic field (see Eq. (2.10)). The fluctuation contribution to the diamagnetic susceptibility in the simplest case of a "zero-dimensional" superconductor (spherical superconducting granule of diameter $d \ll \xi(\epsilon)$) was considered by Shmidt [42]. In this case the order parameter does not depend on the space variables and the free energy can be calculated exactly for all temperatures including the critical region in the same way as was done for the case of the heat capacity in the absence of a magnetic field. Due to the small size of the granule with respect to the GL coherence length the order parameter Ψ does not depend on coordinates. Moreover, due to the smallness of the granule size with respect to the magnetic field penetration depth in superconductor λ, one can assume the equivalence of the average magnetic field in metal $\mathbf{B}$ with the external field $\mathbf{H}$. This allows us to omit the last two terms in Eq. (2.10) since in the assumed approximation they do not depend on fluctuations.

It is why formally the effect of a magnetic field in this case is reduced to the renormalization of the coefficient a, or, in other words, to the suppression of the critical temperature:

$$T_c(H) = T_c(0)(1 - \frac{4\pi^2\xi^2}{\Phi_0^2}\langle\mathbf{A}^2\rangle). \tag{2.47}$$

Here $\Phi_0 = \pi/e$ is the magnetic flux quantum and $\langle \cdots \rangle$ means the averaging over the sample volume. That is why for the granule in a magnetic field one can use the partition function in the same form (2.25) as in the absence of the field but with the renormalized GL parameter $a\,(H) = a + \frac{e^2}{m}\left\langle\mathbf{A}^2\right\rangle$:

$$\begin{aligned} Z_{(0)}\,(H) &= \pi \int d|\Psi_0|^2 \exp\left(-\frac{\left[a + \frac{e^2}{m}\left\langle\mathbf{A}^2\right\rangle\right]|\Psi_0|^2 + \frac{b}{2V}|\Psi_0|^4}{T}\right) \\ &= \sqrt{\frac{\pi^3 VT}{2b}} \exp\left[\frac{a^2\,(H)\,V}{2bT}\right]\left\{1 - \operatorname{erf}\left[a\,(H)\sqrt{\frac{V}{2bT}}\right]\right\}. \end{aligned} \tag{2.48}$$

Such a trivial dependence of the properties of $0D$ samples on the magnetic field immediately allows us to understand its effect on the heat capacity of a granular sample. Indeed, with the growth of the field the temperature dependence of the heat capacity presented in Fig. 2.1 just moves in the direction of lower temperatures.

In the GL region $Gi_{(0)} \lesssim \epsilon$ one can write the asymptotic expression (2.27) for the free energy:

$$F_{(0)}(\epsilon, H) = -T \ln \frac{\pi}{\alpha(\epsilon + \frac{4\pi^2\xi^2}{\Phi_0^2}\langle\mathbf{A}^2\rangle)}.$$

In the case of a spherical particle one has to choose the gauge of the vector-potential $\mathbf{A} = \frac{1}{2}\mathbf{H} \times \mathbf{r}$ yielding $\langle \mathbf{A}^2 \rangle = \frac{1}{40} H^2 d^2$ (calculation of this average value is completely analogous to the calculation of the moment of inertia of a solid sphere). In this way an expression for the $0D$ fluctuation magnetization valid for all fields $H \ll H_{c2}(0)$ can be found:

$$M_{(0)}(\epsilon, H) = -\frac{1}{V}\frac{\partial F_{(0)}(\epsilon, H)}{\partial H} = -\frac{6\pi T \xi^2}{5\Phi_0^2 d}\frac{H}{\left(\epsilon + \frac{\pi^2 \xi^2}{10\Phi_0^2} H^2 d^2\right)}. \tag{2.49}$$

One can see that the fluctuation magnetization turns out to be negative and linear up to some crossover field, which can be called the temperature dependent upper critical field of the granule $H_{c2(0)}(\epsilon) \sim \frac{\Phi_0}{d\xi(\epsilon)} \sim \frac{\xi}{d} H_{c2}(0)\sqrt{\epsilon}$ [13] at which it reaches a minimum. At higher fields $H_{c2(0)}(\epsilon) \lesssim H \ll H_{c2}(0)$ the fluctuation magnetization of the $0D$ granule decreases as $1/H$. In the weak field region $H \ll H_{c2(0)}(\epsilon)$ the diamagnetic susceptibility is:

$$\chi_{(0)}(\epsilon, H) = -\frac{6\pi T \xi_0^2}{5\Phi_0^2 d}\frac{1}{\epsilon} \approx -10^2 \chi_P \left(\frac{\xi}{d}\right)\frac{1}{\epsilon}$$

which coincides with our previous estimate in its temperature dependence but the numerical factor found is very large. Let us underline that the temperature dependence of the $0D$ fluctuation diamagnetic susceptibility turns out to be less singular than the $0D$ heat capacity correction: ϵ^{-1} instead of ϵ^{-2}.

The expression for the fluctuation part of free energy (2.32) is also applicable to the cases of a wire or a film placed in a parallel field: as was already mentioned above all its dependence on the magnetic field is manifested by the shift of the critical temperature (2.47). In the case of a wire in a parallel field the gauge of the vector-potential can be chosen as above what yields $\langle \mathbf{A}^2 \rangle_{(wire,\|)} = H^2 d^2/32$. For a wire in a perpendicular field, or a film in a parallel field, the gauge has to be chosen in the form $\mathbf{A} = (0, Hx, 0)$. One can find $\langle \mathbf{A}^2 \rangle_{(wire,\perp)} = H^2 d^2/16$ for a wire and $\langle \mathbf{A}^2 \rangle_{(film,\|)} = H^2 d^2/12$ for a film.

Calculating the second derivative of Eq. (2.32) with the appropriate magnetic field dependencies of the critical temperature one can find the following expressions for the diamagnetic susceptibility:

$$\chi_{(D)}(\epsilon) = -2\pi \frac{\xi T}{v_F}\chi_P \begin{cases} \frac{1}{\sqrt{\epsilon}}, & \text{wire in parallel field,} \\ \frac{2}{\sqrt{\epsilon}}, & \text{wire in perpendicular field,} \\ \frac{d}{3\xi}\ln\frac{1}{\epsilon}, & \text{film in parallel field.} \end{cases} \tag{2.50}$$

2.3.3 2D *magnetization*

Let us consider the fluctuation contribution to diamagnetic magnetization and susceptibility of the $2D$ system placed in perpendicular magnetic field. This

[13] For a spherical particle $H_{c2(0)}^{sph}(\epsilon) = \frac{\Phi_0}{\pi d \xi}\sqrt{10\epsilon}$.

problem cannot be solved exactly for the entire vicinity of the transition, like it was done above in the case of granule, but restricting consideration to the GL region one can omit the fourth order term in the free energy functional. Moreover, as long as fluctuation effects are comparatively small, the average magnetic field in the metal **B** may be assumed to be equal to the external field **H**. Thus we omit the last two terms in (2.10)and start from the formula for the free energy in the form (2.32). Such simplifications give the possibility of finding solutions valid for a wide range of magnetic fields, which demonstrates the nontrivial behavior of the fluctuation magnetization as the function of magnetic field.

Let us notice that the momentum representation of Eq. (2.32) does not fit anymore for the description of the charged particle (Cooper pair) motion in a uniform magnetic field: the most appropriate thing here turns out to be the Landau representation [47]. The summation over momenta in Eq. (2.32) is nothing other than the calculation of the trace, which is independent of the chosen representation. Therefore, one has to substitute the summation over the $2D$ momentum by that over the degenerate states of each Landau level ($k^2/4m \to \omega_c\,(n+1/2)$):

$$F(\epsilon, H) = -\frac{HS}{\Phi_0} T \sum_{n,k_z} \ln \frac{\pi T}{\alpha T_c \epsilon + \omega_c \left(n+\frac{1}{2}\right)}. \tag{2.51}$$

Here S is the sample cross-section and the degeneracy of Landau levels results in appearance of the number of particle states HS/Φ_0 with the definite quantum numbers n and k_z. For a film of thickness d the integral over k_z in Eq. (2.71) has to be replaced by a summation over the discrete k_z. When $\xi_z(T) \gg d$, only the term with $k_z = 0$ has to be taken into account, which results in the appearance of the coefficient $1/d$.

The expression (2.51) is convenient to rewrite in terms of the dimensionless magnetic field

$$h = \frac{\omega_c}{2\alpha T_c} = \frac{eH}{2m\alpha T_c} = \frac{H}{\widetilde{H}_{c2}(0)} \tag{2.52}$$

and the fluctuation part of the $2D$ superconductor free energy takes the form:

$$F_{(2)}(\epsilon, H) = -\frac{TS}{2\pi\xi_{xy}^2} h \sum_{n=0}^{n_c-1} \ln \frac{\pi}{\alpha\left[\epsilon + 2h\left(n+1/2\right)\right)]}. \tag{2.53}$$

Here we introduce the definition of the second critical field, useful for further consideration

$$\widetilde{H}_{c2}(0) = 2m\alpha T_c/e = \Phi_0/2\pi\xi^2 \tag{2.54}$$

as the linear extrapolation to zero temperature of the GL formula.[14]

[14]Let us remember that the exact definition of $H_{c2}(0)$ contains, in comparison with (2.54), the numerical coefficient $A(0)$ (see Appendix A).

The sum in (2.53) is evidently divergent and in order to regularize it we introduced a formal cut-off parameter n_c, the number of the last Landau level, at which the summation is interrupted. Its value can be evaluated by appealing to the restrictions of the GL theory, which breaks down for short wave length fluctuations and is valid for $k \lesssim k_{\max}$, where

$$k_{\max}^2/4m \sim \alpha T. \tag{2.55}$$

In terms of the Landau level summation this condition can be rewritten as:

$$n_c \sim \alpha T/\omega_c \sim \frac{L_H^2}{\xi_{xy}^2} \sim 1/h, \tag{2.56}$$

where the magnetic length $L_H = (\Phi_0/H)^{1/2}$ is always larger than the coherence length ξ_{xy}.

The formal divergence of the GL free energy does not effect on such observable properties as the heat capacity and the diamagnetic susceptibility. Being the second derivatives of the free energy these physical values are well-defined, the momentum (or Landau levels) summation converges well for them. Nevertheless, the problem of regularization arises for example in calculation of the $3D$ fluctuation magnetization. Being only the first derivative of the free energy it turn out to be also divergent and must be cut-off. Let us stress, that the ultraviolet divergence of the fluctuation part of the GL free energy is the result of the limitedness of the GL functional approach, unapplicable to the description of the short wavelength fluctuations. This problem does not appear at all in the consistent microscopic theory which correctly accounts for the short wavelength fluctuations and does not requires any renormalization procedure.

Coming back to (2.53) we, following [48], will express the $2D$ free energy in terms of the Euler gamma function (see Appendix A). Using its definition in the form of an infinite product (B.2) one can transform the sum in Eq. (2.53) to the product and find

$$\begin{aligned} F_{(2)}(\epsilon, h) &= -\frac{TS}{2\pi\xi_{xy}^2} h \ln \prod_{n=0}^{n_c-1} \frac{\pi/2h\alpha}{\left(n + 1/2 + \frac{\epsilon}{2h}\right)} \\ &= -\frac{TS}{2\pi\xi_{xy}^2} h \left[\ln \frac{n_c!\, n_c^{\epsilon/2h-1/2}}{\Gamma(1/2 + \epsilon/2h)} - n_c \ln \frac{\pi}{2h\alpha}\right]. \end{aligned} \tag{2.57}$$

The natural dimensionless argument of the gamma function appearing here can be expressed in terms of the dimensional variables as

$$\frac{1}{2} + \frac{\epsilon}{2h} = \frac{\epsilon + h}{2h} = \frac{1}{2H}\left(T - T_{c2}(H)\right)\left(-\frac{\partial H_{c2}(T)}{\partial T}\right)\bigg|_{T=T_c-0}. \tag{2.58}$$

Substituting the factorial in (2.57) by its asymptotic expression according to the Stirling's formula (B.3), and taking into account (2.56), one can find for the singular part of the free energy

$$\delta F_{(2)}(\epsilon, h) = -\frac{TS}{2\pi\xi_{xy}^2}\left[h\ln\frac{\Gamma(1/2+\epsilon/2h)}{\sqrt{2\pi}} + \frac{\epsilon}{2}\ln h\right]. \tag{2.59}$$

We have omitted here the terms related to the cut-off procedure which do not contribute to the temperature or magnetic field dependence of the observable values.

In the case of a weak magnetic field $h \ll \epsilon$ one can use the corresponding asymptotic formula (B.4) for the gamma function and find the part of free energy responsible for the singular corrections to heat capacity and susceptibility

$$\delta F_{(2)}(h \ll \epsilon) = -\frac{TS}{2\pi\xi_{xy}^2}\left(\frac{\epsilon}{2}\ln\epsilon - \frac{h^2}{12\epsilon}\right). \tag{2.60}$$

The first term of this expression evidently coincides with the earlier evaluation given by (2.29), the second provides the exact coefficient for Eq. (2.42).

In the opposite case $h \gg \epsilon$

$$\delta F_{(2)}(\epsilon \ll h) = \frac{TS\ln 2}{4\pi\xi_{xy}^2}h. \tag{2.61}$$

Now one can easily find the expression for $2D$ fluctuation magnetization per unit square of the film:

$$\begin{aligned} M_{(2)}(\epsilon, h) &= -\frac{1}{\widetilde{H}_{c2}(0)S}\frac{\partial}{\partial h}\left[\delta F_{(2)}(\epsilon, h)\right] \\ &= \frac{T}{\Phi_0}\left\{\ln\frac{\Gamma(1/2+\epsilon/2h)}{\sqrt{2\pi}} - \frac{\epsilon}{2h}\left[\psi(1/2+\epsilon/2h) - 1\right]\right\}, \end{aligned} \tag{2.62}$$

where $\psi(z)$ is the logarithmic derivative of the Euler gamma function (see Appendix A). Using this formula one can observe the crossover from the weak field linear regime to the saturation of magnetization in strong fields. Really, in a weak field, Eq. (2.62) gives

$$M_{(2)}(h \ll \epsilon) = -\frac{h}{6\epsilon}\left(\frac{T}{\Phi_0}\right), \tag{2.63}$$

while in strong fields the fluctuation magnetization according to the result by Klemm et al. [49] reaches the constant:

$$M_{(2)}(h \gg \epsilon) \to M_\infty = -\frac{\ln 2}{2}\left(\frac{T}{\Phi_0}\right) = -0.346\left(\frac{T}{\Phi_0}\right). \tag{2.64}$$

The further derivation of Eq. (2.63) over magnetic field gives the expression for the diamagnetic susceptibility of a thin film in a weak perpendicular field:

$$\chi_{(2)}(h \ll \epsilon) = \chi_{(film,\perp)} = -\frac{e^2 T}{3\pi d}\frac{\xi_{xy}^2}{\epsilon}. \qquad (2.65)$$

2.4 Layered superconductor in magnetic field

2.4.1 *Lawrence–Doniach model*

Let us pass now to the quantitative analysis of the temperature and field dependencies of the fluctuation magnetization of a layered superconductor. As was already mentioned this system has a great practical importance because of its direct applicability to HTS, where the fluctuation effects are very noticeable. Moreover, the general results obtained will allow us to analyze as limiting cases $3D$ and already familiar $2D$, situations. The effects of a magnetic field are more pronounced for a perpendicular orientation, so let us first consider this case.

The generalization of the GL functional for a layered superconductor (Lawrence–Doniach (LD) functional [50]) in a perpendicular magnetic field can be written as

$$\mathcal{F}_{LD}[\Psi] = \sum_l \int d^2r \left(a\left|\Psi_l\right|^2 + \frac{b}{2}\left|\Psi_l\right|^4 + \frac{1}{4m}\left|\left(\nabla_\parallel - 2ie\mathbf{A}_\parallel\right)\left|\Psi_l\right|^2 \right.$$
$$\left. + \mathcal{J}\left|\Psi_{l+1} - \Psi_l\right|^2 \right), \qquad (2.66)$$

where Ψ_l is the order parameter of the l-th superconducting layer and the phenomenological constant $\mathcal{J}$ is proportional to the energy of the Josephson coupling between adjacent planes. The gauge with $A_z = 0$ is chosen in (2.66). In the immediate vicinity of T_c the LD functional is reduced to the GL one with the effective mass $M = (4\mathcal{J}s^2)^{-1}$ along c-direction, where s is the inter-layer spacing. One can relate the value of $\mathcal{J}$ to the coherence length along the z-direction: $\mathcal{J} = 2\alpha T_c \xi_z^2/s^2$. Since we are dealing with the GL region the fourth order term in (2.66) can be omitted.

As it was already mentioned the Landau representation is the most appropriate for solution of the problems related to the motion of a charged particle in a uniform magnetic field. The fluctuation Cooper pair wave function can be written as the product of a plane wave propagating along the magnetic field direction and a Landau state wave function $\phi_n(\mathbf{r})$. Let us expand the order parameter $\Psi_l(\mathbf{r})$ on the basis of these eigenfunctions:

$$\Psi_l(\mathbf{r}) = \sum_{\mathbf{n},k_z} \Psi_{n,k_z}\phi_n(\mathbf{r})\exp(ik_z l), \qquad (2.67)$$

where $\mathbf{n}$ is the quantum number related to the degenerate Landau state and k_z is the momentum component along the direction of the magnetic field. Substituting this expansion into (2.66) one can find the LD free energy as a functional of the Ψ_{n,k_z} coefficients:

$$\mathcal{F}_{LD}\left[\Psi_{\{n,k_z\}}\right]=\sum_{n,k_z}\left\{\alpha T_c\epsilon+\omega_c\left(n+\frac{1}{2}\right)+\mathcal{J}\left[1-\cos(k_z s)\right]\right\}|\Psi_{n,k_z}|^2. \quad (2.68)$$

In complete analogy with the case of an isotropic spectrum the functional integral over the order parameter configurations Ψ_{n,k_z} in the partition function can be reduced to a product of ordinary Gaussian integrals, and the fluctuation part of the free energy of a layered superconductor in magnetic field takes the form:

$$F(\epsilon,H)=-\frac{SH}{\Phi_0}T\sum_{n,k_z}\ln\frac{\pi T}{\alpha T_c\epsilon+\ \omega_c\left(n+\frac{1}{2}\right)+\mathcal{J}\left[1-\cos(k_z s)\right]}. \quad (2.69)$$

One can see that this expression differs from (2.51) only by presence of the "kinetic energy" along the magnetic field direction in the denominator of the logarithm.

In the limit of weak fields one can carry out the summation over the Landau states by means of the Euler–Maclaurin's transformation:

$$\sum_{n=0}^{N}f(n)=\int_{-1/2}^{N+1/2}f(n)dn-\frac{1}{24}\left[f^{'}(N+1/2)-f^{'}(-1/2)\right] \quad (2.70)$$

and obtain

$$F(\epsilon,H)=F(\epsilon,0)+\frac{\pi STH^2}{24m\Phi_0^2}\int_{-\pi/s}^{\pi/s}\frac{\mathcal{N}sdk_z}{2\pi}\left\{\frac{1}{\alpha T_c\epsilon+\mathcal{J}\left(1-\cos(k_z s)\right)}\right\}. \quad (2.71)$$

Here $\mathcal{N}$ is the total number of layers. Carrying out the final integration over the transversal momentum one gets:

$$F(\epsilon,H)=F(\epsilon,0)+\frac{TV}{24\pi s\xi_{xy}^2}\frac{h^2}{\sqrt{\epsilon(\epsilon+r)}}$$

with the anisotropy parameter defined as[15]

[15]Let us stress the difference between J and $\mathcal{J}$ in the two definitions (2.45) and (2.72) of the anisotropy parameter r. The first one was introduced as the electron tunneling matrix element, while the second one enters in the LD functional as the characteristic Josephson energy for the order parameter. Later on, in the framework of the microscopic theory, it will be demonstrated that, in accordance with our qualitative definition, $r\sim J^2$, while $\mathcal{J}$ also turns out to be proportional to J^2. In the dirty case it depends on the relaxation time of the electron scattering on impurities: $\mathcal{J}\sim\alpha J^2\max\{\tau,1/T\}$. Hence both definitions (2.45), appearing in the qualitative consideration, and (2.72), following from the LD model, are consistent.

$$r = \frac{2\mathcal{J}}{\alpha T} = \frac{4\xi_z^2(0)}{s^2}. \tag{2.72}$$

The diamagnetic susceptibility in a weak field turns out [51, 52] to be

$$\chi_{(layer,\perp)} = -\frac{e^2 T}{3\pi s}\frac{\xi_{xy}^2}{\sqrt{\epsilon(\epsilon + r)}}. \tag{2.73}$$

This result confirms the qualitative estimation (2.44) additionally providing the exact value of the numerical coefficient and the temperature dependence in the crossover region. In the $2D$ limit $r \ll \epsilon$ these results correspond to (2.60) and (2.63), while in the limit $r \gg \epsilon$, Eq. (2.73) transforms into the diamagnetic susceptibility of the $3D$ anisotropic superconductor [53]. Note that Eq. (2.73) predicts a nontrivial increase of diamagnetic susceptibility for clean metals [52]. The usual statement that fluctuations are most important in dirty superconductors with a short electronic mean free path does not hold in the particular case of susceptibility because here ξ turns out to be in the numerator of the fluctuation correction.

2.4.2 *General formula for the fluctuation free energy in magnetic field*

Now we will demonstrate that, besides the crossovers in its temperature dependence, the fluctuation induced magnetization and heat capacity are also nonlinear functions of magnetic field. These nonlinearities, different for various dimensionalities, take place at relatively weak fields. This, strong in comparison with the expected scale of $H_{c2}(0)$, manifestation of the nonlinear regime in fluctuation magnetization and hence, field dependent fluctuation susceptibility, was the subject of the intensive debates in early 1970s [53–62] (see also the old but excellent review of Scokpol and Tinkham [63]) and after the discovery of HTS [64–67]. We will mainly follow here the recent essay of Mishonov and Penev [48] and the paper of Buzdin et al. [68], dealing with the fluctuation magnetization of a layered superconductor, which allows observing in a unique way all variety of the crossover phenomena in temperature and magnetic field.

Let us go back to the general expression (2.69) and evaluate it without taking the magnetic field to small. In order to derive the general formula for the LD free energy in arbitrary magnetic field we will use the already done above summation over the Landau levels (see the passage from (2.51) to (2.57)) and will average Eqs. (2.51)–(2.69) over the transverse to layers Cooper pair motion:

$$\begin{aligned} F_{\mathrm{LD}}(\epsilon, H) &= -\frac{hTS}{2\pi\xi_{xy}^2}\int_{-\pi/s}^{\pi/s}\frac{\mathcal{N}dk_z}{2\pi}\sum_{n=0}^{n_c-1}\ln\frac{\pi/\alpha}{\widetilde{\epsilon}(k_z) + \ 2h\left(n+\frac{1}{2}\right)} \\ &= V\int_0^{2\pi}\frac{d\theta}{2\pi}F_{(2)}(\epsilon(\theta), H), \end{aligned} \tag{2.74}$$

where

$$sk_z = \theta,\ \widetilde{\epsilon}\,(\theta) = \epsilon + \frac{r}{2}\,(1 - \cos\theta) \tag{2.75}$$

and $V = S\mathcal{N}s$ is the sample volume. Using the expression (2.59) one can reduce (2.74) to

$$F_{\mathrm{LD}}(\epsilon, h) = -\frac{TV}{2\pi s\xi_{xy}^2}\left[h\int_0^{2\pi}\frac{d\theta}{2\pi}\ln\frac{\Gamma(1/2 + \widetilde{\epsilon}\,(\theta)\,/2h)}{\sqrt{2\pi}} + \frac{1}{2}\left(\epsilon + \frac{r}{2}\right)\ln h + \mathrm{const}\right]. \tag{2.76}$$

This general expression is valid for any anisotropy parameter and evidently is reduced to the $2D$ Eq. (2.59) for $r \to 0$. In the opposite case $\max\{r, \epsilon\} \gg h$ one can substitute the gamma function by its asymptotic (B.4) and get the expression for the $3D$ singular part of the fluctuation free energy:

$$\delta F_{(3)}(h \ll \max\{\epsilon, r\}) = -\frac{TV}{2\pi s\xi_{xy}^2}\left[(\epsilon + r/2)\ln\frac{\sqrt{(\epsilon + h)} + \sqrt{\epsilon + h + r}}{2\sqrt{2}}\right.$$
$$\left. -\sqrt{(\epsilon + h)\,(\epsilon + h + r)} + \frac{h + \epsilon + r/2}{2} + \frac{h^2}{6\sqrt{(\epsilon + h)\,(\epsilon + h + r)}} + \mathrm{const}\right].$$

One can see that for corresponding limits it reproduces both formulae (2.60) and (2.61).

2.4.3 *Fluctuation magnetization and its crossovers*

Having an analytic result for the LD free energy we can easily derive other thermodynamic variables by its differentiation. For the magnetization, e.g. we have

$$M_{\mathrm{LD}}(\epsilon, h; r) = -\frac{T}{\Phi_0 s}\int_0^{\pi/2}\frac{d\phi}{\pi/2}\left\{\frac{\epsilon + r\sin^2\phi}{2h}\left[\psi\left(\frac{\epsilon + r\sin^2\phi}{2h} + \frac{1}{2}\right) - 1\right]\right.$$
$$\left. -\ln\Gamma\left(\frac{\epsilon + r\sin^2\phi}{2h} + \frac{1}{2}\right) + \frac{1}{2}\ln(2\pi)\right\}.$$

Handling with the Hurvitz zeta functions the general formula for an arbitrary magnetic field in $3D$ case ($\epsilon < r$) can be carried out [57, 48]:

$$M_{(3)}(\epsilon; h \ll r) = 3\frac{T}{\Phi_0 s}\left(\frac{2}{r}\right)^{1/2}\sqrt{h}$$
$$\times\left[\zeta\left(-\frac{1}{2}, \frac{1}{2} + \frac{\epsilon}{2h}\right) - \zeta\left(\frac{1}{2}, \frac{1}{2} + \frac{\epsilon}{2h}\right)\frac{\epsilon}{2h}\right]. \tag{2.77}$$

while in the opposite case of extremely high anisotropy $r < |\epsilon|, h \ll 1$ one obtains the $2D$ result (2.62).

Let us comment on the different crossovers in the $M(\epsilon, H)$ field dependence. Let us fix the temperature $\epsilon \ll r$. In this case the c-axis coherence length exceeds

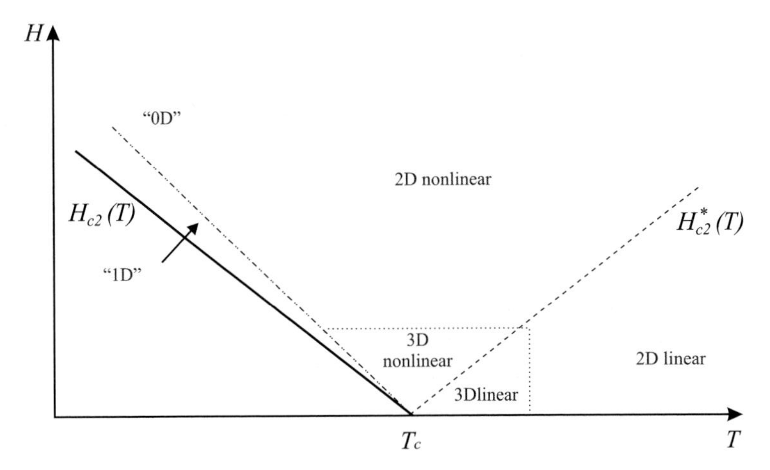

FIG. 2.2. Schematic representation of the different regimes for fluctuation magnetization in the (H, T) diagram. The line $H^*_{c2}(T)$ is mirror-symmetric to the $H_{c2}(T)$ line with respect to a y-axis passing through $T = T_c$. This line defines the crossover between linear and nonlinear behavior of the fluctuation magnetization above T_c [68].

the interlayer distance ($\xi_z \gg s$) and in the absence of a magnetic field the fluctuation Cooper pairs motion has a $3D$ character. For weak fields ($h \ll \epsilon$) the magnetization grows linearly with magnetic field, justifying our preliminary qualitative results:

$$M_{(3)}(\epsilon \ll r, h \to 0) = -\frac{e^2 TH}{6\pi} \xi_{xy}(\epsilon) \,. \tag{2.78}$$

Nevertheless, this linear growth is changed to the nonlinear $3D$ high field regime $M \sim \sqrt{H}$ already in the region of a relatively small fields $H_{c2}(\epsilon) \lesssim H$ ($\epsilon \lesssim h$) (see Fig. 2.2). The further increase of magnetic field at $h \sim r$ leads to the next $3D \to 2D$ crossover in the magnetization field dependence. The limit $\epsilon \ll h$ already is described by Eq. (2.62) and, as we already demonstrated, for large fields the magnetization saturates at the value M_∞.

The substitution of $\epsilon = 0$ gives the result typical of $2D$ superconductors (2.64). Therefore at $h \sim r$ we have a $3D \to 2D$ crossover in $M(H)$ behavior in spite of the fact that all sizes of fluctuation Cooper pair exceed considerably the lattice parameters. Let us stress that this crossover occurs in the region of already strongly nonlinear dependence of $M(H)$ and therefore for a rather strong magnetic field from the experimental point of view in HTS.

Let us mention the particular case of strong magnetic fields $\epsilon \ll h$ Eq. (2.77) reproduces the result by Prange [55] with an anisotropy correction multiplier [48] $\xi_{xy}(0)/\xi_z(0)$[16]

[16]Here we have to use the values for the ζ function

$$M_{(3)}(0,h) = -\frac{0.32T}{\Phi_0^{3/2}} \frac{\xi_{xy}(0)}{\xi_z(0)} \sqrt{H}. \tag{2.80}$$

Near the line of the upper critical field ($h_{c2}(\epsilon) = -\epsilon$) the contribution of the term with $n = 0$ in the sum (2.69) becomes the most important and for the magnetization the expression

$$M(h) = -0.346 \left(\frac{T}{\Phi_0 s} \right) \frac{h}{\sqrt{(h - h_{c2}(\epsilon))(h - h_{c2}(\epsilon) + r)}} \tag{2.81}$$

can be obtained [68]. It contains the already familiar for us "$0D$" regime ($r \ll h - h_{c2} \ll 1$), where the magnetization decreases as $-M(h) \sim \frac{1}{h-h_{c2}}$ (compare with (2.49)), while for $h - h_{c2} << r$ the regime becomes "$1D$" and the magnetization decreases slower, as $-M(h) \sim \frac{1}{\sqrt{h-h_{c2}}}$.

Such an analogy is observed in the next orders in Gi too. In the [69] the analogy was demonstrated for the example of the first eleven terms for the $2D$ case and nine for the $3D$ case. Summation of the series of high order fluctuation contributions to the heat capacity by the Pade–Borel method resulted in its temperature dependence similar to the $0D$ and $1D$ cases without a magnetic field. Nevertheless a considerable difference is not to be forgotten: in the $0D$ and $1D$ cases no phase transition takes place while in the $2D$ and $3D$ cases in a magnetic field a phase transition of first order to the Abrikosov vortex lattice state occurs.

In conclusion, the fluctuation magnetization of a layered superconductor in the vicinity of the transition temperature turns out to be a complicated function of temperature and magnetic field, and it evidently cannot be factorized in these variables. The fit of the experimental data is very sensitive to the anisotropy parameter r and allows determination of the latter with a rather high precision [70,71]. In Fig. 2.3 the successful application of the described approach to fit the experimental data on $YBa_2Cu_3O_7$ is shown [72].

2.4.4 *Fluctuation heat capacity in magnetic field*

Differentiation of the general expression (2.74) for the LD fluctuation free energy with respect to the temperature gives the most singular part of the heat capacity

$$C_{\text{LD}}(\epsilon,h) = -\frac{1}{VT}\frac{\partial^2}{\partial^2\epsilon} F_{\text{LD}}(\epsilon,h) = \frac{1}{4\pi s\xi_{xy}^2}\frac{1}{2h}\int_0^{2\pi}\frac{d\theta}{2\pi}\psi^{(1)}(1/2 + \frac{\widetilde{\epsilon}(\theta)}{2h}).$$

For zero magnetic field one can use the asymptotic (B.16) and easily get

$$C_{\text{LD}}(\epsilon, r, h \to 0) = \frac{1}{4\pi s\xi_{xy}^2(0)} \frac{1}{\sqrt{\epsilon(r+\epsilon)}}. \tag{2.82}$$

This formula evidently contains both $2D$ and $3D$ regimes

$$\zeta(-1/2, 1/2) = \left[-1 + 1/\sqrt{2}\right]\zeta(-1/2) = 0.06. \tag{2.79}$$

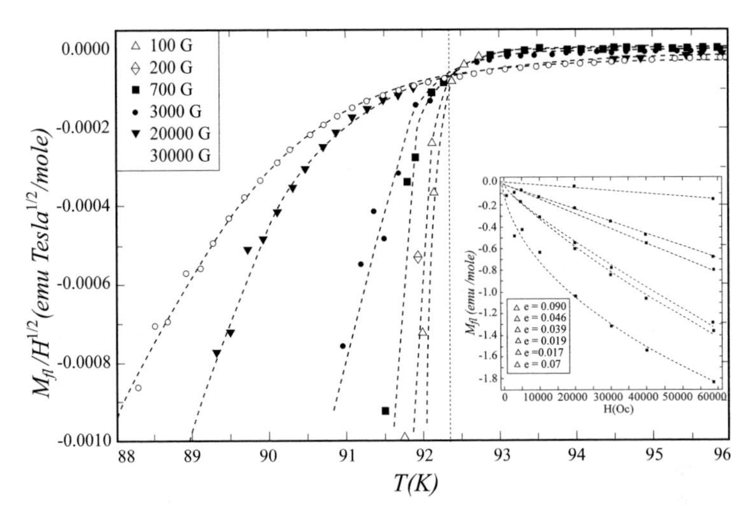

FIG. 2.3. Fluctuation magnetization of a YBaCO123 normalized on $\sqrt{H}$ as the function of temperature in accordance with the described theory shows the crossing of the isofield curves at $T = T_c(0) = 92.3\,K$. The best fit obtained for anisotropy parameter $r = 0.09$. In the inset the magnetization curves as the function of magnetic field are reported.

$$C_{\text{LD}}(\epsilon, r, h \to 0) = \frac{1}{4\pi s \xi_{xy}^2(0)} \begin{cases} \frac{1}{\epsilon}, \text{ when } r \ll \epsilon, \\ \frac{1}{\sqrt{\epsilon r}}, \text{ when } r \ll h. \end{cases} \tag{2.83}$$

At the transition point $T = T_{c0}$

$$\begin{aligned} C_{\text{LD}}(0, r, h) &= \frac{1}{4\pi s \xi_{xy}^2} \frac{1}{2h} \int_0^\pi \psi^{(1)}(1/2 + \frac{r}{2h} \sin^2 \phi) \frac{d\phi}{\pi} \\ &= \frac{1}{4\pi s \xi_{xy}^2} \begin{cases} 1/\sqrt{hr}, \text{ when } \epsilon \ll h \ll r, \\ \pi^2/4h, \text{ when } r \ll h, \end{cases} \end{aligned} \tag{2.84}$$

the heat capacity in magnetic field is finite, while near the line of the upper critical field ($h_{c2}(\epsilon) = -\epsilon$) it diverges (the argument of the function $\psi^{(1)}$ turns zero for $\phi \to 0$). The character of this divergence coincides with that one in (2.81) and can be calculated substituting $\psi^{(1)}$ function by its asymptotic near the pole:

$$\begin{aligned} C_{\text{LD}}(r, h \to h_{c2}(\epsilon)) &= \frac{1}{4\pi s \xi_{xy}^2} \frac{1}{2h} \int_0^\pi \psi^{(1)}(\frac{h - h_{c2}(\epsilon)}{2h} + \frac{r}{2h} \sin^2 \phi) \frac{d\phi}{\pi} \\ &\approx \frac{1}{4\pi s \xi_{xy}^2} \frac{1}{\sqrt{(h - h_{c2}(\epsilon) + r)\,(h - h_{c2}(\epsilon))}}. \end{aligned} \tag{2.85}$$

2.5 Ginzburg–Levanyuk criterion

2.5.1 *Definition of the Gi number from the heat capacity in zero magnetic field*

The fluctuation corrections to the heat capacity obtained above allow us to answer quantitatively the question: where are the limits of applicability of the GL theory?

As we already know this theory is valid not too near the transition temperature and not too far from it. Indeed, far from T_c, when the Cooper pair size $\xi(T)$ does not differ strongly from ξ_0, the gradient expansion used for the construction of the GL functional is not valid anymore: short-wavelength fluctuations become important. Hence the high temperature restriction on the validity of the GL theory is evident: $\epsilon \ll 1$. In the vicinity of the transition point this theory can be applied up to the temperature when the fluctuation corrections become comparable to the value of the corresponding physical values themselves. Historically, for the first time such a criterion was formulated [11, 35] comparing the heat capacity jump at the transition point to the corresponding fluctuation correction found in the framework of the GL theory. Correspondingly, the so-called Ginzburg–Levanyuk number $Gi_{(D)}$ [11,35] is defined as the value of the reduced temperature at which the fluctuation correction (2.34) reaches the value of ΔC (2.9):

$$Gi_{(D)} = \frac{1}{\alpha}\left[\frac{V_D}{V}\vartheta_D b(4m)^{\frac{D}{2}} T_c^{\frac{D}{2}-1}\right]^{2/(4-D)}. \tag{2.86}$$

Substituting into this formula the microscopic values of the GL theory parameters (2.6) one finds

$$Gi_{(D)} = \left[\frac{7\zeta(3)\vartheta_D}{8\pi^2}\left(\frac{V_D}{V}\right)\frac{1}{\nu_D T_c \xi^D}\right]^{2/(4-D)}. \tag{2.87}$$

For future use let us present the Gi number in explicit form for $3D$ and $2D$ cases:

$$Gi_{(3)} = \left[\frac{7\zeta(3)}{64\pi^3}\frac{1}{\nu_3 T_c \xi^3}\right]^2,$$

and

$$Gi_{(2)} = \frac{7\zeta(3)}{32\pi^3}\frac{1}{\nu_2 T_c \xi^2}.$$

Since $\nu_D T_c \sim \nu_D v_F/\xi_c \sim p_F^{D-1}\xi_c^{-1} \sim \mathfrak{a}^{1-D}\xi_c^{-1}$ one can convert Eq. (2.87) to the form

$$Gi_{(D)} \sim \left[\frac{7\zeta(3)\vartheta_D}{8\pi^2}\left(\frac{V_D}{V}\right)\frac{\xi_c \mathfrak{a}^{D-1}}{\xi^D}\right]^{2/(4-D)},$$

where $\mathfrak{a}$ is the interatomic distance. It is worth mentioning that in bulk conventional superconductors, due to the large value of the coherence length ($\xi_c \sim 10^{-6} \div 10^{-4}\, cm$), which drastically exceeds the interatomic distance ($\mathfrak{a} \sim 10^{-8}\, cm$), the fluctuation correction to the heat capacity is extremely small. However, the fluctuation effect increases for small effective sample dimensionality and small electron mean free path. For instance, the fluctuation heat capacity of a superconducting granular system is readily accessible for experimental study.

Using the microscopic expression for the coherence length (A.2), the Gi number (2.87) can be evaluated for different cases of clean (c) and dirty (d) superconductors of various dimensionalities and geometries (film, wire, whisker and granule are supposed to have $3D$ electronic spectra). The corresponding results are given in Table 2.1.

Table 2.1 *The values of the Gi number for different systems*

$Gi_{(3)}$	$Gi_{(2)}$	$Gi_{(1)}$	$Gi_{(0)}$
$80\left(\frac{T_c}{E_F}\right)^4$, (c) $\frac{1.6}{(p_F l)^3}\left(\frac{T_c}{E_F}\right)$, (d)	$\left(\frac{T_c}{E_F}\right)$, (c) $\frac{0.27}{p_F l}$, (d) $\frac{1.3}{p_F^2 l d}$, (d), film	0.5, (c) $1.3\left(p_F^2 S\right)^{-2/3}(T_c\tau)^{-1/3}$, (d), wire $2.3\left(p_F^2 S\right)^{-2/3}$, (c), whisker	$\frac{\sqrt{7\zeta(3)}}{2\pi}\frac{1}{\sqrt{\nu T_c V}}$ $\approx 13.3\frac{T_{c0}}{E_F}\sqrt{\frac{\xi_0^3}{V}}$

One can see that for the $3D$ clean case the result coincides with the original Ginzburg evaluation and demonstrates the negligibility of the superconducting fluctuation effects in clean bulk materials.

Let us mention the relation between the Ginzburg–Levanyuk number found in the dirty $2D$ case $Gi_{(2)}^{(d)}$ with the dimensionless conductance

$$G_{\square} = \frac{\hbar}{e^2 R_{\square}}, \tag{2.88}$$

where $R_{\square}$ is the resistance per unit film square. Really, from the Drude formula for $2D$ conductivity one can easily find:

$$G_{\square}^{(2)} = \nu_{(2)}\mathcal{D}_{(2)} = \frac{p_F l}{2\pi} = \frac{E_F \tau}{\pi}, \tag{2.89}$$

$$G_{\square}^{(\text{film})} = \nu_{(3)}\mathcal{D}_{(3)} = \frac{p_F^2 l d}{3\pi^2}, \tag{2.90}$$

and to discover that both $Gi_{(2,d)}$ numbers of Table 2.1 can be expressed by means of $G_{\square}$ in the unique way:

$$Gi_{(2,d)} \approx \frac{1}{23 G_{\square}}. \tag{2.91}$$

2.5.2 *Other definitions*

The way in which we introduced the Gi number evidently is not unique. For instance, often it is introduced [73] by comparing the fluctuation energy with the magnetic part of energy $\widetilde{H}_{c2}^2(0)\xi^3$:

$$Gi_{(3,H)} = \frac{16\pi^3\kappa^4 T_c^2}{\Phi_0^3 \widetilde{H}_{c2}(0)} = \frac{32 e^4 \lambda^4 T_c^2}{\xi^2}. \tag{2.92}$$

where $\kappa = \lambda/\xi$ is the dimensionless parameter of the GL theory. The magnetic field penetration depth λ is related to the superfluid density and can be expressed in terms of microscopic parameters (see Appendix A)

$$(2e\lambda)^2 = \frac{m}{2\pi n_s} = \frac{7\zeta(3)}{128\pi^3}\frac{1}{\nu_3 T_c^2 \xi^2}, \qquad (2.93)$$

which gives the value for $\widetilde{Gi}_{(3)}$ differing only twice with respect to our definition:

$$\widetilde{Gi}_{(3)} = Gi_{(3,H)} = \frac{2\,(2e\lambda)^4 T_c^2}{\xi^2} = \frac{1}{2}\left(\frac{7\zeta(3)}{8\pi^2}\frac{1}{8\pi}\frac{1}{\nu T_c \xi^3}\right)^2 = \frac{1}{2}Gi_{(3)}\,(0)\,. \qquad (2.94)$$

One more method to define the Gi number is to call in this way the reduced temperature at which the AL correction to conductivity is equal to the normal value of conductivity (as it was done in [74,75]). Such a definition also results in the change of the numerical factor in Gi number:

$$Gi_{(2,\sigma)} \approx 1.44 Gi_{(2)}\,(0)\,. \qquad (2.95)$$

2.5.3 *Broadening of the critical region by a magnetic field*

2.5.3.1 *Vicinity of T_c* Now let us consider how the presence of a strong enough magnetic field changes the fluctuation region. Let us start from general anisotropic $2D$ case and then analyze its $2D$ and $3D$ limits. It is enough to repeat our previous evaluation of the fluctuation contribution to the heat capacity in the presence of magnetic field using the expression for free energy (2.69). We suppose the temperature to be so close to $T_c\,(H)$ that the contribution of only the lowest level of Landau quantization is important (so the magnetic field is supposed to be nonzero):

$$F(\epsilon, H) = -\frac{SH}{\Phi_0}T\int_{-\pi/s}^{\pi/s}\frac{dk}{2\pi}\ln\frac{\pi T}{\alpha T_c\epsilon + \frac{H}{4m\Phi_0} + J\,(1-\cos(k_z s))},$$

$$\delta C(\epsilon, H) = -\frac{1}{VT_c}\left(\frac{\partial^2 F(\epsilon, H)}{\partial \epsilon^2}\right) = \frac{H}{\Phi_0 s}\left(-\frac{\partial}{\partial \epsilon}\right)\frac{1}{\sqrt{\epsilon\,(H)\,[\epsilon\,(H)+r]}}\,.$$

(see Appendix C).

In $3D$ and $2D$ cases one obtains

$$\delta C_{(3)}(\epsilon \ll r, H) = \frac{H}{2\Phi_0 s}\frac{1}{\sqrt{r}}\frac{1}{\epsilon\,(H)^{3/2}}, \qquad (2.96)$$

$$\delta C_{(2)}(r \ll \epsilon, H) = \frac{H}{\Phi_0 s}\frac{1}{\epsilon^2\,(H)}. \qquad (2.97)$$

Comparing (2.97) with the heat capacity jump (2.9) and taking into account (2.6) one finds $Gi_{(2)}\,(H)$

$$\frac{8\pi^2}{7\zeta(3)}\nu_2 T_{\rm c} = \frac{H}{\Phi_0 s}\frac{1}{\epsilon_{cr}^2\left(H\right)}, \tag{2.98}$$

or [76]

$$Gi_{(2)}\left(H\right) = \epsilon_{cr}\left(H\right) = \sqrt{Gi_{(2)}(T_{\rm c},0)\frac{2H}{\widetilde{H}_{c2}(0)}}. \tag{2.99}$$

In the same manner can be done for the $3D$ spectrum

$$\frac{8\pi^2}{7\zeta(3)}\nu_3 T_{\rm c} = \frac{H}{\Phi_0 s}\frac{1}{\sqrt{r}}\frac{1}{\epsilon_{cr}^{3/2}\left(H\right)}, \tag{2.100}$$

$$\begin{aligned}\epsilon_{cr}^{3/2}\left(H\right) &= \left(\frac{7\zeta(3)}{8\pi^2}\frac{1}{8\pi}\frac{1}{\nu_3 T_{\rm c}\xi^3}\right)\frac{H}{\Phi_0 s}\frac{8\pi\xi^3}{\sqrt{r}} \\ &= \frac{2H}{\widetilde{H}_{c2}(0)}\frac{1}{\sqrt{r}}\sqrt{Gi_{(3)}\left(0\right)}\end{aligned}$$

(we used the definition (2.72) for the anisotropy parameter). As the result,

$$Gi_{(3)}\left(H\right) = \left(\frac{2H}{\widetilde{H}_{c2}(0)}\right)^{2/3}\sqrt[3]{Gi_{(3)}\left(T_{\rm c},0\right)}. \tag{2.101}$$

We see that the presence of a magnetic field widens the critical region as $H^{1/2}$ in the $2D$ case and $H^{2/3}$ in $3D$ cases. As it was already mentioned our consideration is for strong enough fields.

2.5.3.2 *Zero temperature: vicinity of $H_{c2}(0)$* In order to have the complete picture of the shape of critical fluctuations domain on $H-T$ phase diagram let us discuss the region of low temperatures $t = T/T_{\rm c} \ll 1$ and high fields $H \sim H_{c2}(0)$. The fluctuations have here a purely quantum nature and we will study them quantitatively below, in Part III. Here we will use the results obtained in [76] only in order to determine the width of the critical region. We will evaluate the Gi number here comparing the first fluctuation correction diamagnetic susceptibility with its jump between the normal and superconducting phases. We speak about the dirty superconductor of strongly second type, so its GL parameter $\varkappa \gg 1$.

The magnetization of such a system was studied in detail as a function of temperature in the entire interval of temperatures in the middle of sixties [77,78]. The corresponding diamagnetic susceptibility is given by the formula [79]

$$\chi\left(T\right) = \frac{\partial M}{\partial H} = -\frac{1}{4\pi\beta_A\left[2\varkappa_2^2\left(T\right)-1\right]} \tag{2.102}$$

where $\beta_A = 1.18$ is the parameter of Abrikosov's theory of superconductors of the second type. The parameter $\varkappa_2\left(T\right)$ was originally introduced by Maki [80] and it is nothing but the generalization of the GL parameter $\varkappa$ for the entire range of temperatures below T_c. For the case interesting to us of dirty superconductors it was shown [78] that

$$\varkappa_2(0) \approx 1.2\varkappa_d = 1.2\frac{3c}{2\pi^2 e v_F l}\left[\frac{7\zeta(3)}{\pi\nu}\right]^{1/2}, \tag{2.103}$$

where the dirty limit of $\varkappa$

$$\varkappa_d = \frac{3c}{2\pi^2 e v_F l}\left[\frac{7\zeta(3)}{\pi\nu}\right]^{1/2} \tag{2.104}$$

was calculated by Gor'kov [81]. Let us remember that for good metal the electron Coulomb energy is of the order of its kinetic one: $e^2/v_F \sim p_F e^2/E_F \sim (e^2/a)/E_F \sim 1$. Hence the jump of the diamagnetic susceptibility at zero temperature between the superconducting and normal phases of a dirty type-II superconductor (near field $H_{c2}(0)$) may be estimated as

$$\Delta\chi(0) \approx \frac{v_F^2}{c^2}(p_F l)^2 \frac{\pi^2}{63\zeta(3)\cdot 2.88\cdot\beta_A}\frac{e^2}{v_F} \approx 0.034\frac{v_F^2}{c^2}(p_F l)^2\frac{e^2}{v_F}. \tag{2.105}$$

Naturally, this value is considerably less than $(4\pi)^{-1}$.

In Part III will be shown that the fluctuation susceptibility of a dirty superconducting film of the thickness d at low temperatures $(t \ll \widetilde{h})$ [76] is given by:

$$\chi_{\text{fl}}\left(t \ll \widetilde{h}\right) = -\frac{e^2}{\pi^2 c^2}\frac{v_F^2\tau}{d}\frac{1}{\widetilde{h}}, \tag{2.106}$$

where the reduced magnetic field

$$\widetilde{h} = \frac{H - H_{c2}(T)}{H_{c2}(T)} \tag{2.107}$$

plays the role of the reduced temperature ϵ in our consideration of fluctuations above T_c. Comparing (2.106) with (2.105): $\chi_{\text{fl}}\left(\widetilde{h}_{cr}\right) = \Delta\chi(0)$, one can find

$$Gi_{(\text{film})}\left(t \ll \widetilde{h}\right) = \widetilde{h}_{cr}\left(t \ll \widetilde{h}\right) = \frac{2.65}{p_F^2 l d} \approx 2Gi_{(\text{film})}(T_c, H = 0). \tag{2.108}$$

One can repeat the same procedure for temperatures t still small but larger than the reduced field $\widetilde{h}$, where (see (9.69))

$$\chi_{\text{fl}}\left(t \gg \widetilde{h}\right) = -\frac{e^2}{\pi^2 c^2}\frac{v_F^2\tau}{d}\frac{\gamma_E t}{\widetilde{h}^2}. \tag{2.109}$$

Comparing this expression with (2.105) one finds for the Ginzburg– Levanyuk number in the region $\widetilde{h} \ll t \ll 1$

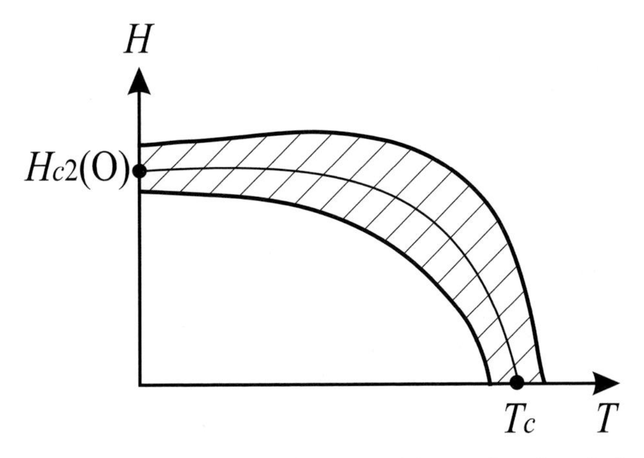

FIG. 2.4. Schematic presentation of the critical region for the whole temperature region below the critical temperature.

$$Gi_{(\text{film})}\left(t \gg \widetilde{h}\right) \approx 2\sqrt{tGi_{(\text{film})}\left(T_{\text{c}}, H=0\right)}. \tag{2.110}$$

Now we are ready to discuss the behavior of the critical region on the phase diagram $H-T$ along the entire interval of temperatures. Near critical temperature T_{c}, in zero magnetic field the width of the critical region is $Gi^{(\text{film})}\left(T_{\text{c}}, H=0\right) = 1.3/\left(p_f^2 ld\right)$. As we move along the line $H_{\text{c}2}(T)$ the critical region is broadened according to Eq. (2.99). One can easily see that in the intermediate region $T_{\text{c}}\left(H\right) \sim T_{\text{c}}/2$ and $H_{\text{c}2}(T) \sim H_{\text{c}2}(0)/2$ formulae obtained from the regions of the high (see (2.99)) and low (see (2.110)) temperatures match each other with the accuracy of $\sqrt{2}$.

At low temperatures, in the vicinity of $H_{\text{c}2}(0)$, the width of the critical region decreases again to (2.108), the value familiar to us from Table 2.1. The sketch of this behavior is presented in Fig. 2.4. Finally one can conclude that the width of the critical region along the curve $H_{\text{c}2}\left(T\right)$ is non-monotonic: it is narrow $\left(Gi_{(\text{film})}\left(T_{\text{c}}, H=0\right)\right)$ near T_{c}, then it broadens up to $\sqrt{Gi_{(\text{film})}\left(T_{\text{c}}, H=0\right)}$ at intermediate temperatures, and finally returns to the same $Gi_{(\text{film})}\left(T_{\text{c}}, H=0\right)$ in the vicinity of $H_{\text{c}2}(0)$.

2.6 Scaling and the renormalization group

In the above study of the fluctuation contribution to heat capacity we have restricted ourselves to the temperature range out of the direct vicinity of the critical temperature: $|\epsilon| \gtrsim Gi_{(D)}$. As we have seen the fluctuations in this region turn out to be weak and neglecting their interaction was justified. In this section we will discuss the fluctuations in the immediate vicinity of the critical temperature ($|\epsilon| \lesssim Gi_{(D)}$) where this interaction turns out to be of great importance.

We will start with the scaling hypothesis, i.e. with the belief that in the immediate vicinity of the transition the only relevant length scale is $\xi(T)$. The temperature dependencies of all other physical quantities can be expressed through $\xi(T)$. This means, for instance, that the formula for the fluctuation part of the free energy (2.29) with the logarithm omitted is still valid in the region of critical fluctuations[17]

$$F_{(D)} \sim -\xi^{-D}(\epsilon), \tag{2.111}$$

the coherence length is a power function of the reduced temperature: $\xi(\epsilon) \sim \epsilon^{-\nu}$. The corresponding formula for the fluctuation heat capacity can be rewritten as

$$\delta C \sim -\frac{\partial^2 F}{\partial \epsilon^2} \sim \epsilon^{D\nu - 2}. \tag{2.112}$$

As was demonstrated in the Introduction, in the GL functional approach, where the temperature dependence of $\xi(T)$ is determined only by the diffusion of the electrons forming Cooper pair, $\xi(\epsilon) \sim \epsilon^{-1/2}$ and $\delta C \sim \epsilon^{-1/2}$. These results are valid for the GL region ($|\epsilon| \gtrsim Gi$) only, where the interaction between fluctuations can be neglected. In the immediate vicinity of the transition (the so-called critical region), where $|\epsilon| \lesssim Gi$, the interaction of fluctuations becomes essential. Here fluctuation Cooper pairs themselves affect the coherence length, changing the temperature dependencies of $\xi(\epsilon)$ and $\delta C(\epsilon)$. In order to find the heat capacity temperature dependence in the critical region one would have to calculate the functional integral with the fourth order term, accounting for the fluctuation interaction, as was done for $0D$ case. For the $3D$ case up to now it is only known how to calculate a Gaussian type functional integral. This was done above when, for the GL region, we omitted the fourth-order term in the free energy functional (2.4).

The first evident step to include in consideration of the critical region would be to develop a perturbation series in b.[18] Any term in this series has the form of a Gaussian integral and can be represented by a diagram, where the solid lines correspond to the correlators $\left\langle \Psi(r)\Psi^*(r')\right\rangle$. The "interactions" b are represented by the points where four correlator lines intersect (see Fig. 2.5).

This series can be written as

$$C \sim \sum_{n=0} c_n \left(\frac{Gi_{(D)}}{\epsilon}\right)^{(4-D)n/2}.$$

For $\epsilon \gtrsim Gi$ it is enough to keep only the first two terms to reproduce the perturbational result obtained above. For $\epsilon \lesssim Gi$ all terms have to be summed. It

[17] The logarithm in (2.29) is essential for the case $D = 2$. This case will be discussed later.

[18] Let us mention that this series is an asymptotic one, i.e. it does not converge even for small b. One can easily see this for small negative b, when the integral for the partition function evidently diverges. This is also confirmed by the exact $0D$ solution (2.25).

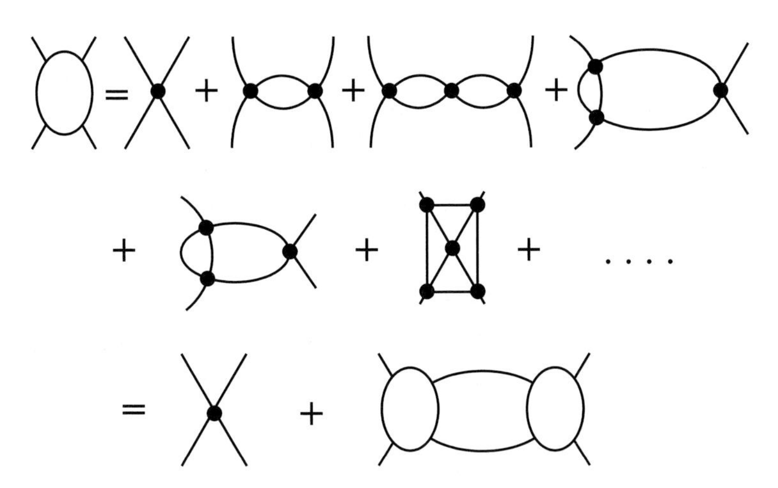

FIG. 2.5. Examples of diagrams for the fluctuation contribution to b.

turns out that the coefficients c_n can be calculated for the space dimensionality $D \to 4$ only. In this case the complex diagrams from Fig. 2.5 (like the diagram similar to an envelope) are smaller by the parameter $\varepsilon = 4 - D$ and in order to calculate c_n it is sufficient to sum the relatively simple "parquet" type diagrammatic series. Such a summation results in the substitution of the "bare" vertex b by some effective interaction $\widetilde{b}$ which diminishes and tends to zero when the temperature approaches the transition point. Such a method was originally worked out in quantum field theory [82–84]. For the problem of a phase transition such a summation was first accomplished in [12]. As result the singularity close to the phase transition in a real $3D$ system was found exactly. It was found that in uni-axial ferroelectrics and ferromagnets with dipole–dipole interactions the specific heat has the singularity:

$$C \sim (-\ln|T - T_{\mathrm{c}}|)^{1/3}. \tag{2.113}$$

Soon this prediction was confirmed experimentally [85].

Let us stress that Eq. (2.113) was carried out for the uni-axial ferromagnetic–paramagnetic phase transition with the simple one-component order parameter, what is already unapplicable in the case of a superconductor with a two-component scalar complex order parameter. In the same paper [12] the effect of the order parameter symmetry on the singularity at the critical point of a an unphysical $4D$ system was considered. It was demonstrated that in the case of transition when the order parameter is an n-component vector the specific heat has singularity:

$$C \sim (-\ln|T - T_{\mathrm{c}}|)^{(4-n)/(n+8)}.$$

Instead of a direct summation of the diagrams it is more convenient and physically obvious to use the method of the multiplicative renormalization group. In the case of quantum field theory it was known long ago [86, 87]. This method is equivalent to that of parquet diagrams summation, but it is simpler and found later applications in different branches of condensed matter theory.

The idea of the renormalization group method consists in separating the functional integration over "fast" ($\psi_{|k|>\Lambda}$) and "slow" ($\psi_{|k|<\Lambda}$) fluctuation modes. If the cut-off Λ is large enough the fast mode contribution is small and the integration over them is Gaussian. After the first integration over fast modes the functional obtained depends on the slow ones only. They can, in their turn, be divided into slow ($|k| < \Lambda_1$) and fast ($\Lambda_1 < |k| < \Lambda$) modes, and the procedure can be repeated. Moving step by step ahead in this way one can calculate the complete partition function.

As an example of the first step of renormalization, the partition function calculation below T_c can be recalled. There we separated the order parameter into the space-independent part $\widetilde{\Psi}$ ("slow" mode) and the fluctuation part $\psi(r)$ ("fast" mode) which was believed to be small in magnitude. Being in the GL region it was enough to average over the fast variables just once, while in the critical region the renormalization procedure requires subsequent approximations.

The cornerstone of the method consists in the fact that in the critical region at any subsequent step the free energy functional has the same form. For D close to 4 this form coincides with the initial free energy GL functional but with the coefficients a_Λ and b_Λ depending on Λ. We will perform these calculations by the method of mathematical induction. Let us suppose that after the $(n-1)$-st step the free energy functional has the form:

$$\mathcal{F}[\widetilde{\Psi}_{\Lambda_{n-1}}] = F_{N,\Lambda_{n-1}} + \int dV \left\{ a_{\Lambda_{n-1}} |\widetilde{\Psi}_{\Lambda_{n-1}}|^2 + \frac{b_{\Lambda_{n-1}}}{2} |\widetilde{\Psi}_{\Lambda_{n-1}}|^4 + \frac{1}{4m} |\nabla \widetilde{\Psi}_{\Lambda_{n-1}}|^2 \right\}. \tag{2.114}$$

Writing $\widetilde{\Psi}_{\Lambda_{n-1}}$ in the form $\widetilde{\Psi}_{\Lambda_{n-1}} = \widetilde{\Psi}_{\Lambda_n} + \psi_{\Lambda_n}$ and choosing Λ_n close enough to Λ_{n-1} it is possible to make ψ_{Λ_n} so small, that one can restrict the functional to the quadratic terms in ψ_{Λ_n} only and perform the Gaussian integration in complete analogy with (2.38). The important property of spaces with dimensionalities close to 4 is the possibility of choosing $\Lambda_n \gg \Lambda_{n-1}$ and still to have $\psi_{\Lambda_n} \ll \widetilde{\Psi}_{\Lambda_n}$. In this case $\widetilde{\Psi}_{\Lambda_n}$ can be taken as coordinate independent, and one can use the result directly following from (2.38):

$$\mathcal{F}[\widetilde{\Psi}_{\Lambda_n}] = F_{N,\Lambda_{n-1}} + \int dV \left\{ a_{\Lambda_n} |\widetilde{\Psi}_{\Lambda_n}|^2 + \frac{b_{\Lambda_n}}{2} |\widetilde{\Psi}_{\Lambda_n}|^4 + \frac{1}{4m} |\nabla \widetilde{\Psi}_{\Lambda_n}|^2 \right\} \tag{2.115}$$

$$-\frac{T}{2} \sum_{\Lambda_n < |\mathbf{k}| < \boldsymbol{\Lambda}_{n-1}} \left\{ \ln \frac{\pi T_c}{(3b_{\Lambda_n} |\widetilde{\Psi}_{\Lambda_n}|^2 + a_{\Lambda_n} + \frac{\mathbf{k}^2}{4m})} + \ln \frac{\pi T_c}{b_{\Lambda_n} |\widetilde{\Psi}_{\Lambda_n}|^2 + a_{\Lambda_n} + \frac{\mathbf{k}^2}{4m}} \right\}.$$

Expanding the last term in (2.115) in a series in $\widetilde{\Psi}_{\Lambda_n}$ one can get for $\mathcal{F}[\widetilde{\Psi}_{\Lambda_n}]$ the same expression (2.114) with the substitution of $\Lambda_{n-1} \to \Lambda_n$. From (2.115), it follows that

$$F_{n,\Lambda_n} = F_{n,\Lambda_{n-1}} - T \sum_{\Lambda_n<|\mathbf{k}|<\mathbf{\Lambda}_{n-1}} \ln \frac{\pi T_\mathrm{c}}{(a_{\Lambda_n} + \frac{\mathbf{k}^2}{4m})},$$

$$a_{\Lambda_n} = a_{\Lambda_{n-1}} + 2T \sum_{\Lambda_n<|\mathbf{k}|<\mathbf{\Lambda}_{n-1}} \frac{b_{\Lambda_n}}{(a_{\Lambda_n} + \frac{\mathbf{k}^2}{4m})}$$

$$b_{\Lambda_n} = b_{\Lambda_{n-1}} - 5T \sum_{\Lambda_n<|\mathbf{k}|<\mathbf{\Lambda}_{n-1}} \frac{b^2_{\Lambda_n}}{(a_{\Lambda_n} + \frac{\mathbf{k}^2}{4m})^2}.$$

Passing to a continuous variable $\Lambda_n \to \Lambda$ one can rewrite these recursion equations as the set of differential equations:

$$\frac{\partial F(\Lambda)}{\partial \Lambda} = -T\mu_D \Lambda^{D-1} \ln \frac{\pi T_\mathrm{c}}{(a(\Lambda) + \frac{\Lambda^2}{4m})}, \tag{2.116}$$

$$\begin{cases} \frac{\partial a(\Lambda)}{\partial \Lambda} = -2T\mu_D \frac{b(\Lambda)\Lambda^{D-1}}{(a(\Lambda)+\frac{\Lambda^2}{4m})} \\ \frac{\partial b(\Lambda)}{\partial \Lambda} = 5T\mu_D \frac{b^2(\Lambda)\Lambda^{D-1}}{(a(\Lambda)+\frac{\Lambda^2}{4m})^2}. \end{cases} \tag{2.117}$$

These renormalization group equations are evidently valid for small enough Λ only, where the transition from discrete to continuous variables is justified. This means at least

$$\Lambda^2/4m \ll T_{\mathrm{c}0} Gi_{(D)}. \tag{2.118}$$

in order to move away from the first approximation.

Let us recall that in the framework of the Landau theory of phase transitions the coefficient $a(T_{\mathrm{c}0}) = 0$ at the transition point and this can be considered as the MFA definition of the critical temperature $T_{\mathrm{c}0}$. The same statement for the function $a = a(\Lambda)$ in the framework of the renormalization group method can be written as the $a(T_{\mathrm{c}0}, \Lambda \sim \xi^{-1}) = 0$. With the decrease of Λ the effect of critical fluctuations is taken into account more and more and the renormalized value of the critical temperature decreases, being defined by the equation: $a(T_\mathrm{c}(\Lambda), \Lambda) = 0$. Finally, after the application of the complete renormalization procedure, one can define the real critical temperature T_c, shifted down with respect to $T_{\mathrm{c}0}$ due to the effect of fluctuations, from the equation:

$$a(T_\mathrm{c}, \Lambda = 0) = 0. \tag{2.119}$$

It is easy to find this shift in the first approximation. Indeed, let us integrate the first equation of (2.117) over Λ in limits $[0, \xi^{-1}]$. The main contribution to the integral will be determined by the region where $a(\Lambda) \ll \Lambda^2/4m$. Being far from the critical point one can assume that the coefficient $b = \text{const}$ and then

$$a(\xi^{-1}) = \alpha\delta T_c = \int_{a(0)}^{a(\xi^{-1})} da = -8mT\mu_D b \int_0^{1/\xi} \Lambda^{D-3} d\Lambda.$$

For the $3D$ case this gives the shift of the critical temperature δT_c duc to fluctuations[19]

$$\frac{\delta T_c^{(3)}}{T_c} \sim -\frac{2mb}{\pi\alpha\xi} = -\frac{b}{2\pi T_c \alpha^2}\frac{1}{\xi^3} = -\frac{7\zeta(3)}{16\pi^3 \nu T_c \xi^3} = -\frac{8}{\pi}\sqrt{Gi_{(3)}}. \tag{2.121}$$

Let us come back to study of properties of the system of equations (2.117). One can find its partial solution at $T = T_c$ in the form:

$$\begin{aligned} a(T_c, \Lambda) &= \frac{4-D}{4m[5+(4-D)]}\Lambda^2 \\ b(T_c, \Lambda) &= \frac{5}{16m^2 T\mu_D}\frac{4-D}{[5+(4-D)]^2}\Lambda^{4-D}. \end{aligned} \tag{2.122}$$

These power solutions are correct in the domain of validity of the system (2.117) itself, i.e. for small enough Λ defined by the condition (2.118). Nevertheless, in a space of dimensionality close to 4 ($D = 4 - \varepsilon, \varepsilon \ll 1$) it is possible to extend their validity up to the GL region and to observe their crossover to the GL results: $a(T_c) = 0$; $b(T_c) = b_0 = \text{const}$.

Indeed, in this case, due to proportionality of $a(\Lambda)$ to $\varepsilon \to 0$, one can omit it in the denominators of the system (2.117) and to write down its second equation in the form:

$$\frac{\partial b}{\partial \ln \Lambda^{-1}} = 5\,(4m)^2\, T\mu_D \frac{b^2}{\Lambda^{4-D}}. \tag{2.123}$$

Let us introduce the dimensionless value γ characterizing the interaction of fluctuations

$$\gamma = b\frac{\Lambda^{4-D}}{5\,(4m)^2\, T\mu_D} \tag{2.124}$$

and will write the equation for it:

[19]In the same way we can analyze the shift of the critical temperature also in $2D$ case and obtain:

$$\frac{\delta T_c^{(2)}}{T_c} = -2Gi_{(2)} \ln \frac{1}{4Gi_{(2)}}. \tag{2.120}$$

As we will show below both $3D$ and $2D$ results for δT_c coincide with those obtained by the analysis of the effect of fluctuations on superconducting density in the perturbation approach.

$$\frac{\partial \gamma}{\partial \ln \Lambda^{-1}} = f(\gamma), \tag{2.125}$$

where in our case

$$f(\gamma) = (4 - D)\gamma - \gamma^2. \tag{2.126}$$

Equation (2.125) was first obtained by Gell-Mann and Low in quantum field theory [86, 87] and it presents the central result of the renormalization group approach. The presence of the negative sign of γ^2 in it results in the appearance at large distances $\Lambda^{-1} \to \infty$ of the fixed point for $\gamma \to \gamma_c = 4 - D$, where $f(\gamma_c) = 0$. For four-dimensional space, $\gamma_c = 0$ and the situation is analogous to the zero-charge situation in quantum electrodynamics [82]. When $4 - D \ll 1$ one can restrict the consideration of the function $f(\gamma)$ to two terms only as we did above (see (2.126)). One can believe that the qualitative picture remains unchanged even for $D = 3$.

The solution of the second equation of the system (2.117) (or (2.125), which is the same) has the form

$$b^{-1}(T_c, \Lambda) = b_0^{-1} + \frac{80m^2}{(4-D)} T\mu_D(\Lambda^{D-4} - \xi^{4-D}). \tag{2.127}$$

We have chosen the constant of integration as $b_0^{-1} - \frac{80m^2}{(4-D)} T\mu_D \xi^{4-D}$ in order to match the renormalization group and GL solutions at the value of $\Lambda = \Lambda_{\max} \sim \xi^{-1}$.

Now let us study the function $a(T, \Lambda)$ for the same interesting case of space dimensionality $D \to 4$ for temperatures slightly different (but still close enough) from T_c, where one can write

$$a(T, \Lambda) = a(T_c, \Lambda) + \alpha(T_c, \Lambda) T_c \epsilon.$$

The first term on the right-hand side is determined by Eq. (2.122). In order to determine $\alpha(T_c, \Lambda)$ let us expand the first equation in (2.117) in terms of ϵ

$$\frac{\partial \alpha(T_c, \Lambda)}{\partial \Lambda} = 2T\mu_D \frac{b(T_c, \Lambda)\Lambda^{D-1}}{\left(a(T_c, \Lambda) + \frac{\Lambda^2}{4m}\right)^2} \alpha(T_c, \Lambda). \tag{2.128}$$

For

$$\Lambda^2 / 4m \gtrsim \alpha(T_c, \Lambda) T_c \epsilon \tag{2.129}$$

we can again use the solution (2.127) for $b(T_c, \Lambda)$ and omit $a(T_c, \Lambda)$ in the denominator of (2.128). The constant of integration, appearing in the process of solution of (2.128), is chosen according to the condition that for $\Lambda = \Lambda_{\max} \sim \xi^{-1}$ we match $\alpha(T_c, \Lambda) = \alpha(T_{c0}, \xi^{-1}) = \alpha_0$ with the GL theory:

$$\alpha(T_c, \Lambda) = \alpha_0 \left[1 + \frac{80m^2 b_0}{(4-D)} T\mu_D(\Lambda^{D-4} - \xi^{4-D})\right]^{-2/5}. \tag{2.130}$$

The condition (2.129) can be written as $\Lambda \gtrsim \xi^{-1}(T)$, where $\xi(T)$ is the generalized coherence length, determined by the equation:

$$\xi^{-2}(T) = 4m\alpha\left(T_c, \xi^{-1}(T)\right) T_c \epsilon.$$

Such a definition is valid at any temperature. For example, far enough from the critical point, in the GL region, $\alpha\left(T_c, \xi^{-1}(T)\right) = \alpha_0$ and one reproduces the result (1.4). Vice versa, in the critical region the main contribution to the right-hand side of Eq. (2.130) results from the second term containing Λ^{D-4} so, putting $\xi^{-1}(T) = \Lambda$, one can rewrite the self-consistent equation for $\xi(T)$ and get

$$\xi(T) = (4m)^{(1-D)/2} \frac{4-D}{20 b_0 T\mu_D \sqrt{T_c \alpha_0}} \epsilon^{-\nu}, \tag{2.131}$$

where $2\nu = [1 - (4-D)/5]^{-1}$. As was already mentioned, strictly speaking this result was carried out for $\varepsilon = 4 - D \ll 1$, therefore it is accurate up to the first order in ε expansion only: $\nu = 1/2 + \varepsilon/10$. Nevertheless, extending it to $\varepsilon = 1$ $(D = 3)$ one obtains $\nu_3 = 3/5$.

Let us calculate of the critical exponent of the heat capacity in the immediate vicinity of the transition. In order to do this one can calculate the second derivative of equation (2.116) with respect to ϵ:

$$\frac{\partial C(\Lambda)}{\partial \Lambda} = T^2 \mu_D \frac{\Lambda^{D-1}\alpha^2(\Lambda)}{(\alpha(\Lambda)T_c\epsilon + \frac{\Lambda^2}{4m})^2}. \tag{2.132}$$

The heat capacity renormalized by fluctuations has the value $C(\Lambda = 0)$ which is the result of the integration over all fluctuation degrees of freedom. Carrying out the integration of (2.132) over all $\Lambda \lesssim \xi^{-1}$ one can divide the domain of integration on the right-hand side in two: $\Lambda \lesssim \xi^{-1}(T)$ and $\xi^{-1}(T) \lesssim \Lambda \lesssim \xi^{-1}$. In the calculation of the integral over the region $\xi > \Lambda \gtrsim \xi^{-1}(T)$ the inequality $\alpha(\Lambda)T_c\epsilon \ll \frac{\Lambda^2}{4m}$ holds, and the function $\alpha(\Lambda)$ can be omitted in the denominator. In the numerator of (2.132) one can use for $\alpha(\Lambda)$ the solution (2.130). In the region $\Lambda \lesssim \xi^{-1}(T)$ one has to use the partial solution (2.122) for $\alpha(\Lambda)$ and can find that the contribution of this domain has the same singularity as that from the region $\Lambda \gtrsim \xi^{-1}(T)$, but with a coefficient proportional to $(4-D)^2 = \varepsilon^2$, hence negligible in our approximation. The result is:

$$C(\Lambda = 0) = \alpha_0^2[(4mT)^2\mu_D]^{\frac{1}{5}} \left[\frac{4}{5}\frac{(4-D)}{b_0}\right]^{4/5} \frac{5}{4-D} \xi^{(4-D)/5}(T). \tag{2.133}$$

Substituting the expression for $\xi(T)$ one can finally find

$$C = 2^{12/5}\alpha_0^2[5\mu_D \frac{m^2T^2}{b_0^4(4-D)}]^{\frac{1}{5}}\epsilon^{-\alpha}, \tag{2.134}$$

confirming the validity of the scaling hypothesis and the relation (2.112). The critical exponent in (2.134) is

$$\alpha = \frac{(4-D)}{10[1-(4-D)/5]} \approx \varepsilon/10.$$

One can see that, generally speaking, the critical exponents ν and α appear in the form of series in powers of ε. More cumbersome calculations allow us to find the next approximations for them in $\varepsilon = 4 - D$. Nevertheless it is worth mentioning that even the first approximation, giving $\nu_3 = 3/5$ and $\alpha_3 = 1/10$ for $\varepsilon = 1$, is already weakly affected by the following steps of the expansion in powers of ε [88].

In conclusion of this section let us make the following note. Depending on the sign of Gell-Mann–Low function the renormalization group equation can have two different kinds of solutions. In quantum electrodynamics, the coupling constant decreases at long distances ("zero charge"). In quantum chromodynamics, the coupling constant is small at short distances ("asymptotic freedom"), but it increases with increasing distance ("confinement"). We demonstrated above that the $4D$ phase transition theory is of the "zero-charge" type and the interaction decreases at long distances.

As we already have seen the four-dimensional case is special for renormalization group theory: the situation becomes logarithmic there. It is interesting that the $3D$ problems with the dipole interaction are equivalent to the solvable $4D$ problem. This is due to the fact that at large distances such interaction depends on the angle between $\mathbf{r}$ and $\mathbf{d}$ and such angle plays the role of the fourth coordinate necessary to make the situation logarithmic. As was mentioned above, such interactions allowed us to resolve the problem of phase transition in ferroelectrics. The same concerns such complex problems as the pinning of a charge density wave where the dipole–dipole interaction helped to resolve the RG equation and to find the exponentially large dielectric susceptibility [89]. As it will be demonstrated below (section 5.2) an analogous situation turns out in the problem of pinning of interacting vortices in superconductor. Here the dipole–dipole interaction takes place at the distances $r \lesssim \lambda$ and for strong enough magnetic fields, where such interaction is relevant, the RG approach allows us to calculate the exponentially small value of the critical current.

2.7 Effect of fluctuations on superfluid density and critical temperature

Among the important thermodynamical properties of a superconductor is the magnetic field penetration depth $\lambda(T)$. It is evidently asymmetric with respect to the critical temperature, growing when the temperature tends to T_c in the superconducting phase and being infinite in the normal phase. The simple London electrodynamics of a superconductor relates the current density $\mathbf{j}$ and $\lambda(T)$ with the superfluid density n_s:

$$\mathbf{j} = -n_s \frac{2e^2}{m}\mathbf{A} = -\frac{1}{4\pi\lambda^2(T)}\mathbf{A},\ \lambda^2(T) = \frac{m}{8\pi n_s e^2}. \tag{2.135}$$

GL theory allows calculating the temperature dependence of $\lambda(T)$ in the vicinity of the critical temperature, identifying the superfluid density with the average value of the order parameter

$$n_s = \widetilde{\Psi}^2 = -\frac{a}{b}$$

and, hence, predicting its inverse square root divergence as a function of the reduced temperature:

$$\lambda(T) = \sqrt{\frac{m}{8\pi e^2}\frac{b}{\alpha T_c|\epsilon|}}. \tag{2.136}$$

Let us study how the appearance of fluctuation Cooper pairs affects the expulsion of the applied magnetic field. The superfluid density goes to zero at the transition point from below and its value determines $\lambda(T)$, so the effect of fluctuations could be well pronounced.

Let us first convince ourselves that the presence of fluctuation Cooper pairs above T_c does not result in the appearance of a superfluid density. One can calculate the fluctuation part of the supercurrent as [90, 91]

$$\begin{aligned}\langle \mathbf{j}_{fl} \rangle &= -\frac{\partial F_{fl}(\mathbf{A})}{\partial \mathbf{A}} = T\frac{\partial \ln Z(\mathbf{A})}{\partial \mathbf{A}} \\ &= -\frac{\int \mathfrak{D}\Psi(\mathbf{r})\mathfrak{D}\Psi^*(\mathbf{r})\frac{\partial \mathcal{F}[\Psi(\mathbf{r}),\mathbf{A}])}{\partial \mathbf{A}}\exp\left(-\frac{\mathcal{F}[\Psi(\mathbf{r}),\mathbf{A}]}{T}\right)}{\int \mathfrak{D}\Psi(\mathbf{r})\mathfrak{D}\Psi^*(\mathbf{r})\exp\left(-\frac{\mathcal{F}[\Psi(\mathbf{r}),0]}{T}\right)}.\end{aligned}$$

This expression can be written in the form of two contributions:

$$\langle \mathbf{j}_{fl} \rangle = \mathbf{j}_1 + \mathbf{j}_2 = -\frac{2e^2}{m}\mathbf{A}\left\langle |\Psi(\mathbf{r})|^2 \right\rangle_{\mathbf{A}=0} - \frac{e}{m}\left\langle \mathrm{Im}[\Psi(\mathbf{r})\nabla\Psi^*(\mathbf{r})] \right\rangle_{\mathbf{A}}. \tag{2.137}$$

The first one, $\mathbf{j}_1$, just reproduces the London expression with the replacement of n_s by $\left\langle |\Psi(\mathbf{r})|^2 \right\rangle$. In the average the GL free energy with $\mathbf{A} = 0$ can be used. The second term, $\mathbf{j}_2$, has a more sophisticated nature. To get a nonzero value for $\mathbf{j}_2$ one has to include the field dependent part of the GL free energy in the process of further averaging in fluctuation fields. Supposing the vector-potential $\mathbf{A}$ to be weak enough one can expand the Gibbs exponent in it and find

$$\mathbf{j}_2 = -\frac{e^2}{m^2T}\left\langle \mathrm{Im}[\Psi(\mathbf{r})\nabla\Psi^*(\mathbf{r})]\int dV_1\,\mathrm{Im}[\Psi(\mathbf{r}_1)\nabla\Psi^*(\mathbf{r}_1)]\mathbf{A}(\mathbf{r}_1)\right\rangle_{\mathbf{A}=0}. \quad (2.138)$$

In the following we will restrict ourselves to a superconductor of the strongly second order type ($\varkappa \gg 1$). In this case the characteristic scale of the space variation of $\mathbf{A}(\mathbf{r}_1)$ is much larger than that of the order parameter, and the vector-potential can be taken out of the integral in (2.138). Thinking about the GL region above T_c, one can expand the order parameter in a Fourier series and reduce the averaging to the expression

$$\mathbf{j}_2 = \frac{2e^2}{m}\sum_k \mathbf{k}(\mathbf{A}\frac{\partial}{\partial\mathbf{k}})\left\langle|\Psi_k|^2\right\rangle,$$

which yields $\langle\mathbf{j}_{\mathrm{fl}}\rangle = 0$, as intuitively expected.

Below T_c the situation is more cumbersome. In order to calculate $\mathbf{j}_2$ here let us separate the equilibrium value of the order parameter and its real and imaginary fluctuating parts: $\Psi(\mathbf{r}) = \widetilde{\Psi} + \psi_r + i\psi_i$. This allows us to calculate the space integral in the expression (2.138):

$$\int dV_1\,\mathrm{Im}[\Psi(\mathbf{r}_1)\nabla\Psi^*(\mathbf{r}_1)] = 2i\sum_{\mathbf{k}}\mathbf{k}\psi_{r\mathbf{k}}\psi_{i,-\mathbf{k}}.$$

Further the functional integration can be carried out in the spirit of the previous calculations for the fluctuation contribution to the heat capacity below T_c, resulting in

$$\langle\mathbf{j}_{\mathrm{fl}}\rangle = -\frac{2e^2}{m}\mathbf{A}\left[\left\langle|\Psi(\mathbf{r})|^2\right\rangle_{\mathbf{A}=0} - \sum_{\mathbf{k}}\frac{T}{(2\alpha T_c|\epsilon| + \mathbf{k}^2/4m)}\right].$$

Let us calculate the remaining $\left\langle|\Psi(\mathbf{r})|^2\right\rangle = \widetilde{\Psi}^2 + \left\langle\psi_r^2\right\rangle + \left\langle\psi_i^2\right\rangle + 2\widetilde{\Psi}\left\langle\psi_r\right\rangle$. The averaging of the first two terms does not create any problem:

$$\left\langle\psi_r^2\right\rangle = \frac{T}{2}\sum_{\mathbf{k}}\frac{1}{2\alpha T_c|\epsilon| + \mathbf{k}^2/4m}, \quad (2.139)$$

$$\left\langle\psi_i^2\right\rangle = \frac{T}{2}\sum_{\mathbf{k}}\frac{1}{\mathbf{k}^2/4m}. \quad (2.140)$$

In order to carry out the $\langle\psi_r\rangle$ term the anharmonic contributions in the GL functional, originating from the fourth order term, have to be taken into account:

$$\langle\psi_r\rangle = -\frac{1}{2\widetilde{\Psi}}\left(3\left\langle\psi_r^2\right\rangle + \left\langle\psi_i^2\right\rangle\right). \quad (2.141)$$

Finally, one finds the expression for the supercurrent, accounting for fluctuations:

$$\langle \mathbf{j}_{\mathrm{fl}} \rangle = -\frac{2e^2}{m}\mathbf{A}\left[\widetilde{\Psi}^2 - 2\sum_{\mathbf{k}}\frac{T}{(2\alpha T_{\mathrm{c}}|\epsilon| + k^2/4m)}\right].$$

One can see that fluctuations suppress it with respect to the MFA result of the GL theory.

The superfluid density renormalized by fluctuations, in its turn, takes the form

$$n_s(T) = \frac{\alpha}{b}\left[T_{\mathrm{c}}|\epsilon| - \frac{2b}{\alpha^2}\sum_{\mathbf{k}}\frac{1}{(2|\epsilon| + k^2/4m\alpha T_{\mathrm{c}})}\right]. \qquad (2.142)$$

In the MFA $\left\langle |\Psi(\mathbf{r})|^2 \right\rangle = \widetilde{\Psi}^2 = n_s(T)$. Calculating $\left\langle |\Psi(\mathbf{r})|^2 \right\rangle = \widetilde{\Psi}^2 - 2\left\langle \psi_r^2 \right\rangle$ one can see that

$$\left\langle |\Psi(\mathbf{r})|^2 \right\rangle = \frac{\alpha}{b}\left[T_{\mathrm{c}}|\epsilon| - \frac{b}{\alpha^2}\sum_{\mathbf{k}}\frac{1}{(2|\epsilon| + k^2/4m\alpha T_{\mathrm{c}})}\right], \qquad (2.143)$$

and hence accounting for fluctuations already in the first order causes $\left\langle |\Psi(\mathbf{r})|^2 \right\rangle$ and $n_s(T)$ to be different.

It is seen that in the $3D$ and $2D$ cases the sum over momenta in (2.142) is formally divergent at large $\mathbf{k}$. This is not the first time that such problem has arisen and we know how to deal with it: this ultraviolet divergence is related to the restrictions on the applicability of the GL functional for $|\mathbf{k}| \gtrsim \xi^{-1}$, therefore the integral has to be cut-off at $\xi \cdot |\mathbf{k}| = C \sim 1$.

For the $3D$ case the transparent way to cut-off the sum is to separate the upper and lower limit contributions of the sum, which have different physical senses, by adding and subtracting the term $\sim \sum |\mathbf{k}|^{-2}$. As a result one obtains two contributions: the first one, $|\epsilon|$-independent, originates from the upper limit cut-off and the second turns out to be proportional to $\sqrt{|\epsilon|}$ and appears from the lower limit one. Using the microscopic relations between the GL parameters $\alpha^2/b = \frac{8\pi^2}{7\zeta(3)}\nu$ and $4m\alpha T_{\mathrm{c}} = \xi^{-2}$ one can finally find

$$n_{s3}(T) = \frac{64m\pi^2}{7\zeta(3)}\nu T_{\mathrm{c}}\xi^2\left[T_{\mathrm{c}} - T - \frac{7\zeta(3)}{32\pi^3\nu\xi^3}C_3 + \frac{7\zeta(3)}{32\pi^2\nu\xi^3}\sqrt{2\frac{T_{\mathrm{c}} - T}{T_{\mathrm{c}}}}\right]. \qquad (2.144)$$

Let us discuss the second term appeared in (2.144). It evidently determines the fluctuation shift of the critical temperature $\delta T_{\mathrm{c}} = T_{\mathrm{c}} - T_{\mathrm{c}0}$, where $T_{\mathrm{c}0}$ is the mean field (BCS) value of the transition temperature. This shift

$$\frac{\delta T_{\mathrm{c}}^{(3)}}{T_{\mathrm{c}0}} \sim -\frac{7\zeta(3)}{16\pi^3\nu T_{\mathrm{c}0}\xi^3} \approx -\frac{8}{\pi}\sqrt{Gi_{(3)}} \qquad (2.145)$$

coincides with that found in the framework of the renormalization group approach (see Eq. (2.121)). In further calculations we will include this shift in $|\epsilon|$

renormalizing the critical temperature to the value of T_c and identifying it with the experimentally observed transition temperature. The accounting for the next order corrections demonstrates that the substitution of $T_{c0} \to T_c$ must be performed also in the last term of (2.144), so for the superfluid density one can finally write

$$n_{s3}(T) = \frac{64m\pi^2}{7\zeta(3)}\nu T_c^2\xi^2 \left[|\epsilon| + 8\sqrt{2Gi_{(3)}|\epsilon|}\right]. \tag{2.146}$$

In spite of the slower decrease of the fluctuation correction to the superfluid density when temperature tends to the critical value, in comparison with the main contribution ($\sqrt{|\epsilon|}$ instead of $|\epsilon|$), it becomes important only at $|\epsilon| \sim Gi_{(3)}$.

In the $2D$ case the problem of the ultraviolet divergence is less important since the cut-off parameter turns out to be involved only in a logarithm:

$$n_{s2}(T) = \frac{32m\pi^2}{7\zeta(3)}\nu T_c\xi^2 \left[T_c|\epsilon| - \frac{7\zeta(3)}{16\pi^3\nu\xi^2}\ln\frac{C_2}{|\epsilon|}\right]. \tag{2.147}$$

This formula, being obtained in the perturbative way, is valid for $|\epsilon| \gtrsim Gi_{(2)}$ and this fact does not allow us to define the position of the true critical temperature more precisely than with the accuracy of the order of $Gi_{(2)}$. Actually we can do this only with the logarithmic accuracy

$$\frac{\delta T_c^{(2)}}{T_{c0}} = -2Gi_{(2)}\ln\frac{C_2}{Gi_{(2)}}. \tag{2.148}$$

The $2D$ superfluid density (2.147) can be also written in terms of the critical temperature renormalized by fluctuations. It is enough to include the cut-off parameter in the T_c-shift (compare with the RG estimation (2.120)): which results in:[20]

$$n_{s2}(T) = \frac{mT_c}{\pi Gi_{(2)}}\left[|\epsilon| - 2Gi_{(2)}\ln\frac{Gi_{(2)}}{|\epsilon|}\right]. \tag{2.149}$$

The specific behavior of the superfluid density in $2D$ system will be discussed in chapter 15 where the interpolation formula for $n_{s2}(T)$ valid for all range of temperatures in vicinity of the true critical temperature will be given.

Let us mention that in $2D$ superconductors not only fluctuations of the superconducting order parameter can play an important role but also the Coulomb potential fluctuations. The latter become important only for very thin films (with $d \ll \xi$), where the screening has $2D$ character. These fluctuations also shift the critical temperature down by the value [92–95]

$$\frac{\delta T_c^{(Cl)}}{T_{c0}} \sim -Gi_{(2)}\ln^3\left(\frac{\xi}{d}\right). \tag{2.150}$$

[20]For the sake of simplicity we still call the reduced temperature $\frac{T_c - T}{T_c}$ as ϵ.

One can see that the obtained fluctuation shift $\delta T_c^{(2)}$ is large with respect to the width of the critical region $Gi_{(2)}T_c$, although Eq. (2.150)) is valid only as long as $\delta T_c^{(2)}$ remains small in comparison with T_c itself.

Let us stress, that calculated here superfluid density n_{sD} appears only below T_c, and namely this fact we used as the definition of the renormalized by fluctuations critical temperature. In contrast to the superfluid density, the local concentration of fluctuation pairs $n_s^{(D)}$ is defined by the average square of the order parameter $\langle|\Psi(\mathbf{r})|^2\rangle$ and it remains nonzero even above transition (see chapter 16). It can be calculated using the inverse Fourier transform of $|\Psi(\mathbf{q})|^2$. For a layered superconductor

$$\begin{aligned} n_s^{(D)} &= \langle|\Psi(\mathbf{r})|^2\rangle = \int \frac{d^3\mathbf{q}}{(2\pi)^3}|\Psi(\mathbf{q})|^2 \exp(-i\mathbf{q}\cdot\mathbf{r}) \\ &= \int \frac{d^2\mathbf{q}_\perp}{(2\pi)^2}\int \frac{dq_z}{2\pi}\frac{1}{\alpha\left(\epsilon+\xi^2\mathbf{q}_\perp^2+\frac{r}{2}(1-\cos q_z s)\right)} \\ &= \frac{1}{4\pi\alpha\xi^2 s}\ln\frac{2}{\sqrt{\epsilon}+\sqrt{\epsilon+r}} = \frac{mT_c}{\pi s}\ln\frac{2}{\sqrt{\epsilon}+\sqrt{\epsilon+r}}, \end{aligned} \tag{2.151}$$

where we have used Eqs. (C.1), (3.5) and (2.6).

Let us note, that the high frequency quantum fluctuations with $\Omega \gg T_c$ also result in a small shift of the critical temperature [96], which is equivalent to the change of a numerical factor at ω_D under the logarithm sign in the corresponding BCS formula.

2.8 Fluctuations of magnetic field

In the above consideration we have discussed fluctuations of the order parameter only supposing the magnetic field to be a constant. At the same time the natural question can arise: what is the effect of the usual black-body radiation fluctuations on the superconductor properties? From the very beginning one can expect this effect to be small at low temperature. The order parameter fluctuation effects in bulk materials turned out to be negligible, too. In the case of type-I superconductors the coherence length ξ can noticeably exceed the magnetic field penetration depth λ. This means that the magnetic field fluctuations will take place on a scale less than the characteristic superconducting one, which can change some intrinsic superconducting properties. In this section we will show that magnetic field fluctuations, being small in amplitude, can change the order of phase transition of a type-I superconductor.

The same GL functional formalism allows us to take into account the fluctuations of a magnetic field [97]. Let us suppose that the external magnetic field $\mathbf{H} = 0$, the system is below the critical temperature, the value of the equilibrium order parameter is $\widetilde{\Psi}$, and we neglect its fluctuations. Further calculations can be done in spirit of the order parameter fluctuation calculations. One can rewrite the expression (2.10) as the functional of the fluctuating vector-potential, choosing the gauge $\nabla \cdot \mathbf{A} = 0$ and writing it in the form of Fourier series:

$$\mathcal{F}[\mathbf{A}(\mathbf{r})] = F_N + \int dV \left[a|\widetilde{\Psi}|^2 + \frac{b}{2}|\widetilde{\Psi}|^4 + \frac{e^2|\widetilde{\Psi}|^2}{m}\mathbf{A}^2 + \frac{\mathbf{k}^2\mathbf{A}^2}{8\pi} \right]$$

$$= F_S[\widetilde{\Psi}, 0] + \sum_{\mathbf{k}} \left(\frac{e^2|\widetilde{\Psi}|^2}{m} + \frac{\mathbf{k}^2}{8\pi} \right) \mathbf{A}_{\mathbf{k}}^2.$$

The corresponding free energy can be written as:

$$F = F_S[\widetilde{\Psi}, 0] + \ln \int \mathfrak{D}A_x \mathfrak{D}A_y \exp\left(-\frac{1}{T} \sum_{\mathbf{k}} \left(\frac{e^2|\widetilde{\Psi}|^2}{m} + \frac{\mathbf{k}^2}{8\pi} \right) \mathbf{A}_{\mathbf{k}}^2 \right) =$$

$$= F_S[\widetilde{\Psi}, 0] + T \sum_{\mathbf{k}} \ln \frac{\pi T}{\frac{e^2|\widetilde{\Psi}|^2}{m} + \frac{\mathbf{k}^2}{8\pi}}. \tag{2.152}$$

The sum in (2.152) in the $3D$ case, evidently, strongly diverges. Namely, the two first terms of the expansion over $|\widetilde{\Psi}|^2$ turn out to be divergent. The most divergent term is nothing other than the consequence of the ultraviolet catastrophe of classical electrodynamics and it, as usual, can be simply attributed to the background value of the free energy. The next, proportional to $|\widetilde{\Psi}|^2$, contribution to the sum in (2.152) turns out to be divergent too. This divergence we treat in the same way as above: renormalize the critical temperature of the superconducting transition with respect to its mean field value. Regularized in this way free energy can be rewritten finally as:

$$F = F_N + a|\widetilde{\Psi}|^2 - \frac{16T\sqrt{\pi}e^3}{m^{3/2}}|\widetilde{\Psi}|^3 + \frac{b}{2}|\widetilde{\Psi}|^4. \tag{2.153}$$

One can see that an unusual cubic term appears in it due to the fluctuations of the electromagnetic field. This contribution, despite being small in magnitude, changes the order of the superconducting phase transition. Indeed, putting $a = 0$ one can see that the minimum of the free energy takes place for a positive order parameter value. This means that already for positive a the free energy has two minima (one at $\Psi = 0$ and a second for positive Ψ) appropriate to the normal and superconducting states. A first-order phase transition takes place when the values of the free energy in both of these minima coincide, i.e. at $\widetilde{a} = 128\pi T^2 e^6/bm^3$, or, using the microscopic expressions for λ and α^2/b, at

$$\widetilde{\epsilon} = \frac{\alpha^2}{b} \frac{2^{13}\pi T^4 e^6}{(4\alpha m T)^3} = 2 \cdot (2\pi)^4 \, Gi_{(3)} \left(\frac{\xi(T)}{\lambda(T)} \right)^6. \tag{2.154}$$

From this result one can see that for a type-I superconductor this region noticeably exceeds the critical one and the phase transition, due to electromagnetic field fluctuations, turns out to be of first order.

For type-II superconductors the region of temperatures $\widetilde{\epsilon}$, where the electromagnetic field fluctuations are important, turns out to lie within the critical one. This means that these fluctuations have to be taken into account in the equations of the renormalization group which results in the conclusion of a first-order phase transition in type-II superconductors too. Nevertheless for temperatures $\epsilon \gtrsim Gi_{(3)}$ these effects are negligible and in the further discussion we will not take them into account.

3

FLUCTUATION CHARGE TRANSPORT

The appearance of fluctuating Cooper pairs above T_c leads to the opening of a "new channel" for charge transfer. In the Introduction the fluctuation Cooper pairs were treated as carriers with charge $2e$ while their lifetime τ_{GL} was chosen to play the role of the scattering time in the Drude formula. Such a qualitative consideration results in the Aslamazov–Larkin (AL) pair contribution to conductivity (1.9) (the so-called paraconductivity.[21]) Below we will present the generalization of the phenomenological GL functional approach to transport phenomena. Dealing with the fluctuation order parameter, it is possible to describe correctly the paraconductivity type fluctuation contributions to the normal resistance and magnetoconductivity, Hall effect, thermoelectric power and thermal conductivity at the edge of the transition. Unfortunately the indirect fluctuation contributions are beyond the possibilities of the description by time-dependent GL (TDGL) approach and they will be calculated in the framework of the microscopic theory (see Part II).

3.1 Time-dependent GL equation

In previous sections, we have demonstrated how the GL functional formalism allows one to accounting for fluctuation corrections to thermodynamic quantities. Let us discuss the effect of fluctuations on the transport properties of a superconductor above the critical temperature.

In order to find the value of paraconductivity, some time-dependent generalization of the GL equations is required. Indeed, the conductivity characterizes the response of the system to the applied electric field. It can be defined as $\mathbf{E} = -\partial\mathbf{A}/\partial t$ but, in contrast to the previous section, $\mathbf{A}$ has to be regarded as being time dependent. The general nonstationary BCS equations are very complicated, even in the limit of slow time and space variations of the field and the order parameter. For our purposes it will be sufficient, following [98–104], to write a model equation in the vicinity of T_c, which in general correctly reflects the qualitative aspects of the order parameter dynamics and in some cases is exact.

Let us revise the GL functional formalism introduced above. One can see that the derived above stationary GL equations do not describe correctly the superconducting properties when a deviation from equilibrium is assumed. Indeed,

[21] This term may have different origins. First of all, evidently, paraconductivity is analogous to paramagnetism and means excess conductivity. Another possible origin is an incorrect onomatopoeic translation from the Russian "paroprovodimost' " that means pair conductivity.

in the absence of equilibrium the order parameter Ψ becomes time-dependent and this in no way was included in the scheme. Nevertheless, the scheme can be improved. For small deviations from the equilibrium it is natural to assume that in the process of order parameter relaxation its time derivative $\partial\Psi/\partial t$ is proportional to the variational derivative of the free energy $\delta\mathcal{F}/\delta\Psi^*$, which is equal to zero at the equilibrium. But this is not all: side by side with the normal relaxation of the order parameter the effect of thermodynamic fluctuations on it has to be taken into account. This can be done by the introduction the Langevin forces $\zeta(\mathbf{r},t)$ in the right-hand side of the equation describing the order parameter dynamics. Finally, gauge invariance requires that $\partial\Psi/\partial t$ should be included in the equation in the combination $\partial\Psi/\partial t + 2ie\varphi\Psi$, where φ is the scalar potential of the electric field. By including all these considerations one can write the model time-dependent GL equation in the form

$$-\gamma_{\mathrm{GL}}\left(\frac{\partial}{\partial t}+2ie\varphi\right)\Psi=\frac{\delta\mathcal{F}}{\delta\Psi^*}+\zeta(\mathbf{r},t) \tag{3.1}$$

with the GL functional $\mathcal{F}$ determined by (2.4), (2.10), (2.66).[22] The dimensionless coefficient γ_{GL} in the left-hand-side of the equation can be related to pair lifetime τ_{GL} (1.2): $\gamma_{\mathrm{GL}}=\alpha T_{\mathrm{c}}\epsilon\tau_{\mathrm{GL}}=\pi\alpha/8$ by the substitution in (3.1) of the first term of (2.4) only.[23]

Neglecting the fourth-order term in the GL functional, Eq. (3.1) can be rewritten in operator form as

$$[\widehat{L}^{-1}-2ie\gamma_{\mathrm{GL}}\varphi(r,t)]\Psi(\mathbf{r},t)=\zeta(\mathbf{r},t) \tag{3.2}$$

with the TDGL operator $\widehat{L}$ and Hamiltonian $\widehat{\mathcal{H}}$ defined as

$$\widehat{L}=\left[\gamma_{\mathrm{GL}}\frac{\partial}{\partial t}+\widehat{\mathcal{H}}\right]^{-1},\ \widehat{\mathcal{H}}=\alpha T_{\mathrm{c}}\left[\epsilon-\widehat{\xi}^2(\widehat{\nabla}-2ie\mathbf{A})^2\right]. \tag{3.3}$$

We have introduced here the formal operator of the coherence length $\widehat{\xi}$ to have the possibility to deal with an arbitrary type of spectrum. For example, in the most interesting case for our applications to layered superconductors, the action of this operator is defined by Eq. (2.66).

In the absence of an electric field one can write the formal solution of equation (3.2) as

[22] An equation of this type was considered by Landau and Khalatnikov [105] in connection with the study of superfluid helium dynamics in early 1950s

[23] It will be shown below that taking into account electron- hole asymmetry leads to the appearance of the imaginary part of γ_{GL} proportional to the derivative $\partial\ln(\rho v^2\tau)/\partial E|_{E_F}\sim\mathcal{O}(1/E_F)$. This is important for such phenomena as fluctuation Hall effect or fluctuation thermopower and, having in mind the writing of the most general formula, we will suppose $\gamma_{\mathrm{GL}}=\pi\alpha/8+i\,\mathrm{Im}\,\gamma_{\mathrm{GL}}$, where necessary.

$$\Psi^{(0)}(\mathbf{r},t) = \widehat{L}\zeta(\mathbf{r},t). \tag{3.4}$$

The correlator of the Langevin forces introduced above must satisfy the fluctuation–dissipation theorem. This means that the correlators $< \Psi_{\mathbf{p}}^{(0)*}(t')\Psi_{\mathbf{p}}^{(0)}(t) >$ at coinciding moments of time have to be the same as $< |\Psi_{\mathbf{p}}|^2 >$, obtained by averaging over fluctuations in thermal equilibrium:

$$< |\Psi_{\mathbf{p}}|^2 >= \frac{\int \mathfrak{D}\Psi_{\mathbf{p}}\mathfrak{D}\Psi_{\mathbf{p}}^*|\Psi_{\mathbf{p}}|^2 \exp\left\{-\alpha(\epsilon+\xi^2\mathbf{p}^2)|\Psi_{\mathbf{p}}|^2\right\}}{\int \mathfrak{D}\Psi_{\mathbf{p}}\mathfrak{D}\Psi_{\mathbf{p}}^* \exp\left\{-\alpha(\epsilon+\xi^2\mathbf{p}^2)|\Psi_{\mathbf{p}}|^2\right\}} = \frac{1}{\alpha(\epsilon+\xi^2\mathbf{p}^2)}. \tag{3.5}$$

This requirement is fulfilled if the Langevin forces $\zeta(\mathbf{r},t)$ and $\zeta^*(\mathbf{r},t)$ are correlated by the Gaussian white-noise law

$$\langle\zeta^*(\mathbf{r},t)\zeta(\mathbf{r}^{'},t^{'})\rangle = 2T\,\mathrm{Re}\,\gamma_{\mathrm{GL}}\delta(\mathbf{r}-\mathbf{r}^{'})\delta(t-t^{'}). \tag{3.6}$$

To show it let us restrict ourselves for sake of simplicity, to the case of $A = 0$ and calculate the correlator

$$\begin{aligned}\langle\ \Psi^*(\mathbf{r},t)\Psi(\mathbf{r}',t)\rangle &= \langle\zeta^*(\mathbf{r},t)\widehat{\widehat{L^*}\widehat{L}}\zeta(\mathbf{r}^{'},t)\rangle = 2T\,\mathrm{Re}\,\gamma_{\mathrm{GL}} \\ &\times\int\frac{d\mathbf{p}}{(2\pi)^D}e^{i\mathbf{p}(\mathbf{r}-\mathbf{r}^{'})}\int_{-\infty}^{\infty}\frac{d\Omega}{2\pi}L^*(\mathbf{p},\Omega)L(\mathbf{p},\Omega).\end{aligned} \tag{3.7}$$

The fundamental solution $L(\mathbf{p},\Omega)$ can be found by making a Fourier transform of (3.3), what gives:

$$L(\mathbf{p},\Omega) = (-i\gamma_{\mathrm{GL}}\Omega + \varepsilon_{\mathbf{p}})^{-1}. \tag{3.8}$$

Substitution of this expression to the previous correlator and calculation of the remaining integral over frequencies complete our demonstration:

$$\langle\Psi^*(\mathbf{r},t)\Psi(\mathbf{r}',t)\rangle_{\mathbf{p}} = 2T\,\mathrm{Re}\,\gamma_{GL}\int_{-\infty}^{\infty}\frac{d\Omega}{2\pi}\frac{1}{|-i\gamma_{GL}\Omega+\varepsilon_{\mathbf{p}}|^2} = \left\langle|\Psi_{\mathbf{p}}|^2\right\rangle. \tag{3.9}$$

Here

$$\varepsilon_{\mathbf{p}} = \alpha T_{\mathrm{c}}(\epsilon+\widehat{\xi}^2\mathbf{p}^2). \tag{3.10}$$

is the fluctuation Cooper pair energy spectrum.

3.2 General expression for paraconductivity

By means of the qualitative consideration based on the Drude formula, we obtained in the Introduction the expression for paraconductivity which correctly reflects its temperature singularity in any dimension. Following this way one could write down some kind of master equation for fluctuation Cooper pairs (this will be done in section 3.7) and obtain indeed the precise expression for paraconductivity. Unfortunately, the applicability of the derived master equation is restricted to relatively weak electric and magnetic fields. For stronger fields $H_{c2}(\epsilon) \lesssim H \ll H_{c2}(0)$ the density matrix has to be introduced and the master equation loses its attractive simplicity. At the same time, as we already know, these fields, quantizing the fluctuation Cooper pair motion, present special interest. That is why in order to include in the scheme the magnetic field and frequency dependencies of the paraconductivity, we return to the analysis of the general TDGL equation (3.1) without the objective to reduce it to a Boltzmann type transport equation.

Let us solve it in the case when the applied electric field can be considered as a perturbation. The method will much resemble an exercise from a course on quantum mechanics. To impose the necessary generality side by side with a formal simplicity of expressions we will introduce a subscript of the kind $\{i\}$ which includes the complete set of quantum numbers and time. By a repeated subscript a summation over a discrete and integration over continuous variables (time in particular) is implied.

We will look for the response of the order parameter to a weak electric field applied in the form

$$\Psi_{k_z}(\mathbf{r},t) = \Psi^{(0)}_{\{i\}} + \Psi^{(1)}_{\{i\}}, \tag{3.11}$$

where $\Psi^{(0)}_{\{i\}}$ is determined by (3.4). Substituting this expression into (3.2) and restricting our consideration to linear terms in the electric field we can write

$$(\widehat{L}^{-1})_{\{ik\}}\Psi^{(1)}_{\{k\}} = 2ie\gamma_{\mathrm{GL}}\varphi_{\{il\}}\Psi^{(0)}_{\{l\}} \tag{3.12}$$

with the solution in the form

$$\Psi^{(1)}_{\{i\}} = 2ie\gamma_{\mathrm{GL}}\widehat{L}_{\{ik\}}\varphi_{\{kl\}}\widehat{L}_{\{lm\}}\zeta_{\{m\}}. \tag{3.13}$$

Let us substitute the order parameter (3.11) in the quantum mechanical expression for current (2.17)

$$\mathbf{j} = 2e\,\mathrm{Re}\left[\Psi^{(0)*}_{\{i\}}\widehat{\mathbf{v}}_{\{ik\}}\Psi^{(1)}_{\{k\}} + \Psi^{(1)*}_{\{i\}}\widehat{\mathbf{v}}_{\{ik\}}\Psi^{(0)}_{\{k\}}\right], \tag{3.14}$$

where $\widehat{\mathbf{v}}_{\{ik\}}$ is the velocity operator which can be expressed by means of the commutator of $\mathbf{r}$ with Hamiltonian (3.3):

$$\widehat{\mathbf{v}}_{\{ik\}} = i\{\widehat{\mathcal{H}},\mathbf{r}\}_{\{ik\}}. \tag{3.15}$$

The second term of (3.14) can be written by means of a transposed velocity operator (which is Hermitian) as the complex conjugated value of the first one:

$$\Psi^{(1)*}_{\{i\}}\widehat{\mathbf{v}}_{\{ik\}}\Psi^{(0)}_{\{k\}} = (\Psi^{(0)*}_{\{k\}}\widetilde{\widehat{\mathbf{v}}}_{\{ik\}}\Psi^{(1)}_{\{i\}})^*, \tag{3.16}$$

which results in

$$\mathbf{j} = 2\,\mathrm{Re}\{\Psi^{(0)*}_{\{i\}}(2e\widehat{\mathbf{v}}_{\{ik\}})\Psi^{(1)}_{\{k\}}\}$$

$$= -8e^2 \operatorname{Im}\{\gamma_{\mathrm{GL}} \widehat{L}^*_{\{ki\}} \widehat{\mathbf{v}}_{\{il\}} \widehat{L}_{\{lm\}} \varphi_{\{mn\}} \widehat{L}_{\{np\}} \zeta^*_{\{k\}} \zeta_{\{p\}}\}. \tag{3.17}$$

Let us average now (3.17) over the Langevin forces, taking into account the property (3.6). Moving the operator $\widehat{L}^*_{\{ki\}}$ from the beginning to the end of the trace one finds

$$\mathbf{j} = -16Te^2 \operatorname{Re}(\gamma_{\mathrm{GL}}) \operatorname{Im}\{\gamma_{\mathrm{GL}} \widehat{\mathbf{v}}_{\{il\}} \widehat{L}_{\{lm\}} \varphi_{\{mn\}} \widehat{L}_{\{np\}} \widehat{L}^*_{\{pi\}}\}. \tag{3.18}$$

Now we choose the representation where the $\widehat{L}_{\{lm\}}$ operator is diagonal (it is evidently given by the eigenfunctions of the Hamiltonian (3.3)):

$$L_{\{m\}}(\Omega) = \frac{1}{\varepsilon_{\{m\}} - i\Omega\gamma_{\mathrm{GL}}}, \tag{3.19}$$

where $\varepsilon_{\{m\}}$ are the appropriate energy eigenvalues. Then we assume that the electric field is coordinate independent but is a monochromatic periodic function of time:

$$\varphi(r,t) = -E^\beta r^\beta \exp(-i\omega t). \tag{3.20}$$

In doing the Fourier transform in (3.18) one has to remember that the time dependence of the matrix elements $\varphi_{\{mn\}}$ results in a shift of the frequency variable of integration $\Omega \to \Omega - \omega$ in both L-operators placed after $\varphi_{\{mn\}}$ or, what is the same, to a shift of the argument of the previous $\widehat{L}_{\{lm\}}$ for ω:

$$\mathbf{j}^\alpha_\omega = 16Te^2 \operatorname{Re}(\gamma_{\mathrm{GL}}) \int \frac{d\Omega}{2\pi} \Re\{\gamma_{\mathrm{GL}} \widehat{\mathbf{v}}^\alpha_{\{il\}} \widehat{L}_{\{l\}}(\Omega+\omega)[-ir^\beta_{\{li\}}]\widehat{L}_{\{i\}}(\Omega)\widehat{L}^*_{\{i\}}(\Omega)\}\mathbf{E}^\beta, \tag{3.21}$$

where $\Re f(\omega) \equiv [f(\omega) + f^*(-\omega)]/2$.

Let us express the matrix element $\mathbf{r}_{\{li\}}$ by means of $\widehat{\mathbf{v}}_{\{li\}}$ using the commutation relation (3.15). One can see that in the representation chosen

$$\widehat{\mathbf{r}}^\beta_{\{li\}} = i \frac{\widehat{\mathbf{v}}^\beta_{\{li\}}}{\varepsilon_{\{i\}} - \varepsilon_{\{l\}}} \tag{3.22}$$

and, carrying out the frequency integration in (3.21), finally write for the fluctuation conductivity tensor ($\mathbf{j}^\alpha_\omega = \sigma^{\alpha\beta}(\omega)\mathbf{E}^\beta$):

$$\sigma^{\alpha\beta}(\epsilon, H, \omega)$$
$$= 8e^2 T \operatorname{Re}(\gamma_{\mathrm{GL}}) \sum_{\{i,l\}=0}^{\infty} \Re\left[\gamma_{\mathrm{GL}} \frac{\widehat{\mathbf{v}}^\alpha_{\{il\}} \widehat{\mathbf{v}}^\beta_{\{li\}}}{\varepsilon_{\{i\}}(\gamma_{\mathrm{GL}}\varepsilon_{\{i\}} + \gamma^*_{\mathrm{GL}}\varepsilon_{\{l\}} - i|\gamma_{\mathrm{GL}}|^2\omega)(\varepsilon_{\{l\}} - \varepsilon_{\{i\}})}\right]. \tag{3.23}$$

This is the most general expression which describes the d.c., galvanomagnetic and high frequency paraconductivity contributions.

The microscopic analysis of the coefficient $\gamma_{\rm GL}$ demonstrates that its imaginary part $\operatorname{Im}\gamma_{\rm GL}$ usually is much smaller than $\operatorname{Re}\gamma_{\rm GL}$. Its origin can be related to the electron–hole asymmetry or other peculiarities of the electron spectrum (see Part III). In the case when one is interested in the diagonal effects only it is enough to accept $\gamma_{\rm GL}$ as real: ($\gamma_{\rm GL} = \operatorname{Re}\gamma_{\rm GL} = \pi\alpha/8$). In this way Eq. (3.24) can be simplified and after symmetrization of the summation variables it takes the form:

$$\sigma^{\alpha\alpha}(\epsilon, H, \omega) = \frac{\pi}{2}\alpha e^2 T \sum_{\{i,l\}=0}^{\infty} \Re\left[\frac{\widehat{\mathbf{v}}^{\alpha}_{\{il\}}\widehat{\mathbf{v}}^{\alpha}_{\{li\}}}{\varepsilon_{\{i\}}\varepsilon_{\{l\}}(\varepsilon_{\{i\}}+\varepsilon_{\{l\}}-i\gamma_{\rm GL}\omega)}\right]. \qquad (3.24)$$

Let us demonstrate the calculation of the d.c. paraconductivity in the simplest case of a metal with an isotropic spectrum. In this case we choose a plane wave representation. By using $\varepsilon_{\mathbf{p}}$ defined by (3.10) one has

$$\widehat{\mathbf{v}}_{\{\mathbf{p}\mathbf{p}'\}} = \mathbf{v}_{\mathbf{p}}\delta_{\mathbf{p}\mathbf{p}'}, \; \mathbf{v}_{\mathbf{p}} = \frac{\partial\varepsilon_{\mathbf{p}}}{\partial\mathbf{p}} = 2\alpha T_{\rm c}\xi^2\mathbf{p}. \qquad (3.25)$$

We do not need to keep the imaginary part of $\gamma_{\rm GL}$, which is necessary to calculate particle-hole asymmetric effects only. As the result one reproduces the AL formula:

$$\sigma^{\alpha\beta}_{(D)} = 2e^2 T \operatorname{Re}\gamma_{\rm GL}\sum_{\mathbf{p}}\frac{\mathbf{v}^{\alpha}_{\mathbf{p}}\mathbf{v}^{\beta}_{\mathbf{p}}}{\varepsilon^3_{\mathbf{p}}} = \delta^{\alpha\beta}\begin{cases} \dfrac{e^2}{32\xi}\dfrac{1}{\sqrt{\epsilon}}, & \text{3D case,} \\ \dfrac{e^2}{16d}\dfrac{1}{\epsilon}, & \text{2D film, thickness}: d \ll \xi, \\ \dfrac{\pi e^2\xi}{16S}\dfrac{1}{\epsilon^{3/2}}, & \text{1D wire, cross-section}: S \ll \xi^2. \end{cases} \qquad (3.26)$$

3.3 Paraconductivity of a layered superconductor

Let us return to the discussion of our general formula (3.24) for the fluctuation conductivity tensor. A magnetic field directed along the c-axis still allows separation of variables even in the case of a layered superconductor. The Hamiltonian in this case can be written as in (2.68), (3.3):

$$\widehat{\mathcal{H}} = \alpha T_{\rm c}\left(\epsilon - \xi^2_{xy}(\nabla_{xy} - 2ie\mathbf{A}_{xy})^2 - \frac{r}{2}(1-\cos(k_z s))\right). \qquad (3.27)$$

It is convenient to work in the Landau representation, where the summation over $\{i\}$ is reduced to one over the ladder of Landau levels $i = 0, 1, 2....$ (each is degenerate with a density H/Φ_0 per unit square) and integration over the c-axis momentum in the limits of the Brillouin zone. The eigenvalues of the Hamiltonian (3.27) can be written in the form

$$\varepsilon_{\{n\}} = \alpha T_c[\epsilon + \frac{r}{2}(1 - \cos(k_z s)) + h(2n+1)] = \varepsilon_{k_z} + \alpha T_c h(2n+1), \qquad (3.28)$$

where $h = eH/2m\alpha T_c$ was already defined by Eq. (2.52). For the velocity operators one can write

$$\widehat{\mathbf{v}}^{x,y} = \frac{1}{2m}(-i\nabla - 2ie\mathbf{A})^{x,y}; \; \widehat{\mathbf{v}}^z = -\frac{\alpha r s}{2} T_c \sin(k_z s). \qquad (3.29)$$

3.3.1 *In-plane conductivity*

Let us start from the calculation of the in-plane components. The calculation of the velocity operator matrix elements requires some special consideration. First of all let us stress that the required matrix elements have to be calculated for the eventuates of a quantum oscillator whose motion is equivalent to the motion of a charged particle in a magnetic field. The commutation relation for the velocity components follows from (3.29) (see [47]):

$$[\widehat{\mathbf{v}}^x, \widehat{\mathbf{v}}^y] = i\frac{eH_z}{2m^2} = \frac{i\alpha T_c}{m} h. \qquad (3.30)$$

In order to calculate the necessary matrix elements let us present the velocity operator components in the form of boson-type creation and annihilation operators $\widehat{a}^+, \widehat{a}$:

$$\langle l|\widehat{a}|n\rangle = \langle n|\widehat{a}^+|l\rangle = \sqrt{n}\delta_{n,l+1},$$

which satisfy the commutation relation $[\widehat{a}, \widehat{a}^+] = 1$. We obtain

$$\widehat{\mathbf{v}}^{x,y} = \sqrt{\frac{\alpha T_c h}{2m}} \begin{pmatrix} \widehat{a}^+ + \widehat{a} \\ i\widehat{a}^+ - i\widehat{a} \end{pmatrix}.$$

One can check that the correct commutation relation (3.30) is fulfilled and see that the only nonzero matrix elements of the velocity operator are

$$\langle l|\widehat{\mathbf{v}}^{x,y}|n\rangle = \sqrt{\frac{\alpha T_c h}{2m}} \begin{pmatrix} \sqrt{l}\delta_{l,n+1} + \sqrt{n}\delta_{n,l+1} \\ i\sqrt{l}\delta_{l,n+1} - i\sqrt{n}\delta_{n,l+1} \end{pmatrix}. \qquad (3.31)$$

Using these relations the necessary product of matrix elements can be calculated:

$$\langle l|\widehat{\mathbf{v}}^x|n\rangle\langle n|\widehat{\mathbf{v}}^x|l\rangle = \frac{\alpha T_c h}{2m}(l\delta_{l,n+1} + n\delta_{n,l+1}). \qquad (3.32)$$

Its substitution to the expression (3.24) gives for the diagonal in-plane component of the paraconductivity tensor

$$\sigma^{xx}(\epsilon, h, \omega) = \frac{\pi\alpha^2 T_c^2 e^2}{4m} h \sum_{\{n,l\}=0}^{\infty} \Re \frac{(l\delta_{l,n+1} + n\delta_{n,l+1})}{\varepsilon_{\{l\}}\varepsilon_{\{n\}}\left[\varepsilon_{\{l\}} + \varepsilon_{\{n\}} - i\gamma_{\mathrm{GL}}\omega\right]}$$

Summation over the subscript $\{l\}$ and accounting of the degeneracy of the Landau levels $H/\Phi_0 = 2m\alpha T_c h/\pi$ (the layer area we assume to be equal one), gives for the diagonal component of the in-plane paraconductivity tensor:

$$\sigma^{xx}(\epsilon, H, \omega) = \frac{e^2 (\alpha T_c)^3 h^2}{2} \int_{-\frac{\pi}{s}}^{\frac{\pi}{s}} \frac{dk_z}{2\pi} \sum_{n=0}^{\infty} \Re \frac{n+1}{\varepsilon_{n+1}\varepsilon_n(\varepsilon_{n+1}+\varepsilon_n - i\gamma_{\mathrm{GL}}\omega)}. \quad (3.33)$$

Expanding the denominator into simple fractions we reduce the problem to the calculation of the c-axis momentum integral, which can be carried out in the general case by use of the identity (C.6). Using it we write the general expression for the diagonal in-plane component of the fluctuation conductivity tensor

$$\sigma^{xx}(\epsilon, h, \omega) = \frac{e^2 h}{8s} \sum_{n=0}^{\infty} (n+1) \left\{ \frac{1}{h - i\widetilde{\omega}} \frac{1}{\sqrt{[\epsilon + h(2n+1)][r + \epsilon + h(2n+1)]}} \right.$$
$$+ \frac{1}{h + i\widetilde{\omega}} \frac{1}{\sqrt{[\epsilon + h(2n+3)][r + \epsilon + h(2n+3)]}}$$
$$\left. - \frac{2h}{h^2 + \widetilde{\omega}^2} \frac{1}{\sqrt{[\epsilon + h(2n+2) - i\widetilde{\omega}][r + \epsilon + h(2n+2) - i\widetilde{\omega}]}} \right\}, \quad (3.34)$$

where $\widetilde{\omega} = \pi\omega/16T_c$.

3.3.2 *Out-of-plane conductivity*

The situation with the out-of-plane component of paraconductivity turns out to be even simpler because of the diagonal structure of the $\widehat{\mathbf{v}}^z_{\{in\}} = -\frac{\alpha r s}{2} T_c \sin(k_z s) \times \delta_{in} \times \delta(k_z - k_{z'})$. Taking into account that the Landau state degeneracy we write

$$\sigma^{zz}(\epsilon, H, \omega) = \frac{1}{2}\pi\alpha e^2 T \sum_{\{i,l\}=0}^{\infty} \Re \left[\frac{\widehat{\mathbf{v}}^z_{\{il\}} \widehat{\mathbf{v}}^z_{\{li\}}}{\varepsilon_{\{i\}}\varepsilon_{\{l\}}(\varepsilon_{\{i\}} + \varepsilon_{\{l\}} - 2i\alpha T_c \widetilde{\omega})} \right] =$$
$$= \frac{\pi e^2 (\alpha T_c)^3}{32} \left(\frac{sr}{\xi_{xy}} \right)^2 h \sum_{n=0}^{\infty} \int_{-\frac{\pi}{s}}^{\frac{\pi}{s}} \frac{dk_z}{2\pi} \Re [\frac{\sin^2(k_z s)}{\varepsilon_n^2(k_z)[\varepsilon_n(k_z) - i\alpha T_c \widetilde{\omega}]}].$$

The following transformations are similar to the calculation of the in-plane component: we expand the integrand into simple fractions and perform the k_z−integration by means of the identity (C.6). The final expression can be written as

$$\sigma^{zz}(\epsilon, H, \omega) = \frac{e^2}{64s} \left(\frac{sr}{\xi_{xy}} \right)^2 h \sum_{n=0}^{\infty} \left(-\frac{\partial}{\partial \lambda} \right)$$
$$\times \Re \left(\frac{1}{\lambda + i\widetilde{\omega}} \right) \left\{ \frac{1}{\sqrt{(\epsilon + h(2n+1) + \lambda)(\epsilon + h(2n+1) + \lambda + r)}} \right.$$

$$- \frac{1}{\sqrt{(\epsilon + h(2n+1) - i\widetilde{\omega})(\epsilon + h(2n+1) - i\widetilde{\omega} + r)}} \Bigg\} |_{\lambda=0} . \quad (3.35)$$

Let us mention, that the formal differentiation has to be done first, then $\lambda = 0$ is substituted.

3.3.3 *Analysis of the limiting cases*

In principle the expressions derived above give an exact solution for the a.c. ($\omega \ll T$) paraconductivity tensor of a layered superconductor in a perpendicular magnetic field $H \ll H_{c2}$ ($h \ll 1$) in the vicinity of the critical temperature ($\epsilon \ll 1$). The interplay of the parameters r, ϵ, ω, h entering into (3.34)–(3.35), as we have seen in the example of fluctuation magnetization, yields a variety of crossover phenomena.

3.3.3.1 *d.c. paraconductivity* The simplest and most important results which can be derived are the components of the d.c. paraconductivity ($\omega = 0$) of a layered superconductor in the absence of magnetic field. Keeping $\omega = 0$ and setting $h \to 0$ one can change the summations over Landau levels into integration and find

$$\sigma^{xx}(\epsilon, h \to 0, \omega = 0) = \frac{e^2}{16s} \frac{1}{\sqrt{[\epsilon(r+\epsilon)]}}, \quad (3.36)$$

$$\sigma^{zz}(\epsilon, h \to 0, \omega = 0) = \frac{e^2 s}{32\xi_{xy}^2} \left(\frac{\epsilon + r/2}{[\epsilon(\epsilon + r)]^{1/2}} - 1 \right). \quad (3.37)$$

3.3.3.2 *Magnetoconductivity* The AL contribution to the magnetoconductivity can be studied by putting $\omega = 0$ and keeping magnetic field as arbitrary. For $2D$ case the sum in (3.34) can be calculated exactly in terms of the ψ-functions:

$$\sigma_{(2)}^{xx}(\epsilon, h) = \frac{e^2}{2s} \frac{1}{\epsilon} F\left(\frac{\epsilon}{2h}\right) = \frac{e^2}{16s} \begin{cases} 1/\epsilon, & h \ll \epsilon \\ 2/h, & \epsilon \ll h \\ 4/(\epsilon + h), & \epsilon + h \to 0 \end{cases}, \quad (3.38)$$

where

$$F(x) = x^2 \left[\psi\left(\frac{1}{2} + x\right) - \psi(x) - \frac{1}{2x} \right]. \quad (3.39)$$

We will not go into the further details and just report in Table 3.1 the results (following [106] with some revision of the coefficient in the $3D$ case):

Here it is worth making an important comment. The proportionality of the fluctuation magnetoconductivity to h^2 is valid when using the parametrization $\epsilon = (T - T_{c0})/T_{c0}$ only. Often the analysis of the experimental data is carried out by choosing as the reduced temperature parameter $\epsilon_h = (T - T_c(H))/T_c(H)$. At that it is important to recognize that the effect of a weak magnetic field on the fluctuation conductivity cannot be reduced to a simple replacement of T_{c0}

Table 3.1 *The asymptotics of the paraconductivity for different magnetic field domains*

	$h \ll \epsilon$	$\epsilon \ll h \ll r$ $(3D)$	$\max\{\epsilon, r\} \ll h$ $(2D)$
σ^{xx}	$\sigma^{xx}(\epsilon, h=0) - \frac{e^2}{2^8 s}\frac{[8\epsilon(\epsilon+r)+3r^2]}{[\epsilon(\epsilon+r)]^{5/2}}h^2$	$\frac{e^2}{4s}\frac{1}{\sqrt{2hr}}$	$\frac{e^2}{8s}\frac{1}{h}$
σ^{zz}	$\sigma^{zz}(\epsilon, h=0) - \frac{e^2 s}{2^8 \xi_{xy}^2}\frac{r^2(\epsilon+r/2)}{[\epsilon(\epsilon+r)]^{5/2}}h^2$	$\frac{3.24 e^2 s}{\xi_{xy}^2}\sqrt{\frac{r}{h}}$	$\frac{7\zeta(3)e^2 s}{2^9 \xi_{xy}^2}\frac{r^2}{h^2}$

by $T_c(H)$ in the appropriate formula without the field. In this parametrization one can get a term in the magnetoconductivity linear in h. Point is that besides the cases of the special specimen geometry, a weak magnetic field shifts the critical temperature linearly. Such linear correction is exactly compensated by the change in the functional dependence of the paraconductivity in magnetic field, and finally it contains the negative quadratic contribution only.

3.3.3.3 *a.c. paraconductivity.* Letting the magnetic field go to zero in expression (3.24) and considering nonzero frequencies of the electromagnetic field one can find general expressions for the components of the a.c. paraconductivity tensor. For the in-plane paraconductivity they are cumbersome enough and in the complete form can be found, for instance, in [107]. We recall here the simplified asymptotics for superconductor in the $2D$ regime only [54, 108]:

$$\operatorname{Re}\sigma_{(2)}^{xx}(r \ll \epsilon, \widetilde{\omega}) = \frac{e^2}{16s}\frac{1}{\epsilon}\left[\frac{2\epsilon}{\widetilde{\omega}}\arctan\frac{\widetilde{\omega}}{\epsilon} - \left(\frac{\epsilon}{\widetilde{\omega}}\right)^2 \ln\left[1+\left(\frac{\widetilde{\omega}}{\epsilon}\right)^2\right]\right]; \quad (3.40)$$

$$\operatorname{Im}\sigma_{(2)}^{xx}(r \ll \epsilon, \widetilde{\omega}) = \frac{e^2}{16s}\frac{1}{\epsilon}\left[\frac{2\epsilon}{\widetilde{\omega}}\left(\arctan\frac{\widetilde{\omega}}{\epsilon} - \frac{\widetilde{\omega}}{\epsilon}\right) - \frac{\epsilon}{\widetilde{\omega}}\ln\left[1+\left(\frac{\widetilde{\omega}}{\epsilon}\right)^2\right]\right]. \quad (3.41)$$

Regarding the out-of-plane component of paraconductivity, it can be easily evaluated from the general expression (3.35)

$$\sigma^{zz}(\epsilon, \omega) = \frac{e^2}{64s}\left(\frac{sr}{\xi_{xy}}\right)^2\left[\frac{1}{\widetilde{\omega}^2}\ln\frac{\sqrt{\epsilon}+\sqrt{\epsilon+r}}{\sqrt{\epsilon-i\widetilde{\omega}}+\sqrt{\epsilon-i\widetilde{\omega}+r}} + \frac{1}{2i\widetilde{\omega}}\frac{1}{\sqrt{\epsilon(\epsilon+r)}}\right]. \quad (3.42)$$

In the $2D$ regime this general expression is reduced to

$$\operatorname{Re}\sigma_{(2)}^{zz}(r \ll \epsilon, \widetilde{\omega}) = \frac{e^2}{2^8 s}\left(\frac{sr}{\xi_{xy}}\right)^2\left(\frac{1}{\widetilde{\omega}}\right)^2 \ln[1+\left(\frac{\widetilde{\omega}}{\epsilon}\right)^2], \quad (3.43)$$

$$\operatorname{Im}\sigma_{(2)}^{zz}(r \ll \epsilon, \widetilde{\omega}) = \frac{e^2}{2^7 s}\left(\frac{sr}{\xi_{xy}}\right)^2\frac{1}{\widetilde{\omega}}\left[\frac{\epsilon}{\widetilde{\omega}}\arctan\frac{\widetilde{\omega}}{\epsilon} - 1\right]. \quad (3.44)$$

The general formulae (3.34)–(3.35) allow one to study the different crossovers in the a.c. conductivity of layered superconductor in the presence of magnetic field of various intensity. We leave this exercise for the reader having some practical interest in the problem.

3.4 Hall paraconductivity

Now we will calculate the paraconductivity contribution to the Hall component σ^{xy} of the conductivity tensor for the layered superconductor placed in a perpendicular magnetic field. Let us start from the general expression (3.23) and apply it to the Hall component. First of all we substitute in it the explicit expressions (3.31) for the matrix elements of the velocity operator. The Hall paraconductivity takes the form:

$$\sigma^{xy}(\epsilon, H, \omega) = \frac{2e^2T}{m}\omega_c \operatorname{Re}\gamma_{\mathrm{GL}} \sum_{\{n,l\}=0}^{\infty} \Re\left[\frac{i\gamma_{\mathrm{GL}}(l\delta_{l,n+1} - n\delta_{n,l+1})\delta_{q_z q_z'}}{\varepsilon_{\{n\}}(\varepsilon_{\{n\}} - \varepsilon_{\{l\}})}\right.$$
$$\left.\cdot\frac{1}{\left[\operatorname{Re}\gamma_{\mathrm{GL}}\left(\varepsilon_{\{n\}} + \varepsilon_{\{l\}}\right) - i|\gamma_{\mathrm{GL}}|^2\omega + i\operatorname{Im}\gamma_{\mathrm{GL}}\left(\varepsilon_{\{n\}} - \varepsilon_{\{l\}}\right)\right]}\right] \tag{3.45}$$

(we have introduced the notation $\omega_c = 2\alpha T_c h$). The principal difference of the expression (3.45) in comparison with the expression for diagonal components (3.24) consists in the imaginarity $< l|\widehat{\mathbf{v}}^y|n >$, that results in the exact cancelation of $\sigma^{xy}(\epsilon, H, 0)$ under the assumption $\operatorname{Im}\gamma_{\mathrm{GL}} = 0$. That is why as the next step we expand the expression (3.45) up to the first order over small $\operatorname{Im}\gamma_{\mathrm{GL}}$ and obtain:

$$\sigma^{xy}(\epsilon, H, \omega) = \frac{2e^2T}{m}\omega_c \operatorname{Im}\gamma_{\mathrm{GL}} \sum_{\{n,l\}=0}^{\infty} \Re\frac{(l\delta_{l,n+1} - n\delta_{n,l+1})\delta_{q_z q_z'}}{\varepsilon_{\{n\}}\left[\left(\varepsilon_{\{n\}} + \varepsilon_{\{l\}}\right) - i\omega\operatorname{Re}\gamma_{\mathrm{GL}}\right]^2}.$$

Calculating the sum over l, shifting the summation index $n+1 \to n'$ in the second term and taking into account that $\varepsilon_{\{n+1\}} - \varepsilon_{\{n\}} = \omega_c$, one can find

$$\sigma^{xy}(\epsilon, H, \omega) = \frac{2e^2T}{\pi}\omega_c^3 \operatorname{Im}\gamma_{\mathrm{GL}} \sum_{n=0}^{\infty}$$
$$\int_{-\frac{\pi}{s}}^{\frac{\pi}{s}} \frac{dq_z}{2\pi} \Re\frac{n+1}{\varepsilon_{\{n\}}\varepsilon_{\{n+1\}}\left(\varepsilon_{\{n\}} + \varepsilon_{\{n+1\}} - i\omega\operatorname{Re}\gamma_{\mathrm{GL}}\right)^2}. \tag{3.46}$$

It is interesting that almost the same sum was already carried out above in the process of calculation of the diagonal component (3.34). Presenting the square in the denominator of (3.46) as the derivative over frequency one can find the useful relation

$$\sigma^{xy}(\epsilon, H, \omega) = 2h\left(\frac{\operatorname{Im}\gamma_{\mathrm{GL}}}{\operatorname{Re}\gamma_{\mathrm{GL}}}\right)\frac{\partial\sigma^{xx}(\epsilon, H, \widetilde{\omega})}{\partial i\widetilde{\omega}} \tag{3.47}$$

(let us remember the notation: $\widetilde{\omega} = \pi\omega/16T_c$). Differentiating (3.34) is cumbersome but trivial. For simplicity, we restrict our analysis of $\sigma^{xy}(\epsilon, H)$ below to the case of direct current ($\omega = 0$). The general formula for Hall paraconductivity of a layered superconductor in an arbitrary magnetic field ($H \ll H_{c2}(0)$) takes the form:

$$\begin{aligned}\sigma^{xy}(\epsilon, H) = \frac{e^2}{4s}\left(\frac{\operatorname{Im}\gamma_{\mathrm{GL}}}{\operatorname{Re}\gamma_{\mathrm{GL}}}\right)\sum_{n=0}^{\infty}(n+1)\Bigg\{&\frac{1}{\sqrt{[\epsilon + h(2n+1)][r+\epsilon+h(2n+1)]}}\\
&-\frac{1}{\sqrt{[\epsilon+h(2n+3)][r+\epsilon+h(2n+3)]}}\\
&-\frac{h}{\sqrt{[\epsilon+h(2n+2)]^3[r+\epsilon+h(2n+2)]}}\\
&-\frac{h}{\sqrt{[\epsilon+h(2n+2)][r+\epsilon+h(2n+2)]^3}}\Bigg\}.\end{aligned} \tag{3.48}$$

Let us analyze, the limiting cases of this result. First of all let us find the general expression $\sigma_{(2)}^{xy}(\epsilon, H)$ in the case of a purely $2D$ spectrum ($r = 0$). In this case the square roots in (3.48) disappear and the summation can be carried out in terms of the familiar ψ-functions [109]:

$$\sigma_{(2)}^{xy}(\epsilon, H) = \frac{e^2}{2s}\left(\frac{\operatorname{Im}\gamma_{\mathrm{GL}}}{\operatorname{Re}\gamma_{\mathrm{GL}}}\right)\frac{1}{h}F_H\left(\frac{\epsilon}{2h}\right), \tag{3.49}$$

where

$$F_H(x) = 4x^2\left[\psi(x) + x\psi'(x) - 1 - \psi\left(\frac{1}{2}+x\right)\right]. \tag{3.50}$$

Using the asymptotics of ψ-function from Appendix B one can find that

$$F_H(x) = \begin{cases} x^2\ln 4\gamma_E, & x\to 0\\ 1/24, & x\to\infty \\ 1/(x+1/2), & x\to -1/2\end{cases} \tag{3.51}$$

and hence

$$\sigma_{(2)}^{xy}(\epsilon, h\to 0) = \frac{e^2}{48s}\left(\frac{\operatorname{Im}\gamma_{\mathrm{GL}}}{\operatorname{Re}\gamma_{\mathrm{GL}}}\right)\frac{h}{\epsilon^2}. \tag{3.52}$$

In the strong field limit:

$$\sigma^{xy}_{(2)}(\epsilon, H) = \frac{e^2}{8s}\left(\frac{\operatorname{Im}\gamma_{\mathrm{GL}}}{\operatorname{Re}\gamma_{\mathrm{GL}}}\right)\frac{\ln 4\gamma_E}{h}. \tag{3.53}$$

In order to analyze the $3D$ case we return to the general expression (3.48). In the case of strong enough magnetic fields ($\epsilon \ll h \ll r$) one can neglect ϵ with respect to h and expand the expression over r:

$$\sigma^{xy}_{(3)}(\epsilon \ll h) = \frac{e^2}{4s}\left(\frac{\operatorname{Im}\gamma_{\mathrm{GL}}}{\operatorname{Re}\gamma_{\mathrm{GL}}}\right)\frac{1}{\sqrt{hr}}\sum_{n=0}^{\infty}(n+1)\left[\frac{1}{\sqrt{2n+1}} - \frac{1}{\sqrt{2n+3}} - \frac{1}{\sqrt{(2n+2)^3}}\right] = \frac{0.022e^2}{s}\left(\frac{\operatorname{Im}\gamma_{\mathrm{GL}}}{\operatorname{Re}\gamma_{\mathrm{GL}}}\right)\frac{1}{\sqrt{hr}}.$$

Vice versa, for weak fields one can leave in Eq. (3.48) only the first three terms and due to the fast convergence of the corresponding sums omit with respect to r not only ϵ but also hn:

$$\sigma^{xy}_{(3)}(h \ll \epsilon) = \frac{e^2}{4s}\left(\frac{\operatorname{Im}\gamma_{\mathrm{GL}}}{\operatorname{Re}\gamma_{\mathrm{GL}}}\right)\frac{1}{\sqrt{r}}\sum_{n=0}^{\infty}(n+1)\left\{\frac{1}{\sqrt{\epsilon+h(2n+1)}} - \frac{1}{\sqrt{\epsilon+h(2n+3)}} - \frac{h}{\sqrt{[\epsilon+h(2n+2)]^3}}\right\}. \tag{3.54}$$

This sum can be easily evaluated by means of the Euler–Maclaurin formula (2.70):

$$\sigma^{xy}(h \ll \epsilon \ll r) = \frac{e^2}{4s}\left(\frac{\operatorname{Im}\gamma_{\mathrm{GL}}}{\operatorname{Re}\gamma_{\mathrm{GL}}}\right)\frac{h}{\epsilon^{3/2}}. \tag{3.55}$$

Summarizing, one can present the $3D$ case results in the form

$$\sigma^{xy}_{(3)}(\epsilon, h) = \frac{e^2}{4s}\left(\frac{\operatorname{Im}\gamma_{\mathrm{GL}}}{\operatorname{Re}\gamma_{\mathrm{GL}}}\right)\frac{1}{\sqrt{r}}\begin{cases} h/\epsilon^{3/2}, & h \ll \epsilon \ll r, \\ 0.09/\sqrt{h}, & \epsilon \ll h \ll r. \end{cases} \tag{3.56}$$

It would be not difficult to obtain the general expression for $\sigma^{xy}(h \ll \epsilon)$ in a weak filed applying the Euler–Maclaurin formula directly to Eq. (3.48). We leave this as the exercise for reader.

As it was mentioned above, the imaginary part $\operatorname{Im}\gamma_{\mathrm{GL}}$ is small and usually is related to the electron–hole asymmetry. The derivation of the microscopic expression for $\operatorname{Im}\gamma_{\mathrm{GL}}$, discussion of other related effects and contributions to Hall effect beyond the GL phenomenology will be given in the Part III.

3.5 Magnetic field angular dependence of paraconductivity[24]

We have seen above that in the case of a geometry with a magnetic field directed along the c-axis many sophisticated fluctuation features of layered superconductors can be studied in the general form. Nevertheless, even the attempt to explore

[24] In this section we base on the results of [111].

the d.c. conductivity in a longitudinal magnetic field (directed in ab plane) [110] or, moreover, with the field directed at some arbitrary angle θ with the z-axis leads to the appearance of the a vector-potential component in the argument of $\cos(k_z s)$ and the problem requires a nontrivial calculation of the matrix elements over the Mathieu functions.

In the vicinity of T_c the problem is simplified. We already learned that at temperatures very near the critical one ($\epsilon \ll r$) the $3D$ anisotropic fluctuation regime takes place. Here the size of the Cooper pairs along the c-axis is so large that the peculiarities of the layered structure do not play any more role. This means that only small values of k_z are important in the k_z-integrations, where the $\cos(k_z s)$ in (2.68) can be expanded and the LD functional is reduced to its GL form written for anisotropic superconductor.

The traditional way to incorporate anisotropy into the phenomenological description of superconductivity is to introduce an anisotropic effective mass tensor into the Ginzburg–Landau or London equations. In the conventional approach one then repeats all the calculations which usually have been done for the isotropic case before. As a result of the appearance of additional parameters and the breaking of spherical symmetry, the corresponding analysis becomes very tedious and thus only few results are known for the general anisotropic case including arbitrary field direction. Below, following [111], we present a scaling approach which provides simple and direct access to the most general anisotropic result by rescaling the anisotropic problem to a corresponding isotropic one on the initial level of GL functional.

To start with, let us consider the Gibbs free energy with an anisotropic effective mass tensor:

$$\mathcal{F}[\Psi] = \int d^3\mathbf{r} \left\{ a|\Psi|^2 + \frac{b}{2}|\Psi|^4 + \sum_{\mu=1}^{3} \frac{1}{4m_\mu} \left| \left(\frac{1}{i}\frac{d}{dx_\mu} - 2eA_\mu \right) \Psi \right|^2 + \frac{B^2}{8\pi} - \frac{\mathbf{H}\cdot\mathbf{B}}{4\pi} \right\}. \tag{3.57}$$

Let us suppose that the external field $\mathbf{H}$ is chosen to lie in the y-z plane and makes angle θ with the z-axis. For sake of simplicity and because the oxide superconductors are within high accuracy uniaxial materials, we choose $m_x = m_y = m^*$, while $m_z^{-1} = 2\alpha T s^2 r$ (compare with (2.68)). The effective anisotropy parameter $\gamma_a^2 = m^*/m_z = 2\alpha T s^2 r m^* < 1$ is introduced. In (3.57), the anisotropy enters only in the gradient term, therefore the simple rescaling of the coordinate axes: $x = \widetilde{x}, y = \widetilde{y}, z = \gamma_a \widetilde{z}$ together with the scaling of the vector-potential: $\mathbf{A} = (\widetilde{A}_x, \widetilde{A}_y, \widetilde{A}_z/\gamma_a)$ will render this term isotropic. The magnetic field evidently is rescaled to $\mathbf{B} = (\widetilde{B}_x/\gamma_a, \widetilde{B}_y/\gamma_a, \widetilde{B}_z)$ and the last three terms in (3.57), describing the magnetic field energy, are transformed to

$$\delta\mathcal{F}[\Psi] = \frac{\gamma_a}{8\pi}\int d^3\widetilde{\mathbf{r}}\left[\frac{1}{4m}\sum_{\mu=1}^{3}\left|\left(\frac{\hbar}{i}\frac{d}{d\widetilde{x}_\mu} - 2e\widetilde{A}_\mu\right)\Psi\right|^2 + \left(\frac{\widetilde{\mathbf{B}}_{xy}^2}{\gamma_a} + \widetilde{B}_z^2\right)\right.$$
$$\left. -2\left(\frac{\widetilde{\mathbf{B}}_{xy}\cdot\mathbf{H}_{xy}}{\gamma_a} + \widetilde{B}_z H_z\right)\right].$$

In short, we have removed the anisotropy from the gradient term but reintroduced it into the magnetic energy term. In general it is not possible to make both terms isotropic in the Gibbs energy simultaneously. However, depending on the physical question addressed, we can neglect fluctuations in the magnetic field, as was mostly done above.

Let us demonstrate how the method works for the example of the d.c. fluctuation conductivity tensor which was calculated above for a magnetic field directed along the z-axis. We restrict our consideration to the $3D$ region ($\epsilon \ll r$). One can write the scaling relations between the electric field and current components before and after the scaling transformation by means of a conductivity tensor and the anisotropy parameter:

$$j_{x,y} = \widetilde{j}_{x,y}, \qquad j_z \sim ev_z \sim \gamma_a \widetilde{j}_z,$$
$$E_{x,y} = \widetilde{E}_{x,y}, \qquad E_z \sim \frac{\partial\varphi}{\partial z} \sim \frac{1}{\gamma_a}\widetilde{E}_z. \tag{3.58}$$

Now let us rewrite the relations between the current and electric field vectors before and after the scale transformation

$$j_\alpha = \sigma_{\alpha\beta}E_\beta, \qquad \widetilde{j}_\alpha = \widetilde{\sigma}_{\alpha\beta}\widetilde{E}_\beta. \tag{3.59}$$

Comparing them with (3.58) and introducing the operator of the direct scaling transformation $T_{\alpha\beta}$

$$T_{\alpha\beta} = \begin{pmatrix} 1 & 0 & 0 \\ 0 & 1 & 0 \\ 0 & 0 & \gamma_a \end{pmatrix},$$

one can write $j_\alpha = T_{\alpha\mu}\widetilde{j}_\mu$, $E_\alpha = (T^{-1})_{\alpha\mu}\widetilde{E}_\mu$ and express the conductivity tensor as

$$\sigma_{\alpha\beta} = T_{\alpha\mu}\widetilde{\sigma}_{\mu\rho}T_{\rho\beta}.$$

Now let us work in the already isotropic coordinate frame. We suppose that initially the magnetic field was directed along the z-axis and now we rotate it in the x-z plane by the angle $\widetilde{\theta}$ with respect to the initial direction. The conductivity tensor will be transformed by the usual law:

$$\widetilde{\sigma}_{\alpha\beta}(\widetilde{\theta}) = R_{\alpha\mu}\widetilde{\sigma}_{\mu\rho}(0)R^T_{\rho\beta} = R_{\alpha\mu}(T^{-1})_{\mu\varsigma}\sigma_{\varsigma\eta}(0)(T^{-1})_{\eta\delta}R^T_{\delta\beta},$$

where

$$R_{\alpha\beta} = \begin{pmatrix} \cos\widetilde{\theta} & 0 & -\sin\widetilde{\theta} \\ 0 & 1 & 0 \\ \sin\widetilde{\theta} & 0 & \cos\widetilde{\theta} \end{pmatrix}$$

and

$$\sigma_{\alpha\beta}(\widetilde{\theta}) = T_{\alpha\gamma}\widetilde{\sigma}_{\gamma\delta}(\widetilde{\theta})T_{\delta\beta} = T_{\alpha\gamma}R_{\gamma\mu}(T^{-1})_{\mu\varsigma}\sigma_{\varsigma\eta}(0)(T^{-1})_{\eta\delta}R^T_{\delta\kappa}T_{\kappa\beta}.$$

Let us remind that the T-matrix is diagonal so $T = T^T$. As a result

$$\left(TRT^{-1}\right)^T = \left(T^{-1}\right)^T (TR)^T = \left(T^{-1}\right)^T R^T T^T = T^{-1}R^T T.$$

The fluctuation conductivity tensor $\sigma_{\alpha\beta}(\theta)$ in the initial tetragonal system with the magnetic field directed at the angle θ with respect to the z-axis can be expressed by means of the effective transformation operator $M_{\alpha\beta}$:

$$\sigma_{\alpha\beta}(\theta) = M_{\alpha\varsigma}(\widetilde{\theta})\sigma_{\varsigma\eta}(0,\widetilde{H})M^T_{\eta\beta}(\widetilde{\theta}),$$

with

$$M_{\alpha\beta}(\widetilde{\theta}) = T_{\alpha\delta}R_{\delta\kappa}T^{-1}_{\kappa\beta} = \begin{pmatrix} \cos\widetilde{\theta} & 0 & -\frac{1}{\gamma_a}\sin\widetilde{\theta} \\ 0 & 1 & 0 \\ \gamma_a\sin\widetilde{\theta} & 0 & \cos\widetilde{\theta} \end{pmatrix}.$$

Let us mention that the conductivity matrix elements $\sigma_{xz} = \sigma_{zx} = \sigma_{yz} = \sigma_{zy} = 0$, what considerably simplifies further calculations. Multiplying the matrices one can find:

$$\sigma_{\alpha\beta}(\widetilde{\theta})$$

$$= \begin{pmatrix} \sigma_{xx}\cos^2\widetilde{\theta} + \frac{\sigma_{zz}}{\gamma_a^2}\sin^2\widetilde{\theta} & \sigma_{xy}\cos\widetilde{\theta} & \left(\frac{\gamma_a}{2}\sigma_{xx} - \frac{1}{2\gamma_a}\sigma_{zz}\right)\sin 2\widetilde{\theta} \\ \sigma_{yx}\cos\widetilde{\theta} & \sigma_{yy} & \sigma_{xx}\gamma_a^2\sin^2\widetilde{\theta} + \sigma_{zz}\cos^2\widetilde{\theta} \\ \left(\frac{\gamma_a}{2}\sigma_{xx} - \frac{1}{2\gamma_a}\sigma_{zz}\right)\sin 2\widetilde{\theta} & \sigma_{xy}\gamma_a\sin\widetilde{\theta} & \sigma_{xx}\gamma_a^4\sin^2\theta + \sigma_{zz}\cos^2\theta \end{pmatrix}.$$

The angle $\widetilde{\theta}$ can be expressed by means the renormalized magnitude of the magnetic field $\widetilde{H} = \sqrt{H_z^2 + \gamma_a^2 H_x^2}$:

$$\cos\widetilde{\theta} = \frac{\widetilde{H}_z}{\widetilde{H}} = \frac{\cos\theta}{\sqrt{\cos^2\theta + \gamma_a^2\sin^2\theta}}; \qquad \sin\widetilde{\theta} = \frac{\gamma_a\sin\theta}{\sqrt{\cos^2\theta + \gamma_a^2\sin^2\theta}}.$$

Finally the paraconductivity tensor of a superconductor with the magnetic field H applied at an arbitrary angle θ with respect to the z-axis is expressed by means of the paraconductivity tensor $\sigma_{\alpha\beta}(0,\widetilde{H})$ placed in the effective field $\widetilde{H}\|\widehat{z}$ in the form:

$$\sigma_{\alpha\beta}(\widetilde{\theta}) = \frac{1}{R^2(\theta,\gamma_a)}$$

$$\times \begin{pmatrix} \sigma_{xx}\cos^2\theta + \sigma_{zz}\sin^2\theta & R(\theta,\gamma_a)\,\sigma_{xy}\cos\theta & \left(\frac{\gamma_a^2}{2}\sigma_{xx} - \frac{1}{2}\sigma_{zz}\right)\sin 2\theta \\ R(\theta,\gamma_a)\,\sigma_{yx}\cos\theta & R^2(\theta,\gamma_a)\,\sigma_{yy} & R(\theta,\gamma_a)\,\sigma_{yx}\gamma_a^2\sin\theta \\ \left(\frac{\gamma_a^2}{2}\sigma_{xx} - \frac{1}{2}\sigma_{zz}\right)\sin 2\theta & R(\theta,\gamma_a)\,\sigma_{xy}\gamma_a^2\sin\theta & \sigma_{xx}\gamma_a^4\sin^2\theta + \sigma_{zz}\cos^2\theta \end{pmatrix}, \tag{3.60}$$

where

$$R(\theta,\gamma_a) = \sqrt{\cos^2\theta + \gamma_a^2\sin^2\theta}.$$

In the simplest case of a longitudinal field $\theta = \pi/2$:

$$\sigma_{\alpha\beta}(\pi/2, H) = \begin{pmatrix} \frac{\sigma_{zz}}{\gamma_a^2} & 0 & 0 \\ 0 & \sigma_{yy} & \sigma_{yx}\gamma_a \\ 0 & \sigma_{xy}\gamma_a & \sigma_{xx}\gamma_a^2 \end{pmatrix}.$$

Obtained general formula allows one to find the paraconductivity of a thin film in a magnetic field of arbitrary orientation. In this case one can put $\gamma_a = 0$ which results in

$$\sigma_{\alpha\beta}^{(\text{film})}(\theta, H) = \sigma_{\alpha\beta}(H\cos\theta).$$

One can see that the paraconductivity of a film placed in magnetic field depends on its perpendicular-to-surface component only.

It worth mentioning that here, as everywhere above through this chapter, we took into account only the orbital effect of magnetic field on fluctuation conductivity. The effect of Zeeman splitting of the forming fluctuation Cooper pair electron spin states have been ignored.

Regarding the regime of applicability we wish to point out that, in spite of starting from a GL-type description, proposed scaling approach is not limited to the regime near T_c. In fact, presented here scaling rules can also be obtained by starting from the London equations, which are valid at any temperature. The scaling approach can be used for the case of layered superconductors as long as the discreteness of the structure is not important. The crossover between quasi-$2D$ and $3D$ anisotropic behavior depends on the physical quantity of interest; however, the regime where the anisotropic description is valid is usually large.

3.6 Paraconductivity of nanotubes[25]

In this section we present the example of use of the general formula (3.24) in application to the very new objects: carbon nanotubes where the superconducting state was detected recently. Carbon nanotubes are mesoscopic systems with a remarkable interplay between dimensionality, interaction and disorder [112]. Recent experiments found that the electron transport in single-wall nanotubes (SWNT) has a $1D$ ballistic behavior [113]. Therefore it may be theoretically described within the model of $1D$ interacting electron system known as Luttinger liquid [114–116]. At the same time, the multiwall nanotubes (MWNT), which are composed of several concentrically arranged graphite shells, show properties which are consistent with the weak-localization features of the diffusive transport in magnetoconductivity and zero-bias anomaly in the tunneling DOS [117]. Similar properties have been observed in ropes of SWNTs [113, 118].

Very recent experimental works [119, 120] have addressed the problem of superconductivity in carbon nanotubes. In the article by Tang *et al.* [119] a superconducting behavior was detected in SWNT's at a mean-field critical temperature evaluated as T_c=15 K. At the same time a pure superconducting state with zero resistance was not found and the authors attribute this fact to the presence of strong fluctuations which alter severely the superconducting order parameter both below and above T_c. In [120] ropes of SWNTs were studied and a truly superconducting transition was discovered at T_c=0.55 K. The suppression of T_c by a magnetic field applied along the tube was also measured.

Let us study the paraconductivity and corresponding magnetoconductivity of a carbon nanotube above T_c. In spite of the nanoscale size of the system we assume the validity of the GL formalism for the description of the fluctuation superconductivity and in conclusion will check the limits of applicability of this assumption evaluating the corresponding Gi number.

In order to describe the one-electron spectrum of carbon nanotubes one has to take into consideration that the electron wavelength around the circumference of a nanotube is quantized due to the periodic boundary conditions and only a discrete number of wavelengths can fit around the tube. Along the tube, however, electronic states are not confined and electrons can move ballistically. Because of the circumferential-mode quantization, the electron states in the tube do not form a single wide energy band but instead they split into a number of $1D$ subbands with band onsets at different energies. Consequently, we assume the electron spectrum in the form

$$\epsilon(\mathbf{p}) = \frac{p_{||}^2}{2m_{||}} + \frac{n^2}{2m_\perp R^2}, \tag{3.61}$$

where $n = -N, ..., N$, $N = [p_F R]$, p_F is the Fermi momentum and R is the nanotube radius. The number N is determined by the value of the chemical

[25] In this section we base on the results of [121].

potential and the distance between the levels. It defines the number of electrons filling the $2N+1$ electron subbands of the nanotube electron spectrum. A typical value for a realistic nanotube is $N \sim 5$ [122].

The GL Hamiltonian for the nanotube geometry is convenient to write in the cylindric frame, choosing the longitudinal coordinate z and the angular variable φ as the natural coordinates for the problem under discussion:

$$\widehat{\mathcal{H}} = \alpha T_{\mathrm{c}} \left[\epsilon - \xi_{\shortparallel}^2 \frac{\partial^2}{\partial z^2} - \xi_{\perp}^2 (\frac{1}{R} \frac{\partial}{\partial \varphi} - 2ieA_{\perp})^2 \right]. \tag{3.62}$$

Here $A_{\perp} = \frac{1}{2}HR = \frac{1}{2eR}\frac{\Phi}{\Phi_0}$ is the tangential component of the vector-potential, $\xi_{\shortparallel} = (4m_{\shortparallel}\alpha T_{\mathrm{c}})^{-1/2}$ and $\xi_{\perp} = (4m_{\perp}\alpha T_{\mathrm{c}})^{-1/2}$ are the longitudinal and the transversal GL coherence lengths. The latter is supposed to be comparable to the nanotube radius: $\xi_{\perp} \sim R$.

The fluctuation order parameter Ψ can be presented as a Fourier series

$$\Psi(z, \varphi, t) = \sum_{n=-\infty}^{\infty} \int_{-\pi/a}^{\pi/a} \frac{dq_{\shortparallel}}{2\pi} \psi_n(q_{\shortparallel}, t) \exp(-in\varphi) \exp(-iq_{\shortparallel} z), \tag{3.63}$$

and the appropriate linearized GL equation for the Fourier component $\psi_n(t)$ can be written as:

$$\left[\epsilon + \xi_{\shortparallel}^2 q_{\shortparallel}^2 + \frac{\xi_{\perp}^2}{R^2} \left(n - \frac{\Phi}{\Phi_0} \right)^2 \right] \psi_n(q_{\shortparallel}, t) = 0. \tag{3.64}$$

When the temperature decreases and the magnetic flux $\Phi \in]-\Phi_0/2\,, \Phi_0/2[$ the superconducting transition occurs at the ψ_0 state.

Let us move to the study of the paraconductivity in a small superconducting cylinder at temperatures above the critical one. We are interested here only in the longitudinal diagonal component of (3.24). The appropriate matrix elements of the velocity operator were already calculated above (see (3.25)). The summation over subscript $\{i\}$ is carried out over the levels of the angular quantization up to the maximal number N and includes the integration over the z-axis momentum. As a result the general formula (3.24) for the longitudinal paraconductivity of a nanotube is given by

$$\sigma^{\shortparallel}(\epsilon, H) = \frac{\pi \alpha e^2}{2S} T \int \frac{dp_{\shortparallel}}{2\pi} \int \frac{dq_{\shortparallel}}{2\pi} \sum_{i,l=-N}^{N} \frac{\widehat{\mathbf{v}}_{il,pq}^{\shortparallel} \widehat{\mathbf{v}}_{li,qp}^{\shortparallel}}{\varepsilon_i(p_{\shortparallel})\, \varepsilon_l(q_{\shortparallel}) \left[\varepsilon_i(p_{\shortparallel}) + \varepsilon_l(q_{\shortparallel}) \right]} =$$

$$= \frac{\pi e^2}{16S} \xi_{\shortparallel} \sum_{n=-N}^{N} \frac{1}{\left[\epsilon + \frac{\xi_{\perp}^2}{R^2} \left(n - \frac{\Phi}{\Phi_0} \right)^2 \right]^{3/2}} \tag{3.65}$$

(here $S = \pi R^2$ is the cross-section area of the nanotube). This formula can be numerically evaluated to obtain the magnetoconductivity. Nevertheless, in order to get a qualitative understanding of the paraconductivity temperature dependence in zero field and its behavior in the presence of a magnetic field at fixed temperature, let us assume $N \gg 1$ and proceed analytically.

3.6.1 *Paraconductivity in zero magnetic field*

In the immediate vicinity of the critical temperature, where $\xi_\perp(\epsilon) \gg R$, only the term with $n = 0$ in Eq. (3.65) contributes to the paraconductivity resulting in the $1D$ behavior of paraconductivity ($\sigma''(\epsilon, 0) \sim \epsilon^{-3/2}$).

Relatively far from the superconducting transition, where $\xi_\perp(\epsilon) = \xi_\perp/\sqrt{\epsilon} \ll R$ (but still $\epsilon \ll 1$), the term with $n = 0$ in Eq. (3.65) can be omitted. The remaining sum in Eq. (3.65) can be replaced with an integral, what results in

$$\sigma''(\epsilon,0) = \frac{\pi e^2}{8S}\xi_{\parallel}\left(\frac{R}{\xi_\perp}\right)^3\left\{\frac{1}{\sqrt{1+\frac{R^2}{\xi_\perp^2(\epsilon)}}\left[1+\sqrt{1+\frac{R^2}{\xi_\perp^2(\epsilon)}}\right]}\right.$$
$$\left.-\frac{1}{\sqrt{N^2+\frac{R^2}{\xi_\perp^2(\epsilon)}}\left[N+\sqrt{N^2+\frac{R^2}{\xi_\perp^2(\epsilon)}}\right]}\right\}. \tag{3.66}$$

Not too far from the transition point $R/N \ll \xi_\perp(\epsilon) \ll R$ the first term in parentheses in Eq. (3.66) dominates over all other contributions and Eq. (3.66) reproduces the $2D$ result $\sigma''(\epsilon, 0) \sim \epsilon^{-1}$. Moving away from the critical region and reaching the temperatures where $\xi_\perp(\epsilon) \ll R/N$, one can find the compensation of the leading-order contributions of two terms in parentheses. The accounting for next approximation in the root expansions gives the contribution of the same singularity in as the first term of Eq. (3.66), but with the enhancement factor $2N+1$. Therefore the system returns to $1D$ behavior upon moving away from the transition point, but in contrast to the immediate vicinity of T_c, the transport occurs within $2N+1$ channels.

All the asymptotics of $\sigma''(\epsilon, 0)$ can be presented in a compact form as:

$$\sigma''(\epsilon,0) = \frac{e^2\xi_{\parallel}}{16R^2}\begin{cases} \dfrac{1}{\epsilon^{3/2}}, & \epsilon \ll \left(\frac{\xi_\perp}{R}\right)^2, \\ \dfrac{R}{\xi_\perp}\dfrac{2}{\epsilon}, & \left(\frac{\xi_\perp}{R}\right)^2 \ll \epsilon \ll \left(\frac{\xi_\perp N}{R}\right)^2, \\ \dfrac{2N+1}{\epsilon^{3/2}}, & \left(\frac{\xi_\perp N}{R}\right)^2 \ll \epsilon. \end{cases} \tag{3.67}$$

The physics of these crossovers is the following. The first one has a geometrical nature: very near to T_c the fluctuation Cooper pairs are so large that they have only one degree of freedom to slide along the tube axis. The first line of Eq. (3.67) exactly reproduces the paraconductivity of a wire with cross-section $S \ll \xi_\perp^2$. This crossover is analogous to the one occurring in thin films of layered superconductors [44].

In the intermediate regime rotations over the tube surface become possible and the paraconductivity temperature dependence transforms into the $2D$ one.

Finally, relatively far from T_c, where $\xi_{\perp}(\epsilon) \sim R/N$, the last, most nontrivial, crossover $2D \to 1D$ in the fluctuations dimensionality takes place. To recognize its physical sense let us recall that the value R/N characterizes the distance between the electron wave-function zeros in the N-th subband. The result of the averaging of the superconducting-type electron correlations in confines of the stripes of this size, parallel to the cylinder's axis, will be evidently nonzero, while the pairing of the one-electron states belonging to different stripes will result in the average out of such contributions. In other words, relatively far from T_c, the value R/N characterizes the width of effective $1D$ channels of the Cooper pair motion on a cylinder surface and the fluctuation Cooper pairs transport takes place in each of such channels separately. That is why the total longitudinal paraconductivity acquires the degeneracy factor $2N+1$ equal to the number of subbands.

3.6.2 *Fluctuation magnetoconductivity*

We now move to the study of paraconductivity in the presence of a magnetic field applied. Due to the Little–Parks effect [123], the critical temperatures $T_c^{(n)}(\Phi)$ are periodic functions of the flux through the tube with period Φ_0. Therefore, we can restrict ourselves to the flux range $-\Phi_0/2 < \Phi < \Phi_0/2$, where

$$T_c(\Phi) = T_c^{(0)}\left[1 - \frac{\xi_{\perp}^2}{R^2}\left(\frac{\Phi}{\Phi_0}\right)^2\right]. \tag{3.68}$$

Evidently, two different regimes can take place: a weak-field one when $\Phi \lesssim \Phi_0 \frac{R}{\xi_{\perp}}\sqrt{\epsilon}$ (which is equivalent to $H \lesssim H_{c2}\left(\frac{R}{\xi_{\perp}}\right)\sqrt{\epsilon}$) and a strong-field regime when $\Phi_0 \frac{R}{\xi_{\perp}}\sqrt{\epsilon} \ll \Phi \ll \Phi_0/2$. In the case of the weak-field regime one can easily see that the main magnetic field dependence comes from the renormalization of the critical temperature, and therefore the three limiting cases are recovered analogously with the preceding discussion of the zero-field case. As the result,

$$\delta\sigma^{\parallel}(\epsilon,\Phi) - \sigma^{\parallel}(\epsilon,\Phi) - \sigma^{\parallel}(\epsilon,0)$$

$$= -\frac{e^2\xi_{\parallel}}{32}\frac{\xi_{\perp}^2}{R^4}\left(\frac{\Phi}{\Phi_0}\right)^2 \begin{cases} \dfrac{3}{\epsilon^{5/2}}, \; \epsilon \ll \left(\frac{\xi_{\perp}}{R}\right)^2, \\ \left(4\dfrac{R}{\xi_{\perp}}\right)^2 \dfrac{2}{\epsilon^2}, \; \left(\frac{\xi_{\perp}}{R}\right)^2 \ll \epsilon \ll \left(\frac{\xi_{\perp}N}{R}\right)^2, \\ 3\dfrac{2N+1}{\epsilon^{5/2}}, \; \left(\frac{\xi_{\perp}N}{R}\right)^2 \ll \epsilon. \end{cases} \tag{3.69}$$

The strong-field regime $\Phi_0 \frac{R}{\xi_{\perp}}\sqrt{\epsilon} \ll \Phi \ll \Phi_0/2$ can be reached (without passing to the next foil of the Little–Parks effect) in the case $R \ll \xi_{\perp}(\epsilon)$ only. In such a situation the main contribution originates from the term with $n = 0$ in Eq. (3.65):

$$\sigma^{\shortparallel}(\Phi) = \frac{e^2}{2^{10}} \frac{R\xi_{\shortparallel}}{\xi_{\perp}^3} \left(\frac{\Phi_0}{\Phi} \right)^3 . \tag{3.70}$$

This result is valid for temperatures $\epsilon \ll (\xi_{\perp}/R)^2$. In the temperature range $(\xi_{\perp}/R)^2 \ll \epsilon \ll (\xi_{\perp}N/R)^2 \ll 1$ (if such interval exists), where in the absence of a magnetic field the fluctuations have a $2D$ character, the effect of the magnetic field is relevant only for fields as high as $\Phi_0 N \frac{R}{\xi_{\perp}} \sqrt{\epsilon} \ll \Phi \ll \Phi_0/2$, but it still is described by the formula (3.70). One can recognize in the effect of the magnetic field on paraconductivity the usual suppression of the effective fluctuation dimensionality, as it happens even in the $3D$ case. Nevertheless we would like to attract the reader's attention to the unusually strong suppression of the nanotube paraconductivity in strong magnetic fields. Its comparison with the corresponding paraconductivity of a layered superconductor shows a remarkable difference in the critical exponent: 3 against 1 (see [106]). This follows from the channel separation in the Cooper pairs motion and hence the effective decreasing of their density in the momentum space. A similar effect is observed in superconducting rings. In its $0D$ regime $\sigma^{(0)}_{\text{ring}}(H) \sim H^{-4}$, (see [124]) instead of $\sigma^{(0)}_{\text{gran}}(H) \sim H^{-2}$ as for superconducting granules.

3.6.3 *Discussion*

Discussing the application of the presented results to recent experimental data concerning realistic nanotubes [113, 117, 119, 120], it is important to remember that the physics of superconductivity in these systems is still controversial and it is very likely to be qualitatively different for systems like MWNTs, the ropes of SWNTs, or individual SWNTs. Namely, the effect of interactions within multi-wall tubes or hopping between neighboring tubes in a rope drives the system away from the one-dimensionality characterizing an individual nanotube. Therefore the physical properties are substantially altered in both normal and superconducting states depending on whether hopping is effective or not. Nevertheless, the above considerations are quite general because the model proposed is based on the GL phenomenology which is independent of the specific pairing mechanism leading to the superconductivity. It is clear that an individual nanotube is rather within the $1D$ limit of Eq. (3.66), $\xi_{\perp}(\epsilon) \gg R$, while for multi-wall tubes or ropes the other regimes may be observed.

The alternative to the presented above interpretation of the various paraconductivity temperature regimes can be given on the basis of comparison of the characteristic fluctuation Cooper pair "binding energy", $T - T_{\text{c}}$, with the angular quantization energy level structure. Here it is necessary to recall that the fluctuation Cooper pairs above the critical temperature are not condensed with the zero energy, like happens below T_{c}, but they are distributed over energy with the rapid decay at $\varepsilon \gtrsim T - T_{\text{c}}$. When $\xi_{\perp}(\epsilon) \gg R$ ($T - T_{\text{c}} \ll 1/2m_{\perp}R^2$) the binding energy is so small that the electrons occupying only the $n = 0$ level can be involved in fluctuation pairing. As the result the $1D$ behavior takes place.

As $T - T_c$ growths ($R/N \lesssim \xi_\perp(\epsilon) \lesssim R$) the electrons from more and more subbands can be involved in pairing (within the same subband) and due to this additional degree of freedom (subband number n) fluctuation behavior becomes $2D$. Finally, when $T - T_c$ exceeds the energy of the last filled level of angular quantization $\varepsilon_N = N^2/2m_\perp R^2$[which means $\xi_\perp(\epsilon) \lesssim R/N$], all $2N+1$ subbands are involved in pairing and each one presents an independent $1D$ channel. Indeed, the corresponding formula differs from the one near T_c by a factor $2N+1$ (see Eqs. (3.67) and (3.65)).

Dealing with superconductivity in such objects of nanoscale size as nanotubes or ultra-small grains, it is necessary to recognize that we already reach the limits of classical superconductivity and the variety of principally new phenomena appear [125]. The GL description of the superconductivity of nanodrops, for instance, is valid until the energy spacing of dimensional quantization (2.26) turns out to be much less than mean field value of the superconducting gap. In nanotubes, however, this criterion seems to be less severe because of the possibility of quasi-continuous motion along the tube axis. We can evaluate the Gi number for the nanotube as it was done above:

$$Gi_{(1)} = \frac{1}{(p_F^2 S)^{2/3}} \sim \frac{1}{\pi^{2/3} N^{4/3}} \ll 1 \tag{3.71}$$

and we see that in the realistic limit $N \gg 1$ some room for quasi-classical description of the system still takes place.

3.7 Transport equation for fluctuation Cooper pairs

In the previous sections we related fluctuations of the superconducting order parameter with the classical fields and in order to study their contribution to the electrical and heat transport of superconductor above T_c we used the formalism of the TDGL equation. Nevertheless it turns out, that for some simple cases, in spite of the wave nature of fluctuation Cooper pairs and evident break down of the quasiparticle concept for their description, some version of the classical electron transport theory could be developed.

In order to demonstrate it let us try to derive the Boltzmann master equation for the fluctuation Cooper pair distribution function

$$n_{\mathbf{p}}(t) = \int \langle \Psi(\mathbf{r},t)\Psi^*(\mathbf{r}',t)\rangle \exp(-i\mathbf{p}(\mathbf{r}-\mathbf{r}'))d(\mathbf{r}-\mathbf{r}'). \tag{3.72}$$

Let us recall that in the state of thermal equilibrium $n_{\mathbf{p}}^{(0)} =< |\Psi_{\mathbf{p}}|^2 >= T/\varepsilon_{\mathbf{p}}$.

We will be interested in electrical conductivity. In order to find the electric field dependence of $n_{\mathbf{p}}$ let us write its time derivative using (3.1)

$$\frac{\partial n_{\mathbf{p}}(t)}{\partial t} = \int d(\mathbf{r}-\mathbf{r}')e^{-i\mathbf{p}(\mathbf{r}-\mathbf{r}')} \left[< \frac{\partial \Psi(\mathbf{r},t)}{\partial t} \Psi^*(\mathbf{r}',t) > + < \Psi(\mathbf{r}',t) \frac{\partial \Psi^*(\mathbf{r},t)}{\partial t} > \right]$$

$$= \int d(\mathbf{r}-\mathbf{r}')e^{-i\mathbf{p}(\mathbf{r}-\mathbf{r}')}\left[2\frac{e}{i}[\varphi(\mathbf{r})-\varphi(\mathbf{r}')] < \Psi(\mathbf{r},t)\Psi^*(\mathbf{r}',t) > \right.$$
$$\left. +\frac{2}{\gamma_{\mathrm{GL}}}\operatorname{Re}\left\langle \Psi(\mathbf{r},t)\frac{\delta\mathcal{F}}{\delta\Psi}(\mathbf{r}',t)\right\rangle + \frac{2}{\gamma_{\mathrm{GL}}}\operatorname{Re}\left\langle \zeta(\mathbf{r},t)\Psi^*(\mathbf{r}',t)\right\rangle\right], \tag{3.73}$$

where $\mathcal{F}$ is determined by (2.10). Expressing the scalar potential by the electric field E one can transform the first term of the last integral into $-2c\mathbf{E}\frac{\partial n_{\mathbf{p}}}{\partial \mathbf{p}}$. The term with the variational derivative can be evaluated by means of (2.30) and expressed in the form $-\frac{2}{\gamma_{\mathrm{GL}}}\varepsilon_{\mathbf{p}}n_{\mathbf{p}}$.

More cumbersome is the evaluation of the last term, containing the Langevin force. Assuming the electric field to be weak enough,[26] to the first approximation, it is possible to use here the order parameter $\Psi^{(0)}(\mathbf{r},t)$ (see (3.4)) unperturbed by the electric field as the convolution $\widehat{L}\zeta(\mathbf{r},t)$. In this way, using (3.6) and (3.7) we calculate the last average in (3.73)

$$\frac{2}{\gamma_{\mathrm{GL}}}\int d(\mathbf{r}-\mathbf{r}')e^{-i\mathbf{p}(\mathbf{r}-\mathbf{r}')}\operatorname{Re}\left\langle \zeta^*(\mathbf{r},t)\widehat{L}\zeta(\mathbf{r}',t)\right\rangle$$
$$= 2T\operatorname{Re}\int_{-\infty}^{\infty}\frac{d\Omega}{2\pi}L(\mathbf{p},\Omega) = \frac{2T}{\gamma_{\mathrm{GL}}} = \frac{2}{\gamma_{\mathrm{GL}}}\varepsilon_{\mathbf{p}}n_{\mathbf{p}}^{(0)}$$

and obtain the transport equation

$$\frac{\partial n_{\mathbf{p}}}{\partial t} + 2e\mathbf{E}\frac{\partial n_{\mathbf{p}}}{\partial \mathbf{p}} = -\frac{2}{\gamma_{\mathrm{GL}}}\varepsilon_{\mathbf{p}}\left(n_{\mathbf{p}} - n_{\mathbf{p}}^{(0)}\right) = -\frac{2}{\tau_{\mathbf{p}}}\left(n_{\mathbf{p}} - n_{\mathbf{p}}^{(0)}\right). \tag{3.74}$$

In absence of a magnetic field $\varepsilon_{\mathbf{p}}$ was determined by (3.10) and the momentum dependent lifetime, corresponding to the GL one, can be introduced:

[26]This requirement implies some restrictions on its value. Indeed, when one neglects the effect of electric field on the order parameter $\Psi^*(\mathbf{r},t)$ correlated with the Langevin force, only the natural temperature decay of the fluctuation Cooper pairs with the characteristic time $\tau_{\mathbf{p}}$ is taken into account. Their additional destruction by a strong enough electric field is lost in this approximation. The break of fluctuation Cooper pair by electric field can be understood qualitatively as follows. The electrons correlated in Cooper pair have almost opposite momenta. Therefore, the same acceleration which they acquire in electric field results in the growth of velocity for one of them and decrease for another, what, in its turn, leads to the increase of the distance between electrons. The pair decays if this distance reached during the pair lifetime τ_{GL} exceeds the coherence length $\xi(T)$. In other words, starting with some characteristic, temperature dependent value of the intensity of electric field E_{cr}, the electron acceleration is so large, that at the distance of the order $\xi(T)$ they change their energy by the value of the order of $T-T_{\mathrm{c}}$, corresponding to the fluctuation Cooper pair "binding energy". The described mechanism results in the additional, field depending, decay of fluctuation pairs and respective deviation of the voltage-current characteristics from the Ohm law. One can see that the threshold electric field E_{cr}, where the nonlinear effects start, tends zero as $\epsilon^{3/2}$ when temperature verges towards T_{c}.

$$\tau_{\mathbf{p}} = \gamma_{\mathrm{GL}}/\varepsilon_{\mathbf{p}} = \frac{\tau_{\mathrm{GL}}(\epsilon)}{1+\xi^2(\epsilon)\mathbf{p}^2}.$$

Let us stress the appearance of the coefficient 2 on the right-hand side of Eq. (3.74). This means that the real Cooper pair lifetime, characterizing its density decay, is $\tau_{\mathbf{p}}/2$.

The effect of a weak electric field on the fluctuation Cooper pair distribution function in the linear approximation is determined by

$$n_{\mathbf{p}}^{(1)} = -\frac{e\mathbf{E}\gamma_{\mathrm{GL}}}{\varepsilon_{\mathbf{p}}}\frac{\partial n_{\mathbf{p}}^{(0)}}{\partial \mathbf{p}} = \frac{eT\gamma_{\mathrm{GL}}}{\varepsilon_{\mathbf{p}}^3}\mathbf{E}\cdot\frac{\partial \varepsilon_{\mathbf{p}}}{\partial \mathbf{p}}. \tag{3.75}$$

Substituting this formula into the expression for the electric current (2.137) side by side with the Cooper pair velocity $\mathbf{v}_{\mathbf{p}} = \partial\varepsilon_{\mathbf{p}}/\partial\mathbf{p}$ one can find

$$\mathbf{j}^{\alpha} = \sum_{\mathbf{p}} (2e\mathbf{v}^{\alpha} n_{\mathbf{p}}) = \sigma^{\alpha\beta} E^{\beta}, \tag{3.76}$$

where the paraconductivity tensor components are:

$$\sigma_{(D)}^{\alpha\beta} = \frac{\pi}{4} e^2 \alpha T \sum_{\mathbf{p}} \frac{\mathbf{v}_{\mathbf{p}}^{\alpha}\mathbf{v}_{\mathbf{p}}^{\beta}}{\varepsilon_{\mathbf{p}}^3}. \tag{3.77}$$

This result evidently coincides with Eq. (3.24) written without magnetic field and frequency. For the case of isotropic spectrum one reproduces Eq. (3.26), which is interesting to compare now with the result carried out in the Introduction from qualitative consideration, based on the Drude formula. Those simple considerations reflect correctly the physics of the phenomenon of paraconductivity but were carried out with the assumption of the momentum independence of the relaxation time $\tau_{\mathbf{p}}$, taken as $\tau_0 = \pi/8(T-T_{\mathrm{c}})$. As we have just seen, in reality $\tau_{\mathbf{p}}$ decreases rapidly with increase of the momentum, the excess "2" appeared in (3.74) because of the wave nature of the fluctuation Cooper pairs; accounting for this circumstance results in the precise coefficients of (3.26), different from (1.9).

For our further use it is convenient to find the formula for paraconductivity in terms of the GL coherence length $\xi(\epsilon)$, so let us express the matrix elements in Eq. (3.77) by means of (3.25) and get

$$\begin{aligned}\sigma_{(D)}^{\alpha\alpha} = \pi e^2 \int \frac{d^D p}{(2\pi)^D}\frac{\xi^4 p_{\alpha}^2}{(\epsilon+\xi^2 p^2)^3} &= \frac{\pi e^2}{2}\xi^{2-D}\mu_D\overline{\cos^2\theta}_{(D)}\int_0^{\infty}\frac{x^{D/2}dx}{(\epsilon+x)^3} \\ &= \frac{e^2}{\xi^2}\frac{\Gamma(2-D/2)}{2^{D+2}\pi^{D/2-1}}\xi^{4-D}(\epsilon).\end{aligned} \tag{3.78}$$

This expression of course, reproduces all the formulae in (3.26).

The paraconductivity takes an especially simple form in the $2D$ case, where, calculated per unit square, it depends on the reduced temperature only:

$$\sigma_{\Box}(T) = \frac{e^2}{16\hbar}\frac{T}{T - T_c}. \tag{3.79}$$

The coefficient in this formula turns out to be a universal constant and is given by the value $\hbar/e^2 = 4.1K\Omega$. For electronic spectra of other dimensionalities this universality is lost, and the paraconductivity comes to depend on the electron mean free path.

Let us compare $\sigma_{\Box}(T)$ with the normal electron Drude part $\sigma_n = n_e e^2\tau/m$ by writing the total conductivity

$$\sigma = \frac{e^2}{\hbar}(\frac{p_F l}{2\pi\hbar} + \frac{1}{16\epsilon}). \tag{3.80}$$

One sees that at $\epsilon_{\text{cr}} = 0.4/(p_F l) \sim Gi_{(2c)}$ the fluctuation correction reaches the value of the normal conductivity. Let us recall that the same order of magnitude for the $2D$ Gi number was obtained above from the heat capacity study. We will discuss the region of applicability of (3.80) in section 8.7.

It is worth mentioning that the results derived here for paraconductivity are valid with the assumption of weak fluctuations: for the temperature range $\epsilon \lesssim Gi_{(D)}$ they are not anymore applicable. Nevertheless, one can see that for not very dirty films, with $p_F^2 ld \gg 1$, a wide region of temperatures $Gi_{(2d)} \ll \epsilon \ll 1$ exists where the temperature dependence of conductivity is determined by fluctuations and in this region the localization effects are negligible.

The transport equation (3.74) was originally derived many years ago by Aslamazov and Larkin [126] but, because of the unavailability of this publication, was forgotten. Recently Mishonov *et al.* [127] rederived it.

4

EFFECT OF FLUCTUATIONS ON THERMOELECTRICITY AND HEAT TRANSPORT

Being interested earlier in the effect of fluctuations on the electric transport coefficients, we expressed corresponding electric current in terms of the general formula for quantum mechanical flow of probability Eq. (3.14) and averaged it over fluctuations. The situation with the heat transport coefficients in an interacting electron system turns out to be much more sophisticated and it has been the subject of almost 50 years discussion (see [128–130, 133, 135]). This is due to the fact that the notion of heat itself is not well defined in the Hamiltonian formalism. Nevertheless, in papers [98, 130] the explicit phenomenological expression for the heat current operator was proposed. The idea of its way of writing reduces to simple substitution of the effective charge of the Cooper pair in the quantum mechanical expression for the operator of the electric current Eq. (3.14) by the value of its effective energy. In spite of its simplicity, such an approach allows us to obtain, in the spirit of the GL formalism of the previous chapter, quantitative results for the direct (paraconductivity like) fluctuation contributions to the Seebek (thermoelectric) and Nernst coefficients.

At the end of the section we will briefly discuss the role of fluctuations in the thermal conductivity of the superconductor above its critical temperature.

4.1 Preliminaries

Thermoelectric effects are difficult both to calculate and to measure if compared with electrical transport properties. At the heart of the problem lies the fact that the thermoelectric coefficients in metals are the small resultant of two opposing currents which almost completely cancel. In calculating the thermoelectric power, one finds that the electrons above the Fermi level carry a heat current that is nearly the negative of that carried by the electrons below E_F. In the model of a monovalent metal in which band structure and scattering probabilities are symmetric about E_F, this cancelation would be exact; in a real metal a small asymmetry survives.

Because of their compensated nature, thermoelectric effects are very sensitive to the characteristics of the electronic spectrum, presence of impurities and peculiarities of scattering mechanisms. The inclusion of many-body effects, such as electron–phonon renormalization, multi-phonon scattering, and drag effect, adds even more complexity to the problem of calculating the thermoelectric power.

Among the effects of interaction, there is the influence of thermodynamical fluctuations on thermoelectric transport in a superconductor above the criti-

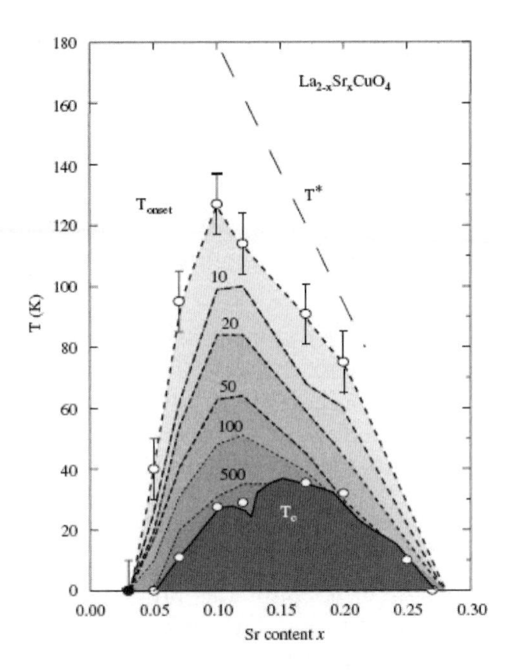

FIG. 4.1. The phase diagram of $La_{2-x}Sr_xCuO_4$ showing T_{onset} of the Nernst effect, the transition temperature T_c, and the pseudogap temperature T^*. The numbers indicate the value of the Nernst coefficient in $nV \cdot T^{-1}K^{-1}$ [149].

cal temperature. The manifestation of the thermoelectric and thermomagnetic phenomena in superconductors is very specific; it is quite different for type one and type two superconductors and differs strongly on those in the normal phase [131, 132]. Accounting for superconducting fluctuations of the Seebek coefficient above the transition temperature was first done by Maki [136], who demonstrated that the direct pair contribution to thermoelectric power is negative, it is less singular than paraconductivity and, in the $2D$ case, diverges only logarithmically. The problem of accounting for fluctuation contributions to the Nernst–Ettingshausen effect attracted the attention of theorists later [103], after the appearance of the first experimental papers [137–141] where a noticeable Nernst signal was detected in HTS samples.

Nevertheless, the Nernst–Ettingshausen effect in superconductors has attracted serious attention recently, when a series of experimental studies has revealed an anomalously strong thermomagnetic signal in the underdoped phase of high-temperature superconductors [142–150]. In a pioneering experiment [142], Xu *et al.* observed a sizeable Nernst effect up to 130 K, well above the transition temperature, T_c, in $La_{2-x}Sr_xCuO_4$ compounds (see Fig. 4.1). This and further similar experiments on cuprates have sparked theoretical interest in thermomagnetic phenomena. Theoretical approaches to the anomalously large Nernst–

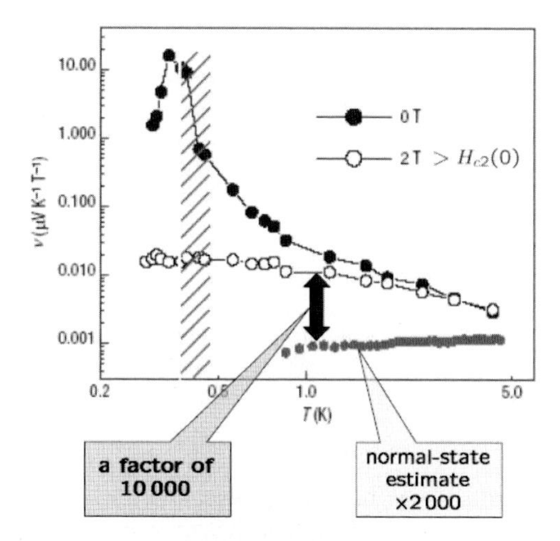

FIG. 4.2. Giant Nernst signal in strongly disordered $Nb_{0.15}Si_{0.85}$ superconducting film of thickness $d = 12.5\,nm$ and with critical temperature $T_c = 0.38\,K$ [157, 158].

Ettingshausen effect currently include models based on the proximity to a quantum critical point [151], vortex motion in the pseudogap phase [143, 152, 153], as well as a superconducting fluctuation scenario [154–156]. While the two former theories are specific to cuprate superconductors, the latter scenario should apply to other more conventional superconducting systems as well.

Very recently, a large Nernst coefficient was observed in the wide range of temperatures in a normal state of disordered superconducting films [157, 158] (see Fig. 4.2). These superconducting films are likely to be well described by the usual BCS model, and hence the new experimental measurements provide an indication that the superconducting fluctuations are likely to be the key to understanding the underlying physics of the giant thermomagnetic response.

Various groups have calculated the fluctuation-induced Nernst coefficient in the vicinity of the classical transition [103, 154–156, 159]. However, these analyses were limited to the case of very weak magnetic fields and temperatures close to the zero-field transition, when Landau quantization of the fluctuating Cooper pair motion can be neglected. In experiment, however, other parts of the phase diagram (in particular, strong fields) are obviously important and how the quantized motion of fluctuating pairs would influence the thermomagnetic response has remained unclear. Below, we will clarify the origin of the giant fluctuating Nernst–Ettingshausen effect, and will present the results of phenomenological GL theory valid for a wide range of magnetic fields above critical temperature.

We start from the definition of transport coefficients. Let us consider a conductor placed in a magnetic field $\mathbf{H}$ and subjected to an applied temperature

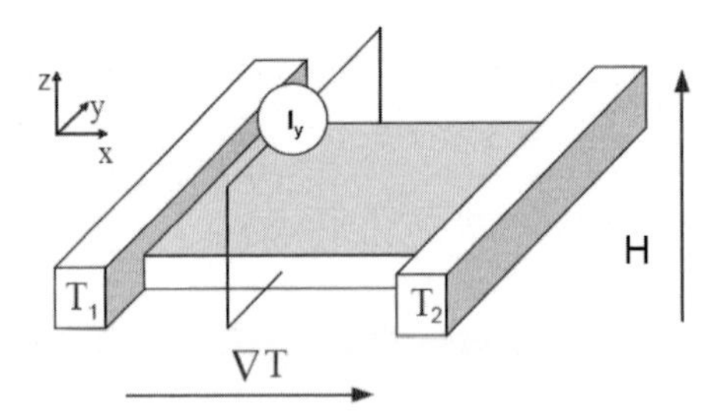

FIG. 4.3. Geometry of the Nernst–Ettingshausen effect: the current flows only along the direction perpendicular to both the magnetic field and the temperature gradient.

gradient ∇T. The electric and heat *transport* currents in it are related to the applied weak-enough electric field and temperature gradient by means of the relations:

$$\mathbf{j}_{\text{tr}}^{(e)\alpha} = \sigma^{\alpha\beta}(\mathbf{H})\,\mathbf{E}^{\beta} + \beta^{\alpha\beta}(\mathbf{H})\,\nabla^{\beta}T, \tag{4.1}$$

$$\mathbf{j}_{\text{tr}}^{(h)\alpha} = \gamma^{\alpha\beta}(\mathbf{H})\,\mathbf{E}^{\beta} - \kappa^{\alpha\beta}(\mathbf{H})\,\nabla^{\beta}T, \tag{4.2}$$

where $\beta^{\alpha\beta}(\mathbf{H}), \gamma^{\alpha\beta}(\mathbf{H})$ and $\kappa^{\alpha\beta}(\mathbf{H})$ are thermoelectricity and heat conductivity tensors. The thermoelectric tensors $\beta^{\alpha\beta}$ and $\gamma^{\alpha\beta}$ are connected by the Onsager relation: $\gamma^{\alpha\beta}(\mathbf{H}) = -T\beta^{\alpha\beta}(-\mathbf{H})$. Let us mention that the validity of the Onsager relation follows from the principle of the symmetry of transport coefficients, which is based on the invariance of the quantum mechanical equations with respect to time-reversal.

Experimentally, it is easier to control the electric current flowing through the specimens instead of electric field in it. Correspondingly, the equations for currents can be rewritten as follows:

$$\mathbf{E}^{\alpha} = \rho^{\alpha\beta}(\mathbf{H})\,\mathbf{j}_{\text{tr}}^{(e)\beta} + S^{\alpha\beta}(\mathbf{H})\,\nabla^{\beta}T, \tag{4.3}$$

$$\mathbf{j}_{\text{tr}}^{(h)\alpha} = \Pi^{\alpha\beta}(\mathbf{H})\,\mathbf{j}_{\text{tr}}^{(e)\beta} - \kappa^{\alpha\beta}(\mathbf{H})\,\nabla^{\beta}T, \tag{4.4}$$

where $\rho^{\alpha\beta}$ is the resistivity tensor, $S^{\alpha\beta} = -\beta^{\alpha\lambda}\left(\sigma^{-1}\right)^{\lambda\beta}$ is the differential thermopower (or the so-called Seebeck coefficient) and $\Pi^{\alpha\beta} = \gamma^{\alpha\lambda}\left(\sigma^{-1}\right)^{\lambda\beta}$ is the Peltier coefficient.

The off-diagonal components of the tensor $\beta^{\alpha\beta}$ in the absence of a magnetic field are equal to zero. When, along with the temperature gradient ∇T, a magnetic field $\mathbf{H}$ is also applied to the sample, the current flows in the circuit along the y axis (or a potential difference V appears when the circuit is broken);

see Fig. 4.3. This so-called Nernst–Ettingshausen effect[27] is well pronounced in semiconductors but is usually small in good metals. It is characterized by the Nernst–Ettingshausen coefficient, which can be expressed by means of the conductivity and thermoelectric tensors as follows:

$$\nu^{\mathrm{NE}} = \frac{E^y}{(-\nabla^x T)\, H} = \frac{1}{H}\frac{\beta^{xy}\sigma^{xx} - \beta^{xx}\sigma^{xy}}{(\sigma^{xx})^2 + (\sigma^{xy})^2}. \tag{4.5}$$

Let us express it in terms of the conductor characteristics. Consider a conductor in the presence of a magnetic field, H_z, and electric field, E_y, directed along the z- and y-axes respectively. The charged carriers (with charge q) subject to these crossed fields acquire a drift velocity $\overline{v}_x = cE_y/H_z$ in the x-direction. The latter would result in the appearance of a transverse current $j_x = nq\overline{v}_x$. When the circuit is broken, no current flows, and the drift of carriers is prevented by the spatial variation of the electric potential: $\nabla_x\varphi = -E_x = (nec/\sigma^{xx})(E_y/H_z)$. Due to electroneutrality, this generates the spatial gradient of the chemical potential: $\nabla_x\mu(n,T) + e\nabla_x\varphi = 0$, which corresponds to the appearance of the transverse temperature gradient $\nabla_x T = \nabla_x\mu(d\mu/dT)^{-1}$ along the x-direction. Hence, the Nernst coefficient can be expressed in terms of the full temperature derivative of the chemical potential:

$$\nu^{NE} = \frac{E_y}{(-\nabla_x T)H_z} = \frac{\sigma}{ne^2c}\frac{d\mu}{dT}. \tag{4.6}$$

E.g., in a degenerate electron gas, the chemical potential

$$\mu(T) = \mu_0 - (\pi^2 T^2/6)(d\ln\nu/d\mu), \tag{4.7}$$

where $\nu(\mu)$ is the density of states, and one easily reproduces the value of the Nernst coefficient in a normal metal [161]:

$$\nu^{NE} = (\pi^2 T/3mc)(d\tau/d\mu), \tag{4.8}$$

where τ is the elastic scattering time. Thus the Nernst effect in metals is small due to the large value of the Fermi energy.

The simple form of Eq. (4.6) suggests that in order to get a large Nernst signal, *a strong temperature dependence of the chemical potential of carriers is required.* As argued below, this remarkably simple and intuitive result alone sheds light on the physics behind the strong Nernst signal often seen in various superconducting compounds as compared to a Fermi liquid. Indeed, a strong temperature-dependence of the chemical potential can be achieved in the vicinity of the superconducting transition where the fluctuating Cooper pairs appear besides the normal electrons. As we know, these excitations are unstable, have the

[27]The Nernst–Ettingshausen effect is strictly related to the Nernst effect, which is just the opposite: it describes the appearance of a temperature gradient in a conductor placed in a magnetic field when an electric current flows through it.

characteristic lifetime of order $\tau_{\rm GL} = \pi/8(T - T_c)$, and form an interacting Bose gas with a variable number of particles. In two dimensions, their concentration is determined by Eq. (2.151). In the vicinity of T_c, the chemical potential of the fluctuating Cooper pairs is defined by their "binding energy" taken with the opposite sign. The chemical potential can be found by identifying its value in the Bose distribution (see Eqs. (1.5)–(1.6)) to give n_s defined by the Eq. (2.151), which leads to

$$\mu^{(2)}_{\rm c.p.}(T) = T_c - T. \tag{4.9}$$

Since $d\mu^{(2)}_{\rm c.p.}/dT = -1$, the fluctuation contribution to the Nernst signal exceeds parametrically the Fermi liquid term. In this sense it is similar to the fluctuation diamagnetism (which also exceeds the Landau/Pauli terms and is effectively a correction to the perfect diamagnetism of a superconductor).

Substituting the values of the paraconductivity $\sigma_{\rm fl}^{xx(D)}$ from Eq. (3.36) one can evaluate (up to logarithmic accuracy) the value of the fluctuation contribution to the Nernst coefficient in the example of a layered superconductor $(D = 2, 3)$:

$$\widetilde{\nu}_{\rm fl}^{\rm NE(D)} = \left(\frac{\partial \mu_{cp}}{\partial T}\right)\left(\frac{\sigma_{\rm fl}^{xx(D)}}{n_s^{(D)}}\right)\frac{1}{(2e)^2 c} \sim -\frac{k_B \hbar}{mcT_{\rm c}}\frac{1}{\sqrt{\epsilon(\epsilon + r)}} \sim -\frac{\mathcal{D}}{cT_{\rm c}}\frac{1}{\sqrt{\epsilon(\epsilon + r)}}. \tag{4.10}$$

Substituting $T_{\rm c} = 0.38\,K$ and $\mathcal{D} = 0.143\,cm^2/s$, we obtain that $\widetilde{\nu}_{\rm fl}^{(2)} \sim -T_{\rm c}/(T - T_{\rm c})\,\mu V \cdot T^{-1}K^{-1}$, which corresponds well to the results of [157] and is three orders more than the value of the Nernst coefficient in typical metals. Hence one can conclude that the giant Nernst effect in a fluctuating superconductor is linked to the very strong dependence of the fluctuation Cooper pairs' chemical potential on temperature.

4.2 Definition of the heat current

Calculation of the transport coefficients related to heat transfer requires knowledge of the explicit form of the heat flow operator. We will start our discussion from its definition in the framework of the phenomenological Ginzburg–Landau approach. Being interested above in the electric current operator, we just expressed it in terms of the general formula for quantum mechanical flow of probability (3.14) and it was enough to calculate correctly the contribution to conductivity related to the order parameter fluctuations. The situation with the heat current turns out to be much more sophisticated due to the fact that the notion of heat itself is not well defined in the Hamiltonian formalism. This is why, in order to be consistent, we start from the basic principles of thermodynamics.

The heat differential for the sample of fixed volume is related to the entropy, number of particles and vector-potential differentials by means of the relation

$$\delta Q\,(t) = T\delta S = \delta\mathcal{E}\,(t) - \mu\delta N\,(t) + \mathbf{j}^{(e)}\,(\mathbf{r}) \cdot \delta\mathbf{A}\,(\mathbf{r}, t)\,. \tag{4.11}$$

We will assume that the vector-potential $\mathbf{A}\,(\mathbf{r}, t)$ consists of two terms. The first one, $\mathbf{A}(\mathbf{r})$, does not depend on time and corresponds to the constant magnetic field $\mathbf{H}$ applied to the sample. The second one, $\mathbf{A}_E(\mathbf{r},t)$, generates the electric field $\mathbf{E} = -\partial \mathbf{A}_E(\mathbf{r},t)/\partial t$ and, along with the latter, is supposed to be small. This is why in Eq. (4.11), instead of the full electric current $\mathbf{j}^{(e)} = \mathbf{j}_{\rm tr}^{(e)} + \mathbf{j}_{\rm magn}^{(e)}$, one can leave only the magnetization current $\mathbf{j}_{\rm magn}^{(e)}(\mathbf{r})$, which does not depend on the time-dependent part $\mathbf{A}_E(\mathbf{r},t)$.[28] Hence $\nabla \cdot \mathbf{j}^{(e)}(\mathbf{r}) = 0$ and we can express $\mathbf{j}^{(e)}$ by means of the sample magnetization $\mathbf{M}$ as follows: $\mathbf{j}^{(e)} = \mathbf{j}_{\rm magn}^{(e)} = \nabla \times \mathbf{M}$. Calculating the time derivative of Eq. (4.11), one can find

$$\frac{\delta Q}{\delta t} = \frac{\delta \mathcal{E}}{\delta t} - \mu \frac{\delta N}{\delta t} + \nabla \times \mathbf{M} \cdot \frac{\delta \mathbf{A}_E(\mathbf{r},t)}{\delta t}. \tag{4.12}$$

Let us define the corresponding currents from the continuity equation

$$\frac{\delta Q}{\delta t} = -\mathrm{div}\, \mathbf{j}^{(Q)}. \tag{4.13}$$

The current $\mathbf{j}^{(Q)}$ is naturally related to the heat flow, the current $\mathbf{j}^{\mathcal{E}}$ ($\mathrm{div}\, \mathbf{j}^{\mathcal{E}} = -\delta\mathcal{E}/\delta t$) is related to the full energy flow, and the current of the ordered (directed) motion of particles is proportional to the electric current $\mathbf{j}_{\rm tr}^{(e)}$. The heat-transport current is defined by the difference of the latter two: $\mathbf{j}_{\rm tr}^{(h)} = \mathbf{j}^{\mathcal{E}} - (\mu/e^*)\,\mathbf{j}_{\rm tr}^{(e)}$. The last term of Eq. (4.12) is related to the equilibrium magnetization current $\mathbf{j}_{\rm magn}^{(h)}$:

$$\nabla \times \mathbf{M} \cdot \mathbf{E} = \mathrm{div}\,(\mathbf{M} \times \mathbf{E}) = -\mathrm{div}\, \mathbf{j}_{\rm magn}^{(h)}, \tag{4.14}$$

where $\mathbf{E}$ is the applied electric field.

Finally, one can express the full heat current as the sum of heat transport and magnetization currents:

$$\mathbf{j}^{(Q)} = \mathbf{j}^{\mathcal{E}} - (\mu/e^*)\,\mathbf{j}_{\rm tr}^{(e)} + \mathbf{j}_{\rm magn}^{(h)} = \mathbf{j}_{\rm tr}^{(h)} + \mathbf{j}_{\rm magn}^{(h)}, \tag{4.15}$$

with $\mathbf{j}_{\rm magn}^{(h)} = -\mathbf{M} \times \mathbf{E}$.[29] The vector of full heat current is related to the vector of electric field by means of some tensor $\widetilde{\beta}^{\alpha\beta}$:

$$\mathbf{j}_{(Q)}^{\alpha} = -T\widetilde{\beta}^{\alpha\beta}\mathbf{E}^{\beta}. \tag{4.16}$$

It turns out that, both in phenomenological and microscopical approaches, it is convenient to calculate namely the value of full heat current. At the same time, the value of the transport heat current $\mathbf{j}_{\rm tr}^{(h)} = \mathbf{j}_{(Q)} - \mathbf{j}_{\rm magn}^{(h)}$ is measured

[28] For example, the current related to the charge transfer due to fluctuation Cooper pairs is proportional to $\mathbf{A}_E(\mathbf{r},t)$, while the magnetization currents in normal phase, precursors of Meissner diamagnetism, do not depend on this part of the vector potential.

[29] One can see that this current is directed perpendicularly to the electric field and it manifests itself in the off-diagonal elements of the thermoelectric tensor only. Nevertheless, its presence provides the fulfillment of the Onsager principle of the symmetry of transport coefficients; this fact was stressed in [134, 135]; in the calculation of the fluctuation contribution to the Nernst effect, accounting for $\mathbf{j}_{\rm magn}^{(h)}$ changes the final coefficient by the factor 3 [154].

by experiment. This is why the observed value of the magneto-thermoelectric tensor $\beta^{\alpha\beta}$, in accordance with the above consideration, is determined by the expression

$$\beta^{\alpha\beta} = \widetilde{\beta}^{\alpha\beta} + \frac{e^{\alpha\beta\gamma}M^{\gamma}}{T}. \tag{4.17}$$

As mentioned by Obraztsov [160], generally speaking, the Nernst coefficient can be considerably renormalized by the effect of magnetization currents that develop in the sample when the magnetic field is applied. This effect is negligible in the case of normal metal in non-quantizing fields, but changes significantly the value of the Nernst coefficient in a fluctuating superconductor [154].

Both temperature and magnetic field dependencies of the fluctuation magnetization $\mathbf{M}$ were studied in detail in chapter 2. This is why our main goal below will be the calculus of the tensor $\widetilde{\beta}^{\alpha\beta}$.

The message of the above discussion consists of the following recipe: in order to calculate the fluctuation contribution to the heat transport current $\mathbf{j}_{\mathrm{tr}}^{(h)}$, which defines the thermoelectric coefficients, one has to average the heat current operator (4.19) over fluctuations and then extract from it the current related to the fluctuation magnetization. Evidently, the last term contributes to the off-diagonal coefficients only, like the Ettingshausen and Nernst effects.

4.3 Fluctuation heat response to the electric field

Let us calculate the fluctuation contribution to the thermoelectric tensor $\widetilde{\beta}^{\alpha\beta}(\mathbf{H})$. We assume the constant arbitrary magnetic field $\mathbf{H}$ to be applied along z-axis, and calculate the heat response function for a weak electromagnetic field $\mathbf{E}$ applied along y-axis, i.e. we will, in fact, study the Nernst effect. Then, by means of the Onsager relation, we will find the explicit expression for the tensor $\widetilde{\beta}^{\alpha\beta}(\mathbf{H})$ and the value of the corresponding Nernst–Ettingshausen coefficient. Since the thermoelectric measurements are performed with direct current, the frequency of the electromagnetic field (ω) will finally tend to zero.

The contribution of the fluctuation Cooper pairs to the full heat current $\mathbf{j}_{(Q)}$ can be found in the framework of the phenomenological Ginzburg–Landau approach, similarly to the case of the fluctuation electric current calculus, discussed earlier (see chapter 3). The explicit phenomenological expression for the full heat current operator was suggested in papers by Schmid [98] and Caroli and Maki [130]. The idea behind its presentation adds up to a simple substitution in the quantum mechanical expression for the electric current operator (2.17) of the effective charge of a Cooper pair by its gauge invariant effective energy:[30]

$$q^* \Rightarrow -i\left(\frac{\partial}{\partial t} + iq^*\varphi\right) = \left(-i\frac{\partial}{\partial t} - 2e\varphi\right). \tag{4.18}$$

[30]Let us stress that, for the charge of a Cooper pair, we assume that $q^* = -2e$, where e is positive. The sign of the charge in this chapter is important, since here we speak about the odd effects in it.

The contribution to the full heat current of the corresponding fluctuation Cooper pairs can be obtained by means of the averaging-out of such an operator over the fluctuating Ginzburg–Landau order parameter:

$$\mathbf{j}_{(Q)} = \operatorname{Re}\left\langle \Psi^* \left(-i\frac{\partial}{\partial t} - 2e\varphi\right)\widehat{\mathbf{v}}\Psi \right\rangle. \tag{4.19}$$

Here $\widehat{\mathbf{v}} = \widehat{\mathbf{p}}/2m = (-i\nabla - 2e\mathbf{A})/2m$ is the Cooper pair velocity operator, $(\varphi, \mathbf{A})$ are electromagnetic field scalar and vector potentials, while Ψ is the order parameter found as the solution of the time-dependent Ginzburg–Landau equation, written down in the presence of the external fields and Langevin forces.

Restricting by the theory of linear response, one can use the expression for the order parameter $\Psi = \Psi^{(0)} + \Psi^{(1)}$ derived in chapter 3 (see Eq. (3.2)):

$$\begin{aligned}\mathbf{j}_{(Q)} &= \operatorname{Re}\langle \Omega\Psi^*\widehat{\mathbf{v}}\Psi\rangle - 2e\operatorname{Re}\langle \Psi^*\widehat{\varphi}\widehat{\mathbf{v}}\Psi\rangle = \\ &= 2\operatorname{Re}\left\langle \Omega\Psi^{(0)*}\widehat{\mathbf{v}}\Psi^{(1)}\right\rangle - 2e\operatorname{Re}\left\langle \Psi^{(0)*}\widehat{\varphi}\widehat{\mathbf{v}}\Psi^{(0)}\right\rangle,\end{aligned} \tag{4.20}$$

where Ω is the Fourier transform of the term with a time derivative in the heat current operator (4.19). Substitution of the values $\Psi^{(0)}$ and $\Psi^{(1)}$, expressed by means of the TDGL operator $\widehat{L}$ (see Eq. (3.8)) in accordance with Eqs. (3.4) and (3.13), into Eq. (4.20) results in

$$\mathbf{j}_{(Q)} = 8eT\operatorname{Re}\gamma_{\mathrm{GL}}\operatorname{Im}\operatorname{tr}\left\{\gamma_{\mathrm{GL}}\Omega\widehat{L}^*\widehat{\mathbf{v}}\widehat{L}\widehat{\varphi}\widehat{L}\right\} - 4eT\operatorname{Re}\gamma_{\mathrm{GL}}\operatorname{Re}\operatorname{tr}\left\{\widehat{L}^*\widehat{\varphi}\widehat{\mathbf{v}}\widehat{L}\right\}. \tag{4.21}$$

The matrix elements of the electric field potential operator $\widehat{\varphi}$ are determined by Eqs. (3.22) and (3.20). In the basis of eigenfunctions of the operator $\widehat{L}$, the operation of trace calculation can be performed explicitly and the expression for the fluctuation contribution to the heat current takes the form

$$\begin{aligned}\mathbf{j}^{\alpha}_{(Q)} =& -8eT\operatorname{Re}\gamma_{\mathrm{GL}}\operatorname{Re}\sum_{\{i,k\}}^{\infty}\left\{\gamma_{\mathrm{GL}}\frac{\widehat{\mathbf{v}}^{\alpha}_{\{ik\}}\widehat{\mathbf{v}}^{\beta}_{\{ki\}}}{\varepsilon_{\{i\}}-\varepsilon_{\{k\}}}\int\frac{\Omega d\Omega}{2\pi}\widehat{L}^*_{\{i\}}\widehat{L}_{\{i\}}\widehat{L}_{\{k\}}\right\}\mathbf{E}^{\beta} - \\ &- 4eT\operatorname{Re}\gamma_{\mathrm{GL}}\operatorname{Im}\sum_{\{i,k\}}^{\infty}\left\{\frac{\widehat{\mathbf{v}}^{\alpha}_{\{ik\}}\widehat{\mathbf{v}}^{\beta}_{\{ki\}}}{\varepsilon_{\{i\}}-\varepsilon_{\{k\}}}\int\frac{d\Omega}{2\pi}\widehat{L}^*_{\{i\}}\widehat{L}_{\{i\}}\right\}\mathbf{E}^{\beta}.\end{aligned} \tag{4.22}$$

Here the matrix elements $\widehat{L}_{\{l\}}$ and $\widehat{\mathbf{v}}^{\alpha}_{\{ik\}}$ are determined by Eqs. (3.19) and (3.15), the eigenvalues of the GL Hamiltonian $\varepsilon_{\{k\}}$ are given by Eq. (3.28), and summation over subscripts $\{i\}$ and $\{k\}$ is performed in accordance with the rules established in chapter 3. The integral over frequencies is carried out by the simple use of the Cauchy theorem:

$$\int\frac{\Omega d\Omega}{2\pi}\widehat{L}_{\{l\}}(\Omega)\widehat{L}_{\{i\}}(\Omega)\widehat{L}^*_{\{i\}}(\Omega) = \frac{i}{2\operatorname{Re}\gamma_{\mathrm{GL}}}\frac{1}{\gamma_{\mathrm{GL}}\varepsilon_{\{i\}}+\gamma^*_{\mathrm{GL}}\varepsilon_{\{l\}}}, \tag{4.23}$$

$$\int \frac{d\Omega}{2\pi} \widehat{L}^*_{\{i\}}(\Omega)\widehat{L}_{\{i\}}(\Omega) = \frac{1}{2\,\mathrm{Re}\,\gamma_{\mathrm{GL}}} \frac{1}{\varepsilon_{\{i\}}}, \tag{4.24}$$

which gives

$$\mathbf{j}^{\alpha}_{(Q)} = -T\widetilde{\beta}^{\alpha\beta}(\epsilon, h)\,\mathbf{E}^{\beta}. \tag{4.25}$$

Therefore, one finds the general expression for the required tensor $\widetilde{\beta}^{\alpha\beta}(\mathbf{H})$:

$$\widetilde{\beta}^{\alpha\beta}(\epsilon, h) = -4e\,\mathrm{Im}\sum_{\{i,k\}}^{\infty} \frac{\widehat{\mathbf{v}}^{\alpha}_{\{ik\}}\widehat{\mathbf{v}}^{\beta}_{\{ki\}}}{\varepsilon_{\{i\}} - \varepsilon_{\{k\}}} \left\{ \frac{\gamma_{\mathrm{GL}}}{\gamma_{\mathrm{GL}}\varepsilon_{\{i\}} + \gamma^*_{\mathrm{GL}}\varepsilon_{\{k\}}} - \frac{1}{2\varepsilon_{\{i\}}} \right\}. \tag{4.26}$$

The latter, together with Eq. (4.17) and the results of section 2.4 for the fluctuation magnetization, describes in the most general form the fluctuation contributions to all magneto-thermoelectric coefficients.

We will see below that, due to the different "imaginarities" of the products of matrix elements in Eq. (4.26) for the diagonal and off-diagonal elements of the magneto-thermoelectric tensor $\beta^{\alpha\beta}$, the magnitudes, magnetic field and temperature behaviors of the former and the latter differ strikingly. Namely, while the diagonal components $\beta^{\alpha\alpha}(H, \epsilon) \sim \mathrm{Im}\,\gamma_{\mathrm{GL}} \ln 1/\epsilon$, the off-diagonal component $\beta^{xy}(H, \epsilon) \sim [H/H_{c2}(0)]\,\epsilon^{-1}$ does not contain the small factor $\mathrm{Im}\,\gamma_{\mathrm{GL}}$, related to the electron–hole asymmetry [103,154]. Moreover, the latter component turns out to be much more singular in a reduced temperature.

4.4 Fluctuation thermoelectric power

Let us start the analysis of a general expression for the thermoelectric tensor (4.17) from its diagonal elements $\beta^{\alpha\alpha}(\epsilon, h)$. First of all, let us notice that the magnetization currents do not contribute to it at all. The next important observation following from Eq. (4.26) is that $\beta^{\alpha\alpha}(\epsilon, h) \sim \mathrm{Im}\,\gamma_{\mathrm{GL}}$, since the product $\widehat{\mathbf{v}}^{\alpha}_{\{il\}}\widehat{\mathbf{v}}^{\alpha}_{\{li\}}$ is real. By the same reason, the second term in Eq. (4.26) disappears. Taking into account the smallness of the imaginary part of the coefficient γ_{GL} with respect to its real part, one can expand Eq. (4.17) in powers of $\mathrm{Im}\,\gamma_{\mathrm{GL}}$, which results in

$$\beta^{\alpha\alpha}(\epsilon, h) = -8e\frac{\mathrm{Im}\,\gamma_{\mathrm{GL}}}{\mathrm{Re}\,\gamma_{\mathrm{GL}}} \sum_{\{i,k\}}^{\infty} \frac{\varepsilon_{\{k\}}\widehat{\mathbf{v}}^{\alpha}_{\{ik\}}\widehat{\mathbf{v}}^{\alpha}_{\{ki\}}}{\left(\varepsilon_{\{i\}} + \varepsilon_{\{k\}}\right)^2\left(\varepsilon_{\{i\}} - \varepsilon_{\{k\}}\right)}. \tag{4.27}$$

Let us apply this formula to our usual object, a layered superconductor, and calculate the contribution of the fluctuation Cooper pairs to the longitudinal component of its thermoelectric power $\beta^{xx}(\epsilon, h)$ above critical temperature. The required product of the matrix elements has already been calculated above (see Eq. (3.32)). Performing a summation over the subscript $\{i\}$, one finds

$$\beta^{xx}(\epsilon, h) = \frac{8e}{\pi s}\frac{\mathrm{Im}\,\gamma_{\mathrm{GL}}}{\mathrm{Re}\,\gamma_{\mathrm{GL}}}(\alpha T h)^2 \sum_{k=0}^{n_c} \int_{-\pi/S}^{\pi/s} \frac{dq_z}{2\pi} \frac{(k+1)}{\left(\varepsilon_{\{k+1\}} + \varepsilon_{\{k\}}\right)^2}. \tag{4.28}$$

The integration in Eq. (4.28) can be performed using the integrals of Appendix C. As a result, the contribution of fluctuating Cooper pairs to the thermoelectric power of a superconductor above critical temperature is presented as the sum over Landau levels as follows:

$$\beta^{xx}(\epsilon, h) = \frac{1}{2s\Phi_0} \frac{\operatorname{Im}\gamma_{\mathrm{GL}}}{\operatorname{Re}\gamma_{\mathrm{GL}}} \sum_{l=0}^{n_c} \frac{(l+1)(l+1+\epsilon/2h+r/4h)}{[(l+1+\epsilon/2h+r/2h)(l+1+\epsilon/2h)]^{3/2}}. \tag{4.29}$$

The problem of formal divergence of sums of this type has already been discussed above: it is related to the inapplicability of Ginzburg–Landau theory to short-wavelength fluctuations with wavelengths larger than $\lambda \sim \xi^{-1}$ (see section 2.4). It can be resolved by means of the cut-off of the summation at $n_c \sim h^{-1}$.

In the $2D$ case ($r = 0$), the summation can be performed exactly and the corresponding result is expressed in terms of the digamma-function and its derivatives (see Appendix B). The general expression for the fluctuation thermoelectric power, valid for an arbitrary magnetic field, takes the form

$$\beta^{xx}_{(2)}(\epsilon, h) = \frac{1}{2s\Phi_0} \frac{\operatorname{Im}\gamma_{\mathrm{GL}}}{\operatorname{Re}\gamma_{\mathrm{GL}}} \left[\ln(1/h) - \psi(1+\epsilon/2h) + \frac{\epsilon}{2h}\psi'(1+\epsilon/2h)\right]. \tag{4.30}$$

One can write down asymptotic expressions in the limits of weak and strong magnetic fields, as well as close to the critical temperature shifted by a magnetic field [159, 136]:

$$\beta^{xx}_{(2)}(\epsilon, h) = \frac{1}{2s\Phi_0} \frac{\operatorname{Im}\gamma_{\mathrm{GL}}}{\operatorname{Re}\gamma_{\mathrm{GL}}} \begin{cases} \ln\frac{1}{\epsilon}, & h \ll \epsilon, \\ \ln\frac{1}{h}, & \epsilon \ll h, \\ \ln\left(\frac{4\gamma_E}{h}\right) + \pi^2/2, & \epsilon + h \ll h. \end{cases} \tag{4.31}$$

In the $3D$ case, the fluctuation part of the thermoelectric power does not contain any divergent in ϵ contribution ($\beta^{xx}_{(3)}(\epsilon) = C_1 - C_2\sqrt{\epsilon}$), and we do not present here the explicit form of its coefficients C_1 and C_2.

As one can see from the results obtained, the diagonal components of the fluctuation magneto-thermoelectric tensor are proportional to $\operatorname{Im}\gamma_{\mathrm{GL}}$. This value is usually related to the derivative $d\ln T_c/d\ln E_F$ and is supposed to be small (let us recall that the thermoelectric power of a normal metal $\beta_n(T)$ turns out to be of the order of T/E_F). Nevertheless, in some special cases, like the Kondo effect, closeness of the Fermi level to the topological singularity of the electron spectrum, this value can grow considerably. Analogously, as it will be demonstrated in section 9.3, in the latter case, $\operatorname{Im}\gamma_{\mathrm{GL}}$ drastically increases in the superconductor, and, as a consequence, the fluctuation contributions to the Hall conductivity and the thermoelectric power become anomalously large.

The microscopic consideration of the problem shows [162, 163] that the indirect fluctuation contribution to the thermoelectric power, related to the decrease of the electron density of states at the Fermi level due to their involvement in fluctuation Cooper pairing, results in the appearance of a contribution of the same sign and the same temperature dependence as shown in Eq. (4.31).

Depending on the concentration of impurities, it can be of the same order or dominate the direct fluctuation contribution (4.31).

4.5 Fluctuation contribution to the Nernst–Ettingshausen coefficient

Let us pass to the discussion of the off-diagonal component of the thermoelectric tensor $\widetilde{\beta}^{xy}(\epsilon, h)$. The product of the matrix elements $\widehat{\mathbf{v}}^{x}_{\{lk\}}\widehat{\mathbf{v}}^{y}_{\{kl\}}$ turns out to be a purely imaginary value (see Eq. (3.45)):

$$\widehat{\mathbf{v}}^{x}_{\{lk\}}\widehat{\mathbf{v}}^{y}_{\{kl\}} = i\frac{\alpha T h}{2m}\left[l\delta_{l,k+1} - k\delta_{k,l+1}\right], \tag{4.32}$$

which results in

$$\widetilde{\beta}^{xy}(\epsilon, h) = -\frac{2e}{\pi s}\alpha T h \sum_{k=0}^{\infty}\int_{-\pi/S}^{\pi/s}\frac{dq_z}{2\pi}(k+1)\left[\frac{2}{\varepsilon_{\{k+1\}}+\varepsilon_{\{k\}}} - \frac{1}{2\varepsilon_{\{k+1\}}} - \frac{1}{2\varepsilon_{\{k\}}}\right]. \tag{4.33}$$

The integration over momentum in Eq. (4.33) is trivial and leads to

$$\widetilde{\beta}^{xy}(\epsilon, h) = \frac{eh}{\pi s}\sum_{k=0}^{\infty}(k+1)\left\{\frac{1}{\sqrt{[\epsilon+h(2k+1)][r+\epsilon+h(2k+1)]}} + \frac{1}{\sqrt{[\epsilon+h(2k+3)][r+\epsilon+h(2k+3)]}} - \frac{2}{\sqrt{[\epsilon+h(2k+2)][r+\epsilon+h(2k+2)]}}\right\}. \tag{4.34}$$

It can be noticed that the same sum that appears in Eq. (4.34) was obtained and analyzed in section 3.4 when studying the longitudinal component of fluctuation magnetoconductivity. This is why we can immediately write down the results.

In the $2D$ case, the summation is performed analytically in terms of digamma-functions, and the exact formula, valid for an arbitrary magnetic field, can be written as follows:

$$\widetilde{\beta}^{xy}_{(2)}(\epsilon, h) = \frac{1}{s\Phi_0}\left(\frac{\epsilon}{2h}\right)\left\{\psi(1/2+\epsilon/2h) - \psi(\epsilon/2h) - \frac{h}{\epsilon}\right\}. \tag{4.35}$$

Naturally, the functional dependence of $\widetilde{\beta}^{xy}_{(2)}(\epsilon, h)$ on temperature and magnetic field completely coincides with that of the fluctuation magnetoconductivity $\sigma^{xx}_{(2)}(\epsilon, h)$ (see Eqs. (3.38) and (3.39)), i.e. one can apply the estimate given by Eq. (4.10) even for arbitrary magnetic fields.

In the limiting cases of a weak and a strong magnetic field, as well as in the vicinity of the value of the critical temperature, shifted by a magnetic field, Eq. (4.35) gives [103]

$$\widetilde{\beta}_{(2)}^{xy}(\epsilon,h)=\frac{1}{2s\Phi_0}\begin{cases} h/2\epsilon, & h\ll\epsilon,\\ 1, & \epsilon\ll h,\\ 2h/(\epsilon+h), & \epsilon+h\ll h.\end{cases} \tag{4.36}$$

In chapter 2 the general expression (2.62) for fluctuation magnetization was obtained. Making use of the asymptotic expression for the logarithm of the gamma-function $\ln\Gamma(x)\approx-\ln x$, one can note down the required asymptotic expressions:

$$M_{(2)}^{z}=-\frac{T}{2s\Phi_0}\begin{cases} h/3\epsilon, & h\ll\epsilon,\\ \ln 2, & \epsilon\ll h,\\ 2h/(\epsilon+h)-\ln\frac{h}{\epsilon+h}, & \epsilon+h\ll h.\end{cases} \tag{4.37}$$

Now we have all of the necessary results to write down the off-diagonal component of the magneto-thermoelectric tensor $\beta_{(2)}^{xy}(\epsilon,h)$. In accordance with the relation (4.17), one has [154]

$$\beta_{(2)}^{xy}=\frac{1}{2s\Phi_0}\begin{cases} h/6\epsilon, & h\ll\epsilon,\\ 1-\ln 2, & \epsilon\ll h,\\ \ln\frac{h}{\epsilon+h}, & \epsilon+h\ll h.\end{cases} \tag{4.38}$$

These expressions demonstrate the importance of accounting for the fluctuation magnetization currents: they at least change the numerical coefficient of the off-diagonal component of the magneto-thermoelectric tensor in comparison to the corresponding value $\widetilde{\beta}_{(2)}^{xy}$. Close to the transition point shifted by the magnetic field its becomes crucial: it changes the very functional dependence of $\beta_{(2)}^{xy}$ on temperature in comparison with the corresponding dependence of $\widetilde{\beta}_{(2)}^{xy}$. Moreover, performing the microscopic calculation of the effect of quantum fluctuations on the value of the Nernst effect at zero temperature (above $H_{c2}(0)$), one finds that $\widetilde{\beta}_{(2)}^{xy}$ diverges as T^{-1}, while, in accordance with the third law of thermodynamics, the Nernst effect should disappear. It is the accounting for the fluctuation magnetization currents in these conditions that results in the compensation of the divergent and even the next, constant, terms originating from $\widetilde{\beta}_{(2)}^{xy}$ in Eq. (4.17) and allows $\beta_{(2)}^{xy}$ to satisfy the general requirements of thermodynamics.

The value of $\widetilde{\beta}_{(\text{layer})}^{xy}(h\ll\epsilon)$ for a layered superconductor in a weak magnetic field that one can obtain by assuming that $h\ll\epsilon,r$ in Eq. (4.34) is

$$\widetilde{\beta}_{(\text{layer})}^{xy}(h\ll\epsilon)=\frac{h}{4\Phi_0 s}\frac{1}{\sqrt{\epsilon(\epsilon+r)}}. \tag{4.39}$$

The value of the fluctuation magnetization $\mathbf{M}$ in the same limit $h\ll\epsilon,r$ turns out to be (see Eq. (2.73))

$$M_{(\text{layer})}^{z}(h\ll\epsilon)=-\frac{Th}{6\Phi_0 s}\frac{h}{\sqrt{\epsilon(\epsilon+r)}}. \tag{4.40}$$

Therefore one can make certain that, in a weak field, due to accounting for magnetization currents, the cancelation of 2/3 of the value of $\widetilde{\beta}_{(2)}^{xy}$ takes place not only for $2D$, but also in the general case:

$$\beta^{xy}_{(\text{layer})}(h \ll \epsilon) = \frac{h}{12\Phi_0 s}\frac{1}{\sqrt{\epsilon(\epsilon+r)}}. \tag{4.41}$$

In the $3D$ case, the last formula takes the following form:

$$\beta^{xy}_{(3D)}(h \ll \epsilon) = \frac{h}{12\Phi_0 s}\frac{1}{\sqrt{\epsilon r}} = \frac{e^2 H}{3\pi}\frac{\xi^2_{xy}}{\xi_z}\frac{1}{\sqrt{\epsilon}}. \tag{4.42}$$

One can make certain that the component of the magneto-thermoelectric tensor β^{xy} determines the contribution of the fluctuation pairs to the Nernst–Ettingshausen signal. Indeed, since $\sigma^{xy} \ll \sigma^{xx}$, one can write down the expression for the fluctuation part of the Nernst–Ettingshausen coefficient (4.31):

$$\nu^{\text{NE}}_{\text{fl}} = \frac{1}{H}\frac{\beta^{xy}_{\text{fl}}\sigma^{xx}_n + \beta^{xy}_n\sigma^{xx}_{\text{fl}} - \beta^{xx}_{\text{fl}}\sigma^{xy}_n - \beta^{xx}_n\sigma^{xy}_{\text{fl}}}{(\sigma^{xx}_n)^2}. \tag{4.43}$$

The last term in the numerator is the most singular in ϵ. Nevertheless, being proportional to the product of two small factors (T/E_F) and $\operatorname{Im}\gamma_{\text{GL}}$, determining the degree of electron–hole asymmetry, it turns out to be negligible. The third term is small by the same reason, and besides it is not so singular in ϵ. The second term has the same singularity in temperature as the first one but, being proportional to the value β^{xy}_n, it can be neglected in comparison with the first one by the parameter T/E_F. In such a way one can see that the fluctuation part of the Nernst–Ettingshausen coefficient is determined by the value β^{xy}:

$$\nu^{\text{NE}}_{\text{fl}}(h,\epsilon) = \frac{1}{H}\frac{\beta^{xy}_{\text{fl}}}{\sigma^{xx}_n} = \frac{\pi\hbar k_B}{24mcT}\frac{1}{\sqrt{\epsilon(\epsilon+r)}}, \tag{4.44}$$

which coincides with our qualitative evaluation (4.10).

4.6 Fluctuation heat conductivity

The contribution of fluctuation Cooper pairs to the heat transfer can, in principle, be calculated in the frameworks of the developed approach, or a microscopic one, as the response at the temperature gradient applied [169,170]. Nevertheless, we will skip here such a quantitative analysis of the fluctuation heat conductivity. The reason is that, in contrast to all fluctuation corrections discussed above, the direct fluctuation Cooper pair contribution to the heat conductivity does not contain any characteristic singularity close to the transition temperature, but just tends zero. Moreover, the microscopic analysis of the problem [170] also demonstrates that indirect quantum fluctuation contributions (DOS and MT contributions) do not contribute to heat transfer. This is why we will just demonstrate qualitatively the absence of the temperature singularity in the contribution of the fluctuation Cooper pairs to the heat transfer (heat paraconductivity $\kappa^{\alpha\beta}$) close to transition.

In the Introduction we demonstrated how the electric paraconductivity can be evaluated using the Drude-like formula: $\sigma \sim e^2\mathcal{N}\tau^*/m$, where $\mathcal{N}$, τ^* and m are the concentration, the lifetime and the mass of the fluctuation Cooper pair.

The only difference of such an evaluation for the case of heat paraconductivity consists in the substitution of the pair's effective charge $2e$ by its effective energy $\varepsilon^* \sim T - T_c$. As a result, one can easily find

$$\kappa(\epsilon) - \kappa(0) \sim \frac{(\epsilon^*)^2}{(2e)^2} \Delta\sigma(\epsilon) \sim \begin{cases} \epsilon^{3/2}, & D = 3, \\ \epsilon \ln \frac{1}{\epsilon}, & D = 2, \\ \epsilon^{1/2}, & D = 1, \end{cases} \tag{4.45}$$

and ensure the absence of the singular in ϵ contribution to the heat conductivity for any dimensionality. This qualitative consideration is confirmed by the microscopic analysis [170, 171].

5

FLUCTUATIONS IN VORTEX STRUCTURES

The properties of the vortex state of type-II superconductors have been described in detail in handbooks devoted to superconductivity (see, e.g. example, [131,172, 132]) and review articles [73,173,174]. That is why the main goal of the present chapter will be only the discussion of the effect of fluctuations on them. We will restrict our consideration to simple estimations which nevertheless allow us to recognize the qualitative picture of the phenomena.

There are two different types of fluctuations which affect the properties of the vortex lattice considerably. The first one is quenched disorder. By this definition we assume the time-independent structure fluctuations, related to nonhomogeneous distribution of impurities, dislocations and other defects of the crystalline lattice. The second type of fluctuations changing the properties of the vortex lattice qualitatively are the thermal fluctuations of the order parameter already studied above.

5.1 Vortex lattice and magnetic flux resistivity

What is a vortex lattice? In 1949 Onsager introduced the notion of quantum vortices in superfluid helium. In 1957 Abrikosov discovered similar defects in superconductors in magnetic fields and showed that the magnetic field $H_{c1} \leq H \leq H_{c2}$ penetrates in type-II superconductors in the form of a regular vortex lattice. The lower critical field H_{c1} is determined by the formation in the system of the first isolated vortex of the normal phase in the superconducting state. The upper one, H_{c2}, corresponds to the formation of the first equilibrium superconducting nuclei in the volume of the bulk normal state (neglecting fluctuations). These conclusions were made on the basis of the analysis of the system of GL equations

$$\xi^2\left[\nabla+\frac{2\pi i}{\Phi_0}\mathbf{A}\right]^2\Psi+\left(1-\frac{|\Psi|^2}{|\Psi_0|^2}\right)\Psi=0, \tag{5.1}$$

$$\lambda^2\frac{|\Psi_0|^2}{|\Psi|^2}\nabla\times(\nabla\times\mathbf{A})+\mathbf{A}=-\frac{\Phi_0}{2\pi}\nabla\varphi. \tag{5.2}$$

In the conditions defined above this system has the periodic solution in the form of the triangular lattice of vortices with the period

$$a_\triangle=\left(\frac{2}{\sqrt{3}}\right)^{1/2}\left(\frac{\Phi_0}{B}\right)^{1/2} \tag{5.3}$$

($B = H + 4\pi M$). The coordinate-dependent order parameter of the isolated vortex ($\xi \ll a$) can be approximately presented as

$$\Psi(r,\theta) = \Psi_0 \frac{r}{\sqrt{r^2 + 2\xi^2}} e^{i\theta}. \tag{5.4}$$

One can see that the amplitude of the order parameter is depressed in the vortex core region of the size $\sim \xi$, while its phase changes by 2π when one circles the vortex.

If there is no quenched disorder in a superconductor (there are no pinning centers) an applied current causes the drift of the vortex lattice and dissipation takes place [175]. Indeed, when the transport current of density $\mathbf{j}_{\text{tr}}$ runs through the superconductor the Lorentz force

$$F_L = j_{\text{tr}} \Phi_0 / c \tag{5.5}$$

applied to the unit length of a unique vortex appears. In homogeneous superconductors under the effect of this force the vortices start to move with velocity v_v. This velocity is related to the Lorentz force by means of the viscosity coefficient η_v:[31]

$$F_{\text{L}} = \eta_v v_v. \tag{5.6}$$

The corresponding flow of the magnetic flux induces, according to the Faraday law, the electric field $E = (v_v/c)\,B$ and energy dissipation of the density $\text{j}_{\text{tr}} E = \text{j}_{\text{tr}} v_v B/c$. This dissipation is equal to the power losses of the Lorentz force. The resistivity ρ_f appearing in the process of the magnetic flux flow in the direction perpendicular to the transport current is called the flux flow resistivity:

$$\rho_f = E/j_{\text{tr}} = \frac{B\Phi_0}{\eta_v c^2}. \tag{5.7}$$

The energy dissipation takes place only in the region of the normal vortex core; thus the value of ρ_f can be evaluated as the normal metal resistivity multiplied on the portion of the area occupied by vortices [175–178]:

$$\rho_f \sim \rho_n \frac{\xi^2}{a_\triangle^2} \sim \rho_n \frac{B}{B_{c2}}. \tag{5.8}$$

Comparison of Eqs. (5.7) – (5.8) gives

$$\eta_v\,(T \ll T_{\text{c}}) = \frac{\Phi_0^2}{2\pi c^2 \rho_n \xi^2}. \tag{5.9}$$

Close to T_{c} more rigorous treatment of the problem is possible. It takes into account the renormalization of the normal excitation spectrum inside the core and for fields not very close to B_{c2} results in [174]

[31] η_v is the force acting on unit length of the vortex moving with unit velocity.

$$\rho_f \sim 0.2\rho_n \frac{B}{B_{c2}(T)} \sqrt{1 - T/T_c} \left[1 + \frac{2}{\tau_\varepsilon (T_c - T)} \right]. \qquad (5.10)$$

Here τ_ε is the energy relaxation time. The corresponding viscosity is

$$\eta_v (T \to T_c) \sim \eta_v (T \ll T_c) \frac{5\sqrt{1 - T/T_c}}{\left[1 + \frac{2}{\tau_\varepsilon (T_c - T)} \right]}. \qquad (5.11)$$

All the results reported in this section were obtained without taking into account two different fluctuation effects, which we will discuss now.

5.2 Collective pinning

In order to have a quiet superconducting life in our laboratory system we should prevent the motion of vortices, i.e. introduce the dry friction of vortices. So, we should create, either naturally or artificially, centers which would pin the vortex lattice. Thus, to obtain the global superconductivity, we should destroy superconductivity at rare random positions, or, in other words, introduce quenched disorder. Let us mention that such disorder is static by nature and manifests itself as the fluctuations of the GL functional parameters a, b, c. It can appear due to the structure inhomogeneities of the initial crystalline lattice (dislocations, accumulations of impurities, separation of grains of the other phase). Usually these fluctuations are small and they weakly affect the properties of a superconductor without vortices. Nevertheless even small structural fluctuations change qualitatively the properties of the vortex structure. This is due to the fact that being the centers of pinning they break down the Galilean invariance and result in the appearance of friction.

Let us start our consideration from the case of the interaction of an isolated vortex with some pinning center. As the example of the pinning center we assume the normal region to lie in a superconducting environment with the small size in comparison with the vortex core size (ξ). The local critical temperature T_c we suppose to be lower than the average one. The GL coefficient a can be presented in the form $a = a_0 + \delta a(\mathbf{r})$. In the first order in δa the vortex energy in the field of such a pinning center is

$$E_p = |\Psi(\mathbf{r})|^2 \int \delta a(\mathbf{r}')\, d\mathbf{r}', \qquad (5.12)$$

where r is the distance between the center of pinning and the vortex axis. If $\delta a(\mathbf{r}') > 0$ then the vortex energy is minimal when its axis passes through the pinning center. The displacement of the vortex from this position results in the appearance of an attractive force

$$|\mathbf{f}| = |\partial E_p / d\mathbf{r}| \sim E_p/\xi. \qquad (5.13)$$

The problem of summation of such forces induced by randomly distributed pinning centers presents the central problem of the pinning theory [73,179,180]. Such an average force would be equal to zero if the vortex lattice were undeformable.

The presence of strong pinning centers in a deformable vortex lattice can cause its plastic deformation [181] and give rise to dry friction. However, such types of pinning centers exist only in strongly disordered low temperature superconductors (rigid superconductors) and do not help to resolve the problem for others (for example, HTS). An analogous desperate situation had formed in Russia as a result of individual terrorism at the end of nineteenth century, when Lenin said: "We shall go another way."

Such another way for pinning was found, the so-called collective pinning. A single weak center causes only a weak elastic deformation, but the collective effect of a large number of weak centers destroys the lattice. A common belief is that a weak force leads to only a small distortion of a system. The result of the action of many weak forces applied at many different points of the system may depend on the system: it may lead either to a weak or to a strong distortion. If you accidentally step on the foot of another person, you say "Excuse me" and it is OK. But what would happen if many people stepped on the feet of other people? The result would depend on the system. For example, when a lot of people participate in an overcrowded party some of them may occasionally step on the feet of others. They would not be too disturbed if these were their friends. Yet if the offenders were all strangers, it could lead to disorder. A similar situation occurs in solid state physics.

A small concentration of impurities does not destroy the crystalline lattice. The long-range order remains, because these impurities belong to the lattice and move with it. If the concentration of impurities is small, then even a large number of them in a macroscopic body does not destroy the crystalline lattice.

But there are other types of lattices, such as the Abrikosov lattice of vortices in superconductors. Here impurities (centers of pinning) belong to our laboratory frame and do not move together with vortices. Such pinning centers destroy the long-range order in the lattice. This results in the appearance of a friction force, a critical current, a hysteresis. Short-range order exists only at distances shorter than the correlation length L_c [182], to which discussion we pass now.

5.2.1 *Correlation length*

How does the destruction of long-range order happen? A single vortex displaced by a weak pinning center transmits its displacement to other vortices by elastic forces. The displacement of vortices u caused by a force f_i due to a single pinning center i leads to displacements which only slowly decay with the distance r. Writing down the equation of elasticity

$$C\Delta u\left(\mathbf{r}\right) = f_i \delta\left(\mathbf{r} - \mathbf{r}_i\right)$$

one finds its solution

$$u = \frac{f_i}{C|\mathbf{r} - \mathbf{r}_i|}, \tag{5.14}$$

where C is elastic modulus and we ignore its tensor nature.[32] Due to the long-range character of the interaction (5.14) every vortex of the lattice receives information from a million other vortices: "A pinning center from another laboratory system has stepped on my foot". As a result, this vortex displaces strongly, and the long-range order in the vortex lattice disappears at distances $L \gg L_c$. At shorter distances $L \ll L_c$, the crystalline order is destroyed weakly and all vortices are displaced by the same distance u.

One can find the correlation length L_c as follows. Inside this region the pinning centers affect vortices independently in various directions; therefore we have to average its square value. Thus, the square of the vortex displacement $\langle u^2 \rangle$ in a crystal of size L can be estimated using formula (5.14) as the square of the typical force multiplied on the number of pinning centers in volume L^3:

$$\langle u(L)^2 \rangle \sim \left(\frac{f}{CL} \right)^2 nL^3. \tag{5.15}$$

n here is the density of pinning centers. As the size of the system grows, $\langle u^2 \rangle \propto L$ increases. When $\sqrt{\langle u^2 \rangle}$ reaches the size of the vortex core ξ, vortices can be assumed to be uncorrelated and the formula (5.14) works no more. Thus L_c, defined by the condition $u(L_c) = \xi$ [182], or

$$L_c = \frac{C^2 \xi^2}{n f^2},$$

can be chosen as the correlation length of the vortex lattice.

The order parameter in a superconductor with the regular Abrikosov vortex lattice varies periodically in space. It can be compared with an antiferromagnet. As a result of collective pinning the order parameter Δ becomes a random function of position in space, similar to the spins in spin glasses. The vortex lattice behaves here like a glass. The glassy phase properties are not yet fully studied.

The existence of the long-range order of the Abrikosov lattice, as in normal crystals, means that δ-function peaks have to be observed at the Bragg wave vectors in the scattering amplitudes of the neutrons and X-rays. Let us remember the that presence of a small concentration of impurities in the usual crystal does not destroy the long-range order; it decreases the amplitude of such peaks but does not smear it out. In contrast to the usual crystal lattice, the collective pinning in the Abrikosov lattice leads to the destruction of the long-range order in vortices positions. Nevertheless a weak logarithmic singularity in the form-factor at the Bragg wave vectors remains [183–185].

[32] Let us recall that in normal crystals impurities move together with the lattice, the displacement u falls down quicker, $u \sim 1/r^2$, and the collective effect does not arise.

5.2.2 *Critical current*

Each region of size of the order of L_c finds its position of equilibrium at a minimum of the random pinning potential. In order to move these regions from those positions a force is needed. This force of dry friction F_{fr} is equal to the Lorentz force $j_c B$ at equilibrium; therefore to evaluate the critical current let us estimate F_{fr}. Each pinning center produces a randomly directed force f_i. In order to find the average pinning force per volume L_c^3 one has to calculate the mean square value of these random forces

$$F_{fr} = \frac{\sqrt{\sum_i f_i^2}}{L_c^3} \sim \frac{f\sqrt{n}}{L_c^{3/2}}. \tag{5.16}$$

Now, comparing this result with the Lorentz force, we obtain for the critical current [179, 180]

$$j_c = \frac{1}{B}\,\frac{n^2 f^4}{\xi^3 C^3}. \tag{5.17}$$

One can notice that for a hard lattice with a large elastic modulus C the friction is less. This news would not be surprising to the experienced driver: it is well known that strongly pumped tires grip less in braking than weakly pumped.

Let us remember that formula (5.17) was obtained omitting the tensor nature of the elasticity. In reality, even an isotropic crystal is characterized by several constants: by the moduli of compression C_{11}, tilt, C_{44}, and shear, C_{66}. Being the components of the dynamical matrix of the theory of elasticity, these values, generally speaking, are some functions of the quasi-momentum $\mathbf{k}$. Nevertheless, this dispersion has to be taken into account only when the correlation length L_c turns out less than the penetration depth λ.

When $L_c \lesssim \lambda$, it is more energetically favorable to deform the vortex line without the deformation of magnetic field lines. Indeed, in the absence of pinning centers the vortex line coincides with the magnetic field one. The introduced pinning center drags the core of the vortex line and, in principle, the magnetic field line could be deformed together with the vortex. Nevertheless, the magnetic field energy related to the large scale field distortion exceeds the local gain of the elastic energy near the pinning center and the vortices locally deviate from magnetic field lines. In this case the critical current depends exponentially on the concentration of the pinning centers [174]. Even more important is that in these conditions the critical current j_c grows exponentially when the magnetic field tends to H_{c2} (the so-called peak effect [186]).

When $L_c \gtrsim \lambda$ the peak effect near H_{c2} still takes place but it has another origin. This is due to the fact that close to T_c the vortex lattice becomes soft and even one weak pinning center can generate plastic deformation and pin the lattice [180]. Analogously to the theory of phase transitions (see section 2.6) the RG method turns out to be useful in the pinning theory [89]. In contrast to the former the Gell–Mann–Low function here has an opposite sign and the effective charge (pinning force) grows with the distance increase (compare with the confinement in quantum chromodynamics). One can recognize the physical sense of this statement as follows. In the pinning theory the effective charge is equal to the averaged force of disordered pinning centers. We have demonstrated above that this force increases with the distance (see (5.16)) and becomes large at a distance of the order of L_c.

Anther important difference between the RG methods in the theories of pinning and phase transition consists in the object for which the Gell–Mann–Low equation is written down. Instead of the Gell–Mann–Low function depending only on the effective charge (phase transition theory) in the pinning theory the Gell–Mann–Low functional has to be introduced [187]. This functional depends on the correlator of pinning forces, which is in its turn the function of the displacement u. Such an RG description is especially convenient when $L_c \lesssim \lambda$, inter-vortex interaction decreases slowly, elastic moduli have an anomalous dispersion and the problem is equivalent to the four-dimensional one. One can recognize this statement recalling the above described situation with the RG application to the ferroelectric with the long-range dipole–dipole forces, where the angular dependence of the interaction transforms the problem into effectively a $4D$ one which allows us to resolve the RG equations [12]. The two loop approximation of the functional RG allows us to find the prefactor in the exponential dependencies of L_c and j_c on magnetic field [188].

5.2.3 *Collective pinning in other systems*

Collective pinning is a generic phenomenon which arises in a number of physical situations. For instance, for the charge density wave [89, 189–191] the role of critical current is played by the critical electric field E_{cr}.

The collective pinning phenomenon arises when two rough elastic surfaces are brought into contact. The renormalization group method allowed us to find the force of dry friction [192].

In a similar way the destruction of long-range order occurs in ferromagnets [193]. In this problem the magnetic rigidity plays the role of elastic modulus, and the random magnetic field plays the role of pinning centers.

5.2.4 *Thermal depinning*

The value of critical current (5.17) is effected by the space variations of the GL coefficient $\delta a(\mathbf{r})$, determined by the specifics of the quenched disorder profile. Moreover it depends on such characteristics of the Abrikosov vortex lattice as $\Psi_0(T)$, $\xi(T)$ and $C(T)$ (see Eqs. (5.4), (5.12), (5.13)). All latter values considerably depend on temperature only in the scale $T \sim T_c$, so one could expect the critical current saturation at temperatures as low as $T \ll T_c$. However, considerable temperature dependence of critical current was found experimentally even at so low temperatures.

One more source of the critical current temperature dependence consists in the fluctuation oscillations of the vortices themselves, which sometimes can become important even far below the critical temperature. Indeed, the average square of the vortex displacement of elastic lattice depends on temperature

$$\langle u_T^2 \rangle = \sum_{\mathbf{k}} \frac{T}{Ck^2}. \tag{5.18}$$

Due to the numerical smallness of the triangular Abrikosov lattice shear modulus C_{66}, $u(T)$ can exceed ξ even noticeably below T_c. The thermal vibrations of the vortex leads to the increase of its effective size and when the displacement $u(T) \gtrsim \xi$ one has to substitute the value ξ in the formula (5.17) by $u(T)$. This will result in a decrease of critical current $j_c(T)$ in comparison with the value $j_c(0)$, which can take place even at temperatures $T \ll T_c$.

5.3 Creep

Quenched fluctuations transform the Abrikosov lattice into some kind of glass. Omitting thermal and quantum fluctuations one can find that the persistent superconducting currents with density less than j_c can still run in this glass state. Nevertheless it is necessary to remember that the current carrying state in such a hard type-II superconductor is only metastable and thus tends to decay. The accounting for thermal and quantum fluctuations shows that under their influence the flux lines can overcome their pinning barriers and seldom jump to the neighbor favorable pinning valley. This leads to a thermally activated motion of vortices known as creep [194, 195].

The average velocity of such vortex creep is determined by the Arrenius-type thermal activation process probability:

$$v = v_c \exp\left(-\frac{U(j)}{T}\right), \tag{5.19}$$

where the "pinning barrier" height $U(j)$ depends on the carrying current. Equation (5.19) determines the nonlinear diffusion equation for the current density

$$\frac{dj}{dt} \approx -\frac{j_c}{\tau} \exp\left(-\frac{U(j)}{T}\right), \tag{5.20}$$

which can be solved with the logarithmic accuracy [196]:

$$U(j) = T \ln\left(1 + \frac{t}{t_0}\right), \tag{5.21}$$

where t_0, the same as τ, depends on the sample size.

The characteristic property of a glass system is the existence of a huge number of metastable states separated by energy barriers of different heights. During the time of experiment t the system succeeds in overcoming only relatively low barriers of the heights $U < T \ln t/t_0$ and still stays in some metastable state separated from the more energy favorable ones by higher barriers.

The maximal barrier height depends on the flowing current but the exact form of this dependence is still unknown. One can only say that $U(j) \to 0$ when

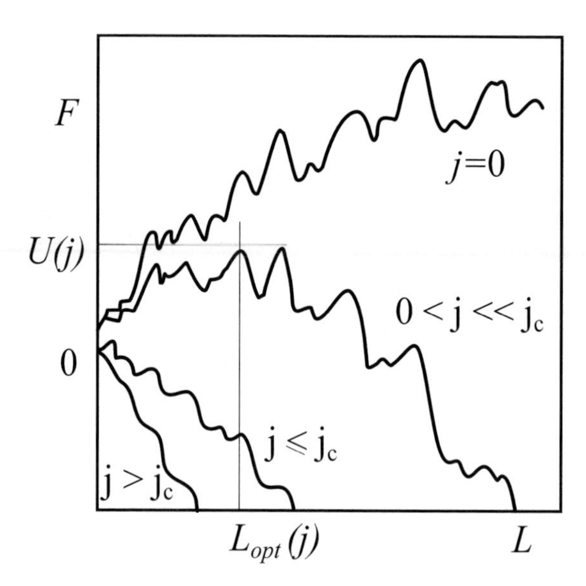

FIG. 5.1. Free energy functional versus length L of the hopping segment, which plays the role of a generalized coordinate. The free energy is made up of the two terms, the barrier energy, growing slowly with L, and the energy gain due to the Lorentz force, with a small prefactor but a more rapid growth in L. For a fixed driving current density j, a minimal segment of length $L_{opt}(j)$ has to overcome the barrier height $U(j)$ in order to reach the next favorable metastable state.

$j \to j_c$ and $U(j) \to \infty$ when $j \to 0$. The interpolation formula can be written as

$$U(j) \simeq \frac{U_c}{\mu}\left[\left(\frac{j_c}{j}\right)^{\mu} - 1\right]. \tag{5.22}$$

The energy scale for the pinning barrier U is determined by the collective pinning energy U_c.

The qualitative character of the dependence of the energy of the vortex system on some generalized coordinate at different values of flowing current j can be seen in Fig.5.1. Evidently, at $j > j_c$ no more metastable states take place; barriers do not exist anymore. When $j_c - j \ll j_c$ the height of the barriers is small. The reason for the barrier height growth with the decrease of current can be recognized looking at Fig. 5.2.

When $j = 0$ the vortex (or the system of vortices) is located in one of the most profound valleys and the neighbor valleys are higher in energy. The switch of the current means the appearance of the Lorentz force; hence the formation of the common slope in the "mountains chain" and formation of the low energy valleys. Nevertheless for small values of j such valleys will appear far enough from

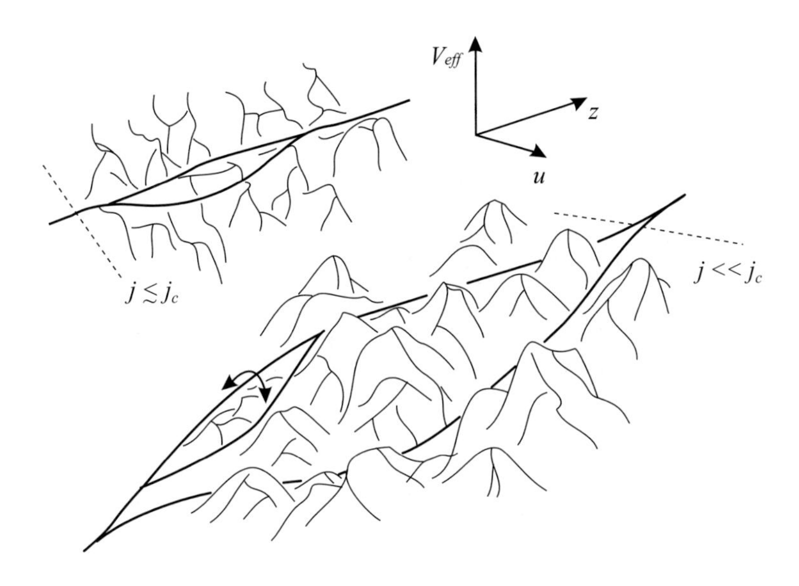

FIG. 5.2. Effective tilted random potential felt by the flux line in the presence of quenched disorder and of a driving transport current density j. The elastic vortex line relaxes into a low-lying metastable state.

the initial position of the chosen vortex and the displacement only of the long enough vortex segment to the new valley can result in the energy decrease. To reach the new valley this long segment will overcome a lot of high tops. That is the reason for the barrier height growth with the increase of the segment length, i.e. with the current decrease.

The time evolution of the screening current can be easily found from (5.21) knowing the explicit form of $U(j)$ (5.22):

$$j(t) = j_{\rm c}\left[1 + \frac{\mu T}{U_{\rm c}} \ln\left(1 + \frac{t}{t_0}\right)\right]^{-1/\mu}. \tag{5.23}$$

Under the theory of collective creep, the dynamical critical exponent μ is related to the one for static properties (ζ) and the space dimensionality D [197]:

$$\mu = \frac{2\zeta + D - 2}{2 - \zeta}. \tag{5.24}$$

We see that the temporal decay of the transport current is thus determined by the ratio $T/U_{\rm c}$, which can be found experimentally by measuring the relaxation of the diamagnetic moment of a sample in the critical state. The activation energy $U_{\rm c}$ is therefore an experimentally accessible quantity and it provides one test for the validity of the weak collective pinning theory. Typical experimental results for the activation energy $U_{\rm c}$, obtained in magnetic relaxation experiments at low temperatures, are in the range $U_{\rm c} \sim 100 - 1000\,K$ [198–204]. For conventional hard type-II superconductors $T/U_c \sim 10^{-3}$ [205,206] and the creep phenomenon can be observed only for currents close to the critical one. Expanding (5.23) for small T/U_c we reproduce the famous logarithmic time decay of the diamagnetic current [194]:

$$j(t) = j_c \left[1 - \frac{T}{U_c} \ln\left(1 + \frac{t}{t_0}\right)\right]. \tag{5.25}$$

In the new oxide superconductors, however, the corresponding decay coefficients turn out to be much larger, reaching the value 10% at $T \sim 0.5T_c$. These large logarithmic decay rates are a result of various factors, such as the high temperatures available in an experiment, the small pinning energies U_c, which in turn are a consequence of the small coherence length ξ, and the large anisotropy of oxides. Combining the large decay coefficients with a typical logarithmic time factor $\ln(t/t_0)$ of the order of 20 (waiting time $t \approx 1$ min, the characteristic t_0 may be evaluated as 10^{-6} s) we have to conclude that the experimentally measured current density j has been roughly halved due to creep, as compared with the critical current density j_c even at such low temperatures as $T \sim 10\,K$. Therefore, it is important to realize that the determination of the critical current density in the oxides is always affected by the presence of creep, and the condition $j_c - j \ll j_c$ is no longer fulfilled. The expression "giant creep" was therefore introduced [198] to describe the phenomenon of very large creep rates characteristic of the oxide superconductors.

From a fundamental science point of view, the case $j \ll j_c$ and, in particular, the limit $j \to 0$ is very interesting, too: wishing to probe the thermodynamic state of the vortex structure, we should perturb the system only infinitesimally and record its response. For a truly superconducting state we would expect to observe the vanishing resistivity ρ_f in the limit $j \to 0$ or, to put it somewhat differently, to see a sublinear "glassy" response of the vortex structure. In fact, as is seen from (5.22) barriers $U(j)$ against creep diverge algebraically with vanishing current density j, which implies strongly subohmic current–characteristics of the form [207]

$$V \sim \exp\left[-\frac{U_c}{T}\left(\frac{j_c}{j}\right)^{\mu}\right]. \tag{5.26}$$

This exponential and strongly nonlinear current–voltage characteristic means that the glassy properties show up not only in the structure of the vortex lattice, but also in its dynamics [207].

We have considered above the creep generated by thermal fluctuations. At low temperatures the quantum tunneling of vortices under the potential barriers is more likely [208–213]. The probability of such tunneling is determined by the action of the underbarrier action $S(j)$:

$$v = v_c \exp\left(-\frac{S(j)}{\hbar}\right). \tag{5.27}$$

Repeating the same considerations as in the case of thermal creep one finds:

$$j(t) = j_c \left[1 - \frac{\hbar}{S_c} \ln\left(1 + \frac{t}{t_0}\right)\right]. \tag{5.28}$$

5.4 Melting of the vortex lattice

Without accounting for quenched disorder and thermal fluctuations in a type-II superconductor, at the magnetic field $H = H_{c2}$, the phase transition of the second order to the Abrikosov periodic structure takes place. The question arises: how does the character of this transition change due to fluctuations? Let us start our consideration from thermal fluctuations only. Similarly to the situation in standard crystals the thermal fluctuations result in the melting of the vortex lattice [214]. This effect turns out to be especially important for HTS.

As we already know the region of strong fluctuations is determined by the Gi number hence one can expect the melting of the vortex lattice at temperatures of the order GiT_c below critical temperature (the magnetic field is supposed to be fixed). The quantitative consideration shows that the large numerical factor appears in our evaluation that results in the vortex lattice melting at temperatures noticeably below the critical one. For instance in normal solids the only energy parameter which can be constructed from the fundamental values is the Bohr energy, which is of the order of $10^5\,K$. Nevertheless, the energy of atomic interactions by numeric reasons turns out one order less than the Bohr energy, while the melting temperature is yet 1–2 orders less than the interaction energy. Another example: the melting of the Wigner crystal takes place at the temperature $T = 10^{-2}e^2/r$, where r is the inter-electron distance.

The melting temperature is determined by the equality of the free energies of solid and liquid phases. Due to the above mentioned numerical smallness the lattice free energy can be calculated in the harmonic approximation, while the calculation of the liquid phase free energy is an extremely difficult task.

The Gi parameter of conventional superconductors is very small and the melting of the vortex lattice takes place in the immediate vicinity of T_c (or H_{c2}). In this case, the calculations of the fluctuation free energy of the vortex liquid can be simplified as has been done in the approximation of the lowest Landau level (LLL) [215–220]. It is based on the fact that in a strong magnetic field the momentum perpendicular to the magnetic field direction is quantized and there appears a region where the lowest Landau level plays a dominant role. Such calculations were carried out up to the 9-th order in Gi for the $3D$ case and up to the 11-th order for the $2D$ case. Carrying out the summation of these series by the Borel–Pade' method one can find the free energy of the liquid phase. Comparing it with the free energy of the lattice one finds

$$T_m - T_c(H) = yGi(H)\,T_c(H), \tag{5.29}$$

where $y \approx -10$ for the $2D$ case and $y \approx -7$ for the $3D$ case and the $Gi(H)$ number in magnetic field is determined by Eqs. (2.99) and (2.101).

In the case of HTS the Gi number is already not too small. That is why the melting field H_m is not too close to H_{c2} and the LLL approximation no longer works. In the general case the melting temperature T_m is determined by the empirical Lidemann criterion which relates the mean-squared thermal displacement u_T determined by Eq. (5.18) to the lattice parameter a:

$$\sqrt{\langle u_T^2 \rangle} = c_L a. \tag{5.30}$$

The empirical Lidemann parameter c_L belongs to the interval 0.1–0.2. Coming back to the vortex lattice one can say that the smallness of the melting temperature is related to the smallness both of the value of Lidemann's parameter and the shear modulus of the triangular lattice.

Calculating the elastic moduli of the vortex lattice one can find from (5.18) the temperature and magnetic field dependence of $\sqrt{\langle u_T^2 \rangle}$. Finally substituting it into the Lidemann criterion (5.30) one can find the temperature dependence of the melting field $B_{\mathrm{m}}(T)$ [221]:

$$\frac{\sqrt{b_{\mathrm{m}}(t)}}{1 - b_{\mathrm{m}}(t)} \frac{t}{\sqrt{1-t}} \left[\frac{4\left(\sqrt{2}-1\right)}{\sqrt{1 - b_{\mathrm{m}}(t)}} + 1 \right] = \frac{2\pi c_L^2}{\sqrt{Gi}}. \tag{5.31}$$

(Here we use the dimensionless variables $b_{\mathrm{m}}(t) = B_{\mathrm{m}}(T)/B_{c2}(T)$ and $t = T/T_{\mathrm{c}}$.)

In fields close to $B_{c2}(T)$ one can get

$$\left[1 - b_{\mathrm{m}}(t)\right]^{3/2} = \frac{0.26}{c_L^2} \sqrt{Gi} \frac{t}{\sqrt{1-t}}. \tag{5.32}$$

Comparing this expression with (5.29) and taking into account that $Gi(H)$ is related to Gi by means of the relation (2.101) we find that (5.29) and (5.32) are equivalent in the case $c_L = 0.14$.

In the opposite limit $B_{\mathrm{m}}(T) \ll B_{c2}(T)$ from the general equation (5.31) one finds

$$B_{\mathrm{m}}(t) = \frac{1-t}{t^2} \frac{\pi^2 c_L^4}{8Gi} B_{c2}(T). \tag{5.33}$$

Comparison of this equation with the numerical results of [222, 223] shows their coincidence with $c_L = 0.25$.

The mechanism of the influence of thermal fluctuations on the vortex lattice can be easily realized in the assumptions of a strongly type-II superconductor ($\kappa \gg 1$) and relatively weak fields $B \ll B_{c2}(T)$. The large value of κ means that the order parameter disappears in very narrow (of the order of $\xi \ll a_\triangle$) domains only, so the vortex lattice resembles the normal crystal one with the only difference being that it is constructed of lines instead of particles. Temperature tilts of the vortex lines can result in the melting of the lattice analogously to the melting of a crystal lattice due to the thermal oscillations of its atoms. A precise analytic theory of the lattice melting does not exist, but with good accuracy for evaluation the Lidemann criterion is used.

In the $2D$ case in the narrow region between the liquid and lattice can exist the so-called hexatic phase [224] where the translational invariance no longer exists although it still conserves the orientational order in the vortices positioning.

Weak quenched disorder changes weakly T_{m} and the melting latent heat. But when the vortex displacements induced by the quenched disorder become comparable to the thermal displacements the phase transition can change its order from the first to the second [225]. Close to the second-order liquid–glass transition the scaling theory was developed [226]. One has to remember that this is the transition from vortex liquid to vortex glass, where there are many metastable states. The transition time between such states can exceed the experiment duration time. This can result in the identification of the true second-order phase transition (in the thermodynamic sense, when the experiment time can be supposed to equal infinity) as the first order one studied in real experiments with finite freezing time [227].

In conventional superconductors where Gi is small, quenched disorder can smear out the phase transition more strongly than thermal fluctuations. In this case the transition has a percolation character and will be discussed in chapter 12.

PART II

BASIC NOTIONS OF THE MICROSCOPIC THEORY

6

MICROSCOPIC DERIVATION OF THE TDGL EQUATION

6.1 Preliminaries

We have seen above how the phenomenological approach based on the GL functional allows one to describe fluctuation Cooper pairs (Bose particles) near the superconducting transition and to accounting for their contribution to different thermodynamical and transport characteristics of the system. Now we will consider of the microscopic description of fluctuation phenomena in superconductors. The development of the microscopic approach is necessary for the following reasons:

- This description allows microscopic determination of the values of the phenomenological parameters of the GL theory.
- This method is more powerful than the phenomenological GL approach and allows treatment of fluctuation effects quantitatively even far from the transition point and for magnetic fields strong as H_{c2}, taking into account the contributions of dynamical and short wavelength fluctuations.
- The electron energy relaxation times in metals are relatively large ($\tau_\varepsilon \gg \hbar/T$) which causes the electron low-frequency dynamics to be sensitive to the nearness to the superconducting transition. That is why the temperature dependence of fluctuation corrections can be determined generally speaking not only by the Cooper pair motion but also by changes in the single-electron properties.
- There are some fluctuation phenomena in which the direct Cooper pair contribution is considerably suppressed, or is even absent altogether. Among them we can mention the nuclear magnetic relaxation rate, tunnel conductivity, c-axis transport in strongly anisotropic layered metals, thermoelectric power and heat conductivity where the fluctuation pairing manifests itself by means of the indirect influence on the properties of the single-particle states of electron system.

In Part I the integration over the electron degrees of freedom in process of writing down the GL functional was performed implicitly (it was included in the GL coefficients). In its turn the averaging over the superconducting order parameter has been accomplished explicitly, by means of a functional integration over all possible configurations of the corresponding bosonic fields. In this description we have dealt with the fluctuation Cooper pair related effects only and the method of the functional integration turned out to be simple and effective for their description. In order to take into account more delicate consequences of the fluctuation pairing in the system of interacting electrons we will develop in the following sections the diagrammatic method of Matsubara temperature Green's functions since it is more adequate for the description of the properties of a Fermi system of interacting electrons.

6.2 The Cooper channel of electron–electron interaction: the fluctuation propagator

Let us start the microscopic description of fluctuation phenomena in a superconductor from the electron Hamiltonian. We will choose it in the simple BCS form:[33]

$$\mathcal{H} = \sum_{\mathbf{p},\sigma} \xi(\mathbf{p}) \widetilde{\psi}^{+}_{\mathbf{p},\sigma} \widetilde{\psi}_{\mathbf{p},\sigma} - g \sum_{\mathbf{p},\mathbf{p}',\mathbf{q},\sigma,\sigma'} \widetilde{\psi}^{+}_{\mathbf{p}+\mathbf{q},\sigma} \widetilde{\psi}^{+}_{-\mathbf{p},-\sigma} \widetilde{\psi}_{-\mathbf{p}',-\sigma'} \widetilde{\psi}_{\mathbf{p}'+\mathbf{q},\sigma'}. \tag{6.1}$$

The momentum conservation law side by side with singlet pairing is already taken into account in the interaction term. Here $\xi(\mathbf{p}) = E(\mathbf{p}) - E_F$ is the quasiparticle energy measured from the Fermi level; $-g$ is the negative constant of electron–electron attraction which is supposed to be momentum independent and different from zero in a narrow domain of momentum space in the vicinity of the Fermi surface where

$$p_F - \frac{\omega_D}{v_F} < |\mathbf{p}|, |\mathbf{p}'| < p_F + \frac{\omega_D}{v_F}.$$

$\widetilde{\psi}^{+}_{\mathbf{p},\sigma}$ and $\widetilde{\psi}_{\mathbf{p},\sigma}$ are the creation and annihilation field operators in the Heisenberg representation, so the first term is just the kinetic energy of the noninteracting Fermi gas. The interaction term is chosen in the traditional form characteristic of the electron-phonon mechanism of superconductivity.[34]

For the description of the properties of an interacting electron system with the Hamiltonian (6.1) we will use the formalism of the Matsubara temperature diagrammatic technique. The state of a noninteracting quasiparticle is described by its Green function

$$G(\mathbf{p}, \varepsilon_n) = \frac{1}{i\varepsilon_n - \xi(\mathbf{p})}, \tag{6.2}$$

where $\varepsilon_n = (2n+1)\pi T$ is a fermion Matsubara frequency.

The effective electron–electron attraction leads to a reconstruction of the ground state of the electron system which formally manifests itself in the appearance at the critical temperature of a pole in the two particle Green function

$$\mathcal{L}(p, p', q) = \langle T_\tau [\widetilde{\psi}_{p+q,\sigma} \widetilde{\psi}_{-p,-\sigma} \widetilde{\psi}^{+}_{p'+q,\sigma'} \widetilde{\psi}^{+}_{-p',-\sigma'}] \rangle, \tag{6.3}$$

where T_τ is the time ordering operator and $4D$ vector notations are used [228]. The two particle Green function can be expressed in terms of the vertex part

[33] We suppose that reader is familiar with the BCS formulation of the theory of superconductivity (see e.g. [228]).

[34] Fluctuations in the framework of more realistic Eliashberg [229] model of superconductivity were studied by Narozhny [230]. He demonstrated that the strong coupling does not change drastically the results of the weak coupling approximation.

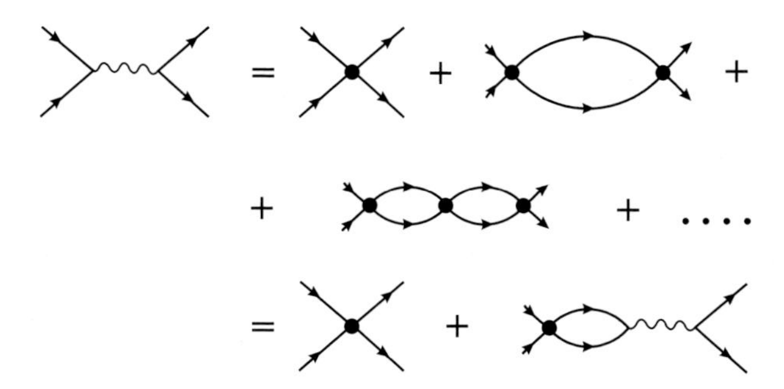

FIG. 6.1. The Dyson equation for the fluctuation propagator (wavy line) in the ladder approximation. Solid lines represent one-electron Green's functions, bold points correspond to the model electron–electron interaction.

[228]. In the case under consideration, it is the vertex part of the electron–electron interaction in the Cooper channel $L(\mathbf{q},\Omega_k)$ which will be called the fluctuation propagator below. The Dyson equation for $L(\mathbf{q},\Omega_k)$, accounting for the e–e attraction in the ladder approximation, is represented graphically in Fig. 6.1. It can be written down analytically as

$$L^{-1}(\mathbf{q},\Omega_k) = -g^{-1} + \Pi(\mathbf{q},\Omega_k), \tag{6.4}$$

where the polarization operator $\Pi(\mathbf{q},\Omega_k)$ is defined as a loop of two single-particle Green's functions in the particle–particle channel:

$$\Pi(\mathbf{q},\Omega_k) = T\sum_{\varepsilon_n}\int \frac{d^3\mathbf{p}}{(2\pi)^3} G(\mathbf{p}+\mathbf{q},\varepsilon_{n+k})G(-\mathbf{p},\varepsilon_{-n}). \tag{6.5}$$

Let us emphasize, that the two quantities introduced above, $\mathcal{L}\,(p,p',q)$ and $L(q)$, are closely connected with each other. The former being integrated over momenta p and p' becomes an average of the product of the two order parameters:

$$\int dpdp'\mathcal{L}\,(p,p',q) = \frac{1}{g^2}\left\langle \Delta_q\Delta_q^*\right\rangle, \tag{6.6}$$

where Δ_q is the superconducting gap proportional to the condensate wave function Ψ. The value (6.6) may be written down by means of the polarization operator introduced above as

$$\int dpdp'\mathcal{L}\,(p,p',q) = -\frac{\Pi}{1-g\Pi} = \frac{\Pi}{g}L. \tag{6.7}$$

Comparing this equation with Eq. (6.4) for the fluctuation propagator we see that the corresponding expressions are very similar. After analytic continuation to the real frequencies the fluctuation propagator $L(q,i\Omega)$ coincides (up to a constant) with the quantity defined by Eq. (6.6).

One can calculate the propagator (6.4) using the one-electron Green functions of the normal metal (6.2). For sake of convenience of future calculations let us define the correlator of two one-electron Green functions

$$\mathcal{P}(\mathbf{q}, \varepsilon_1, \varepsilon_2) = \int \frac{d^3\mathbf{p}}{(2\pi)^3} G(\mathbf{p}+\mathbf{q}, \varepsilon_1) G(-\mathbf{p}, \varepsilon_2)$$
$$= 2\pi\nu\theta(-\varepsilon_1\varepsilon_2) \left\langle \frac{1}{|\varepsilon_1 - \varepsilon_2| + i\Delta\xi(\mathbf{q}, \mathbf{p})|_{\epsilon(\mathbf{p})=E_F}} \right\rangle_{F.S.}, \tag{6.8}$$

where $\theta(x)$ is Heavyside step function, ν is the one-electron density of states, $<>_{\text{F.S.}} = \int \frac{d\Omega_{\mathbf{p}}}{4\pi}$ means the averaging over the Fermi surface,

$$\Delta\xi(\mathbf{q}, \mathbf{p})|_{\epsilon(\mathbf{p})=E_F} = [\xi(\mathbf{q}+\mathbf{p}) - \xi(-\mathbf{p})]|_{\epsilon(\mathbf{p})=E_F} \approx (\mathbf{v_p q})_{\xi(\mathbf{p})=0}. \tag{6.9}$$

The last approximation is valid not too far from the Fermi surface, i.e. when $(\mathbf{v_p q})_{\xi(\mathbf{p})=0} \ll E_F$.

It is impossible to carry out the angular averaging in (6.8) for an arbitrary anisotropic spectrum. Nevertheless, in the following calculations of fluctuation effects in the vicinity of critical temperature only small momenta $\mathbf{v_p q} \ll T$ will be involved in the integrations, so we can restrict our consideration here to this region, where one can expand the integrand in powers of $\mathbf{v_p q}$. Indeed, the presence of $\theta(-\varepsilon_1\varepsilon_2)$ leaves the difference of the two fermionic frequencies in (6.8) to be of the order of the temperature which allows this expansion. The first term in $\mathbf{v_p q}$ will evidently be averaged out, so with quadratic accuracy one can find:

$$\mathcal{P}(\mathbf{q}, \varepsilon_1, \varepsilon_2) = 2\pi\nu \frac{\theta(-\varepsilon_1\varepsilon_2)}{|\varepsilon_1 - \varepsilon_2|} \left(1 - \frac{\langle(\mathbf{v_p q})^2\rangle_{F.S.}}{|\varepsilon_1 - \varepsilon_2|^2}\right). \tag{6.10}$$

Now one can calculate the polarization operator $\Pi(\mathbf{q}, \Omega_k)$ which in the discussed case of a clean superconductor is just the correlator (6.8) summed over the electronic Matsubara frequencies ε_n:

$$\Pi(\mathbf{q}, \Omega_k) = T \sum_{\varepsilon_n} \mathcal{P}(\mathbf{q}, \varepsilon_{n+k}, \varepsilon_{-n}) = \nu \left[\sum_{n \geq 0} \frac{1}{n + 1/2 + \frac{|\Omega_k|}{4\pi T}} \right.$$
$$\left. - \frac{\langle(\mathbf{v_p q})^2\rangle_{F.S.}}{(4\pi T)^2} \sum_{n=0}^{\infty} \frac{1}{\left(n + 1/2 + \frac{|\Omega_k|}{4\pi T}\right)^3} \right]. \tag{6.11}$$

The calculation of the sums in (6.11) can be carried out in terms of logarithmic derivatives of the Γ function $\psi^{(n)}(x)$ already familiar to us (see Appendix B).

It worth mentioning that the first sum is well-known in the BCS theory, one can recognize in it the so-called "Cooper logarithm"; its logarithmic divergence at the upper limit ($\psi(x \gg 1) \approx \ln x$) is cut-off by the Debye energy ($N_{\max} = \omega_D/2\pi T$) and one gets:

$$\frac{1}{\nu}\Pi(\mathbf{q},\Omega_k) = \psi\left(\frac{1}{2}+\frac{|\Omega_k|}{4\pi T}+\frac{\omega_D}{2\pi T}\right) - \psi\left(\frac{1}{2}+\frac{|\Omega_k|}{4\pi T}\right) + \frac{\langle(\mathbf{v_p q})^2\rangle_{F.S.}}{2(4\pi T)^2}\psi''\left(\frac{1}{2}+\frac{|\Omega_k|}{4\pi T}\right). \tag{6.12}$$

The critical temperature in the BCS theory is determined as the temperature T_c at which the pole of $L(0,0,T_c)$ occurs

$$L^{-1}(\mathbf{q}=0,\Omega_k=0,T_c) = -g^{-1} + \Pi(0,0,T_c) = 0.$$

Taking into account that $\psi\left(\frac{1}{2}\right) = -C - 2\ln 2$, where $C = 0.577...$ is the Euler constant, one can express the critical temperature in the famous BCS form:

$$T_c = \frac{2\gamma_E}{\pi}\omega_D \exp\left(-\frac{1}{\nu g}\right). \tag{6.13}$$

Here $\gamma_E = e^{C_{\text{Euler}}} = 1.78...$

Introducing the reduced temperature $\epsilon = \ln(T/T_c)$, one can write the propagator as

$$L^{-1}(\mathbf{q},\Omega_k) = -\nu\left[\epsilon + \psi\left(\frac{1}{2}+\frac{|\Omega_k|}{4\pi T}\right) - \psi\left(\frac{1}{2}\right) - \frac{\langle(\mathbf{v_p q})^2\rangle_{F.S.}}{2(4\pi T)^2}\psi''\left(\frac{1}{2}+\frac{|\Omega_k|}{4\pi T}\right)\right]. \tag{6.14}$$

We found (6.14) for bosonic imaginary Matsubara frequencies $i\Omega_k = 2\pi iTk$. These frequencies are necessary for the calculation of fluctuation contributions to any thermodynamic characteristics of the system.

In the vicinity of the transition point one can restrict oneself in summations of the expressions with $L(\mathbf{q},\Omega_k)$ over Matsubara frequencies to the so-called static approximation, taking into account the term with $\Omega_k = 0$ only, which turns out to be the most singular term in $\epsilon \ll 1$. This approximation physically means that the product of Heisenberg field operators $\widetilde{\psi}_{p,\sigma}\psi_{-p,-\sigma}$ appears here like a classical field Ψ, which in the phenomenological approach corresponds to the Cooper pair wave function and in the vicinity of critical temperature is proportional to the fluctuation order parameter. Having in mind the GL region of temperatures we restricted ourselves above to the assumption of small momenta $\mathbf{v_p q} \ll T$. In these conditions the static propagator reduces to

$$L(\mathbf{q},0) = -\frac{1}{\nu}\frac{1}{\epsilon + \xi^2\mathbf{q}^2}. \tag{6.15}$$

With an accuracy of a numerical factor and the total sign this correlator coincides with the expression (3.5) for $\langle|\Psi_{\mathbf{q}}|^2\rangle$. By this expression we also have finally obtained the microscopic value of the coherence length ξ for a clean superconductor with an isotropic D-dimensional Fermi surface which was often mentioned previously (compare with (A.3))

$$\xi_{(D)}^2 = \frac{7\zeta(3)\mathbf{v}_\mathbf{F}^2}{16D\pi^2T^2}. \tag{6.16}$$

In order to describe the fluctuation contributions to transport phenomena one has to start from the analytic continuation of the propagator (6.14) from the discrete set of $\Omega_k \geq 0$ to the whole upper half-plane of imaginary frequencies. The analytic properties of $\psi^{(n)}(x)$ functions (which have poles at $x = 0, -1, -2, ...$) allow one to obtain the retarded propagator $L^R(\mathbf{q}, -i\Omega)$ by simple substitution $i\Omega_k \to \Omega$. For small $|\Omega| \ll T$ the $\psi-$functions can be expanded in $-i\Omega/4\pi T$ and the propagator acquires the simple pole form:

$$L^R(\mathbf{q},\Omega) = -\frac{1}{\nu}\frac{1}{-\frac{i\pi}{8T}\Omega + \epsilon + \xi^2\mathbf{q}^2} = \frac{8T}{\pi\nu}\frac{1}{i\Omega - \left(\tau_{\mathrm{GL}}^{-1} + \frac{8T}{\pi}\xi^2\mathbf{q}^2\right)}. \tag{6.17}$$

This expression provides us with the microscopic value of the GL relaxation time $\tau_{\mathrm{GL}} = \frac{\pi}{8(T-T_\mathrm{c})}$, widely used above in the phenomenological theory. Moreover, comparison of the microscopically derived Eq. (6.17) with the phenomenological expressions (3.3), (3.8) and (3.19) shows that $\alpha T_\mathrm{c} = \nu$ and $\gamma_{\mathrm{GL}} = \pi\alpha/8 = \pi\nu/8T_\mathrm{c}$ (see Appendix A).

In evaluating $L(\mathbf{q}, \Omega_k)$ we neglected the effect of fluctuations on the one-electron Green functions. This is correct when fluctuations are small, i.e. not too near the transition temperature. The exact criterion of this approximation will be discussed in the following.

6.3 Diagrammatic representation of fluctuation corrections

The propagator (3.8) can be considered as the effective inter-electron interaction, therefore using the rules of the diagrammatic technique [228] one can easily write down the series of the perturbation theory diagrams for the thermodynamical potential of the system. These diagrams are presented by the closed loops of increasing complexity; the corresponding first order correction is demonstrated in Fig. 6.2 (a). Drawing the diagrams with fluctuation propagator one has to remember the rule concerning the arrows directions: the diagrams represent the interaction in the particle–particle channel, hence every couple of solid lines entering or exiting the propagator must be directed contrarily. This rule restrains the existence of some types of diagrams. For example, the second-order diagram with the overlapping wavy lines cannot be realized.

The fluctuation corrections to the magnetization, entropy and other first derivatives of the thermodynamical potential can be obtained by direct graphical differentiation of the corresponding diagram (Fig. 6.2a). Indeed, the Green function derivatives in general case can be expressed by means of the Ward

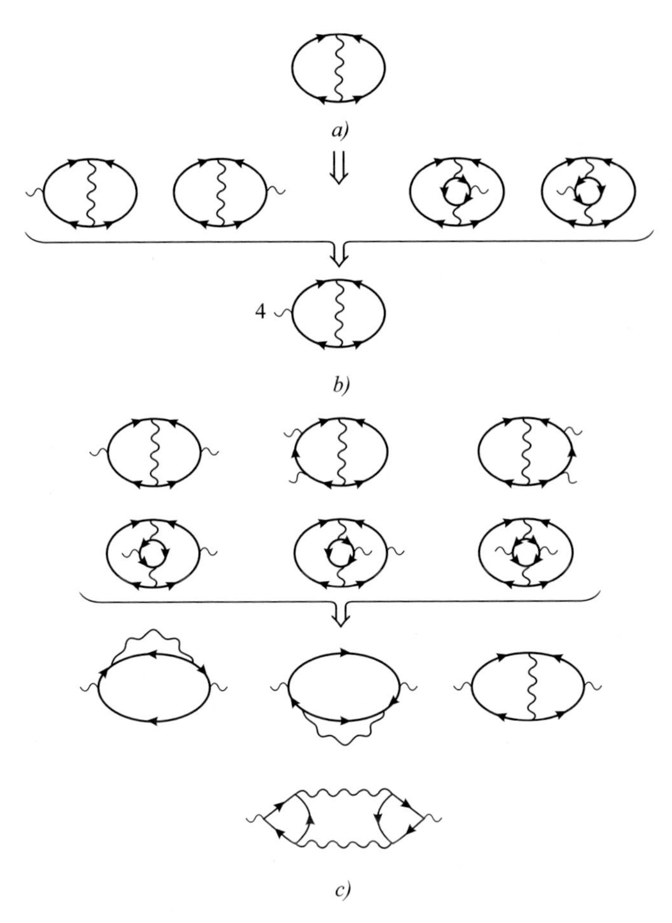

FIG. 6.2. The diagrammatic representation of the first-order fluctuation corrections to: (a) thermodynamical potential; (b) magnetization, entropy; (c) diamagnetic susceptibility, heat capacity. Solid and wavy lines correspond to bare one-electron Green functions and fluctuation propagators respectively. The external vertices correspond to the differentiation procedure.

identities [228] relating them with some convolutions of the Green function's square and corresponding vertex parts. In our case, dealing with the bare Green functions, the situation is trivial. For example,

$$\frac{\partial G^{(0)}(\mathbf{p},\varepsilon_n)}{\partial \mathbf{p}} = \frac{\partial}{\partial \mathbf{p}} \frac{1}{i\varepsilon_n - \xi(\mathbf{p})} = G^{(0)}(\mathbf{p},\varepsilon_n)\, \mathbf{v} G^{(0)}(\mathbf{p},\varepsilon_n). \tag{6.18}$$

Hence, graphically the differentiation means the interruption of the Green function by some vertex (in the case of the differentiation over momentum this vertex is the velocity operator). One has to remember, that the fluctuation propagator itself contains the loops of the bare Green functions, so looking for the complete set of the fluctuation diagrams in the certain order of perturbation theory it is necessary to extract explicitly these loops from fluctuation propagator and to differentiate them too. Such a procedure is demonstrated in Fig. 6.2(b). The first two diagrams are the result of the direct differentiation of the first-order diagram for the free energy. The third and fourth diagrams appear as the derivatives of the propagator, but as is seen, topologically they are equivalent to the first two.

In such a way one can draw the diagrams corresponding to the first-order fluctuation corrections for diamagnetic susceptibility and heat capacity. They are presented in Fig. 6.2(c). The first three follow directly from Fig. 6.2(b), as the next order derivatives of the first two diagrams. The differentiation of the hidden in fluctuation propagator electron loop results in the appearance of the nontrivial last diagram in Fig. 6.2(c). Interrupting of the wavy line in the diagram results in the increase of the number of propagators in it and hence increases its order in the perturbation series. One can see that the diagram equivalent to the last one in Fig. 6.2(c) can be obtained as the second derivative of the second diagram in Fig. 6.2(a).

Collecting carefully the equivalent diagrams appearing in the process of differentiation one can define the corresponding combinatorial factor.

6.4 Superconductor with impurities

6.4.1 *Accounting for electron scattering by impurities*

In order to study fluctuations in real systems like superconducting alloys one has to perform an impurity average in the graphical equation for the fluctuation propagator (see Fig. 6.1). This procedure can be done in the framework of the Abrikosov–Gor'kov approach [228], which we shortly recall below.

The one-particle Green function can be presented in the form of the expansion over products of the exact eigenfunctions of the Hamiltonian $\widehat{H}$:

$$G(\mathbf{r},\mathbf{r}',\varepsilon_n) = \sum_i \frac{\psi_i(\mathbf{r})\,\psi_i^*(\mathbf{r}')}{i\varepsilon_n - E_i}, \tag{6.19}$$

where $\psi_i(\mathbf{r})$ are wave-functions corresponding to the exact electron states of this Hamiltonian. In order to accounting for the effect of impurities with potential $U(\mathbf{r})$ let us start from the corresponding equation for the electron Green function:

$$\left(E - U(\mathbf{r}) - \widehat{H}\right) G_E(\mathbf{r},\mathbf{r}') = \delta(\mathbf{r}-\mathbf{r}'). \tag{6.20}$$

If we solve this equation using the perturbation theory for the impurity potential and average the solution, then the average product of two Green functions, can be presented as series, each term of which is associated with a graph drawn according the rules of a diagrammatic technique (see Fig. 6.3). In this technique solid lines correspond to bare Green functions and dashed lines to random potential correlators. We assume that the impurity system random potential $U(r)$

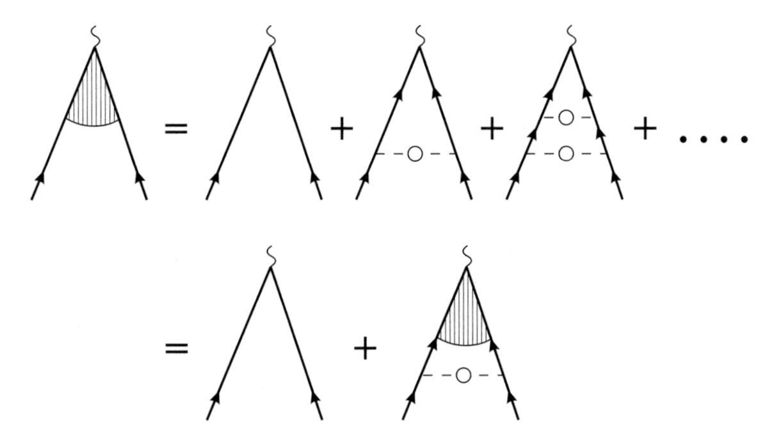

FIG. 6.3. The equation for the vertex part $\lambda(q, \omega_1, \omega_2)$ in the ladder approximation. Solid lines correspond to bare one-electron Green functions and dashed lines to the impurity random potential correlators.

is distributed according to the Gauss law. In this case the high rank correlators can be represented as the products of pair correlators. The latter for impurities with short-range potential can be assumed as δ-functions:

$$\langle U(r)\rangle = 0, \ \langle U(r)U(r')\rangle = \left\langle U^2\right\rangle \delta(r - r'), \tag{6.21}$$

where the angle brackets denote averaging over the impurity configuration. Equation (6.21) corresponds to the Born approximation for the electron interaction with short range impurities, and $\left\langle U^2\right\rangle = n_i \left(\int V(\mathbf{r})d\mathbf{r}\right)^2$ where n_i is the impurity concentration and $V(\mathbf{r})$ is the potential of the single impurity.

In conductors (far enough from the metal–insulator transition) the mean free path is much greater than the electron wavelength $l \gg \lambda = 2\pi/p_F$ (which in practice means the mean free path up to tens of interatomic distances). In the case of the electron spectra with dimensionality $D > 1$ the angular integration in momentum space reduces considerably the contribution of the diagrams with intersecting impurity lines that allows us to omit them up to the leading approximation in $(p_F l)^{-1}$ [228]. In this approximation the one-electron Green function keeps the same form as the bare one (6.2) with the only substitution

$$\varepsilon_n \Rightarrow \widetilde{\varepsilon}_n = \varepsilon_n + \frac{1}{2\tau}\mathrm{sign}(\varepsilon_n), \tag{6.22}$$

where $1/\tau = 2\pi\nu \left\langle U^2\right\rangle$ is the frequency of elastic collisions.

Another effect of the coherent scattering on the same impurities by both electrons forming a Cooper pair is the renormalization of the vertex part $\lambda(\mathbf{q}, \varepsilon_1, \varepsilon_2)$ in the particle–particle channel. Let us demonstrate the details of its calculation. The renormalized vertex $\lambda(\mathbf{q}, \varepsilon_1, \varepsilon_2)$ is determined by a graphical equation of the ladder type (see Fig. 6.3). Here after the averaging over the impurity configurations the value $\left\langle U^2\right\rangle = 1/2\pi\nu\tau$ is associated with the dashed line.

The corresponding integral equation is

$$\lambda(\mathbf{q}, \varepsilon_1, \varepsilon_2) = 1 + \frac{1}{2\pi\nu\tau} \int \frac{d^3\mathbf{p}}{(2\pi)^3} \lambda(\mathbf{q}, \varepsilon_1, \varepsilon_2) G(\mathbf{p}+\mathbf{q}, \widetilde{\varepsilon}_1) G(-\mathbf{p}, \widetilde{\varepsilon}_2), \tag{6.23}$$

which, in the momentum representation, is reduced to the algebraic one

$$\lambda^{-1}(\mathbf{q}, \varepsilon_1, \varepsilon_2) = 1 - \frac{1}{2\pi\nu\tau} \mathcal{P}(\mathbf{q}, \widetilde{\varepsilon}_1, \widetilde{\varepsilon}_2), \tag{6.24}$$

The correlator $\mathcal{P}(\mathbf{q}, \widetilde{\varepsilon}_1, \widetilde{\varepsilon}_2)$ was defined by Eq. (6.8) but now it depends on the arguments $\widetilde{\varepsilon}_1, \widetilde{\varepsilon}_2$.

Now one has to perform a formal averaging of the general expression (6.8) over the Fermi surface ($\langle\ ...\rangle_{\mathrm{F.S.}}$). Restricting consideration to small momenta

$$\Delta\xi(\mathbf{q}, \mathbf{p})|_{|\mathbf{p}|=p_F} \ll |\widetilde{\varepsilon}_1 - \widetilde{\varepsilon}_2|, \tag{6.25}$$

the calculation of $\lambda(\mathbf{q}, \omega_1, \omega_2)$ for the practically important case of an arbitrary spectrum can be done analogously to (6.10). Indeed, expanding the denominator of (6.8) one can find

$$\lambda(\mathbf{q}, \varepsilon_1, \varepsilon_2) = \frac{|\widetilde{\varepsilon}_1 - \widetilde{\varepsilon}_2|}{|\varepsilon_1 - \varepsilon_2| + \frac{\langle(\Delta\xi(\mathbf{q},\mathbf{p})|_{|\mathbf{p}|=p_F})^2\rangle_{\mathrm{F.S.}}}{\tau|\widetilde{\varepsilon}_1 - \widetilde{\varepsilon}_2|^2} \theta(-\varepsilon_1\varepsilon_2)}. \tag{6.26}$$

The assumed restriction on momenta is not too severe and is almost always satisfied in calculations of fluctuation effects at temperatures near T_c. In this region of temperatures the effective propagator momenta are determined by $|\mathbf{q}|_{\mathrm{eff}} \sim [\xi^{\mathrm{GL}}(T)]^{-1} = \xi^{-1}\sqrt{\epsilon} \ll \xi^{-1}$, while the Green function $\mathbf{q}$-dependence becomes important for much larger momenta $q \sim \min\{l_T^{-1}, l^{-1}\}$, [35] which is equivalent to the limit of the condition (6.25).

The average in (6.26) can be calculated for some particular types of spectra. For example, in the cases of $2D$ and $3D$ isotropic spectra it is expressed in terms of the diffusion coefficient $\mathcal{D}_{(D)}$:

$$\langle(\Delta\xi(\mathbf{q}, \mathbf{p})|_{|\vec{p}|=p_F})^2\rangle_{\mathrm{F.S.}(D)} = \tau^{-1}\mathcal{D}_{(D)}q^2 = \frac{v_F^2 q^2}{D}. \tag{6.27}$$

6.4.2 *Propagator for superconductor with impurities*

In section 6.1, in the process of the microscopic derivation of the TDGL equation, the fluctuation propagator was introduced. This object is of first importance for the microscopic fluctuation theory and it has to be generalized for the case of an impure metal with an anisotropic electron spectrum. This is easy to do using the averaging procedure presented in the previous section. Formally, it is enough

[35] $l_T = \sqrt{\mathcal{D}_{(D)}/T}$ is so-called "temperature length". In clean superconductor it coincides with the accuracy of the numerical factor with ξ.

to use in Eq. (6.4) the polarization operator $\Pi(\mathbf{q}, \Omega_k)$ averaged over impurity positions, which can be expressed in terms of $\mathcal{P}(\mathbf{q}, \widetilde{\varepsilon}_{n+k}, \widetilde{\varepsilon}_{-n})$ introduced above:

$$\Pi(\mathbf{q}, \Omega_k) = T \sum_{\varepsilon_n} \lambda(q, \varepsilon_{n+k}, \varepsilon_{-n}) \mathcal{P}(\mathbf{q}, \widetilde{\varepsilon}_{n+k}, \widetilde{\varepsilon}_{-n})$$
$$= T \sum_{\varepsilon_n} \frac{1}{\mathcal{P}^{-1}(\mathbf{q}, \widetilde{\varepsilon}_{n+k}, \widetilde{\varepsilon}_{-n}) - 1/(2\pi\nu\tau)}. \tag{6.28}$$

For relatively small $\mathbf{q}$ $(\Delta\xi(\mathbf{q}, \mathbf{p})|_{|E(\mathbf{p})|=E_F} \ll |\widetilde{\varepsilon}_{n+k} - \widetilde{\varepsilon}_{-n}| \sim \max\{T, \tau^{-1}\})$ and $\Omega \ll T$ one can find an expression for the fluctuation propagator, which can be useful in studies of fluctuation effects near T_c $(\epsilon \ll 1)$ for the dirty and intermediate but not very clean case $(T\tau \ll 1/\sqrt{\epsilon})$. Substituting in Eq. (6.28) Eq. (6.10) for $\mathcal{P}(\mathbf{q}, \widetilde{\varepsilon}_{n+k}, \widetilde{\varepsilon}_{-n})$ and expanding in powers of $\langle(\mathbf{v_p q})^2\rangle_{\text{F.S.}}/|\widetilde{\varepsilon}_{n+k} - \widetilde{\varepsilon}_{-n}|^2$ one can find

$$\Pi(\mathbf{q}, \Omega_k) = 2\pi\nu T \sum_{\varepsilon_n} \frac{\theta(\varepsilon_{n+k}\varepsilon_n)\left(1 - \frac{\langle(\mathbf{v_p q})^2\rangle_{F.S.}}{|\widetilde{\varepsilon}_{n+k} - \widetilde{\varepsilon}_{-n}|^2}\right)}{|\widetilde{\varepsilon}_{n+k} - \widetilde{\varepsilon}_{-n}| \left[1 - \frac{1}{\tau}\frac{1}{|\widetilde{\varepsilon}_{n+k} - \widetilde{\varepsilon}_{-n}|} + \frac{\langle(\mathbf{v_p q})^2\rangle_{F.S.}}{\tau|\widetilde{\varepsilon}_{n+k} - \widetilde{\varepsilon}_{-n}|^3}\right]}$$
$$= 4\pi\nu T \left[\sum_{n=0} \frac{1}{2\varepsilon_n + \Omega_k} - \sum_{n=0} \frac{\langle(\mathbf{v_p q})^2\rangle_{F.S.}}{\left(2\varepsilon_n + \Omega_k + \frac{1}{\tau}\right)(2\varepsilon_n + \Omega_k)^2}\right]. \tag{6.29}$$

Being in the vicinity of the transition temperature we can omit the Ω_k -dependence of the term proportional to small $\langle(\mathbf{v_p q})^2\rangle_{\text{F.S.}}$. Performing the remaining summation over Matsubara frequencies we find the result very similar Eq. (6.12) for the polarization operator in clean case:

$$\frac{1}{\nu}\Pi(\mathbf{q}, \Omega_k) = \ln \frac{\omega_D}{2\pi T} - \psi\left(\frac{1}{2} + \frac{|\Omega_k|}{4\pi T}\right) + \tau^2 \langle(\mathbf{v_p q})^2\rangle_{\text{F.S.}}$$
$$\times \left[\psi\left(\frac{1}{2} + \frac{1}{4\pi T\tau}\right) - \psi\left(\frac{1}{2}\right) - \frac{1}{4\pi T\tau}\psi'\left(\frac{1}{2}\right)\right]. \tag{6.30}$$

Now one can write the fluctuation propagator $L^R(q, \Omega)$ in the vicinity of the transition and for the small arguments essential for our needs, $\Omega \ll T$, $\xi(T\tau)q \ll 1$ in the form completely coinciding with Eq. (6.17):

$$L^R(\mathbf{q}, \Omega) = -\frac{1}{\nu} \frac{1}{\epsilon - i\frac{\pi\Omega}{8T} + \xi_{(D)}^2(T\tau)\mathbf{q}^2} \tag{6.31}$$
$$= -\frac{1}{\nu} \frac{1}{\epsilon + \frac{\pi}{8T}\left(-i\Omega + \widehat{\mathcal{D}}\mathbf{q}^2\right)}. \tag{6.32}$$

The only difference with respect to the clean case is in the appearance of a natural dependence of the effective coherence length on the elastic relaxation time. In the isotropic D-dimensional case it can be written as

$$\xi_{(D)}^2(T\tau) = \frac{\pi}{8T}\widehat{\mathcal{D}} = (4m\alpha T)^{-1} = \eta_{(D)}$$

$$= -\frac{\tau^2 v_F^2}{D}\left[\psi\left(\frac{1}{2}+\frac{1}{4\pi T\tau}\right)-\psi\left(\frac{1}{2}\right)-\frac{1}{4\pi T\tau}\psi'\left(\frac{1}{2}\right)\right] \quad (6.33)$$

(we introduced here the parameter $\eta_{(D)}$ frequently used in the microscopic theory).[36] Let us stress that the phenomenological coefficient γ_{GL} turns out to be equal to the same value $\pi\nu/8T$ as in clean case, and hence does not depend on the impurity concentration.

We have introduced here the generalization of the notion of diffusion coefficient $\widehat{\mathcal{D}}$:

$$\widehat{\mathcal{D}} = \begin{cases} \tau v_F^2/D, T\tau \ll 1, \\ v_F^2/2DT, T\tau \gg 1, \end{cases} \quad (6.34)$$

which in the clean case is formed by means of the substitution $\tau \to 1/2T$. In the case of a complex spectrum $\widehat{\mathcal{D}}$ can become operator (see the following section).

One has to remember that Eq. (6.32) was derived under the assumption of small momenta $\Delta\xi(\mathbf{q},\mathbf{p})|_{|E(\mathbf{p})|=E_F} \ll |\widetilde{\varepsilon}_{n+k} - \widetilde{\varepsilon}_{-n}| \sim \max\{T, \tau^{-1}\}$, so the range of its applicability is restricted to the region of long wavelength fluctuations ($\epsilon = \ln(T/T_c) \ll 1$), where the integrands of diagrammatic expressions have singularities at small momenta of the Cooper pair center of mass.

Another useful formula for the polarization operator can be carried over from Eq. (6.29) for the dirty case ($T\tau \ll 1$). In this approximation the summation can be performed without the expansion over small q and general expression for $\Pi(\mathbf{q},\Omega_k)$ valid for $\max\left\{\Omega_k, \mathcal{D}q^2\right\} \ll \omega_D$ reads as

$$\Pi(\mathbf{q},\Omega_k) = 4\pi\nu T \sum_{\varepsilon_n>0} \frac{1}{2\varepsilon_n + |\Omega_k| + \tau\langle(\mathbf{v_p q})^2\rangle_{F.S.}}$$
$$= \nu\left[\ln\frac{\omega_D}{2\pi T} - \psi\left(\frac{1}{2}+\frac{|\Omega_k| + \mathcal{D}q^2}{4\pi T}\right)\right]. \quad (6.35)$$

This formula will be useful for study of fluctuation phenomena far from transition temperature and other cases.

Finally let us express the Gi parameter for the important $2D$ case in terms of the microscopic parameter $\eta_{(2)}$. In accordance with (2.87) and the definition (6.33):

$$Gi_{(2)}(T\tau) = \frac{7\zeta(3)}{16\pi^2}\frac{1}{mT_c\eta_{(2)}(T\tau)}. \quad (6.36)$$

One can see that this general definition in the limiting cases of clean and dirty metals results in the same values $Gi_{(2c)}$ and $Gi_{(2d)}$ as were reported in Table 2.1.

[36] Let us recall that its square determines the product of the GL parameter α and the Cooper pair mass entering in the GL functional. In the clean case we supposed the latter equal to two free electron masses and defined α in accordance with (A.12). As we just have seen in the case of the impure superconductor ξ depends on impurity concentration and this dependence, in principle, can be attributed both to α or m. For our further purposes it is convenient to leave α in the same form (A.12) as in the case of a clean superconductor. The Cooper pair mass in this case becomes dependent on the electron mean free pass what physically can be attributed to the diffusion motion of the electrons forming the pair.

6.5 Layered superconductor

The main object for applications of the microscopic theory of superconducting fluctuations in this book will be a layered superconductor. We have already studied above the manifestation of fluctuations in some of its properties in the framework of the phenomenological Lawrence–Doniach model and were convinced in its richness. The quasi-$2D$ fluctuation Cooper pair spectrum allows us to obtain in a unique way both $3D$ and $2D$ results, to study crossovers between them, to elucidate the role of anisotropy, to get directly results for thin superconducting films, to study the out-of-plane transport, to apply corresponding results to the tunneling and Josephson systems etc. Finally, the class of HTS most investigated during the last two decades consists mainly of layered systems and it will be potential target of our consideration.

Intending to operate in the framework of the microscopic theory we begin our discussion from the properties of the quasiparticle normal state energy spectrum and the choice of the appropriate model. While models of layered superconductors with several conducting layers per unit cell and with either intralayer or interlayer pairing have been considered [231], it has been shown [232] that all of these models give rise to a Josephson pair potential that is periodic in k_z, the wave-vector component parallel to the c-axis, with period $2\pi/s$ where s is the c-axis repeat distance. While such models differ in their superconducting densities of states, they all give rise to qualitatively similar fluctuation propagators, which differ only in the precise definitions of the parameters and in the precise form of the Josephson coupling potential. Ignoring the rather unimportant differences between such models in the Gaussian fluctuation regime above T_c, we therefore consider the simplest model of a layered superconductor, in which there is one layer per unit cell, with intralayer pairing.

Some remarks regarding the normal–state quasiparticle momentum relaxation time are necessary. In the "old" layered superconductors the materials were generally assumed to be in the dirty limit (like TaS_2(pyridine)$_{1/2}$). In the HTS cuprates, however, both single crystals and epitaxial thin films are nominally in the "intermediate" regime, with $l/\xi_{xy} \approx 2 - 5$. In addition, the situation in the cuprates is complicated by the presence of phonons for $T \simeq T_c \simeq 100\,K$, the nearly localized magnetic moments on the Cu^{2+} sites, and by other unspecified inelastic processes. We will suppose the impurities to be located in conducting layers and the electron scattering to be elastic.

Below we will mainly restrict our consideration by the local limit of the fluctuation pairs motion. This means that we consider the case of not too clean superconductors, keeping the impurity concentration n_i and reduced temperature such that the resulting electron mean-free path satisfies the requirement $l < \xi_{xy}(T) = \xi_{xy}/\sqrt{\varepsilon}$ and the impurity vertex can be taken in the local form (6.26) with $\langle(\Delta\xi(\mathbf{q},\mathbf{p})|)^2\rangle_{\text{F.S.}}$ determined by (6.38). What concerns the phase-breaking time τ_φ it will be supposed much larger than τ.

The important example is already familiar case of quasi-$2D$ electron motion in a layered metal:

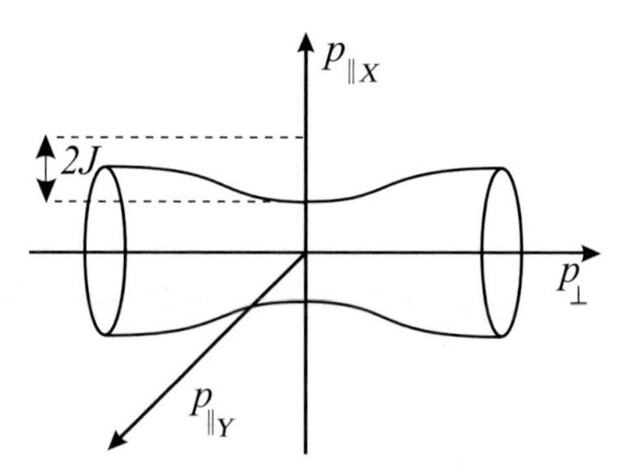

FIG. 6.4. The Fermi surface in the form of a corrugated cylinder.

$$\xi(\mathbf{p}) = E(\mathbf{p}_{\|}) + J\cos(p_z s) - E_F, \tag{6.37}$$

where $E(\mathbf{p}_{\|}) = \mathbf{p}_{\|}^2/(2m)$, $\mathbf{p} \equiv (\mathbf{p}_{\|}, p_z)$, $\mathbf{p}_{\|} \equiv (p_x, p_y)$, J is the effective nearest-neighbor interlayer hopping energy for quasiparticles. We note that J characterizes the width of the band in the c-axis direction taken in the strong-coupling approximation and can be identified with the effective energy of electron tunneling between planes (see Eqs. (2.45), (2.45) and footnote therein). The Fermi surface, defined by the condition $\xi(\mathbf{p}) = 0$, is a corrugated cylinder (see Fig. 6.4). In this case the average (6.27) is written in a more sophisticated form:

$$\langle (\Delta\xi(\mathbf{q},\mathbf{p})|_{|\vec{p}|=p_F})^2 \rangle_{F.S.} = \frac{1}{2}\left(v_F^2 \mathbf{q}^2 + 4J^2 \sin^2(q_z s/2)\right) = \tau^{-1}\widehat{\mathcal{D}}q^2, \tag{6.38}$$

where we have introduced the definition of the generalized diffusion coefficient operator $\widehat{\mathcal{D}}$ in order to deal with an arbitrary anisotropic spectrum. The generalization of (6.32) for the case of a layered superconductor with a quasi $2D$ electronic spectrum is evident:

$$L^R(q,\Omega) = -\frac{1}{\nu}\frac{1}{\epsilon - i\frac{\pi\Omega}{8T} + \eta_{(2)}\mathbf{q}_{\|}^2 + r\sin^2(q_z s/2)}. \tag{6.39}$$

Let us stress, that the anisotropy parameter $r = \frac{J^2}{T} = \frac{4\xi_z^2(0)}{s^2}$ more than once used in the phenomenological description of fluctuations (see Eq. (2.45)) appears here in the natural way. The coefficient $\eta_{(2)}$ in its turn is nothing else as the in-plane coherence length ξ_{xy} square (see Eq. (6.33)).

7

MICROSCOPIC DERIVATION OF THE GL FUNCTIONAL

7.1 GL functional of a conventional superconductor

In this chapter we will demonstrate how the GL functional itself can be carried out from the microscopic theory of superconductivity. For this aim we will use the method of functional integration alternative to the diagrammatic technique approach.

Let us start again from the BCS Hamiltonian (6.1) and write it in the form:

$$\mathcal{H} = \mathcal{H}_0 + \mathcal{H}_{\text{int}}, \tag{7.1}$$

where

$$\mathcal{H}_0 = \widetilde{\psi}^+(\mathbf{r}) \left(-\frac{\nabla^2}{2m} + U(\mathbf{r}) - \mu \right) \widetilde{\psi}(\mathbf{r}), \tag{7.2}$$

$$\mathcal{H}_{\text{int}} = -g \int \widetilde{P}^+(\mathbf{r}) \widetilde{P}(\mathbf{r})\, d\mathbf{r}, \tag{7.3}$$

and

$$\widetilde{P}(\mathbf{r}) = \widetilde{\psi}_\downarrow(\mathbf{r}) \widetilde{\psi}_\uparrow(\mathbf{r}) \tag{7.4}$$

is the operator of the Cooper pair annihilation in the Heisenberg representation. The potential $U(\mathbf{r})$ describes, as above, the interaction of electrons with impurities.

Our task now consists of the calculation of the partition function

$$Z = \operatorname{tr} \exp\left(-\int_0^\beta \mathcal{H} d\tau \right) = \operatorname{tr} \exp\left(-\frac{\mathcal{H}}{T} \right), \tag{7.5}$$

but in contrast to Eq. (2.1) of chapter 1, with the microscopic Hamiltonian (7.1) instead of the phenomenological functional (2.4).

In the ideal gas $g = 0$, and $\mathcal{H}_0$ is just the quadratic form of the Heisenberg electron field operators $\widetilde{\psi}(\mathbf{r})$ and $\widetilde{\psi}^+(\mathbf{r})$. One can diagonalize it and carry out the trace operation in (2.1) to calculate the partition function. In order to calculate the partition function of the interacting electron gas with Hamiltonian (7.1) one has in advance to present the operator $\mathcal{H}_{\text{int}}$ as the quadratic form over the field operators.

Let us separate the exponent with $\mathcal{H}_{\text{int}}$ in the expression for the partition function in the form of multiplier. The problem is nontrivial because the operators $\mathcal{H}_0$ and $\mathcal{H}_{\text{int}}$ do not commute. Nevertheless, introducing the operator of the imaginary time ordering T_τ and the interaction representation for the Hamiltonian $\mathcal{H}$ [228] one can present the partition function in the form:

$$Z = \mathrm{tr}\left\langle \exp\left(-\int_0^\beta \mathcal{H}_0 d\tau\right) T_\tau \exp\left(-\int_0^\beta \mathcal{H}_{\mathrm{int}}(\tau)\, d\tau\right)\right\rangle. \tag{7.6}$$

Now let us present the $\mathcal{H}_{\mathrm{int}}$ in the form of the quadratic form over the field operators by means of the Hubbard–Stratonovich transformation. For this purpose let us write

$$\exp\left(-\int_0^\beta \mathcal{H}_{\mathrm{int}}(\tau)\, d\tau\right) = \prod_x \exp\left(g\widetilde{P}^+(x)\widetilde{P}(x)d^4x\right), \tag{7.7}$$

where $x = (\mathbf{r}, \tau)$. Each multiplier in the product (7.7) can be presented in the form of the integral[37]

$$e^{g\widetilde{P}^+(x)\widetilde{P}(x)d^4x}, = \int d^2\Delta(x) \exp\left[\left(-\frac{|\Delta(x)|^2}{g}\right.\right.$$
$$\left.\left. -\Delta^*(x)\widetilde{P}(x) - \Delta(x)\widetilde{P^+}(x)\right) d^4x\right]. \tag{7.8}$$

Here $d^2\Delta(x) = d(\mathrm{Im}\,\Delta(x))\, d(\mathrm{Re}\,\Delta(x))$. The product of all these multipliers is the functional integral over $\mathfrak{D}\Delta(\mathbf{r},\tau)\mathfrak{D}\Delta^*(\mathbf{r},\tau)$. Finally [233],

$$Z = tr\left\langle T_\tau \int \mathfrak{D}\Delta(\mathbf{r},\tau)\mathfrak{D}\Delta^*(\mathbf{r},\tau) \exp\left(\int_0^\beta -\frac{|\Delta(\mathbf{r},\tau)|^2}{g}\right.\right.$$
$$\left.\left. -\Delta^*(\mathbf{r},\tau)\widetilde{\psi}_\downarrow(\mathbf{r})\,\widetilde{\psi}_\uparrow(\mathbf{r}) - \Delta(\mathbf{r},\tau)\widetilde{\psi}_\downarrow^+(\mathbf{r})\,\widetilde{\psi}_\uparrow^+(\mathbf{r}) - \mathcal{H}_0 d\tau\right)\right\rangle. \tag{7.9}$$

Now, contained in the exponent form is quadratic over the operators $\widetilde{\psi}$ and the problem is reduced to the description of one electron motion in the fluctuation field $\Delta(\mathbf{r},\tau)$ being an arbitrary function of the coordinate $\mathbf{r}$ and imaginary time τ.

Let us introduce the normal $G_\Delta(\mathbf{r},\tau,\mathbf{r}',\tau') = -\left\langle T_\tau\widetilde{\psi}_\downarrow(\mathbf{r},\tau)\,\widetilde{\psi}_\downarrow^+(\mathbf{r}',\tau')\right\rangle$ and anomalous $F_\Delta(\mathbf{r},\tau,\mathbf{r}',\tau') =< T_\tau\widetilde{\psi}_\downarrow(\mathbf{r},\tau)\,\widetilde{\psi}_\uparrow(\mathbf{r}',\tau') >$ electron Green functions of superconducting state which can be written in the matrix Gor'kov–Nambu representation

$$\widehat{G}_\Delta = \begin{pmatrix} G_\Delta(\mathbf{r},\mathbf{r}',\tau,\tau') & F_\Delta(\mathbf{r},\mathbf{r}',\tau,\tau') \\ F_\Delta^+(\mathbf{r},\mathbf{r}',\tau,\tau') & -G_\Delta^+(\mathbf{r}',\mathbf{r},\tau',\tau) \end{pmatrix}. \tag{7.10}$$

They satisfy the Gor'kov equation written in the form:

[37] We omit here the unimportant coefficient.

$$\left[\tau_z \frac{\partial}{\partial \tau} + \tau_0 \widehat{\xi} + i\tau_y \operatorname{Re}\Delta(\mathbf{r},\tau) + i\tau_x \operatorname{Im}\Delta(\mathbf{r},\tau)\right] \widehat{G}_\Delta = \delta(\mathbf{r}-\mathbf{r}')\,\delta(\tau-\tau'). \tag{7.11}$$

Here $\tau_i (i = x, y, z)$ are the Pauli matrices, τ_0 is the unit matrix, $\widehat{\xi} = -\frac{\nabla^2}{2m} + U(\mathbf{r}) - \mu$, the magnetic field is supposed to be equal zero.

Calculating the trace over electron field operators $\widetilde{\psi}_\downarrow, \widetilde{\psi}_\uparrow^+$ one finds the corresponding contribution to the partition function

$$Z_\Delta = e^{\mathcal{S}_0} \int \mathfrak{D}\Delta(\mathbf{r},\tau)\mathfrak{D}\Delta^*(\mathbf{r},\tau) \exp\{-\mathcal{S}[\Delta(\mathbf{r},\tau)]\} \tag{7.12}$$

with

$$\mathcal{S}[\Delta(\mathbf{r},\tau)] = -\int d\mathbf{r} \int_0^\beta d\tau \left[tr\left\{\ln \widehat{G}_\Delta^{-1}\right\} - \frac{|\Delta(\mathbf{r},\tau)|^2}{g}\right] = \mathcal{S}_0 + \mathcal{S}_\Delta. \tag{7.13}$$

The trace here is supposed to be carried out over the Nambu–Gor'kov subscripts. The independent on the order parameter part of action

$$\mathcal{S}_0 = -\int d\mathbf{r} \int_0^\beta d\tau \left[\operatorname{tr}\left\{\ln \widehat{G}_{\Delta=0}^{-1}\right\}\right] \tag{7.14}$$

determines the ideal gas partition function and is introduced in order to provides the natural normalization $Z_\Delta(\Delta = 0) = 1$, what corresponds $\mathcal{S}_\Delta(\Delta = 0) = 0$.

In the case when $\Delta(\mathbf{r},\tau) = \text{const}$ one can calculate the matrix $\widehat{G}^{-1}$ by means of the Fourier transform of Eq. (7.11):

$$\widehat{G}_\Delta(\mathbf{p},\varepsilon_n) = -\frac{i\tau_z\varepsilon_n + \tau_0\xi(\mathbf{p}) - i\tau_y \operatorname{Re}\Delta - i\tau_x \operatorname{Im}\Delta}{\Delta^2 + \varepsilon_n^2 + \xi^2(\mathbf{p})}. \tag{7.15}$$

Here ε_n are the fermionic Matsubara frequencies and $\xi(\mathbf{p})$ are the eigenvalues of the $\widehat{\xi}$–operator. In the absence of impurities $\xi(\mathbf{p}) = p^2/2m - \mu$. Recalling that the free energy is related to the action as $F = T\mathcal{S}$, and substituting these values in Eq. (7.12), one can find

$$-\frac{T\mathcal{S}_\Delta}{V} = T\sum_{\varepsilon_n} \nu \int d\xi \ln\left(1 + \frac{\Delta^2}{\varepsilon_n^2 + \xi^2}\right) - \frac{|\Delta|^2}{g}. \tag{7.16}$$

The mean field approximation as usual corresponds to the calculation of the functional integral in Eq. (7.12) by the steepest descend method. One can see that the saddle point condition ($\partial\mathcal{S}/\partial\Delta = 0$) coincides with the BCS self-consistency equation

$$1 = gT\nu \sum_{\varepsilon_n} \int d\xi \frac{1}{\varepsilon_n^2 + \xi^2 + \Delta^2}. \tag{7.17}$$

Let us notice that Eq. (7.16) describes not only the mean field approximation but also the order parameter fluctuations independent on the imaginary time and space variables. Expanding its right hand side in the vicinity of the critical temperature in the series over the order parameter powers one can reproduce the famous expression for the free energy in the Landau form

$$\frac{F}{V} = A\Delta^2 + \frac{B}{2}\Delta^4 \tag{7.18}$$

with the coefficients defined by means of the microscopic characteristics of superconductor:

$$A = \nu\left(\frac{1}{g} - \ln\frac{\omega_D}{2\pi T}\right) = \nu \ln\frac{T}{T_c}, \tag{7.19}$$

$$B = \frac{\nu T}{2}\sum_{\varepsilon_n}\int\frac{d\xi}{\left(\varepsilon_n^2 + \xi^2\right)^2} = \frac{7\zeta(3)}{8\pi^2 T^2}\nu. \tag{7.20}$$

It worth mentioning that these coefficients, and therefore the critical temperature T_c, do not depend on the concentration of nonmagnetic impurities which do not break the time reversal symmetry. This statement remains correct until the interaction (7.3)–(7.4) has the point character, and corresponding order parameter is coordinate independent. The situation turns out to be certainly different for the order parameter with p- or d-wave symmetry. In these cases even nonmagnetic impurities suppress the superconducting state [234]. Below we will derive the GL functional for these cases.

When the order parameter depends on $\mathbf{r}$ and τ the fluctuation part of action $\mathcal{S}_\Delta$ can be presented in the form of the series over $\Delta(\mathbf{r},\tau)$. It takes the form:

$$\begin{aligned} -\mathcal{S}_\Delta = & \int \Delta^*(\mathbf{r}_1,\tau_1)\,\widehat{L}^{-1}(\mathbf{r}_1,\tau_1,\mathbf{r}_2,\tau_2)\,\Delta(\mathbf{r}_2,\tau_2)\,d\mathbf{r}_1 d\tau_1 d\mathbf{r}_2 d\tau_2 \\ & -\frac{1}{2}\int \Delta^*(\mathbf{r}_1,\tau_1)\,\Delta^*(\mathbf{r}_2,\tau_2)\,\widehat{B}(\mathbf{r}_1,\tau_1,\mathbf{r}_2,\tau_2,\mathbf{r}_3,\tau_3,\mathbf{r}_4,\tau_4) \\ & \times\Delta(\mathbf{r}_3,\tau_3)\,\Delta(\mathbf{r}_4,\tau_4)\,d\mathbf{r}_1 d\tau_1 d\mathbf{r}_2 d\tau_2 d\mathbf{r}_3 d\tau_3 d\mathbf{r}_4 d\tau_4 \end{aligned} \tag{7.21}$$

In Eq. (12.19), operator $\widehat{L}$ is the fluctuation propagator: $\hat{L}_\omega = \left[-g^{-1} + \hat{\Pi}_\omega\right]^{-1}$; polarization operator $\hat{\Pi}_\omega$ in the coordinate representation has the form

$$\Pi_\omega(\mathbf{r},\mathbf{r}') = T\sum_\varepsilon \Pi_\omega(\mathbf{r},\mathbf{r}';\varepsilon) = T\sum_\varepsilon G_\varepsilon(\mathbf{r},\mathbf{r}')G_{\omega-\varepsilon}(\mathbf{r},\mathbf{r}'), \tag{7.22}$$

where G_ε is the one-electron Green function of a normal metal at Matsubara frequency. Nonlinear operator $\widehat{B}$ in Eq. (12.19) corresponds to the product of the four such Green functions.

The condition for saddle point definition $\delta\mathcal{S}_\Delta/\delta\Delta(\mathbf{r},t) = 0$ in the case of a gapless superconductor results in the TDGL equation [104].

Close to T_c the coordinate dependence of the fluctuating order parameter is smooth while the dependence of $\Delta(\mathbf{r},\tau)$ on τ can be omitted at all. As the result the function B remains constant while L^{-1} can be expanded over the powers of q^2 (see (6.32)). In result the free energy (7.16) is presented in the form of the GL series (A.6) with the coefficients A and B determined by Eqs. (7.19)–(7.20). The q^2 coefficient C is related to the square of coherence length by Eq. (A.9) (see also Eq. (6.33)).

Let us mention that for most problems $\Delta(\mathbf{r}, t)$ is the smooth self-averaging function, hence only the coefficient C has to be averaged over impurities configuration. After such averaging the expression for propagator L in the frequency-momentum representation is determined by Eqs. (6.4) and (6.30). Nevertheless for some problems of the mesoscopic character (for instance, see the problem of the optimal fluctuation in Part IV), where the disorder fluctuations are important themselves, it is necessary first to integrate over the order parameter fluctuations and only subsequently to perform the impurity average.

In order to get the match between the developed microscopic and phenomenological theory results (7.18) and (2.4) the correspondence between the phenomenological order parameter Ψ and the microscopic fluctuation field intensity Δ has to be established. Here, it is necessary to make the following comment. The choice of the coefficient of this proportionality is the delicate procedure where some arbitrariness takes place. The most common are two following choices. In the first one the identity $\Delta \equiv \Psi$ is postulated. In this case $a = A, b = B, 1/4m = C$. Such a choice is convenient since in this case the value of Δ is equal to the value of one-particle spectrum gap, following from the microscopic theory. Yet such a choice becomes embarrassing in the case of an impure superconductor, where the Cooper pair mass $2m$ ceases to be universal anymore and starts to depend on the impurity concentration. The second choice assumes the Cooper pair mass to be fixed and equal to two free electron masses. In this case Δ, a and b are determined by Eqs. (A.10) and (7.1).

Due to the short range nature of the interaction in the BCS model we succeeded, applying Hubbard–Stratonovich transformation, to separate it and in this way to reduce the problem to the accounting for only long wave length fluctuations of Δ. In some problems (such as those discussed in [92, 95], where the inter-electron interaction is considered in strongly disordered superconductors) the long range character of the Coulomb interaction, or, in other words, the electric field fluctuations, is of the first importance. Accounting for such interaction means the appearance in Hamiltonian (7.1) of the additional term

$$\mathcal{H}_Q = \frac{\rho(\mathbf{r})\,\rho(\mathbf{r}_1)}{|\mathbf{r} - \mathbf{r}_1|}, \tag{7.23}$$

with $\rho(\mathbf{r}) = e\,|\Psi|^2$. After the Hubbard–Stratonovich transformation in action appears the additional term

$$-\mathcal{S}_e = -\frac{(\nabla\varphi)^2}{8\pi} + \varphi\rho, \tag{7.24}$$

where $\varphi(\mathbf{r})$ is the electric field potential. Strong Coulomb interaction results in the smallness of the Debye radius ($|r_D|^2 = e^2\nu$) with respect to the Cooper pair size (coherence length ξ), while the plasma frequency $\omega_p^2 = 4\pi e^2 n/m$ is large in comparison with Δ. This is the reason why in the most part of the problems instead of functional integration over the fields $\varphi(\mathbf{r})$ one can restrict himself to accounting for only neutrality constrain $\rho = 0$.

7.2 GL functional in the case of a nontrivial order parameter symmetry[38]

Let us derive the GL functional for the general case of the anisotropic order parameter. The inter-electron interaction can be written in the most general form as

$$V_{ee}(\mathbf{p}, \mathbf{p}_1, \mathbf{q}) = V_{\mathbf{p},\mathbf{p}_1,\mathbf{q}} \widetilde{\psi}^+_{\mathbf{p}} \widetilde{\psi}^+_{-\mathbf{p}+\mathbf{q}} \widetilde{\psi}_{\mathbf{p}_1} \widetilde{\psi}_{-\mathbf{p}_1+\mathbf{q}}. \tag{7.25}$$

We will be interested in fluctuations with the wave vectors q much smaller than p_F and it is natural to suppose that both electron momenta $\mathbf{p}, \mathbf{p}_1$ belong to the vicinity of the Fermi surface. In this case the matrix elements $V_{\mathbf{p},\mathbf{p}_1,\mathbf{q}}$ weakly depend on q while their dependence on $\mathbf{p}, \mathbf{p}_1$ is reduced to the function of their directions only. For example, in the case of a $2D$ isotropic metal $V_{\mathbf{p},\mathbf{p}_1,\mathbf{q}}$ depends only on the angle, $\varphi = \varphi - \varphi_1$, between the vectors $\mathbf{p}$ and $\mathbf{p}_1$. Let us expand this function in the Fourier series:

$$V(\varphi) = -\sum_l g_l (\cos l\varphi \cos l\varphi_1 + \sin l\varphi \sin l\varphi_1). \tag{7.26}$$

Each positive g_l in principle can generate the pairing with corresponding order parameter symmetry, but that one conformable to the largest g_l is realized. For example, in the case when the coefficient g_0 is the largest one the s-pairing takes place. If the g_2 is larger than all other g_l the d-wave pairing is realized in the system.

In the crystalline lattice

$$V_{\mathbf{p},\mathbf{p}_1} = \sum_l g_l \chi_l(\mathbf{p}) \chi_l(\mathbf{p}_1). \tag{7.27}$$

Here the quantum number l does not have anymore the rigorous sense of the angular momentum but just shows how many zeros the function χ_l has. The habitual presentation for d-wave case $\chi_2 = p_x^2 - p_y^2$ means that the function χ_2 has four zeros which take place in directions $p_x = \pm p_y$. We will ignore the exotic case when two interaction constants g_l and g_m are almost equal and will suppose the interaction to be weak. So in the series (7.27) one can leave only the term

[38] In this section we base on the results of [234].

with the largest g_ρ. The corresponding interaction Hamiltonian will be described by the same Eq. (7.3) but the operators (7.4) must be written in the form

$$\widetilde{P}_\rho(\mathbf{r}) = \sum_{\mathbf{p},\mathbf{q}} \chi_\rho(\mathbf{p})\, \widetilde{\psi}_{\mathbf{p}+\mathbf{q}/2}(\mathbf{r})\, \widetilde{\psi}_{\mathbf{p}-\mathbf{q}/2}(\mathbf{r}) \exp(i\mathbf{q}\mathbf{r}). \tag{7.28}$$

The formulae (7.7)–(7.21) will be valid with the simple substitution of Δ by the anisotropic gap $\Delta(\mathbf{p}) = \Delta\chi_\rho(\mathbf{p})$ while the function $\chi_\rho(\mathbf{p})$ is normalized according to

$$\overline{\chi_\rho^2} = \frac{1}{\nu}\int \chi_\rho^2(\mathbf{p})\,\delta(\varepsilon_\mathbf{p} - \mu)\, d\mathbf{p} = 1. \tag{7.29}$$

The GL functional maintains its form (A.6) but the coefficients B and C must me substituted by their anisotropic values

$$B_\rho = B\overline{\chi_\rho^4}, \tag{7.30}$$

$$C_\rho = C\overline{\left(\chi_\rho^2(\mathbf{p})\,\mathbf{v}_\mathbf{p}^2\right)}/v_F^2. \tag{7.31}$$

Namely by this reason all physical consequences of the GL phenomenology are applicable to the p- and d-superconductors ($\rho = 1, 2$).

The important difference of an anisotropic case in comparison with the isotropic one consists in the role of impurities. This is due to the fact that in an anisotropic superconductor the presence of nonmagnetic impurities changes, with respect to the clean case, not only coefficient C but also A, B and T_c. Let us demonstrate this.

Due to the scattering by impurities the electron momentum changes with time in such a way that the correlator

$$\langle \mathbf{p}(t)\,\mathbf{p}(0)\rangle = p_F^2 e^{-t/\tau_p}. \tag{7.32}$$

In the case of p-pairing $\chi_p(\mathbf{p}) \sim \mathbf{p}$ and the same relaxation law can be written for $\langle \chi_p(\mathbf{p},t)\,\chi_p(\mathbf{p},0)\rangle \sim \exp(-t/\tau_p)$. The corresponding average value $\overline{\chi_p} = 0$.[39] One can check that the same law but with $\tau_d = \tau_p/2$ is valid also for d-pairing: $\langle \chi_d(\mathbf{p},t)\,\chi_d(\mathbf{p},0)\rangle \sim \exp(-t/\tau_d)$,and again $\overline{\chi_d} = 0$. These statements are valid for any type of pairing with $\rho \neq 0$. The exception is the case of s-pairing when in the BCS approximation $\overline{\chi_s} = 1$. Nevertheless even in the case of s-pairing but with the anisotropic electron spectrum $\overline{\chi_s}$ can differ from 1. Let us see to what consequences this will lead for the critical temperature. In the general case $\overline{\chi_\rho} \neq 0$ correlator of the parameter $\chi_l(\mathbf{p})$ relaxes to its average value as

$$\langle \chi_\rho(\mathbf{p},t)\,\chi_\rho(\mathbf{p},0)\rangle = e^{-t/\tau_\rho} + \left(\overline{\chi_\rho(\mathbf{p})}\right)^2 \left(1 - e^{-t/\tau_\rho}\right). \tag{7.33}$$

Performing the Fourier transform of expression (7.33) one can find

[39] The sign $\overline{(\cdots)}$ means the averaging over the state with the energy close to the Fermi level.

$$\langle \chi_\rho(\mathbf{p},t)\,\chi_\rho(\mathbf{p},0)\rangle_\omega = \frac{1-\left(\overline{\chi_\rho(\mathbf{p})}\right)^2}{i\omega + 1/\tau_\rho} + \frac{\left(\overline{\chi_\rho(\mathbf{p})}\right)^2}{i\omega} \tag{7.34}$$

On can calculate this correlator for electron with energy ξ_l in the basis of the electron eigenfunctions of the Hamiltonian with the exact impurity potential

$$\langle \chi_\rho(\mathbf{p},t)\,\chi_\rho(\mathbf{p},0)\rangle = \theta(t)\sum_{m,k}\left|\left(\chi_\rho(\mathbf{p})\right)_{km}\right|^2 \exp\left(i\omega_{mk}t\right)\delta\left(\xi_k\right), \tag{7.35}$$

where $\omega_{mk} = \xi_k - \xi_m$. After the Fourier transform one finds

$$\langle \chi_\rho(\mathbf{p},t)\,\chi_\rho(\mathbf{p},0)\rangle_\omega = \sum_{m,k}\frac{\left|\left(\chi_\rho(\mathbf{p})\right)_{km}\right|^2}{\omega - \omega_{mk}}\delta\left(\xi_k\right). \tag{7.36}$$

In the same basis of the electron eigenfunctions one can write the equation for the critical temperature which follows from (7.17):

$$\begin{aligned} 1 &= -g_\rho T_\mathrm{c}\sum_{\varepsilon_n}\sum_{m,k}\left|\left(\chi_\rho(\mathbf{p})\right)_{mk}\right|^2 \frac{1}{i\varepsilon_n - \xi_m}\frac{1}{i\varepsilon_n + \xi_k} \\ &= -g_\rho T_\mathrm{c}\sum_{\varepsilon_n}\sum_{m,k}\frac{\left|\left(\chi_\rho(\mathbf{p})\right)_{mk}\right|^2}{2\varepsilon_n - i\omega_{mk}}\left(\frac{1}{\varepsilon_n + i\xi_m} + \frac{1}{\varepsilon_n - i\xi_k}\right). \end{aligned} \tag{7.37}$$

The summation over eigen-states can be performed in the following way. Recalling that $\varepsilon_n \sim T \ll E_F$ one can substitute the last parenthesis as the Lorentzian $2\varepsilon_n\left(\xi_k^2 + \varepsilon_n^2\right)^{-1}$, which, in its turn, to take as the delta function $\delta\left(\xi_k\right)$ and finally to use Eq. (7.36):

$$\begin{aligned} 1 &= g_\rho \nu T_\mathrm{c}\sum_{\varepsilon_n}\sum_{m,k}\frac{\left|\left(\chi_\rho(\mathbf{p})\right)_{mk}\right|^2}{-2i\varepsilon_n - \omega_{mk}}\delta\left(\xi_k\right) = \\ &. = g_\rho \nu T_\mathrm{c}\sum_{\varepsilon_n}\left[\frac{\left(\overline{\chi_\rho(\mathbf{p})}\right)^2}{2|\varepsilon_n|} + \frac{1-\left(\overline{\chi_\rho(\mathbf{p})}\right)^2}{2|\varepsilon_n| + 1/\tau_\rho}\right]. \end{aligned} \tag{7.38}$$

For the traditional case of s-pairing in not very anisotropic superconductor $\overline{\chi_\rho(\mathbf{p})} = 1$ and Eq. (7.38) is reduced to Eq. (7.17) resulting in the BCS value of the critical temperature T_{c0}.

Here it is necessary to underline that in the case of s-pairing in superconductor with the strongly anisotropic electron spectrum $\overline{\chi_\rho(\mathbf{p})} \neq 1$. In result, the presence of even nonmagnetic impurities reduces the critical temperature according to the relation:

$$\frac{1}{\nu g_\rho} = \ln\frac{\omega_D}{2\pi T_\mathrm{c}} - \psi\left(\frac{1}{2}\right) + \left[\left(\overline{\chi_\rho(\mathbf{p})}\right)^2 - 1\right]\left[\psi\left(\frac{1}{2} + \frac{1}{4\pi T_\mathrm{c}\tau_\rho}\right) - \psi\left(\frac{1}{2}\right)\right].$$

Introducing the critical temperature of the corresponding clean superconductor T_{c0} one can express the shifted critical temperature T_c as [234]:

$$\ln \frac{T_c}{T_{c0}} = \left[\left(\overline{\chi_\rho(\mathbf{p})} \right)^2 - 1 \right] \left[\psi \left(\frac{1}{2} + \frac{1}{4\pi T_c \tau_\rho} \right) - \psi \left(\frac{1}{2} \right) \right]. \tag{7.39}$$

Nevertheless the anisotropy in conventional superconductors usually is small and corresponding renormalization of the critical temperature is weak and ignored (the so-called Anderson theorem) [235, 236].

In the cases of pairing with higher momenta $\overline{\chi}_\rho = 0$ and the corresponding equation for critical temperature takes the form

$$\ln \frac{T_c}{T_{c0}} = \psi(1/2) - \psi \left(\frac{1}{2} + \frac{1}{4\pi T_c \tau_\rho} \right). \tag{7.40}$$

That is the nonmagnetic impurities for p- and d-superconductors produce the same effect on critical temperature as the magnetic impurities do for the conventional s-wave superconductor.

Evidently exists some critical impurities concentration ($\tau_\rho^{(cr)} \sim T_{c0}^{-1}$) when impurities completely destroy the state of superconductivity. Let us find the corresponding $\tau_\rho^{(cr)}$. For this purpose one can expand the function ψ in Eq. (7.40) according to Eq. (B.16) and, putting $T_c = 0$, to obtain

$$\ln \frac{1}{4\pi \tau_\rho^{(cr)} T_{c0}} = \psi \left(\frac{1}{2} \right) = -\ln 4\gamma_E, \tag{7.41}$$

what means that the critical concentration corresponds to

$$\tau_\rho^{(cr)} = \frac{\gamma_E}{\pi T_{c0}}. \tag{7.42}$$

8

MICROSCOPIC THEORY OF FLUCTUATION CONDUCTIVITY

8.1 Qualitative discussion of the different fluctuation contributions

In chapter 3, the direct fluctuation effect on conductivity, related to the charge transfer by means of fluctuation Cooper pairs, was discussed in detail. Nevertheless, in this section we return to its discussion and will demonstrate the corresponding calculations by means of the microscopic theory. This will be done in order to prepare the basis for studies of the AL contribution to a variety of physical values like magnetoconductivity near the upper critical field, conductivity far from transition point and in ultra-clean limit, Hall conductivity, etc.

The microscopic approach allows us also to calculate the above-cited indirect fluctuation effects such as the so-called DOS and MT contributions. We will start now from their qualitative discussion.

The important consequence of the presence of fluctuating Cooper pairs above T_c is the decrease in the one-electron density of states at the Fermi level. Indeed, if some electrons are involved in pairing they cannot simultaneously participate in charge transfer and energy absorbtion as single-particle excitations. Nevertheless, the total number of the electronic states cannot be changed by the Cooper interaction, and only a redistribution of the levels along the energy axis is possible [169, 237]. In this sense one can speak of the opening of a fluctuation pseudogap at the Fermi level. The decrease of the one-electron DOS at the Fermi level leads to a reduction of the Drude conductivity. This, indirect, fluctuation correction to the conductivity is called the DOS contribution and it appears side by side with the paraconductivity (or AL contribution). It has the opposite (negative) sign and turns out to be much less singular in $(T-T_c)^{-1}$ in comparison with the AL contribution, so that in the vicinity of T_c it was usually omitted. However, in many cases [52, 108, 238–241], when for some special reasons the main, most singular, corrections are suppressed, the DOS correction becomes of major importance. Such a situation is realized in study of fluctuation quasiparticle current in tunnel structures, of fluctuation c-axis transport in strongly anisotropic HTS, fluctuation corrections to the NMR relaxation rate or thermoelectric power.

The correction to the normal state conductivity above the transition temperature related to the fluctuation DOS renormalization can be evaluated qualitatively. Indeed, the fact that some electrons ($\Delta\mathcal{N}_e$ per unit volume) participate in fluctuation Cooper pairings means that the effective number of carriers taking part in one-electron charge transfer diminishes leading to a decrease of conductivity (we deal here with the longitudinal component):

$$\delta\sigma_{xx}^{DOS} = -\frac{\Delta\mathcal{N}_e e^2\tau}{m_e} = -\frac{2n_s e^2\tau}{m_e}, \tag{8.1}$$

where n_s is the fluctuation Cooper pairs concentration above transition point. The latter was already calculated in chapter 2 and is given by the Eq. (2.151). Let us recall, that the mass m in Eq. (2.151) represents the effective Cooper pair mass and generally speaking should not coincide with the doubled electron mass. Substituting the Eq. (2.151) to the Eq. (8.1) and using the explicit form for the Ginzburg–Landau coefficient α in accordance with Eq. (A.12) one can find that

$$\delta\sigma_{xx}^{DOS} \sim -\frac{e^2}{s}\ln\frac{2}{\sqrt{\epsilon}+\sqrt{\epsilon+r}}, \tag{8.2}$$

what coincides up to the accuracy of numerical coefficient with the microscopic expression (8.28), which will be carried out below in result of much more cumbersome microscopic analysis.

The third, purely quantum, fluctuation contribution is generated by the coherent scattering of the electrons forming a Cooper pair on the same elastic impurities. This is the so called anomalous Maki–Thompson (MT) contribution [40,41] which can be treated as the result of Andreev reflection of the electron by fluctuation Cooper pairs. This contribution appears only in transport coefficients and often turns out to be important. Its temperature singularity near T_c is similar to that of the paraconductivity, although being extremely sensitive to electron phase-breaking processes and to the type of orbital symmetry of pairing it can be suppressed. Let us evaluate it.

The physical origin of the MT correction consists in the fact that the Cooper interaction of electrons with nearly opposite momenta changes the mean free path (diffusion coefficient) of electrons. As we have already seen in the previous section the amplitude of this interaction increases drastically when $T \to T_c$:

$$g_{\text{eff}} = \frac{g}{1-\nu g\ln\frac{\omega_D}{2\pi T}} = \frac{1}{\ln\frac{T}{T_c}} \approx \frac{T}{T-T_c} = \frac{1}{\epsilon}.$$

What is the reason for this growth? One can say that the electrons scatter one at another in a resonant way with the virtual Cooper pair formation. Or, it is possible to imagine that the electrons undergo Andreev reflection by fluctuation Cooper pairs, binding in the Cooper pairs themselves. The probability of such induced pair irradiation (let us remember that Cooper pairs are Bose particles) is proportional to their number in the final state, i.e. $n(p)$ (1.5). For small momenta $n(p) \sim 1/\epsilon$.

One can ask why such an interaction does not manifest itself considerably far from the transition point? This is due to the fact that just a small number of electrons with the total momentum $q \lesssim \xi^{-1}(T)$ interacts so intensively. In accordance with the Heisenberg principle the minimal distance between such electrons is of the order of $\sim \xi(T)$. On the other hand, such electrons, in order to interact, have to approach one another approximately up to a distance of the Fermi length $\lambda_F \sim 1/p_F$. The probability of such event may be estimated in the spirit of the self-intersecting trajectories contribution evaluation in the weak-localization theory [242, 131].

In the process of diffusion motion the distance between two electrons increases with time according to the law: $R(t) \sim (\mathcal{D}t)^{1/2}$. Hence the scattering probability

$$W \sim \int_{t_{\min}}^{t_{\max}} \frac{\lambda_F^{D-1}}{R^D(t)} v_F dt.$$

The lower limit of the integral can be estimated from the condition $R(t_{\min}) \sim \xi(T)$ (only such electrons interact in the resonant way). The upper limit is determined by the phase-breaking time τ_φ since for larger time intervals the phase coherence, necessary for the pair formation, is broken. As the result the relative correction to conductivity due to such processes is equal to the product of the scattering probability on the effective interaction constant: $\delta\sigma^{MT}/\sigma = W\, g_{\text{eff}}$. In the $2D$ case

$$\delta\sigma^{MT} \sim \frac{e^2}{8\epsilon} \ln \frac{\mathcal{D}\tau_\varphi}{\xi^2(T)}.$$

This result will be confirmed below in the framework of the microscopic consideration.

8.2 The electromagnetic response operator

The most general relation between the current density $\mathbf{j}(\mathbf{r},t)$ and vector-potential $\mathbf{A}(\mathbf{r}',t')$ is given through the so-called electromagnetic response operator $Q_{\alpha\beta}$ [228]:

$$\mathbf{j}_\alpha(\mathbf{r},t) = -\int Q_{\alpha\beta}(\mathbf{r},\mathbf{r}',t,t')\mathbf{A}_\beta(\mathbf{r}',t')d\mathbf{r}'dt'. \tag{8.3}$$

Assuming space and time homogeneity, one can take the Fourier transform of this relation and compare it with the definition of the conductivity tensor $j_\alpha = \sigma_{\alpha\beta}E_\beta$. This allows us to express the conductivity tensor in terms of the retarded electromagnetic response operator

$$\sigma_{\alpha\beta}(\omega) = -\frac{1}{i\omega}[Q_{\alpha\beta}]^R(\omega)\ . \tag{8.4}$$

The electromagnetic response operator $Q_{\alpha\beta}(\omega_\nu)$, defined on Matsubara frequencies $\omega_\nu = 2\nu\pi T$, can be presented as the correlator of two one-electron Green's functions [228] averaged over impurity positions and accounting for interactions, in our case the particle–particle interactions in the Cooper channel. The appropriate diagrams corresponding to the first order of perturbation theory in the fluctuation amplitude can be drawn according to the rules of the diagrammatic technique (see Fig. 6.2) and they are shown in Fig. 8.1.

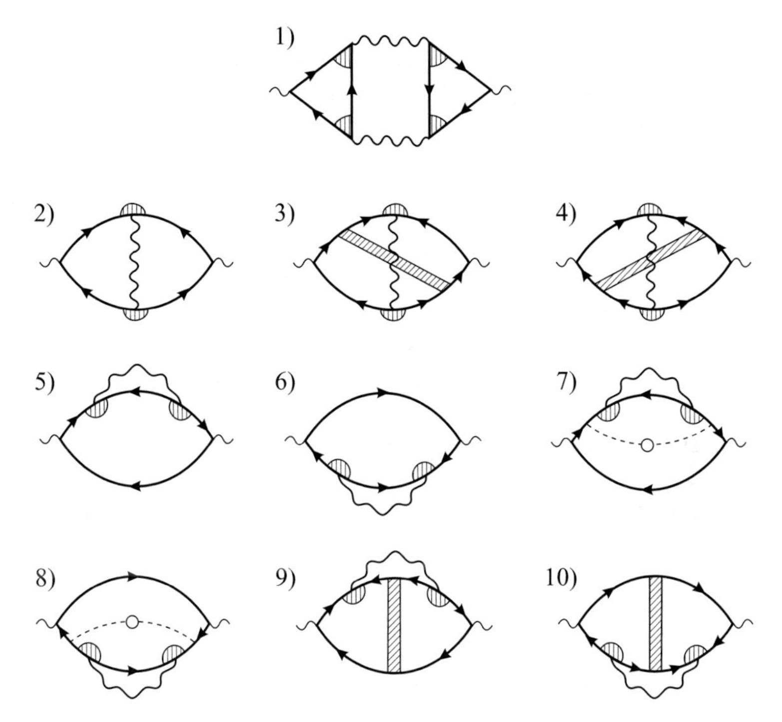

FIG. 8.1. Feynman diagrams for the leading-order contributions to the fluctuation conductivity. Wavy lines are fluctuation propagators, thin solid lines with arrows are impurity-averaged normal state Green's functions, shaded semicircles are renormalized by scattering of electrons by impurities vertex functions, dashed lines with central crosses are additional impurity renormalizations and shaded rectangles are impurity ladders. Diagram 1 represents the AL term; diagrams 2–4 represent the MT type contributions; diagrams 5–10 arise from corrections to the normal state DOS.

With each electromagnetic field component A_α we associate the external vertex $ev_\alpha(p) = e\partial\xi(p)/\partial p_\alpha$. For the longitudinal conductivity tensor elements (parallel to the layers, for which $\alpha = x, y$), the resulting vertex is simply ep_α/m. For the c-axis conductivity, the vertex is given by

$$ev_z(p) = e\frac{\partial\xi(p)}{\partial p_z} = -eJs\sin(p_z s). \tag{8.5}$$

Each solid line in the diagrams represents a one-electron Green's function averaged over impurities (6.2), a wavy line represents a fluctuation propagator $L(\mathbf{q}, \Omega_k)$ (6.32), three-leg vertices were defined by Eq. (6.26). The four-leg impurity vertex, appearing in diagrams 3–4, 9–10 of the Fig. 8.1, is called the Cooperon in the weak localization (WL) theory (see, e.g. [243]). It differs from the above three-leg vertex only by the additional factor $(2\pi\nu\tau)^{-1}$. We do not renormalize the current vertex since this renormalization only leads to the substitution of the scattering time τ by the transport one $\tau_{\rm tr}$ in the final results (see [228]). We integrate over the internal Cooper pair momentum $\mathbf{q}$ and electron momentum $\mathbf{p}$ and sum over the internal fermionic and bosonic Matsubara frequencies, with momentum and energy conservation at each vertex (fluctuation propagator endpoint) in the analytic expressions for the diagrams presented in Fig. 8.1.

After these necessary introductory remarks and definitions we will consider the microscopic calculation of the different fluctuation contributions.

8.3 Fluctuation conductivity of a layered superconductor in the vicinity of $T_{\rm c}$

Let us consider the microscopic calculation of the fluctuation conductivity in the framework of the microscopic approach on the actual example of a layered superconductor (superconductor with the quasi-2D electron spectrum being in its normal state).

8.3.1 *AL contribution*

We first examine the AL paraconductivity (diagram 1 of Fig. 8.1). Actually this contribution was already studied in section 3.2 in the framework of the TDGL equation but, in order to demonstrate how the method works, we will carry out here the appropriate calculations in the microscopic approach, as was originally done by Aslamazov and Larkin [36].

The AL contribution to the electromagnetic response operator tensor has the form:

$$Q_{\alpha\beta}^{AL}(\omega_\nu) = -4e^2 T \sum_{\Omega_k} \int \frac{d^3\mathbf{q}}{(2\pi)^3} B_\alpha(\mathbf{q},\Omega_k,\omega_\nu) L(\mathbf{q},\Omega_k) \times B_\beta(\mathbf{q},\Omega_k,\omega_\nu) L(\mathbf{q},\Omega_k+\omega_\nu), \tag{8.6}$$

where

$$B_\alpha(\mathbf{q},\Omega_k,\omega_\nu) = T\sum_{\varepsilon_n} \lambda(\mathbf{q},\varepsilon_{n+\nu},\Omega_k-\varepsilon_n)\lambda(\mathbf{q},\varepsilon_n,\Omega_k-\varepsilon_n) \times \int \frac{d^3\mathbf{p}}{(2\pi)^3} v_\alpha(\mathbf{p}) G(\mathbf{p},\varepsilon_{n+\nu}) G(\mathbf{p},\varepsilon_n) G(\mathbf{q}-\mathbf{p},\Omega_k-\varepsilon_n) \tag{8.7}$$

is the block of three Green's functions with the impurity vertex integrated over the electron momentum and summed over the fermionic frequency.

In the vicinity of $T_{\rm c}$, due to the pole structure of the fluctuation propagators in (8.6), the leading contribution to the electromagnetic response operator $Q_{\alpha\beta}^{\rm AL(R)}$ arises from them rather than from the weak frequency dependence of the blocks B_α, so we can neglect the Ω_k- and ω_ν-dependencies of the Green functions blocks and use the expression for $B_\alpha(\mathbf{q},0,0)$ valid for small $\mathbf{q}$ only:

$$B_\alpha(\mathbf{q}) = B_\alpha(\mathbf{q},0,0)$$

$$= -\nu T \sum_{\varepsilon_n} \lambda^2(\mathbf{q}, \varepsilon_n, -\varepsilon_n) \int_{-\infty}^{\infty} \frac{d\xi}{(i\widetilde{\varepsilon}_n - \xi)^2} \left\langle \frac{v_\alpha(\mathbf{p})}{(i\widetilde{\varepsilon}_n + \xi - \mathbf{vq})} \right\rangle_{FS} . \tag{8.8}$$

Taking into account that only small values of q, defined by the poles of propagator will be involved in the further q-integration, one can perform the angular integration expanding over q the last Green's function. The q-dependence of the impurity vertices can be ignored since it occurs for $\widehat{\mathcal{D}}q^2 \sim \varepsilon_n \sim T$. As a result the first term is averaged out while the second gives:

$$\begin{aligned} B_\alpha(\mathbf{q}) &= -\nu T \left\langle v_\alpha v_\beta q_\beta \right\rangle_{FS} \sum_{\varepsilon_n} \frac{|\widetilde{\varepsilon}_n|^2}{|\varepsilon_n|^2} \int_{-\infty}^{\infty} \frac{d\xi}{(\xi^2 + \widetilde{\varepsilon}_n^2)^2} \\ &= -\nu \left\langle v_\alpha v_\beta q_\beta \right\rangle_{\mathrm{FS}} \pi T \sum_{n=0}^{\infty} \frac{1}{(\varepsilon_n + 1/2\tau)\, \varepsilon_n^2}, \end{aligned} \tag{8.9}$$

where $\langle v_\alpha v_\beta q_\beta \rangle_{FS} = v_F^2 q_\alpha / D$ in the case of D dimensional isotropic spectrum. The last sum already was carried out in evaluation of the (6.28) and finally (8.9) turns out to be proportional to the square of coherence length (6.33). In the case of the isotropic spectrum

$$\begin{aligned} B_\alpha(\mathbf{q}) &= \frac{2\nu\tau^2 v_F^2}{D} \left[\psi\left(\frac{1}{2} + \frac{1}{4\pi T\tau}\right) - \psi\left(\frac{1}{2}\right) - \frac{1}{4\pi T\tau} \psi'\left(\frac{1}{2}\right) \right] q_\alpha \\ &= -2\nu \xi_{(D)}^2 (T\tau) q_\alpha . \end{aligned} \tag{8.10}$$

In the other case important for our consideration, the layered superconductor, the Green function block in the vicinity of the transition temperature takes the form

$$B_\alpha(\mathbf{q}) = -2\nu \frac{\eta_{(2)}}{v_F^2} \begin{cases} v_F^2 q_\alpha, \ \alpha = x, y, \\ sJ^2 \sin q_z s, \ \alpha = z. \end{cases} \tag{8.11}$$

Let us return to the study of the general expression (8.6) for $Q_{\alpha\beta}^{\mathrm{AL}}(\omega_\nu)$. The Ω_k-summation in it can be transformed into a contour integral, using the identity [244]

$$T \sum_{\Omega_k} f(\Omega_k) = \frac{1}{4\pi i} \oint_{C_0} dz \coth \frac{z}{2T} f(-iz), \tag{8.12}$$

where $z = i\Omega_k$ is a variable in the plane of complex frequency and the contour C_0 encloses all bosonic Matsubara frequencies over which the summation is carried out (see Fig. 8.2(a)). Applying this transformation to Eq. (8.6) one can write

$$Q_{\alpha\beta}^{AL}(\omega_\nu) = -\frac{e^2}{\pi i} \int \frac{d^3\mathbf{q}}{(2\pi)^3} B_\alpha(\mathbf{q}) B_\beta(\mathbf{q}) \oint_C dz \coth \frac{z}{2T} L(\mathbf{q}, -iz) L(\mathbf{q}, -iz + \omega_\nu), \tag{8.13}$$

where the integration contour C is some continuously deformed contour C_0, which we choose for convenience in further integration. In order to do this let us

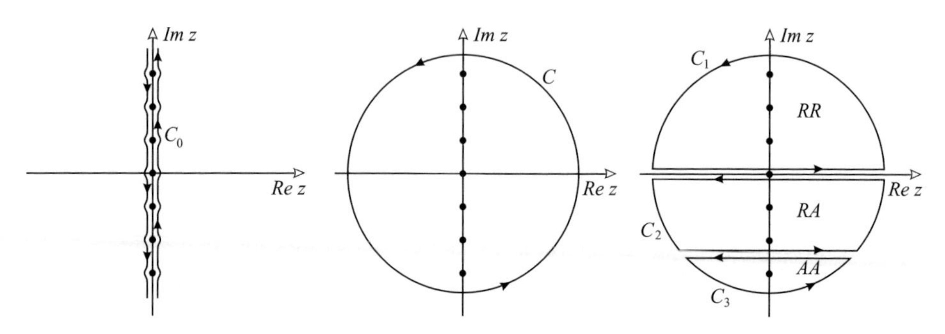

FIG. 8.2. The contour of integration in the plane of complex frequencies.

notice that the integrand function in (8.13) has the breaks of analyticity at the lines $\operatorname{Im} z = 0$ and $\operatorname{Im} z = -i\omega_\nu$. Indeed, the fluctuation propagator $L(\mathbf{q}, \Omega_k)$ was defined on the bosonic Matsubara frequencies only, while now we have to use it as the function of the continuous variable z. As is known from the properties of Green's functions in the complex plane z, two analytic functions, related to $L(\mathbf{q}, \Omega_k)$, can be introduced. The first one, $L^R(\mathbf{q}, -iz)$ (retarded), is analytic in the upper half-plane ($\operatorname{Im} z > 0$), while the second one, $L^A(\mathbf{q}, -iz)$ (advanced), has no singularities in the lower half-plane ($\operatorname{Im} z < 0$). In accordance with this observation let us cut the z-plane along the lines $\operatorname{Im} z = 0$ and $\operatorname{Im} z = -i\omega_\nu$ (see Fig. 8.2(b)) and choose the contour C in (8.13) as the aggregate of three closed contours $C_1 + C_2 + C_3$ (see Fig. 8.2(c)). Each of them already encloses a domain of well-defined analyticity of the integrand function.

Hence the integrand function is analytic along each of these contours and the corresponding integral, according to the Cauchy theorem, is determined by the sum of residues in the poles of $\coth(z/2T)$ (see the formula (8.12)). If the modulus of function $f(-iz)$ on the large circle $|z| = R \to \infty$ tends to zero as $\sim R^{-1-\gamma} (\gamma > 0)$ the value of the integral over this circle in (8.12) tends to zero too. Finally, the contour integral in (8.13) is reduced to the four integrals along the horizontal lines:

$$\begin{aligned} I^{AL}_{\alpha\beta}(\mathbf{q}, \omega_\nu) &= \oint_{C_1+C_2+C_3} dz \coth \frac{z}{2T} L(\mathbf{q}, -iz) L(\mathbf{q}, -iz + \omega_\nu) \\ &= \int_{-\infty}^{\infty} dz \coth \frac{z}{2T} L^R(\mathbf{q}, -iz + \omega_\nu) \left[L^R(\mathbf{q}, -iz) - L^A(\mathbf{q}, -iz) \right] \\ &+ \int_{-\infty - i\omega_\nu}^{\infty - i\omega_\nu} dz \coth \frac{z}{2T} L^A(\mathbf{q}, -iz) \left[L^R(\mathbf{q}, -iz + \omega_\nu) - L^A(\mathbf{q}, -iz + \omega_\nu) \right]. \end{aligned} \tag{8.14}$$

Now one can shift the variable in the last integral to $z = z' - i\omega_\nu$, take into account that $i\omega_\nu$ is the period of $\coth \frac{z}{2T}$ and get an expression analytic in $i\omega_\nu \to \omega$. Finally:

$$Q_{\alpha\beta}^{AL(R)}(\omega) = -\frac{2e^2}{\pi}\int \frac{d^3\mathbf{q}}{(2\pi)^3} B_\alpha(\mathbf{q})B_\beta(\mathbf{q})$$
$$\times \int_{-\infty}^{\infty} dz \coth\left(\frac{z}{2T}\right)\left[L^R(\mathbf{q},-iz-i\omega)+L^A(\mathbf{q},-iz+i\omega)\right]\operatorname{Im} L^R(\mathbf{q},-iz).$$

Being interested here in the d.c. conductivity one can expand the integrand function in ω. It is possible to show that the zeroth-order term is canceled by the same type contributions from all other diagrams (this cancelation confirms the absence of anomalous diamagnetism above the critical temperature). The remaining integral can be integrated by parts and then carried out taking into account that the contribution most singular in ϵ comes from the region $z \sim \epsilon \ll T$:

$$\sigma_{\alpha\beta}^{AL} = \frac{e^2}{2\pi T}\int \frac{d^3\mathbf{q}}{(2\pi)^3} B_\alpha(\mathbf{q})B_\beta(\mathbf{q})\int_{-\infty}^{\infty}\frac{dz}{\sinh^2\frac{z}{2T}}\left[\operatorname{Im} L^R(\mathbf{q},-iz)\right]^2. \quad (8.15)$$

Coming back to the case of the layered superconductor one can reproduce the Lawrence–Doniach expression (3.36) for the in-plane paraconductivity. Indeed, separating the imaginary part of Eq. (6.39) and substituting it together with (8.11) to (8.15) one can find:

$$\sigma_{xx}^{AL} = \frac{\pi e^2\eta_{(2)}^2}{8T}\int \frac{q_x^2 d^2\mathbf{q}}{(2\pi)^2}\int_{-\pi/s}^{\pi/s}\frac{dq_z}{2\pi}\int_{-\infty}^{\infty}\frac{dz}{\left[\left(\eta_{(2)}\mathbf{q}^2+\epsilon+r\sin^2\frac{q_z s}{2}\right)^2+\left(\frac{\pi z}{8T}\right)^2\right]^2}$$
$$= \frac{\pi e^2\eta_{(2)}^2}{2s}\int \frac{d^2\mathbf{q}}{(2\pi)^2}\int_{-\pi}^{\pi}\frac{d\theta}{2\pi}\frac{q^2}{\left[\eta_{(2)}\mathbf{q}^2+\epsilon+\frac{r}{2}\left(1-\cos\theta\right)\right]^3},$$

where the Lawrence–Doniach anisotropy parameter r [50] was already defined by (2.45). The integral over θ can be at first presented as the complete second derivative and then expressed by means of the standard for the theory of layered superconductors integral (C.3). Carrying out the remaining integration over x by parts one can find

$$\sigma_{xx}^{AL} = \frac{e^2}{16s}\int_0^\infty x\frac{d^2}{dx^2}\left[\frac{1}{\sqrt{(\epsilon+x)(\epsilon+x+r)}}\right]dx$$
$$= \frac{e^2}{16s}\frac{1}{[\epsilon(\epsilon+r)]^{1/2}} \to \frac{e^2}{16s}\begin{cases}1/\sqrt{\epsilon r}, & \epsilon\ll r,\\ 1/\epsilon, & \epsilon\gg r.\end{cases} \quad (8.16)$$

In the same way one can evaluate the AL contribution to the transverse fluctuation conductivity and get the familiar from the phenomenological consideration Eq. (3.37) [238, 245, 246]:

$$\sigma_{zz}^{AL} = \frac{\pi e^2 s r^2}{32}\int \frac{d^2\mathbf{q}}{(2\pi)^2}\frac{1}{\left[(\eta_{(2)}\mathbf{q}^2+\epsilon)(\eta_{(2)}\mathbf{q}^2+\epsilon+r)\right]^{3/2}} \quad (8.17)$$

$$= \frac{e^2 s}{32\xi_{xy}^2}\left(\frac{\epsilon + r/2}{[\epsilon(\epsilon+r)]^{1/2}} - 1\right) \to \frac{e^2 s}{64\xi_{xy}^2}\begin{cases}\sqrt{r/\epsilon}, \epsilon \ll r, \\ (r/2\epsilon)^2. \epsilon \gg r\end{cases}$$

Note that contrary to the case of in-plane conductivity, the critical exponent for σ_{zz} above the Lawrence–Doniach crossover temperature T_{LD} (for which $\epsilon(T_{\text{LD}}) = r$) is 2 instead of 1, so the crossover occurs from the $0D$ to $3D$ regimes. This is related to the tunneling (so from the band structure point of view effectively zero dimensional $0D$) character of electron motion along the c-axis.

8.3.2 *Contributions from fluctuations of the DOS*

In original paper of Aslamazov and Larkin [36] the most singular AL contributions to conductivity, heat capacity and other properties of a superconductor above the critical temperature were considered. The diagrams of the type 5–6 were pictured and correctly evaluated as less singular in ϵ. Nevertheless the specific form of the AL contribution to the transverse conductivity of a layered superconductor, which may be considerably suppressed for small interlayer transparency, suggested to re-examine the contributions from diagrams 5–10 of Fig. 8.1 which are indeed less divergent in ϵ, but turn out to be of lower order in the transmittance and of the opposite sign with respect to the AL one [238,239]. These, so-called DOS, diagrams describe the changes in the normal Drude-type conductivity due to fluctuation renormalization of the normal quasiparticles DOS above the transition temperature (see section 10.1). In the dirty limit, the calculation of contributions to the longitudinal fluctuation conductivity σ_{xx} from such diagrams was discussed in [241,247]. Contrary to the case of the AL contribution, the in-plane and out-of-plane components of the DOS contribution differ only in the square of the ratio of effective Fermi velocities in the parallel and perpendicular directions. This allows us to calculate both components simultaneously. The contribution to the fluctuation conductivity due to diagram 5 (diagram 6 gives an identical contribution) is

$$Q_{\alpha\beta}^{(5)}(\omega_\nu) = 2e^2 T \sum_{\Omega_k} \int \frac{d^3\mathbf{q}}{(2\pi)^3} L(\mathbf{q}, \Omega_k) \sum\nolimits_{\alpha\beta}^{(5)}(\mathbf{q}, \Omega_k), \tag{8.18}$$

where

$$\sum\nolimits_{\alpha\beta}^{(5)}(\mathbf{q}, \Omega_k, \omega_\nu) = T \sum_{\varepsilon_n} \lambda^2(\mathbf{q}, \varepsilon_n, \Omega_k - \varepsilon_n) I_{\alpha\beta}^{(5)}(\mathbf{q}, \varepsilon_n, \Omega_k, \omega_\nu) \tag{8.19}$$

and

$$\begin{aligned} I_{\alpha\beta}^{(5)}(\mathbf{q}, \varepsilon_n, \Omega_k, \omega_\nu) &= \int \frac{d^3\mathbf{p}}{(2\pi)^3} v_\alpha(\mathbf{p}) v_\beta(\mathbf{p}) G^2(\mathbf{p}, \varepsilon_n) G(\mathbf{q}-\mathbf{p}, \Omega_k - \varepsilon_n) G(\mathbf{p}, \varepsilon_{n+\nu}) \\ &= \int \frac{d^3\mathbf{p}}{(2\pi)^3} \frac{1}{(i\widetilde{\varepsilon}_n - \xi_{\mathbf{p}})^2} \frac{v_\alpha(\mathbf{p}) v_\beta(\mathbf{p})}{(i\widetilde{\varepsilon}_{n+\nu} - \xi_{\mathbf{p}})} \frac{1}{\left(i(\widetilde{\Omega_k - \varepsilon_n}) - \xi_{\mathbf{q}-\mathbf{p}}\right)}. \end{aligned} \tag{8.20}$$

Evaluation of the integral (8.20) can be performed passing from momentum to the ξ-integration with the further averaging of the result obtained over the Fermi surface. Let us start from the ξ-integration. It can be accomplished with the use of the Cauchy theorem (see Fig. 8.3). One can see that depending on the sign of the frequencies six principally different dispositions of the integral poles are possible. Two of them do not contribute to $I_{\alpha\beta}^{(5)}$: when all three poles are located from the same side of the real axis the integral over ξ turns to be zero. In the cases of the other four poles dispositions (each one is accounted by means of the appropriate Heavyside theta-function) and one can find:

1 + 2) $\theta(\varepsilon_n \varepsilon_{n+\nu}) = 1, \qquad \theta(\varepsilon_n \varepsilon_{n-k}) = 1;$

$$I_{\alpha\beta}^{(5)(1+2)}(\mathbf{q}, \varepsilon_n, \Omega_k, \omega_\nu) = -2\pi\nu \left\langle \frac{v_\alpha(\mathbf{p}) v_\beta(\mathbf{p})}{(\widetilde{\varepsilon}_{n-k} + \widetilde{\varepsilon}_{n+\nu} + i\mathbf{v}\mathbf{q})} \times \frac{\theta(\varepsilon_n \varepsilon_{n-k})\, \theta(\varepsilon_n \varepsilon_{n+\nu})\, \mathrm{sign}\varepsilon_n}{(\widetilde{\varepsilon}_{n-k} + \widetilde{\varepsilon}_n + i\mathbf{v}\mathbf{q})^2} \right\rangle_{\mathrm{FS}}.$$

3) $\theta(-\varepsilon_n \varepsilon_{n+\nu}) = 1, \qquad \theta(-\varepsilon_n \varepsilon_{n-k}) = 1;$

4) $\theta(-\varepsilon_n \varepsilon_{n+\nu}) = 1, \qquad \theta(\varepsilon_n \varepsilon_{n-k}) = 1.$

$$I_{\alpha\beta}^{(5)(3+4)}(\mathbf{q}, \varepsilon_n, \Omega_k, \omega_\nu) = -2\pi\nu \left\langle \frac{v_\alpha(\mathbf{p}) v_\beta(\mathbf{p})}{(\widetilde{\varepsilon}_{n-k} + \widetilde{\varepsilon}_{n+\nu} + i\mathbf{v}\mathbf{q})} \left[\frac{\theta(-\varepsilon_n \varepsilon_{n+\nu})}{(\omega_\nu + 1/\tau)^2} - \frac{\theta(-\varepsilon_n \varepsilon_{n+\nu})\, \theta(\varepsilon_n \varepsilon_{n-k})}{(\widetilde{\varepsilon}_{n-k} + \widetilde{\varepsilon}_n + i\mathbf{v}\mathbf{q})^2} \right] \right\rangle_{\mathrm{FS}}.$$

Collecting all contributions

$$I_{\alpha\beta}^{(5)}(\mathbf{q}, \varepsilon_n, \Omega_k, \omega_\nu) = -2\pi\nu \left\langle \frac{v_\alpha(\mathbf{p}) v_\beta(\mathbf{p})}{(\widetilde{\varepsilon}_{n-k} + \widetilde{\varepsilon}_{n+\nu} + i\mathbf{v}\mathbf{q})} \left[\frac{\theta(\varepsilon_n \varepsilon_{n-k})\, \theta(\varepsilon_n \varepsilon_{n+\nu})\, \mathrm{sign}\varepsilon_n}{(\widetilde{\varepsilon}_{n-k} + \widetilde{\varepsilon}_n + i\mathbf{v}\mathbf{q})^2} + \frac{\theta(-\varepsilon_n \varepsilon_{n+\nu})}{(\omega_\nu + 1/\tau)^2} - \frac{\theta(-\varepsilon_n \varepsilon_{n+\nu})\, \theta(\varepsilon_n \varepsilon_{n-k})}{(\widetilde{\varepsilon}_{n-k} + \widetilde{\varepsilon}_n + i\mathbf{v}\mathbf{q})^2} \right] \right\rangle_{FS}. \tag{8.21}$$

This result is written in the most general form, which we will use later while discussing the case of temperatures far from the critical one. Being in the vicinity of the transition, in order to obtain the leading singular behavior, it suffices to restrict consideration to the term with $\Omega_k = 0$ [241]. This approximation corresponds to neglecting the contribution of dynamical fluctuations. Moreover, due to the singular structure of the propagator $L(\mathbf{q}, 0)$ in the vicinity of T_c one can neglect the Green function q-dependence (as it was already mentioned the characteristic momenta for the propagator are $q_{\mathrm{eff}}^{(\mathrm{pr})} \sim \xi_{\mathrm{GL}}^{-1}(\epsilon)$, while for the Green

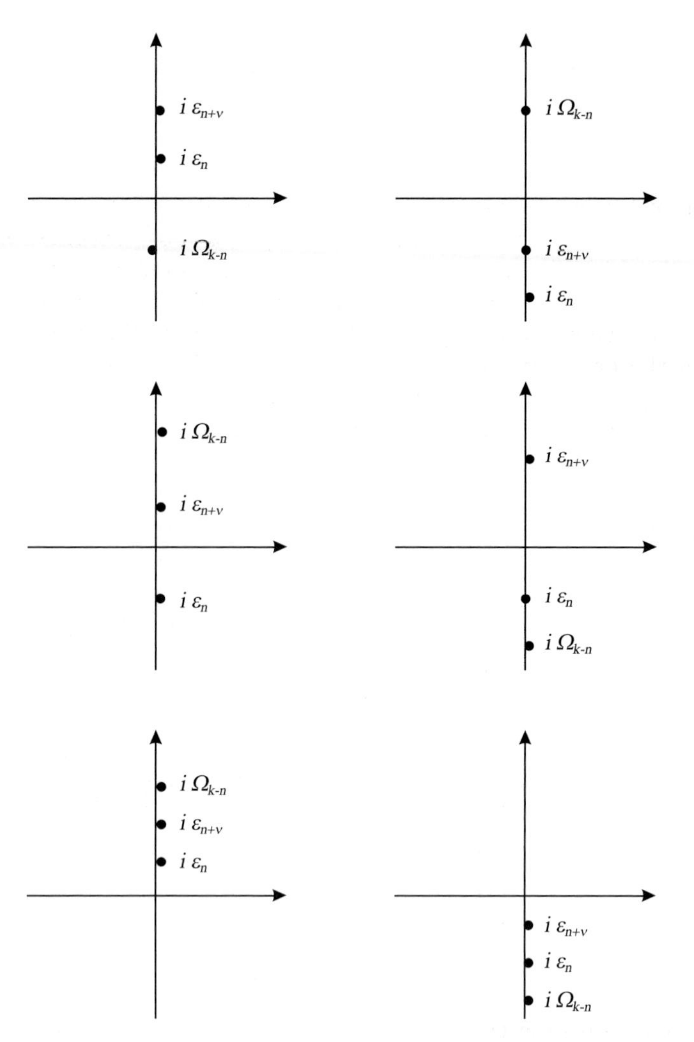

FIG. 8.3. The ξ-integration in (8.7): six possible dispositions of the poles in the complex plane of ξ.

function $q_{\text{eff}}^{(\text{GF})} \sim \max\left\{l^{-1}, l_T^{-1}\right\}$). The expression (8.19) can be now considerably simplified

$$\sum\nolimits_{\alpha\beta}^{(5)}(\mathbf{q},0,\omega_\nu) = -\pi\nu A_{\alpha\beta}v_F^2 T\left(\sum_{-\infty}^{-\nu-1} + \sum_{0}^{\infty}\right)\frac{1}{\left[|2\varepsilon_n| + \widehat{\mathcal{D}}q^2\right]^2}\frac{\operatorname{sign}\varepsilon_n}{(\widetilde{\varepsilon}_n + \widetilde{\varepsilon}_{n+\nu})}$$
$$-\pi\nu A_{\alpha\beta}v_F^2 T\sum_{-\nu}^{-1}\frac{|2\widetilde{\varepsilon}_n|^2}{\left[|2\varepsilon_n| + \widehat{\mathcal{D}}q^2\right]^2}\left[\frac{1}{(\omega_\nu + 1/\tau)^2} - \frac{1}{(2\widetilde{\varepsilon}_n)^2}\right],$$

where $A_{xx} = A_{yy} = 1, A_{zz} = (sJ/v_F)^2, A_{\alpha\neq\beta} = 0$ are the components of the tensor

$$A_{\alpha\beta} = 2\frac{\langle v_\alpha(\mathbf{p})v_\beta(\mathbf{p})\rangle_{FS}}{v_F^2},$$

calculated for the layer superconductor with spectrum (6.37).

The summation over the fermionic frequency ε_n can be performed in terms of the derivatives of the function ψ (see (B.11)). The summation over the negative n in the first and last sums can be transformed into a summation over the positive numbers by the substitution $n \to -n'$. Moreover further shift of the summation variable in the first sum by ν excludes it from the lower limit. Then one can decrease the power in the denominators presenting the corresponding fractions in the form of the derivatives and to perform the expansion in simple multipliers. As a result:

$$\begin{aligned}
T\sum\nolimits_{\alpha\beta}^{(5)}(\mathbf{q},0,\omega_\nu) = {} & \frac{\nu A_{\alpha\beta}v_F^2}{4}\left(\frac{\partial}{\partial\widehat{\mathcal{D}}q^2}\right)\frac{1}{\tau^{-1}-\omega_\nu-\widehat{\mathcal{D}}q^2} \\
& \times\left[\psi\left(\frac{1}{2}+\frac{\omega_\nu}{4\pi T}+\frac{1}{4\pi T\tau}\right)-\psi\left(\frac{1}{2}+\frac{\omega_\nu}{2\pi T}+\frac{\widehat{\mathcal{D}}q^2}{4\pi T}\right)\right] \\
& +\frac{\nu A_{\alpha\beta}v_F^2}{4}\left(\frac{\partial}{\partial\widehat{\mathcal{D}}q^2}\right)\frac{1}{\tau^{-1}+\omega_\nu-\widehat{\mathcal{D}}q^2} \\
& \times\left[\psi\left(\frac{1}{2}+\frac{\omega_\nu}{4\pi T}+\frac{1}{4\pi T\tau}\right)-\psi\left(\frac{1}{2}+\frac{\widehat{\mathcal{D}}q^2}{4\pi T}\right)\right] \\
& +\frac{\nu A_{\alpha\beta}v_F^2}{4\left(\omega_\nu+\tau^{-1}\right)^2}\left(1-\left(\omega_\nu+2\tau^{-1}-\widehat{\mathcal{D}}q^2\right)\frac{\partial}{\partial\widehat{\mathcal{D}}q^2}\right) \\
& \times\left[\psi\left(\frac{1}{2}+\frac{\omega_\nu}{2\pi T}+\frac{\widehat{\mathcal{D}}q^2}{4\pi T}\right)-\psi\left(\frac{1}{2}+\frac{\widehat{\mathcal{D}}q^2}{4\pi T}\right)\right].
\end{aligned}$$

The obtained expression is already analytic function of the ω_ν. Hence one can make analytic continuation $\omega_\nu \to -i\omega \to 0$ and expand the result over the small ω :

$$\begin{aligned}
T\sum\nolimits_{\alpha\beta}^{(5)R}(\mathbf{q},0,\omega) = {} & T\sum\nolimits_{\alpha\beta}^{(5)}(\mathbf{q},0,0)+i\omega\frac{\nu A_{\alpha\beta}v_F^2}{8\pi T}\left(\frac{\partial}{\partial\widehat{\mathcal{D}}q^2}\right)\frac{1}{\tau^{-1}-\widehat{\mathcal{D}}q^2} \\
& \times\left[\psi'\left(\frac{1}{2}+\frac{\widehat{\mathcal{D}}q^2}{4\pi T}\right)-\psi'\left(\frac{1}{2}+\frac{1}{4\pi T\tau}\right)\right] \\
& -i\omega\frac{\nu A_{\alpha\beta}v_F^2\tau^2}{8\pi T}\left[\psi'\left(\frac{1}{2}+\frac{\widehat{\mathcal{D}}q^2}{4\pi T}\right)\right. \\
& \left.-\frac{2\tau^{-1}-\widehat{\mathcal{D}}q^2}{4\pi T}\psi''\left(\frac{1}{2}+\frac{\widehat{\mathcal{D}}q^2}{4\pi T}\right)\right],
\end{aligned} \tag{8.22}$$

where

$$T\sum_{\alpha\beta}^{(5)}(\mathbf{q},0,0)=\frac{\nu A_{\alpha\beta}v_F^2}{2}\left(\frac{\partial}{\partial\widehat{\mathcal{D}}q^2}\right)\frac{1}{\tau^{-1}-\widehat{\mathcal{D}}q^2}$$
$$\times\left[\psi\left(\frac{1}{2}+\frac{1}{4\pi T\tau}\right)-\psi\left(\frac{1}{2}+\frac{\widehat{\mathcal{D}}q^2}{4\pi T}\right)\right].$$

It is possible to show [108] that the part independent on ω is canceled by the corresponding contributions coming from other diagrams (we already observed how such a term appeared in the AL contribution). This cancelation has a simple physical sense: above the critical temperature $Q(\omega=0)\equiv 0$. Calculating derivatives in (8.22) and then tending $\widehat{\mathcal{D}}q^2\to 0$ (our local approximation for Cooperons supposes $\widehat{\mathcal{D}}q^2\ll\tau^{-1}$) one finds:

$$T\sum_{\alpha\beta}^{(5)}(\mathbf{q},0,\omega)=-\frac{i\omega\nu\tau^2v_F^2}{8\pi T}A_{\alpha\beta}\left\{\psi'\left(\frac{1}{2}+\frac{1}{4\pi T\tau}\right)-\frac{3}{4\pi\tau T}\psi''\left(\frac{1}{2}+\frac{\widehat{\mathcal{D}}q^2}{4\pi T}\right)\right\}. \tag{8.23}$$

Substituting this expression to (8.18), performing the q_z-integration according to the relation (C.3) and cutting the logarithmic divergence at $\widehat{\mathcal{D}}q^2\sim T$ we obtain the contribution to conductivity from the DOS type diagram 5 [238, 240]:

$$Q_{\alpha\beta}^{(5)}(\omega_\nu)=i\omega\kappa_1(T\tau)\frac{\pi\eta_{(2)}e^2}{4s}A_{\alpha\beta}\int_{|\mathbf{q}|\leq\xi^{-1}}\frac{d^2\mathbf{q}}{(2\pi)^2}\frac{1}{\sqrt{\left(\epsilon+\eta_{(2)}q^2\right)\left(\epsilon+r+\eta_{(2)}q^2\right)}}$$
$$=i\omega\frac{e^2\kappa_1}{16s}A_{\alpha\beta}\ln\left(\frac{2}{\epsilon^{1/2}+(\epsilon+r)^{1/2}}\right), \tag{8.24}$$

where

$$\kappa_1=\frac{\tau^2v_F^2}{\pi^2\eta}\left[\psi'\left(\frac{1}{2}+\frac{1}{4\pi T\tau}\right)-\frac{3}{4\pi T\tau}\psi''\left(\frac{1}{2}\right)\right]. \tag{8.25}$$

In order to cut off the ultraviolet divergence in the q-integration we have introduced here a cut-off parameter $q_{\max}\sim\xi^{-1}=\eta_{(2)}^{-1/2}$ in complete agreement with section 2.7. Let us stress that in the framework of the phenomenological GL theory we attributed this cut-off to the breakdown of the GL approach at momenta as large as $q\sim\xi^{-1}$. The microscopic approach developed here allows to see how this cut-off appears in a natural way: the divergent short-wavelength contribution arising from GL-like fluctuation propagators is automatically restricted by the q-dependencies of the impurity vertices and Green's functions, which manifest themselves at the scale $q\sim l^{-1}$.

The topologically equivalent diagram 6 gives the same contribution as that from the diagram 5, hence:

$$\sigma_{\alpha\beta}^{(5+6)}=-\frac{e^2\kappa_1}{8s}A_{\alpha\beta}\ln\left(\frac{2}{\epsilon^{1/2}+(\epsilon+r)^{1/2}}\right). \tag{8.26}$$

In a similar manner, the equal contributions from diagrams 7 and 8 sum to

$$\sigma_{\alpha\beta}^{(7+8)} = -\frac{\pi e^2}{2s} A_{\alpha\beta}\kappa_2\eta_{(2)} \int_{|\mathbf{q}|\leq q_{\max}} \frac{d^2\mathbf{q}}{(2\pi)^2} \frac{1}{\left[(\epsilon+\eta_{(2)}\mathbf{q}^2)(\epsilon+r+\eta_{(2)}\mathbf{q}^2)\right]^{1/2}}$$
$$\approx -\frac{e^2\kappa_2}{8s} A_{\alpha\beta} \ln\left(\frac{2}{\epsilon^{1/2}+(\epsilon+r)^{1/2}}\right), \tag{8.27}$$

$$\kappa_2 = \frac{(v_F\tau)^2}{2\pi^3\eta T\tau}\psi''\left(\frac{1}{2}\right).$$

Comparing (8.26) and (8.27), we see that in the clean limit, the main contributions from the DOS fluctuations arise from diagrams 5 and 6. In the dirty limit, diagrams 7 and 8 are also important, having $-\frac{1}{3}$ times the value of diagrams 5 and 6, for both σ_{xx} and σ_{zz}. Diagrams 9 and 10 are not singular in $\epsilon << 1$ at all and can be neglected. The total DOS contribution to the in-plane and c-axis conductivity is therefore

$$\sigma_{\alpha\beta}^{\mathrm{DOS}} = -\frac{e^2}{2s}\kappa(T\tau)A_{\alpha\beta}\ln\left(\frac{2}{\epsilon^{1/2}+(\epsilon+r)^{1/2}}\right), \tag{8.28}$$

where

$$\kappa(T\tau) = \kappa_1+\kappa_2 = \frac{-\psi'\left(\frac{1}{2}+\frac{1}{4\pi\tau T}\right)+\frac{1}{2\pi\tau T}\psi''\left(\frac{1}{2}\right)}{\pi^2\left[\psi\left(\frac{1}{2}+\frac{1}{4\pi\tau T}\right)-\psi\left(\frac{1}{2}\right)-\frac{1}{4\pi\tau T}\psi'\left(\frac{1}{2}\right)\right]}$$
$$\to \begin{cases} 56\zeta(3)/\pi^4 \approx 0.691, & T\tau \ll 1 \\ 8\pi^2(T\tau)^2/[7\zeta(3)] \approx 9.384\,(T\tau)^2, & 1 \ll T\tau \ll 1/\sqrt{\epsilon} \end{cases} \tag{8.29}$$

is a function of τT only. We will see in the next section that in the case of in-plane conductivity it is doubled by the regular MT contribution. As it will be shown below, at the upper limit of the locality of the fluctuation theory ($T\tau \sim 1/\sqrt{\epsilon}$) the DOS contribution reaches the value of the more singular anomalous MT one and in the limit of $T\tau \to \infty$ they exactly eliminate each other.

8.3.3 *MT contribution*

We now consider another quantum correction to fluctuation conductivity which is called the MT contribution (diagram 2 of Fig. 8.1). It was firstly discussed by Maki [40] in a paper which appeared almost simultaneously with the paper of Aslamazov and Larkin [36]. Both of these articles gave rise to the microscopic theory of fluctuations in superconductor. Maki found that, in spite of the seeming weaker singularity of diagram 2 with respect to the AL one (it contains one propagator only, while the AL one contains two of them) it can contribute to conductivity comparably or even stronger than AL one.

In his original paper Maki found that in $3D$ case his fluctuation correction is four times larger than the AL one. In the $2D$ case the result was striking: the found contribution to fluctuation conductivity simply diverged. Later Thompson recognized [41] that the infrared divergence of the Cooper pair center of mass momentum integration in Maki diagram can be cut off by electron phase-breaking scattering on paramagnetic impurities τ_s. Today we know that there are many sources of such pair-breaking: paramagnetic impurities, phonons, fluctuation Cooper pairing itself, etc. We will discuss them in the next subsection while below we will demonstrate the technical details of the MT contribution calculation in the assumption of the presence of the finite phase-breaking time $\tau_\varphi \gg \tau$. and will compare it with other contributions. We will see, that the divergence of Maki diagram can also be removed by the minimal quasi-two-dimensionality of the electron spectrum.

Although the MT contribution to the in-plane conductivity is expected to be important in the case of weak pair-breaking, experiments on HTS have shown that the excess in-plane conductivity can usually be explained in terms of the fluctuation paraconductivity alone. Two possible explanations can be found for this fact. The first one is that the pair-breaking in these materials is not weak. The second is related to the d-wave symmetry of pairing which kills the anomalous MT process [248, 249]. We will consider below the case of s-pairing, where the MT process is well pronounced.

The appearance of the anomalously large MT contribution is nontrivial and worth being discussed. We consider the scattering lifetime τ and the pair-breaking lifetime τ_φ to be arbitrary, but satisfying $\tau_\varphi > \tau$. In accordance with diagram 2 of Fig.8.1 the analytic expression for the MT contribution to the electromagnetic response tensor can be written as

$$Q_{\alpha\beta}^{\mathrm{MT}}(\omega_\nu) = 2e^2 T \sum_{\Omega_k} \int \frac{d^3\mathbf{q}}{(2\pi)^3} L(\mathbf{q}, \Omega_k) I_{\alpha\beta}(\mathbf{q}, \Omega_k, \omega_\nu), \tag{8.30}$$

where

$$I_{\alpha\beta}^{\mathrm{MT}}(\mathbf{q}, \Omega_k, \omega_\nu) = T \sum_{\varepsilon_n} \lambda(\mathbf{q}, \varepsilon_{n+\nu}, \Omega_{k-n-\nu}) \lambda(\mathbf{q}, \varepsilon_n, \Omega_{k-n}) \int \frac{d^3\mathbf{p}}{(2\pi)^3} v_\alpha(\mathbf{p}) v_\beta(\mathbf{q}-\mathbf{p})$$
$$\times G(\mathbf{p}, \varepsilon_{n+\nu}) G(\mathbf{p}, \varepsilon_n) G(\mathbf{q}-\mathbf{p}, \Omega_{k-n-\nu}) G(\mathbf{q}-\mathbf{p}, \Omega_{k-n}). \tag{8.31}$$

In the vicinity of T_c, it is possible to restrict consideration to the static limit of the MT diagram, simply by setting $\Omega_k = 0$ in (8.30). Although the dynamic effects can be important for the longitudinal fluctuation conductivity faraway the critical temperature, the static approximation is correct close to T_c, as was shown in [108, 271]. The main q-dependence in (8.30) arises from the propagator and vertices λ. That is why we can assume $q = 0$ in the Green functions and to calculate the electron momentum integral passing, as usual, to a $\xi(\mathbf{p})$ integration:

$$I_{\alpha\beta}^{\mathrm{MT}}(q, 0, \omega_\nu) = \pi\nu \left\langle v_\alpha(p) v_\beta(q-p) \right\rangle_{\mathrm{FS}} T \sum_{\varepsilon_n} \frac{1}{\left(|2\varepsilon_{n+\nu}| + \widehat{\mathcal{D}} q^2\right)}$$

$$\times \frac{1}{\left(|2\varepsilon_n| + \widehat{\mathcal{D}}q^2\right)} \frac{1}{|\widetilde{\varepsilon}_{n+\nu}| + |\widetilde{\varepsilon}_n|}. \tag{8.32}$$

In evaluating the sum over the Matsubara frequencies ε_n in (8.32) it is useful to split it into the two parts. In the first ε_n belongs to the domains $]-\infty, -\omega_\nu[$ and $[0, \infty[$, which finally give two equal contributions. This gives rise to the *regular* part of the MT diagram. The second, *anomalous*, part of the MT diagram arises from the summation over ε_n in the domain $[-\omega_\nu, 0[$. The further analytic continuation of the contribution of this interval to $Q^{\mathrm{MT}}_{\alpha\beta}(\omega_\nu)$ leads to the appearance, side by side with already present fluctuation propagator, of an additional diffusive pole. Namely this pole, increasing the temperature singularity, generates the famous anomalous MT contribution to conductivity comparable in its temperature singularity with the AL one. Let us stress, that as we will show below, this anomalous interval contributes to conductivity not only by the anomalous MT term, but it changes the contribution of the regular interval too.

In order to demonstrate the details of calculation let us introduce the notations

$$I^{\mathrm{MT}}_{\alpha\beta}(q, 0, \omega_\nu) = \nu \left\langle v_\alpha(\mathbf{p}) v_\beta(\mathbf{q}-\mathbf{p}) \right\rangle_{FS} \left[\Sigma^{(\mathrm{reg1})} + \Sigma^{(an)} + \Sigma^{(\mathrm{reg2})}\right],$$

where

$$\Sigma^{(\mathrm{reg1})} = 2\pi T \sum_{n=0}^{\infty} \frac{1}{\left(2\varepsilon_{n+\nu} + \widehat{\mathcal{D}}q^2\right)} \frac{1}{\left(2\varepsilon_n + \widehat{\mathcal{D}}q^2\right)} \frac{1}{2\varepsilon_n + \omega_\nu + \tau^{-1}}$$

and

$$\begin{aligned}
\Sigma^{(\mathrm{an})} + \Sigma^{(\mathrm{reg2})} &= \frac{\pi T}{\omega_\nu + \tau^{-1}} \sum_{n=-\nu}^{-1} \frac{1}{\left(2\varepsilon_{n+\nu} + \widehat{\mathcal{D}}q^2\right)} \frac{1}{\left(-2\varepsilon_n + \widehat{\mathcal{D}}q^2\right)} \\
&= \frac{\pi T}{\omega_\nu + \tau^{-1}} \frac{1}{2\left(\omega_\nu + \widehat{\mathcal{D}}q^2\right)} \sum_{n=-\nu}^{-1} \left[\frac{1}{2\varepsilon_{n+\nu} + \widehat{\mathcal{D}}q^2} + \frac{1}{-2\varepsilon_n + \widehat{\mathcal{D}}q^2}\right] \\
&= \frac{\pi T}{\omega_\nu + \tau^{-1}} \frac{1}{2\left(\omega_\nu + \widehat{\mathcal{D}}q^2\right)} \left[\sum_{n=0}^{\nu-1} \frac{1}{2\varepsilon_n + \widehat{\mathcal{D}}q^2} + \sum_{n=-\nu}^{-1} \frac{1}{-2\varepsilon_n + \widehat{\mathcal{D}}q^2}\right].
\end{aligned} \tag{8.33}$$

The limits of summation in the first sum do not depend on ω_ν, so it is an analytic function of this argument and can be continued to the upper half-plane of the complex frequency. Let us calculate it expanding in simple multipliers:

$$\Sigma^{(\mathrm{reg1})}(q, \omega_\nu) = \frac{\pi T}{\omega_\nu} \left[\frac{1}{\omega_\nu + \tau^{-1} - \widehat{\mathcal{D}}q^2} \sum_{n=0}^{\infty} \frac{1}{\left(2\varepsilon_n + \widehat{\mathcal{D}}q^2\right)}\right.$$

$$
\left. - \frac{1}{-\omega_\nu + \tau^{-1} - \widehat{\mathcal{D}}q^2} \sum_{n=0}^{\infty} \frac{1}{\left(2\varepsilon_{n+\nu} + \widehat{\mathcal{D}}q^2\right)} \right]
+ \frac{2\pi T}{\left(\tau^{-1} - \widehat{\mathcal{D}}q^2\right)^2 - \omega_\nu^2} \sum_{n=0}^{\infty} \frac{1}{2\varepsilon_n + \omega_\nu + \tau^{-1}}.
$$

Carrying out the summation in terms of the already familiar digamma function one can easily find:

$$
\Sigma^{(\mathrm{reg1})} = \frac{1}{4\omega_\nu} \left[\frac{\psi\left(\frac{1}{2} + \frac{\omega_\nu}{2\pi T} + \frac{\widehat{\mathcal{D}}q^2}{4\pi T}\right)}{-\omega_\nu + \tau^{-1} - \widehat{\mathcal{D}}q^2} - \frac{\psi\left(\frac{1}{2} + \frac{\widehat{\mathcal{D}}q^2}{4\pi T}\right)}{\omega_\nu + \tau^{-1} - \widehat{\mathcal{D}}q^2} \right]
- \frac{1}{2} \frac{\psi\left(\frac{1}{2} + \frac{\omega_\nu}{4\pi T} + \frac{1}{4\pi T\tau}\right)}{\left(\tau^{-1} - \widehat{\mathcal{D}}q^2\right)^2 - \omega_\nu^2}. \tag{8.34}
$$

Now one can make an analytic continuation $\omega_\nu \to -i\omega$, expand the result over ω/T and, as $\omega \to 0$, find

$$
\Sigma^{(\mathrm{reg1})} = \frac{1}{2} \frac{\psi\left(\frac{1}{2} + \frac{\widehat{\mathcal{D}}q^2}{4\pi T}\right) - \psi\left(\frac{1}{2} + \frac{1}{4\pi T\tau}\right)}{\left(\tau^{-1} - \widehat{\mathcal{D}}q^2\right)^2} + \frac{1}{8\pi T} \frac{\psi'\left(\frac{1}{2} + \frac{\widehat{\mathcal{D}}q^2}{4\pi T}\right)}{\tau^{-1} - \widehat{\mathcal{D}}q^2}
+ \frac{i\omega}{8\pi T} \frac{1}{\left(\tau^{-1} - \widehat{\mathcal{D}}q^2\right)^2} \left[\psi'\left(\frac{1}{2} + \frac{1}{4\pi T\tau}\right) - \psi'\left(\frac{1}{2} + \frac{\widehat{\mathcal{D}}q^2}{4\pi T}\right)
\left. - \frac{1}{4\pi T} \left(\tau^{-1} - \widehat{\mathcal{D}}q^2\right) \psi''\left(\frac{1}{2} + \frac{\widehat{\mathcal{D}}q^2}{4\pi T}\right) \right]. \tag{8.35}
$$

It is a time to recall that for evaluation of (8.32) we have used the impure vertices taken in the local approximation ($q \ll l^{-1}$). This means that conservation of the value $\widehat{\mathcal{D}}q^2$ side by side with τ^{-1} in Eq.(8.35) would be exceeding of the accuracy of our calculations, hence it can be omitted. The value of characteristic momentum, which defines the domain of convergeability of the final integral of the propagator $L(\mathbf{q}, 0)$ in (8.30) (analogously to (8.27) is determined by the pole of propagator. As a result

$$
\Sigma^{(\mathrm{reg1})} = -\frac{\tau^2}{2} \left[\psi\left(\frac{1}{2} + \frac{1}{4\pi T\tau}\right) - \psi\left(\frac{1}{2}\right) - \frac{1}{4\pi\tau T}\psi'\left(\frac{1}{2}\right) \right]
+ \frac{i\omega\tau^2}{8\pi T} \left[\psi'\left(\frac{1}{2} + \frac{1}{4\pi T\tau}\right) - \psi'\left(\frac{1}{2}\right) - \frac{1}{4\pi T\tau}\psi''\left(\frac{1}{2}\right) \right]. \tag{8.36}
$$

The appearance of the constant in $Q_{\alpha\beta}(\omega_\nu)$ was already discussed in the case of the AL contribution and, as was mentioned there, it is canceled with the similar contributions of the other diagrams [108]; we will not consider it any more.

Now let us calculate the sum in Eq. (8.33). Expanding the fraction in the simple multipliers, shifting the variable of summation $n \to n+\nu$ in the first sum and passing to the new summation variable $n' \to -n$ in the second, one can find that their contributions are equal and finally get

$$\Sigma^{(\mathrm{an})} + \Sigma^{(\mathrm{reg2})} = \frac{\pi T}{\omega_\nu + \tau^{-1}} \frac{1}{2\left(\omega_\nu + \widehat{\mathcal{D}}q^2\right)} \sum_{n=-\nu}^{-1} \left[\frac{1}{2\varepsilon_{n+\nu} + \widehat{\mathcal{D}}q^2} + \frac{1}{-2\varepsilon_n + \widehat{\mathcal{D}}q^2} \right]$$

$$= \frac{\pi T}{\omega_\nu + \tau^{-1}} \frac{1}{\left(\omega_\nu + \widehat{\mathcal{D}}q^2\right)} \sum_{n=0}^{\nu-1} \frac{1}{2\varepsilon_n + \widehat{\mathcal{D}}q^2} = \frac{1}{4}\left(\frac{1}{\omega_\nu + \tau^{-1}}\right)$$

$$\times \frac{1}{\omega_\nu + \widehat{\mathcal{D}}q^2} \left[\psi\left(\frac{1}{2} + \frac{2\omega_\nu + \widehat{\mathcal{D}}q^2}{4\pi T}\right) - \psi\left(\frac{1}{2} + \frac{\widehat{\mathcal{D}}q^2}{4\pi T}\right) \right]. \quad (8.37)$$

Making the analytic continuation $i\omega_\nu \to \omega \to 0$

$$\Sigma^{(\mathrm{an})} + \Sigma^{(\mathrm{reg2})} = -\frac{i\omega\tau}{8\pi T} \frac{1}{-i\omega + \widehat{\mathcal{D}}q^2} \psi'\left(\frac{1}{2} + \frac{\widehat{\mathcal{D}}q^2}{4\pi T}\right) \quad (8.38)$$

and taking into account that in the further q-integration of $I_{\alpha\beta}^{(an)R}(q, \omega \to 0)$, the important range of momenta is $\widehat{\mathcal{D}}q^2 \ll T$, one can the part of the sum corresponding to the so-called anomalous (singular) MT contribution

$$\Sigma^{(\mathrm{an})\mathrm{R}}(q,\omega) = -\frac{i\pi\omega\tau}{16T} \frac{1}{-i\omega + \widehat{\mathcal{D}}q^2}. \quad (8.39)$$

Let us attract the readers' attention to the fact, that the expansion of the digamma function (8.38) over $\widehat{\mathcal{D}}q^2/4\pi T$ contributes to the regular part:

$$\Sigma^{(\mathrm{reg2})} = -\frac{i\omega\tau}{32\pi^2 T^2} \psi''\left(\frac{1}{2}\right). \quad (8.40)$$

Finally

$$\Sigma^{(\mathrm{reg})\mathrm{R}}(q,\omega) = \Sigma^{(\mathrm{reg1})} + \Sigma^{(\mathrm{reg2})} = \frac{i\omega\tau^2}{8\pi T} \left[\psi'\left(\frac{1}{2} + \frac{1}{4\pi T\tau}\right) \right.$$

$$\left. -\psi'\left(\frac{1}{2}\right) - \frac{1}{2\pi T\tau}\psi''\left(\frac{1}{2}\right) \right]. \quad (8.41)$$

and we can proceed with the angular averaging and lastly the q-integration. Because of the considerable difference in the angular averaging of the different tensor components we discuss the MT contribution to the in-plane and out of plane conductivities separately.

Taking into account that $\langle v_x(\mathbf{p})v_x(\mathbf{q}-\mathbf{p})\rangle_{\mathrm{FS}} = -v_F^2/2$ one can find that the calculation of the regular part of MT diagram to the in-plane conductivity is completely similar to the corresponding DOS contribution and here we list the final result [240] only:

$$\sigma_{xx}^{\mathrm{MT(reg)}} = -\frac{e^2}{2s}\tilde{\kappa}\ln\left(\frac{2}{\epsilon^{1/2}+(\epsilon+r)^{1/2}}\right), \tag{8.42}$$

where

$$\tilde{\kappa}(T\tau) = \frac{-\psi'\left(\frac{1}{2}+\frac{1}{4\pi\tau T}\right)+\psi'\left(\frac{1}{2}\right)+\frac{1}{2\pi T\tau}\psi''\left(\frac{1}{2}\right)}{\pi^2\left[\psi\left(\frac{1}{2}+\frac{1}{4\pi\tau T}\right)-\psi\left(\frac{1}{2}\right)-\frac{1}{4\pi\tau T}\psi'\left(\frac{1}{2}\right)\right]}$$
$$\to\begin{cases} 56\zeta(3)/\pi^4\approx 0.691, & T\tau\ll 1,\\ \dfrac{8}{\pi}T\tau, & 1\ll T\tau\ll 1/\sqrt{\epsilon}, \end{cases} \tag{8.43}$$

is a function only of τT. We note that this regular MT term is negative, as is the overall DOS contribution. Moreover, one can find, that in the dirty case $\tilde{\kappa}(T\tau\ll 1)=\kappa(T\tau\ll 1)$, hence the correct accounting for the regular MT contribution doubles the total DOS contribution (originating from all four diagrams 5–8).

For the anomalous part of the in-plane MT contribution we have:

$$\sigma_{xx}^{\mathrm{MT(an)}} = 8e^2\eta_{(2)}T\int\frac{d^3q}{(2\pi)^3}\frac{1}{[1/\tau_\varphi+\widehat{\mathcal{D}}q^2][\epsilon+\eta_{(2)}\mathbf{q}^2+\frac{r}{2}(1-\cos q_z s)]}$$
$$=\frac{e^2}{4s(\epsilon-\gamma_\varphi)}\ln\left(\frac{\epsilon^{1/2}+(\epsilon+r)^{1/2}}{\gamma_\varphi^{1/2}+(\gamma_\varphi+r)^{1/2}}\right), \tag{8.44}$$

where, according to [41], the infrared divergence for the purely $2D$ case ($r=0$) is cut-off at $\mathcal{D}q^2\sim 1/\tau_\varphi$. The dimensionless parameter

$$\gamma_\varphi = \frac{2\eta}{v_F^2\tau\tau_\varphi}\to\frac{\pi}{8T\tau_\varphi}\begin{cases}1, & T\tau\ll 1,\\ 7\zeta(3)/\left(2\pi^3T\tau\right), & 1\ll T\tau\ll 1/\sqrt{\epsilon},\end{cases}$$

is introduced for simplicity. If $r\neq 0$ the MT contribution turns out to be finite even with $\tau_\varphi=\infty$. Comparison of the expressions (8.16) and (8.44) indicates that in the weak pair-breaking limit, the MT diagram makes an important contribution to the longitudinal fluctuation conductivity: it is four times larger than the AL contribution in the $3D$ regime, and even logarithmically exceeds it in the $2D$ regime above $T_{\mathrm{LD})}$. For finite pair-breaking, however, the MT contribution is greatly reduced in magnitude.

We now consider the calculation of the MT contribution to the transverse conductivity. The explicit expressions for $v_z(p)$ and $v_z(q-p)$ (see Eq. (8.5)), result in $\langle v_x(p)v_x(q-p)\rangle_{FS}=\frac{1}{2}J^2s^2\cos q_z s$. We take the limit $J\tau<<1$ in evaluating the remaining integrals, which may then be performed exactly.

The regular part of the MT contribution to the transverse conductivity is

$$\sigma_{zz}^{MT(\text{reg})} = -\frac{e^2 s^2 \pi r \tilde{\kappa}(T\tau)}{4} \int \frac{d^3 q}{(2\pi)^3} \frac{\cos q_z s}{\epsilon + \eta_{(2)} \mathbf{q}^2 + \frac{r}{2}(1 - \cos q_z s)}$$
$$= -\frac{e^2 s}{16 \xi_{xy}^2} \tilde{\kappa}(T\tau) \left[(\epsilon + r)^{1/2} - \epsilon^{1/2} \right]^2 .$$

This term is smaller in magnitude than is the DOS one, and therefore makes a relatively small contribution to the overall fluctuation conductivity. In the $3D$ regime (which takes place below $T_{\text{LD})}$) it is proportional to J^2, while in the $2D$ regime (above $T_{\text{LD})}$), it is proportional to J^4.

For the anomalous part of the MT contribution one can find

$$\sigma_{zz}^{MT(\text{an})} = \frac{\pi e^2 J^2 s^2 \tau}{4} \int \frac{d^3 q}{(2\pi)^3} \frac{\cos q_z s}{[1/\tau_\varphi + \widehat{\mathcal{D}} q^2][\epsilon + \eta_{(2)} \mathbf{q}^2 + \frac{r}{2}(1 - \cos q_z s)]}$$
$$= \frac{\pi e^2 s}{4(\epsilon - \gamma_\varphi)} \int \frac{d^2 \mathbf{q}}{(2\pi)^2} \left[\frac{\gamma_\varphi + \eta_{(2)} \mathbf{q}^2 + r/2}{\left[(\gamma_\varphi + \eta_{(2)} \mathbf{q}^2)(\gamma_\varphi + \eta_{(2)} \mathbf{q}^2 + r) \right]^{1/2}} - \frac{\epsilon + \eta_{(2)} \mathbf{q}^2 + r/2}{\left[(\epsilon + \eta_{(2)} \mathbf{q}^2)(\epsilon + \eta_{(2)} \mathbf{q}^2 + r) \right]^{1/2}} \right]$$
$$= \frac{e^2 s}{16 \xi_{xy}^2} \left(\frac{\gamma_\varphi + r + \epsilon}{[\epsilon(\epsilon + r)]^{1/2} + [\gamma_\varphi(\gamma_\varphi + r)]^{1/2}} - 1 \right). \tag{8.45}$$

In examining the asymptotics of the expression (8.45), it is useful to consider the cases of weak ($\gamma_\varphi \ll r, \Longleftrightarrow J^2 \tau \tau_\varphi \gg 1$) and strong ($\gamma_\varphi \gg r, \Longleftrightarrow J^2 \tau \tau_\varphi \ll 1$) pair-breaking separately.[40] For weak pair-breaking, we have

$$\sigma_{zz}^{MT(\text{an})} \to \frac{e^2 s}{16 \xi_{xy}^2} \begin{cases} \sqrt{r/\gamma_\varphi}, \ \epsilon \ll \gamma_\varphi \ll r \\ \sqrt{r/\epsilon}, \ \gamma_\varphi \ll \epsilon \ll r \\ r/(2\epsilon), \ \gamma_\varphi \ll r \ll \epsilon \end{cases} .$$

In this case, there is the usual $3D$ to $2D$ crossover in the anomalous MT contribution at T_{LD}. There is an additional crossover at T_φ (where $T_c < T_\varphi < T_{\text{LD}}$), characterized by $\epsilon(T_\varphi) = \gamma_\varphi$, below which the anomalous MT term saturates. Below T_{LD}, in the $3D$ regime, the MT contribution is proportional to J, while above T_{LD},in the $2D$ regime, it is proportional to J^2.

For strong pair-breaking

$$\sigma_{zz}^{\text{MT(an)}} \to \frac{e^2 s}{32 \xi_{xy}^2} \begin{cases} r/\gamma_\varphi, & \epsilon \ll r \ll \gamma_\varphi, \\ r^2/(4\gamma_\varphi \epsilon), & r \ll \min\{\gamma_\varphi, \epsilon\}. \end{cases}$$

[40] Physically the value $J^2 \tau$ characterizes the effective interlayer tunneling rate [240, 250]. When $1/\tau_\phi \ll J^2 \tau \ll 1/\tau$, the quasiparticle scatters many times before tunneling to the neighboring layer, and the pairs live long enough for them to tunnel coherently. When $J^2 \tau \ll 1/\tau_\phi$, the pairs decay before both paired quasiparticles tunnel.

In this case, the $3D$ regime (below T_{LD}) is not singular at all, and the anomalous MT contribution is proportional to J^2, rather than J for weak pair-breaking. In the $2D$ regime, it is proportional to J^4 for strong pair-breaking, as opposed to J^2 for weak pair-breaking. In addition, the overall magnitude of the anomalous MT contribution with strong pair-breaking is greatly reduced from that for weak pair-breaking.

Let us now compare the regular and anomalous MT contributions. Since these contributions are opposite in sign, it is important to determine which will dominate. For the in-plane resistivity, the situation is straightforward: the anomalous part always dominates over the regular and the latter can be neglected. The case of c-axis conductivity requires more discussion. Since we expect $\tau_\varphi \geq \tau$, strong pair-breaking is likely in the dirty limit. When the pair-breaking is weak, the anomalous term is always of lower order in J than the regular term, so the regular term can be neglected. This is true for both the clean and dirty limits. The most important regime for the regular MT term is the dirty limit with strong pair-breaking. In this case, when $\tau_\varphi T \sim 1$, the regular and anomalous terms are comparable in magnitude. In short, it is usually a good approximation to neglect the regular term, except in the dirty limit with relatively strong pair-breaking and only for the out-of-plane conductivity.

Finally let us mention that the contributions from the two other diagrams of the MT type (diagrams 3 and 4 of Fig. 8.1) in the vicinity of critical temperature can be omitted: one can check that they have an additional square of the Cooper pair center of mass momentum q in the integrand of q-integration with respect to diagram 2 and hence turn out to be less singular in ϵ with respect to the anomalous contribution. Regarding the contribution of the diagrams 3 and 4 in comparison with $\sigma_{xx}^{\mathrm{MT(reg)}}$ it has the same singularity in ϵ but in the dirty case turns out to be smaller by the parameter $\tau T \ll 1$. In the clean case it does not play any essential role in comparison with the DOS contribution.

8.3.4 *Phase-breaking time τ_φ*

Since the moment of its discovery the MT contribution became the subject of intense controversy. Maki himself found the infrared divergence of the contribution to fluctuation conductivity proposed by him. This paradox was, at least at the level of recipe, resolved by Thompson [41]: he proposed to cut-off the infrared divergence in the momentum integration by introduction of the finite length l_s of electron scattering on paramagnetic impurities. In the further papers of Patton [251], Keller and Korenmann [252] it was cleared up that the presence of paramagnetic impurities or other external phase-breaking sources is not necessary: the fluctuation Cooper pairing of two electrons results in a change of the quasiparticle phase itself and the corresponding phase-breaking time τ_φ appears as the natural cut-off parameter of the MT divergence in the strictly $2D$ case.

Let us come back to the discussion of the physical origin of the MT correction to conductivity. As was already mentioned this correction appears above T_c due to the Andreev reflection of electrons on virtual Cooper pairs. In order to form a virtual Cooper pair two electrons have to first transit to almost the same orbital quantum states (their spin quantum states must be opposite), i.e. form a Cooperon. Any process which withdraws electrons from such quantum states or at least partially destroys their coherence (breaks down the wave function phase) aggravates the virtual Cooper pair formation.

The same Cooperons are responsible for appearance of the weak-localization correction (WL) to conductivity of a normal impure metal [253]. The transition from the MT correction to the WL one at temperatures $T \sim T_c$ will be discussed below (see 8.6). In the $2D$ case the WL correction diverges at zero temperature in the same logarithmic way as does the MT one near T_c, phase-breaking in the same way removes this divergence. The intensity of such processes is characterized by the phase-breaking time τ_φ.

Among the mechanisms of such phase-breaking we can mention the electron scattering by paramagnetic impurities; by phonons; Coulomb interaction; electron–electron interaction in Cooper channel (scattering by superconducting fluctuations); Andreev reflection from superconducting drops in normal metal. The scattering of electron by elastic impurities changes its quantum number (e.g. momentum) in the confines of the initial degenerated quantum state but it does not result in a transition between two states with the different energy hence does not contribute to τ_φ.

Magnetic impurities induce the transitions between electron states with different spin projections. In the absence of a magnetic field such transitions do not change the electron energy (or change it weakly [254]) but destroy the phase coherence of the electron state. The corresponding τ_φ coincides with the spin flip collision time [255]:

$$1/\tau_\varphi^{(s)} = \tau_s^{-1} = \frac{n_i m p_F}{(2\pi)^2} \left(\frac{S(S+1)}{3} \right) \int |U(\mathbf{p} - \mathbf{p}')|^2 d\Omega_{p'}.$$

(compare with Eq. (6.21)).

The electron–phonon scattering results in transitions between the states corresponding to different energy of electron. The electron–phonon phase-breaking time is determined by the formula

$$1/\tau_\varphi^{(ph)} = \tau_{e-ph}^{-1} = \frac{7\zeta(3)\pi g_{e-ph} T^3}{2(sp_F)^2} \sim \frac{T^3}{\omega_D^2},$$

where s is the sound velocity and g_{e-ph} is the constant of electron–phonon interaction.

The Coulomb interaction of electrons also changes both their phases and energies. The corresponding values are related:

$$\delta\varphi = \int \delta\varepsilon dt.$$

From this relation one can see that due to the long range character of the Coulomb interaction corresponding scattering time is large and therefore the

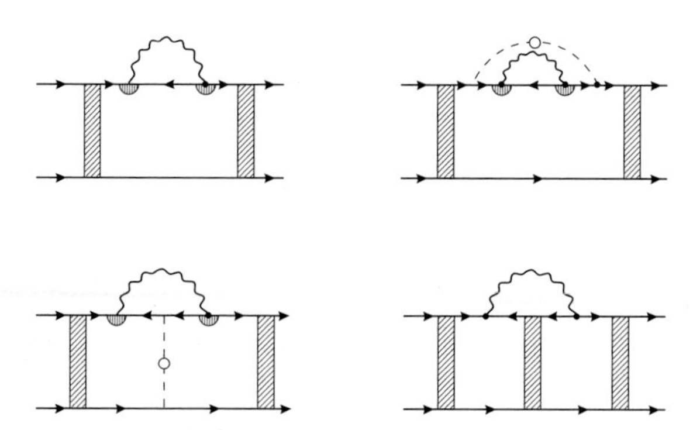

FIG. 8.4. Quantum corrections to Cooperon vertex C.

phase changes faster than energy. The thorough study of this problem [256] demonstrated that in $2D$ case the characteristic energy relaxation time $\tau_{\varepsilon}^{(\mathrm{Coul})} \sim p_F l/T$ logarithmically exceeds $\tau_{\varphi}^{(\mathrm{Coul})}$:

$$1/\tau_{\varphi}^{(\mathrm{Coul})} = \frac{T}{2\pi\hbar G_{\square}} \ln \pi G_{\square} = \frac{T}{p_F l} \ln \frac{\pi p_F l}{2}, \tag{8.46}$$

where the dimensionless conductance $G_{\square}$ is determined by Eqs. (2.88)–(2.89).

In superconducting metal at temperatures close to T_c the scattering by superconducting fluctuations becomes the main phase-breaking mechanism [251,252]. The process of Andreev reflection increases the phase relaxation. Indeed, in the process of Andreev reflection electron is scattered in the form of hole and in this way is excluded from the interference process. It again comes into the game only after its inverse transformation to electron as the result of one of the subsequent Andreev reflections. The increase of WL can be also treated as the Andreev reflection, in the same way as it was done in the qualitative treatment of the MT contribution at the beginning of this chapter. The fact that the order parameter inhomogeneities operate as magnetic impurities will be confirmed later (see 12.3).[41]

The effect of superconducting fluctuations on the phase-breaking time was studied in [257]. The authors took into account first fluctuation corrections to the Cooperon (see diagrams in Fig. 8.4). The corresponding expression was found as

$$1/\tau_{\varphi}^{(\mathrm{fl})}(\epsilon, \varepsilon) = -8\pi \left(\nu\tau^2\right)^2 \int \frac{d^D q}{(2\pi)^D} \int_{-\infty}^{+\infty} d\Omega \coth \frac{\Omega}{2T} \operatorname{Im} L^R(q, i\Omega)$$

[41] Andreev reflection on the static fluctuations of the order parameter renormalizes the Cooperon and diffuson in different ways. In the case of the diffuson it contributes both to the Green function and to vertex renormalization. As a result the diffusion conserves, the MT process only renormalizes the diffusion coefficient. In the case of the Cooperon the fluctuation correction topologically can be constructed only for the Green function, the corresponding vertex is not renormalized by fluctuations.

$$\times \lambda^2 (q, \varepsilon_n, i\Omega - \varepsilon_n) \left(i\Omega + \mathcal{D}q^2\right)\Big|_{\varepsilon_n \to -i\varepsilon}, \tag{8.47}$$

where ε is the electron energy counted from the Fermi level (we have omitted here the irrelevant terms). In the limit of a weak pair-breaking ($1/\tau_\varphi^{(\mathrm{fl})} \ll T\epsilon \ll T$)

$$1/\tau_\varphi^{(\mathrm{fl})}(\epsilon) = 1/\tau_\varphi^{(\mathrm{fl})}(\epsilon, \varepsilon \to 0) = \frac{16}{\pi} \frac{T^2 \tau_{\mathrm{GL}}}{p_F l} \ln 2. \tag{8.48}$$

Close to transition $T\tau_{\mathrm{GL}} \gg 1$ hence $\tau_\varphi^{(\mathrm{fl})} \ll \tau_\varphi^{(\mathrm{Coul})} \sim p_F l / T$. As far as the temperature tends to the transition point $\tau_\varphi^{(\mathrm{fl})}$, being proportional to $T - T_c$, decreases: superconducting fluctuations become significant sources of the phase relaxation. Such growth of the phase relaxation was found experimentally [258, 259].

As it follows from Eq. (8.44) the MT contribution saturates at $T\epsilon\tau_\varphi \sim 1$ and does not grow anymore with the further decrease of temperature. Equation (8.44) was obtained by Thompson in the assumption of the constant phase-breaking rate, generated by paramagnetic impurities. The statement concerning the saturation of the MT contribution close to T_c ($\epsilon \lesssim 1/\tau_\varphi$) qualitatively remains also valid for the case of phase relaxation induced by electron scattering by superconducting fluctuations. Yet the explicit form of the functional dependence $\sigma^{(\mathrm{fl})}(\epsilon)$ in the region $T\epsilon\tau_\varphi \lesssim 1$ is modified and there are two reasons for this change.

The first one consists in the dependence of $\tau_\varphi^{(\mathrm{fl})}$ on the electron energy ε_n [260]. Equation (8.48) was derived as the result of analytic continuation over energy with the further assumption that $i\varepsilon_n \to \varepsilon \ll T\epsilon$. Let us understand which ε are essential for calculation of $\Sigma^{(\mathrm{an})}$ in Eq. (8.37). In order to conduct a self-consistent treatment of the problem it would be necessary to add in Eq. (8.37) $1/\tau_\varphi^{(\mathrm{fl})}$ side by side with $\mathcal{D}q^2$. After the analytic continuation one can see that the essential $\varepsilon \sim \mathcal{D}q^2 \sim 1/\tau_\varphi^{(\mathrm{fl})}$. It is why in the region $T\epsilon\tau_\varphi \gg 1$ one can omit the energy dependence of $\tau_\varphi^{(\mathrm{fl})}(\epsilon, \varepsilon)$ and the validity of the formula (8.48) is justified. But it is not valid when temperature becomes so close to T_c that $T\epsilon\tau_\varphi \lesssim 1$.

The second reason briefly can be formulated as follows. It is the value $\tau_\varphi^{(\mathrm{fl})}(\epsilon)$ which leads to saturation of the MT correction close to T_c. But one can see that the value of $\tau_\varphi^{(\mathrm{fl})}$ itself is determined by the same Andreev reflection process as is the MT correction, so being treated in the self-consistent way $\tau_\varphi^{(\mathrm{fl})}(\epsilon)$ saturates and does not decrease anymore when $\epsilon \to 0$ in Eq. (8.48). The self-consistent procedure [257] consists of substitution of τ_{GL} by $\tau_\varphi^{(\mathrm{fl})}$ in the right-hand-side of Eq. (8.48) when $\tau_\varphi^{(\mathrm{fl})}(\epsilon)$ reaches the value of τ_{GL}. As a result

$$1/\tau_\varphi^{(\mathrm{fl})}(\epsilon \to 0) = T\sqrt{Gi}. \tag{8.49}$$

We see that indeed growth of the MT contribution side by side with growth of the $1/\tau_\varphi^{(\mathrm{fl})}(\epsilon)$ is saturated at $\epsilon \sim \sqrt{Gi}$. This statement, being correct qualitatively, allows us only to estimate the saturated value of the MT contribution. This lack of accuracy is due to the fact that the fluctuation contribution to $1/\tau_\varphi^{(\mathrm{fl})}(\epsilon)$ presents the example of the nonlinear fluctuation effect. As it will be demonstrated below these effects cannot be reduced to the simple scheme of the introduction of the self-consistent $\tau_\varphi^{(\mathrm{fl})}(\epsilon)$. The picture turns out to be more sophisticated since due to superconducting fluctuations the pseudogap $\Delta_{\mathrm{pg}} \sim T\sqrt{Gi}$ opens in the quasiparticle spectrum. The quantitative theory of strong nonlinear fluctuation effects still does not exist.

As a milestone in the study of such phenomena it is necessary mention the paper [256], where the authors succeeded in accounting for the contribution of strong long-wavelength fluctuations of the electric field to τ_φ of the normal metal. This is due to the fact that the electric field fluctuations mainly change the wave function phase and the Green function can be calculated exactly (the first order differential equation solution always can be presented in some integral form). Unfortunately the same scheme cannot be applied to the case of superconducting fluctuations since the long-wavelength fluctuations of the order parameter effect not only on wave function phase but also change the quasiparticle spectrum and the problem formally is reduced to solution of two differential equations, that cannot be done in general form.

As we already know and will also see below the fluctuation effects considerably grow in granular systems. This increase is particularly seen in the phase-breaking time. The authors of [261] calculated temperature dependence of electron phase-breaking time $\tau_\varphi(T)$ for a disordered metal with small concentration n_i of superconducting grains. One can say that these grains play the role of quenched superconducting fluctuations, so the authors of [261] in some sense revisited the problem of the effect of superconducting fluctuations on phase-breaking time, but they did this in a nonperturbative way. As a result it was found that above the global superconducting transition line, when electrons in the metal do not have yet gap in spectrum, Andreev reflection from the grains leads to appearance a nearly temperature-independent dephasing rate. In a broad temperature range (even noticeably far from T_{c}, see Fig. 8.5) the value of $\tau_\varphi^{-1}(T)$ strongly exceeds the prediction of the classical theory of dephasing in normal disordered conductors (see Eq. (8.46)).

8.3.5 *Comparison with the experiment*

Although the in-plane and out-of-plane components of the fluctuation conductivity tensor of a layered superconductor contain the same fluctuation contributions, their temperature behavior may be qualitatively different. In fact, for $\sigma_{xx}^{(\mathrm{fl})}$, the negative contributions are considerably less than the positive ones in the entire experimentally accessible temperature range above the transition, and it is a positive monotonic function of the temperature. Moreover, for HTS compounds, where the pair-breaking is strong and the anomalous MT contribution is in the saturated regime, it is almost always enough to take into account only the paraconductivity to fit experimental data. Some examples of the experimental findings for in-plane fluctuation conductivity of HTS materials on can see

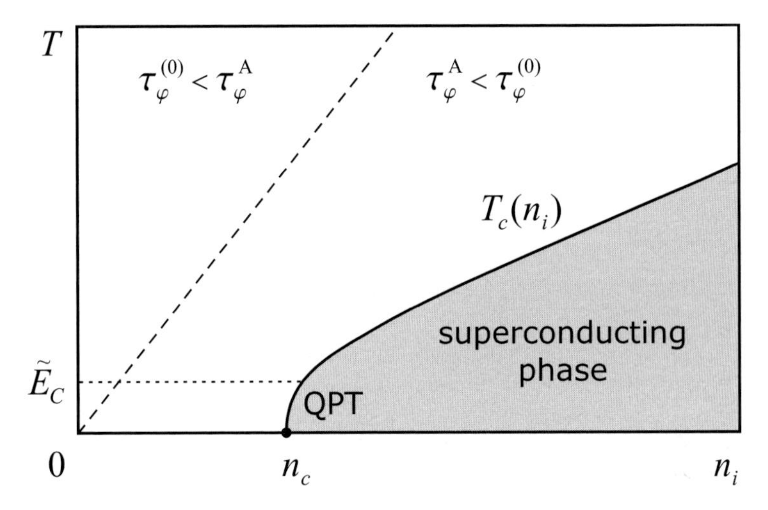

FIG. 8.5. Schematic (n_i, T) phase diagram of a metal with superconducting grains. The dephasing time due to Andreev reflection is shorter than in a broad range above $T_c(n_i)$.

in [262–269].

In Fig. 8.6, the fluctuation part of in-plane conductivity $\sigma_{xx}^{(\mathrm{fl})}$ is plotted as a function of $\epsilon = \ln T/T_c$ on a double logarithmic scale for three HTS samples (the solid line represents the $2D$ AL behavior $(1/\epsilon)$, the dotted line represents the $3D$ one: $3.2/\sqrt{\epsilon}$) [270]. One can see that paraconductivity of the less anisotropic YBCO compound asymptotically tends to the $3D$ behavior $(1/\epsilon^{1/2})$ for $\epsilon < 0.1$, showing the LD crossover at $\epsilon \approx 0.07$; the curve for more anisotropic 2223 phase of BSCCO starts to bend for $\epsilon < 0.03$ while the most anisotropic 2212 phase of BSCCO shows a $2D$ behavior in the whole temperature range investigated. All three compounds show a universal $2D$ temperature behavior above the LD crossover up to the limits of the GL region. It is interesting that around $\epsilon \approx 0.24$ all the curves bend down and follow the same asymptotic $1/\epsilon^3$ behavior (dashed line). Finally at the value $\epsilon \approx 0.45$ all the curves fall down indicating the end of the observable fluctuation regime.

Reggiani et al. [271] extended the $2D$ AL theory to the high temperature region by taking into account the short-wavelength and dynamic fluctuations. The following universal formula for $2D$ paraconductivity of a clean $2D$ superconductor as a function of the generalized reduced temperature $\epsilon = \ln T/T_c$ was obtained:[42]

$$\sigma_{xx}^{(\mathrm{fl})} = \frac{e^2}{16s} f(\epsilon)$$

with $f(\epsilon) = \epsilon^{-1}$, $\epsilon \ll 1$ and $f(\epsilon) = \epsilon^{-3}$, $\epsilon \gtrsim 1$.

[42] In section 9.4 we will demonstrate how such a dependence $\left(1/\ln^3(T/T_c)\right)$ appears by accounting for short-wavelength fluctuations for the $2D$ fluctuation diamagnetic susceptibility.

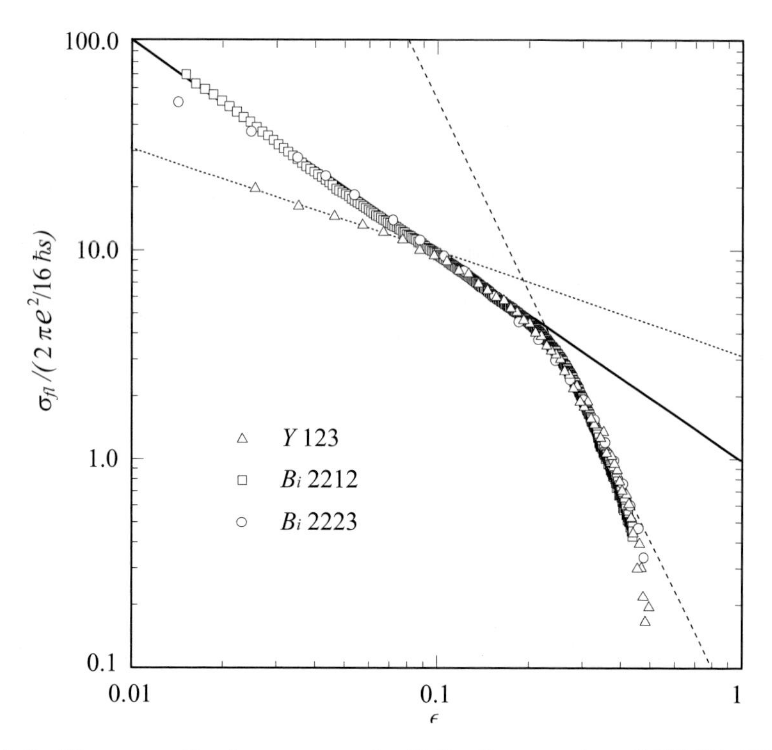

FIG. 8.6. The normalized excess conductivity for samples of YBCO-123 (triangles), BSSCO-2212(squares) and BSSCO-2223 (circles) plotted against $\epsilon = \ln T/T_c$ on a ln-ln plot as described in [270]. The dotted and solid lines are the AL theory in $3D$ and $2D$ respectively. The dashed line is the extended theory of [271].

In the case of the out-of-plane conductivity the situation is quite different. Both positive contributions (AL and anomalous MT) are suppressed here by the necessity of the interlayer tunneling, what results in a competition between positive and negative terms. Such concurrence can lead to formation of a maximum in the temperature dependence of the c-axis resistivity. This maximum belongs to the domain of the $2D$ behavior of fluctuations (we discuss the case $J\tau \ll 1, r\kappa \ll 1$ and $\gamma_\varphi \kappa > 1$) and its position is determined by the formula:

$$\epsilon_m/r \approx \frac{1}{(8r\kappa)^{1/2}} - \frac{1}{8\kappa}\left[\tilde{\kappa} - \frac{1}{2\gamma_\varphi}\right].$$

This nontrivial effect of fluctuations on the transverse resistance of a layered superconductor allows a successful fit to the data observed on optimally doped and overdoped HTS samples (see, for instance, Fig. 8.7) where the growth of the resistance still can be treated as a correction. The fluctuation mechanism of the growth of the transverse resistance can be easily understood in a qual-

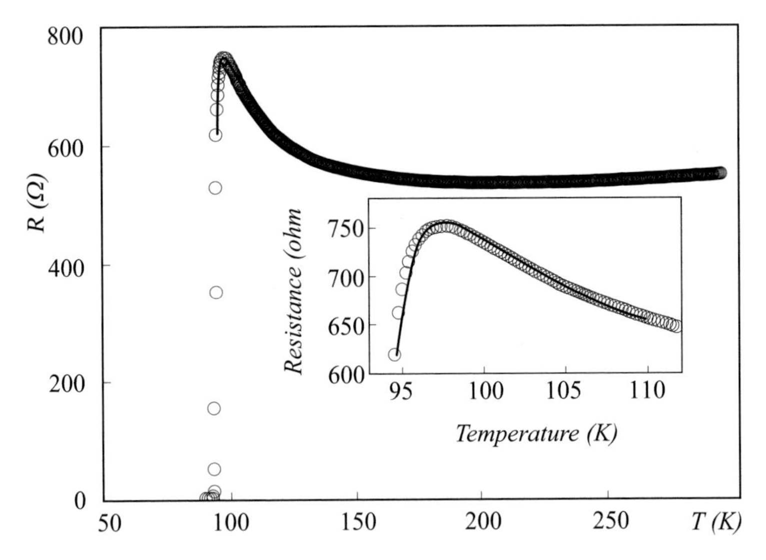

FIG. 8.7. Fit of the temperature dependence of the transverse resistance of a slightly underdoped BSCCO c-axis oriented film with the results of the fluctuation theory [272]. The inset shows the details of the fit in the temperature range between T_c and 110 K.

itative manner. Indeed, to modify the in-plane result (3.79) for the case of c-axis paraconductivity one has to take into account the hopping character of the electronic motion in this direction. If the probability of one-electron interlayer hopping is $\mathcal{P}_1$, then the probability of coherent hopping for two electrons during the fluctuation Cooper pair lifetime τ_{GL} is the conditional probability of these two events: $\mathcal{P}_2 = \mathcal{P}_1(\mathcal{P}_1\tau_{\mathrm{GL}})$. The transverse paraconductivity may thus be estimated as $\sigma_\perp^{\mathrm{AL}} \sim \mathcal{P}_2\sigma_\parallel^{\mathrm{AL}} \sim \mathcal{P}_1^2\frac{1}{\epsilon^2}$, in complete accordance with (8.17). We see that the temperature singularity of $\sigma_\perp^{AL}$ turns out to be stronger than that in $\sigma_\parallel^{\mathrm{AL}}$, however, for a strongly anisotropic layered superconductor $\sigma_\perp^{\mathrm{AL}}$ is considerably suppressed by the square of the small probability of interlayer electron hopping which enters in the prefactor. It is this suppression which leads to the necessity of taking into account the DOS contribution to the transverse conductivity. The latter is less singular in temperature but, in contrast to the paraconductivity, manifests itself in the first, not the second, order in the interlayer transparency $\sigma_\perp^{\mathrm{DOS}} \sim -\mathcal{P}_1 \ln 1/\varepsilon$. The DOS fluctuation correction to the one-electron transverse conductivity is negative and, being proportional to the first order of $\mathcal{P}_1$, can completely change the traditional picture of fluctuations just rounding the resistivity temperature dependence around transition. The shape of the temperature dependence of the transverse resistance mainly is determined by the competition between the opposite sign contributions: the paraconductivity and MT term, which are strongly temperature dependent but are suppressed by the square of the barrier transparency and the DOS contribution which has a weaker temperature dependence but depends only linearly on the barrier transparency.

8.4 Fluctuation conductivity in a.c. field

The a.c. conductivity of a superconductor can be expressed by the same analytically continued electromagnetic response operator $Q_{\alpha\beta}^{(R)}(\omega)$ (see Eq. (8.4)) but in contrast to the d.c. conductivity case, calculated without the assumption $\omega \to 0$. Let us recall that the paraconductivity tensor in an a.c. field was already studied in chapter 3 in the framework of the TDGL equation [54] and the most interesting asymptotics for our discussion (3.40)–(3.43), valid for $\omega \ll T$ in the $2D$ regime, were calculated there. The microscopic calculation of the AL diagram [108] shows that in the vicinity of $T_{\rm c}$ and for $\omega \ll T$ the leading singular contribution to the response operator $Q_{\alpha\beta}^{\rm AL(R)}$ arises from the fluctuation propagators rather than from the B_α blocks, which confirms the TDGL results. Nevertheless the DOS and MT corrections can be calculated only by the microscopic method, as was done in [108, 273].

Let us note that the external frequency ω_ν enters in the expression for the DOS contribution to $Q_{\alpha\beta}(\omega)$ only by means of the Green's function $G(\mathbf{p}, \omega_{n+\nu})$ and it is not involved in q integration. So, near $T_{\rm c}$, even in the case of an arbitrary external frequency, we can restrict our consideration to the static limit, taking into account only the propagator frequency $\Omega_k = 0$, and to get [273]:

$$\operatorname{Re}\sigma_{\alpha\beta}^{\rm DOS}(\omega) = -\frac{e^2}{2\pi s}\hat{\kappa}\left(\omega, T, \tau\right) A_{\alpha\beta} \ln\left[\frac{2}{\sqrt{\epsilon + r} + \sqrt{\epsilon}}\right],$$

where the anisotropy tensor $A_{\alpha\beta}$ was introduced in (8.26). Let us stress that, in contrast to the AL frequency dependent contribution, this result has been found with only the assumption $\epsilon \ll 1$, so it is valid for any frequency, and impurity concentration. The function $\hat{\kappa}\left(\omega, T, \tau\right)$ was calculated in [273] exactly but we present here only its asymptotics for the clean and dirty cases:

$$\hat{\kappa}_{\rm d}\left(\omega, T \ll \tau^{-1}\right) = \frac{8}{\pi}\begin{cases} \frac{7\zeta(3)}{2\pi^2}, & \omega \ll T \ll \tau^{-1}, \\ \left(\frac{T}{\omega}\right)^2, & T \ll \omega \ll \tau^{-1}, \\ -\frac{\pi T^2}{\omega^3 \tau}, & T \ll \tau^{-1} \ll \omega. \end{cases}$$

$$\hat{\kappa}_{\rm cl}\left(\omega, T \gg \tau^{-1}\right) = \frac{\pi^3}{28\zeta(3)}\begin{cases} \left(T\tau\right)^2, & \omega \ll \tau^{-1} \ll T, \\ \left(\frac{T}{\omega}\right)^2, & \tau^{-1} \ll \omega \ll T, \\ -4\left(\frac{T}{\omega}\right)^3, & \tau^{-1} \ll T \ll \omega. \end{cases}$$

The general expression for the MT contribution is too cumbersome, so we restrict ourselves here to the important $2D$ saturated regime ($r \ll \epsilon \leq \gamma_\varphi$):

$$\sigma_{zz}^{\rm MT(an)(2D)}(\omega) = \frac{e^2 s}{2^7 \eta_{(2)}}\frac{r^2}{\gamma_\varphi \epsilon}\begin{cases} 1, & \omega \ll \tau_\varphi^{-1}, \\ \left(\frac{8T_{\rm c}\gamma_\varphi}{\pi\omega}\right)^2, & \omega \gg \tau_\varphi^{-1}. \end{cases}$$

$$\sigma_{xx}^{\mathrm{MT(an)(2D)}}(\omega) = \frac{e^2}{8s}\begin{cases} \frac{1}{\gamma_\varphi}\ln\frac{\gamma_\varphi}{\epsilon}, & \omega \ll \tau_\varphi^{-1}, \\ \left(\frac{8T_\mathrm{c}\gamma_\varphi}{\pi\omega}\right)^2, & \omega \gg \tau_\varphi^{-1}. \end{cases}.$$

Let us discuss the results obtained. Because of the large number of parameters entering the expressions we restrict our consideration to the most interesting c-axis component of the fluctuation conductivity tensor in the $2D$ region (above the Lawrence–Doniach crossover temperature).

The AL contribution describes the fluctuation condensate response to the applied electromagnetic field. The current associated with it can be treated as the precursor phenomenon of the screening currents in the superconducting phase. As was demonstrated above the characteristic "binding energy " of fluctuation Cooper pair is of the order of $T - T_\mathrm{c}$, so it is not surprising that the AL contribution decreases when the electromagnetic field frequency exceeds this value. Indeed $\omega^{\mathrm{AL}} \sim T - T_\mathrm{c}$ is the only relevant scale for σ^{AL}: its frequency dependence does not contain T, τ_φ and τ. The independence from the latter is due to the fact that elastic impurities do not present obstacles for the motion of Cooper pairs. The interaction of the electromagnetic wave with the fluctuation Cooper pairs resembles, in some way, the anomalous skin-effect where the reflection is determined by the interaction with the free electron system.

The anomalous MT contribution also is due to fluctuation Cooper pairs, but this time they are formed by electrons moving along self-intersecting trajectories. Being the contribution related to the Cooper pair electric charge transfer it does not depend on the elastic scattering time but turns out to be extremely sensitive to the phase-breaking mechanisms. So two characteristic scales turn out to be relevant in its frequency dependence: $T - T_\mathrm{c}$ and τ_φ^{-1}. In the case of HTS, where τ_φ^{-1} has been estimated as at least $0.1T_\mathrm{c}$, for temperatures up to $5 \div 10\ K$ above T_c the MT contribution is already saturated, it is almost temperature independent and is determined by the value of τ_φ.

The DOS contribution to $\mathrm{Re}\,\sigma(\omega)$ is quite different from those above. In the wide range of frequencies $\omega \ll \tau^{-1}$ the lack of electron states at the Fermi level leads to the opposite sign effect in comparison with the AL and MT contributions: $\mathrm{Re}\,\sigma^{\mathrm{DOS}}(\omega)$ turns out to be negative and this means an increase of the surface impedance, or, in other words, decrease of the reflectance. Nevertheless, at very high frequencies $\omega \sim \tau^{-1}$ the applied electromagnetic field starts to affect the electron distribution over energies and the DOS contribution changes its sign. It is interesting that the DOS contribution, as a one-electron effect, depends on the impurity scattering time in a similar manner to the normal Drude conductivity. The decrease of $\mathrm{Re}\,\sigma^{DOS}(\omega)$ starts at frequencies $\omega \sim \min\{T, \tau^{-1}\}$ which for HTS are much higher than $T - T_\mathrm{c}$.

The ω-dependence of $\mathrm{Re}\,\sigma_{zz}^{\mathrm{tot}}$ with the most natural choice of parameters ($T_\mathrm{c}r \ll T_\mathrm{c}\epsilon \leq \tau_\varphi^{-1} \ll \min\{T, \tau^{-1}\}$) is presented in Fig. 8.8. Let us discuss it referring to a strongly anisotropic layered superconductor. The positive AL and MT contributions to $\sigma_{zz}^{\mathrm{tot}}$, being suppressed by the square of the interlayer transparency, are small in magnitude and they vary in the low frequency region

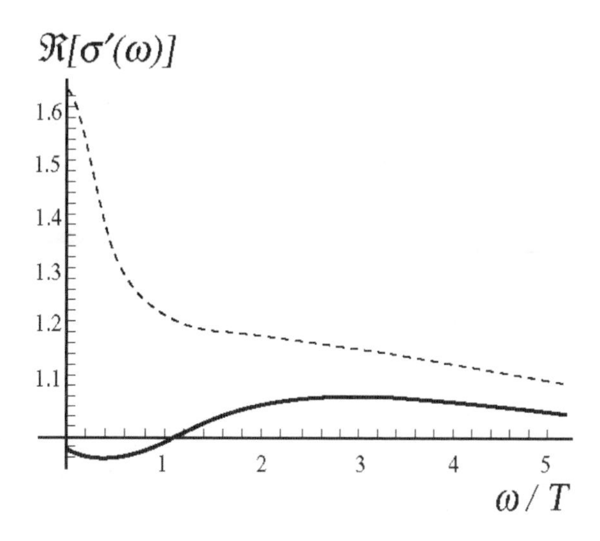

FIG. 8.8. The theoretical dependence [273] of the real part of the conductivity, normalized by the Drude normal conductivity, on ω/T, $\Re\left[\sigma'(\omega)\right] = \mathrm{Re}\left[\sigma(\omega)\right]/\sigma^{\mathrm{n}}$. The dashed line refers to the ab-plane component of the conductivity tensor whose Drude normal conductivity is $\sigma_{\parallel}^{\mathrm{n}} = N(0)e^2\tau v_F^2$. The solid line refers to the c-axis component whose Drude normal conductivity is $\sigma_{\perp}^{\mathrm{n}} = \sigma_{\parallel}^{\mathrm{n}} J^2 s^2/v_F^2$. In this plot we have put $T\tau = 0.3, E_F/T = 50, r = 0.01, \epsilon = 0.04, T\tau_{\varphi} = 4$.

$\omega \sim \min\{T - T_{\mathrm{c}}), \tau_{\varphi}^{-1}$. The DOS contribution remains almost invariable in this region and being proportional only to the first order of transparency can quite prevail over the AL and MT contributions. With a further increase of frequency $\min\{T - T_{\mathrm{c}}, \tau_{\varphi}^{-1}\} \lesssim \omega$ the AL and MT contributions decay; $\mathrm{Re}\,\sigma_{\perp}$ remains negative up to $\omega \sim \min\{T, \tau^{-1}\}$, then it changes its sign at $\omega \sim \tau^{-1}$, reaches maximum and rapidly decreases. The following high frequency behavior is governed by the Drude law. So one can see that the characteristic pseudogap-like behavior in the frequency dependence of the c-axis a.c. conductivity takes place: a transparency window appears in the range $\omega \in [T - T_{\mathrm{c}}, \tau^{-1}]$.

In the case of the ab-plane a.c. conductivity the two first positive contributions are not suppressed by the interlayer transparency, and exceed considerably the negative DOS contribution in a wide range of frequencies. Any pseudogap-like behavior is therefore unlikely in $\sigma_{xx}^{\mathrm{tot}}(\omega)$: the reflectivity will be of the metallic kind.

8.5 Fluctuation conductivity in ultra-clean superconductors

8.5.1 *Preliminaries*

Everywhere in the above consideration the presence of the electron scattering by impurities has been taken into account by use of the averaged over impurity positions electron Green functions in the standard form (6.2) with (6.22 and the impurity vertex function (Cooperon) taken in the local approximation (6.26). The used Green's function expression has a wide region of applicability from the clean metal up to the almost amorphous material ($E_F\tau \gg 1$). At that the smallness of the momentum of the center of mass of fluctuation Cooper pair q was widely used in calculations. Namely, in order to expand the quasiparticle energy over q we assumed that the condition $|\mathbf{q}| \ll p_F$ is fulfilled.

The restrictions on momenta were much more rigid for the Cooperon taken in the form (6.26). This is due to the fact, that in order to obtain the Cooperon in the form (6.26) the expansion over $lq \ll 1$ was made in (6.24). In the vicinity of T_c the effective propagator's momenta are determined by the value $|\mathbf{q}|_{\text{eff}} \sim [\xi_{\text{GL}}(T)]^{-1} = \xi^{-1}\sqrt{\epsilon}$. Hence, for validity of the used above local expression (6.26) for Cooperon, the condition $|\mathbf{q}|_{\text{eff}} l \sim \frac{l}{\xi}\sqrt{\epsilon} \ll 1$ had to be satisfied. This restriction is valid for the cases, considered above, of the dirty ($l \ll \xi_0$) and clean ($\xi_0 \ll l \ll \xi_{\text{GL}} = \xi_0/\sqrt{\epsilon}$) superconductor, where the motion of electrons forming fluctuation Cooper pair has a diffusive character. In a very clean superconductor, where the electron mean free path considerably exceeds the effective size of the fluctuating Cooper pair $l \gg \xi_{\text{GL}}(T)$, the electron motion has a quasi-ballistic character and the traditional picture of the fluctuation phenomena manifestation above the critical temperature significantly changes [274, 275]. In particular, the nontrivial cancelation of the contributions, previously divergent in $T\tau$ (see, e.g. (8.29)), will be shown in this section. This circumstance results that in the ultra-clean case the AL term remains the only fluctuation correction to conductivity. We will restrict our consideration below to the most interesting case of a $2D$ electron spectrum.

In a normal metal the transition from the diffusive to the quasi-ballistic regime is controlled by the relation between the electron elastic mean free path l and the diffusive length $l_T = \sqrt{\mathcal{D}/T}$ ($\mathcal{D} = v_F^2\tau/D$ is the diffusion coefficient) or, in other words, by the ratio of the temperature T and the elastic scattering rate. This ratio is given by the value of $T\tau$. In a superconductor being close to the transition temperature, due to the presence of the additional length scale ξ_0, the range of impurity concentrations of a clean metal ($\xi_0 \ll l \Leftrightarrow T\tau \gg 1$) can be divided into clean ($\xi_0 \ll l \ll \xi_{GL}(T)$) and ultra-clean ($\xi_{\text{GL}}(T) \ll l$) regimes. In terms of the reduced temperature scale, the narrow range $\epsilon \ll 1/(T\tau)^2$ (which includes both the diffusive and clean regimes) can still be described by the local fluctuation theory, while the study of the fluctuation transport in the most interesting temperature interval $1/(T\tau)^2 \ll \epsilon \ll 1$ (ultra-clean regime) requires a nonlocal treatment. The latter case may be particularly relevant in the case of the HTS cuprates, where the parameter $T\tau$ for the high-quality samples is estimated to reach 5-10,[43] and almost all the range of the experimentally accessible

[43]To be more precise let us remember that being in the vicinity of the critical temperature we operate with the coherence length in the form (6.33), so the parameter which governs the clean and dirty regimes is not $T\tau$ but $4\pi T\tau$. This means that even for $T\tau \sim 1$ the effective value $4\pi T\tau \gg 1$.

reduced temperature ϵ belongs to the ultra-clean limit.

Indeed in a sufficiently clean superconductor ($T\tau \gg 1$), the locality condition $\epsilon \ll 1/(T\tau)^2$ almost contradicts to the $2D$ thermodynamical Ginzburg-Levanyuk criterion ($Gi \sim \frac{T_c}{E_F} \ll \epsilon$). Therefore, here almost is no room for applicability of the standard fluctuation theory. Moreover, as we will see a little bit later, the high order corrections for the transport coefficients become comparable to first order perturbation theory results much before than those for the thermodynamical ones, namely at $\epsilon \sim \sqrt{Gi}$ [276, 277]. Hence, being interested in the study of the fluctuations in clean superconductors, *de facto*, one can speak only about their nonlocal theory.

8.5.2 *Nonlocal fluctuation conductivity in quasi-ballistic regime*

8.5.2.1 *The absence of the total DOS and MT contributions in ballistic regime (superconductor without impurities)* Let us start our discussion of the nonlocal fluctuations from the case of a perfectly clean superconductor, not taking into account the presence of impurities at all. In this case the electromagnetic response operator is determined by the four diagrams 1, 2, 5, 6 from Fig. 8.1 without any impurity averaging. In order to get as general a result as possible let us not restrict our consideration to the static fluctuations (see the previous section) and will consider the contribution from nonzero bosonic frequencies Ω_k to conductivity side by side with $\Omega_k = 0$. Using in the expression (8.18) the vertex not renormalized by impurities $\lambda(q, \varepsilon_1, \varepsilon_2) \equiv 1$ and taking the Green functions in the form (6.2), one can carry out the momentum integration of the product of four Green functions without any simplifying assumptions on frequencies, as usually, passing to the ξ-integration. As a result six different pole dispositions can be turned out. Applying the Cauchy theorem and carrying out the summation over the fermionic frequency one can find [271]:

$$\sigma^{(5+6)} = \left(\frac{iv_F^2\nu}{\omega^3}\right)\left(\frac{e^2}{4s}\right)\int\frac{d^2\mathbf{q}}{(2\pi)^2}\int_{-\infty}^{\infty} d\Omega \coth\left(\frac{\Omega}{2T}\right)\operatorname{Im} L^R(\mathbf{q}, -i\Omega)$$
$$\times\left\{\psi\left(\frac{1}{2}+\frac{i\omega}{4\pi T}-\frac{i\Omega}{4\pi T}\right)+\psi\left(\frac{1}{2}+\frac{i\Omega}{4\pi T}\right)+\psi\left(\frac{1}{2}-\frac{i\omega}{4\pi T}-\frac{i\Omega}{4\pi T}\right)\right.$$
$$+\psi\left(\frac{1}{2}-\frac{i\omega}{2\pi T}+\frac{i\Omega}{4\pi T}\right)-2\psi\left(\frac{1}{2}-\frac{i\Omega}{4\pi T}\right)-2\psi\left(\frac{1}{2}-\frac{i\omega}{4\pi T}+\frac{i\Omega}{4\pi T}\right)$$
$$\left.-\frac{i\omega}{2\pi T}\left[\psi'\left(\frac{1}{2}+\frac{i\Omega}{4\pi T}\right)-\psi\left(\frac{1}{2}-\frac{i\omega}{4\pi T}+\frac{i\Omega}{4\pi T}\right)\right]\right\}. \tag{8.50}$$

The calculation of the MT diagram is even more cumbersome, but one can evaluate it also using the general expressions (8.30)–(8.31) with $\lambda(q, \varepsilon_1, \varepsilon_2) \equiv 1$. The anomalous MT contribution in this case naturally does not even appear, and the final contribution from the MT diagram to conductivity takes the form

$$\sigma^{(2)} = -\left(\frac{iv_F^2\nu}{\omega^3}\right)\left(\frac{e^2}{4s}\right)\int\frac{d^2\mathbf{q}}{(2\pi)^2}\int_{-\infty}^{\infty} d\Omega \coth\left(\frac{\Omega}{2T}\right)\operatorname{Im} L^R(\mathbf{q}, -i\Omega)$$

$$\times \left\{ \psi\left(\frac{1}{2}+\frac{i\omega}{4\pi T}+\frac{i\Omega}{4\pi T}\right)-\psi\left(\frac{1}{2}+\frac{i\Omega}{4\pi T}\right)+\psi\left(\frac{1}{2}-\frac{i\omega}{4\pi T}+\frac{i\Omega}{4\pi T}\right)\right.$$
$$-\psi\left(\frac{1}{2}+\frac{i\Omega}{4\pi T}\right)-\psi\left(\frac{1}{2}+\frac{i\omega}{4\pi T}-\frac{i\Omega}{4\pi T}\right)+\psi\left(\frac{1}{2}-\frac{i\Omega}{4\pi T}\right)$$
$$\left.+\psi\left(\frac{1}{2}-\frac{i\Omega}{4\pi T}\right)-\psi\left(\frac{1}{2}-\frac{i\omega}{4\pi T}-\frac{i\Omega}{4\pi T}\right)\right\}. \tag{8.51}$$

For simplicity we assumed $q = 0$ in the Green functions. Expanding the ψ-functions in (8.50) and (8.51) in ω up to the fourth order one can find that the first three orders in this expansion are canceled out and total DOS and MT contribution in the ballistic regime starts from the term proportional the electromagnetic field frequency: $\sigma^{(5+6)}+\sigma^{(2)}\sim\omega$. This means that in the direct current measurements with clean sample the only relevant contribution remains the AL one.

8.5.2.2 *Cancelation of the DOS and MT terms in quasi-ballistic regime*[44] In the previous section, we observed the exact cancelation of the MT and DOS contributions to conductivity for the case of superconductor without impurities. This result at the first glance contradicts the conclusions of our previous consideration: the DOS contribution (8.29) obtained by standard diagrammatic technique calculations [238] strongly diverges when $T\tau\to\infty$, while the anomalous MT contribution formally turns out to be τ-independent (see (8.44)). Nevertheless, one can notice [275] that at the upper limit of the clean case, when $T\tau\sim 1/\sqrt{\epsilon}$, both the DOS (8.29) and anomalous MT (8.44) contributions turn out to be of the same order of magnitude but have opposite signs. Moreover, the anomalous MT contribution, being induced by the pairing on the Brownian diffusive trajectories [107], in some way must be dependent on $T\tau$.[45]

Let us remember that in the derivation of mentioned results the local form of the fluctuation propagator and Cooperons were used. That is why the direct extension of their validity for $T\tau\gg 1/\sqrt{\epsilon}\to\infty$ is incorrect. Hence one can suspect that in the case of a correct procedure of impurity averaging valid for the quasi-ballistic regime ($T\tau\gtrsim 1/\sqrt{\epsilon}$) the large negative DOS contribution is strongly compensated by the positive anomalous MT one and finally, according to the results of the previous subsection, in the ballistic regime ($T\tau\to\infty$) the total sum of the last eight diagrams from Fig. 8.1 tends to zero.

Let us start our revision from the expression for the impurity vertex. We can start from Eqs. (6.24) and (6.8) where the expansion over small lq still was not performed. In the case of a $2D$ electron spectrum the integral in (6.8) can be calculated exactly for the case of an arbitrary electron mean free path:

$$\mathcal{P}(\mathbf{q},\varepsilon_n,\varepsilon_m)=\int\frac{d^2p}{(2\pi)^2}G(\mathbf{p}+\mathbf{q},\varepsilon_n)G(-\mathbf{p},\varepsilon_m)=\frac{2\pi\nu\theta(-\varepsilon_n\varepsilon_m)}{\sqrt{v_F^2q^2+(\widetilde{\varepsilon}_n-\widetilde{\varepsilon}_m)^2}}.$$

[44] In this section we base on the results of [275].

[45] In [274, 278] it was shown that the anomalous MT contribution is τ-independent up to $T\tau\sim 1/\sqrt{\epsilon}$ only, then for $T\tau\gg 1/\sqrt{\epsilon}$ it diverges as $T\tau\ln(T\tau)$.

As a result the nonlocal expression for the Cooperon takes the form

$$\lambda(\mathbf{q},\varepsilon_n,\varepsilon_m)=\left(1-\frac{\theta(-\varepsilon_n\varepsilon_m)}{\tau\sqrt{v_F^2q^2+(\tilde{\varepsilon}_n-\tilde{\varepsilon}_m)^2}}\right)^{-1}. \tag{8.52}$$

One can see that it is reduced to (6.26) in the case of $v_F q \ll |\tilde{\varepsilon}_1-\tilde{\varepsilon}_2|$. Let us stress that this result was carried out without any expansion over the Cooper pair center of mass momentum and is valid in the $2D$ case for an arbitrary lq.

In the same way, using Eq. (6.28), the nonlocal expression for the fluctuation propagator for the $2D$ superconductor with an arbitrary mean free path can be written:

$$-[\nu L(\mathbf{q},\Omega_k)]^{-1}=\ln\frac{T}{T_c}+\sum_{n=0}^{\infty}\left\{\frac{1}{n+1/2}\right.$$
$$\left.-\frac{1}{\sqrt{\left(n+\frac{1}{2}+\frac{\Omega_k}{4\pi T}+\frac{1}{4\pi T\tau}\right)^2+\frac{v_F^2\mathbf{q}^2}{16\pi^2T^2}}-\frac{1}{4\pi T\tau}}\right\}. \tag{8.53}$$

Near T_c $\ln\frac{T}{T_c}\approx\epsilon$ and in the local limit, when only small momenta $lq\ll 1$ are involved in the final integrations, Eq. (8.53) can be expanded in $v_F q/\max\{T,\tau^{-1}\}$ and reduces to the appropriate local expression.

Let us demonstrate the specifics of the nonlocal calculations for the example of the MT contribution. We restrict our consideration to the vicinity of the critical temperature, where the static approximation is valid. Using the nonlocal expression for the Cooperon after integration over electronic momentum one can find:

$$Q^{(\mathrm{MT})}(\omega_v)=-4\pi\nu v_F^2e^2T^2\sum_{\varepsilon_n}\int\frac{d^2\mathbf{q}}{(2\pi)^2}L(\mathbf{q},0)$$
$$\times\left[\mathcal{M}\left(\tilde{\varepsilon}_n,\tilde{\varepsilon}_{n+\nu},\mathbf{q}\right)+\mathcal{M}\left(\tilde{\varepsilon}_{n+\nu},\tilde{\varepsilon}_n,\mathbf{q}\right)\right], \tag{8.54}$$

where

$$\mathcal{M}(\alpha,\beta,\mathbf{q})=\frac{R_q(2\alpha)R_q(\alpha+\beta)-\theta(\alpha\beta)R_q(2\alpha)R_q(2\beta)}{(\beta-\alpha)^2\left(R_q(2\alpha)-\frac{1}{\tau}\right)\left(R_q(2\beta)-\frac{1}{\tau}\right)R_q(\alpha+\beta)}$$

and $R_q(x)=\sqrt{x^2+v_F^2\mathbf{q}^2}$. The analogous consideration of the DOS diagrams results in similar expression:

$$Q^{(5+6)}(\omega_v)=4\pi\nu v_F^2e^2T^2\sum_{\varepsilon_n}\int\frac{d^2\mathbf{q}}{(2\pi)^2}L(\mathbf{q},0)\ \left[\mathcal{D}(\tilde{\varepsilon}_n,\tilde{\varepsilon}_{n+\nu},\mathbf{q})+\mathcal{D}(\tilde{\varepsilon}_{n+\nu},\tilde{\varepsilon}_n,\mathbf{q})\right] \tag{8.55}$$

with

$$\mathcal{D}(\alpha,\beta,\mathbf{q}) = (\beta-\alpha)^{-2}\left(R_q(2\alpha)-\frac{1}{\tau}\right)^2\left[\frac{R_q^2(2\alpha)+2\alpha(\alpha-\beta)}{R_q(2\alpha)} - \frac{\theta\left(\alpha\beta\right)R_q^2(2\alpha)}{\left(R_q(\alpha+\beta)-\frac{1}{\tau}\right)}\right].$$

One can see that, after analytic continuation over the external frequency $\omega_\nu \to -i\omega$ and the consequent tending $\omega \to 0$, each of the DOS or MT type diagrams is written in the form of a Laurent series of the type $C_{-2}(T\tau)^2 + C_{-1}(T\tau) + C_0 + C_1(T\tau)^{-1} + \cdots$ and is divergent at $T\tau \to \infty$ according to (8.29). Nevertheless the expansion in a Laurent series of the sum of these diagrams leads to the exact cancelation of all divergent contributions. The leading order of the sum of the MT and DOS contributions in the limit of $T\tau \gg 1$ turns out to be proportional only to $(T\tau)^{-1}$ and in accordance with the previous section disappears in the ballistic regime $\tau \to \infty$.

8.5.2.3 *AL contribution in quasi-ballistic and ballistic regimes* [46] In the previous sections we had been studied in details the paraconductivity tensor. One of the most impressive results found was the formula for $2D$ paraconductivity (3.79) which turned out to be independent of the electron mean free path both in dirty and clean cases.

In the qualitative considerations of section 8.1 this statement was justified as natural: the electron elastic relaxation time τ in the Drude formula was substituted by the fluctuation Cooper pair lifetime τ_{GL} and the former disappeared from our consideration forever. Not much more could elucidate situation the phenomenological TDGL approach: here the information concerning the impurity scattering is concealed in the phenomenological coefficients and the limits of its applicability was not discussed. Finally, the result (3.79) was carried out in the framework of the microscopic theory valid both for clean and dirty superconductors but substantially based on the local approximation. As was recalled at the beginning of this section in very clean superconductors, where the electron mean free path considerably exceeds the fluctuation Cooper pair size, the results of the local theory, obtained with the Cooperon in the form (6.26), are not valid more, so the validity of the universal formula (3.79) has to be revised too.

One can try to follow the scheme of the previous subsection and to calculate the Green function block (8.7) completely ignoring the electron scattering by impurities (see details in the next section). Below we will see that in such, extremely nonlocal regime the block $B_\alpha(\mathbf{q},\Omega_k,\omega_\nu \neq 0) \equiv 0$ for any $\omega_\nu \neq 0$ and only for $\omega_\nu = 0$ it turns different from zero. This is due to the fact, that for any $\omega_\nu \neq 0$, besides the trivial situation in (8.9) with two poles $\xi_{1,2} = \pm i\varepsilon_n$, appears the possibility of the poles disposition $\xi_1 = i\varepsilon_{n+\nu}$ above the real axis and $\xi_2 = i\varepsilon_n$ below it (see Fig. 8.3).

The corresponding contribution is $\sim \theta\left(-\varepsilon_n\varepsilon_{n+\nu}\right)$. Important, that after the fermion frequency summation it exactly cancels the contribution appearing from the situation when both poles are situated at the same side of the real axis $\sim \theta\left(\varepsilon_n\varepsilon_{n+\nu}\right)$ [271] and we obtain the discouraging result $B_\alpha(\mathbf{q},\Omega_k,\omega_\nu \neq 0) \equiv$

[46] In this section we base on the results of [109].

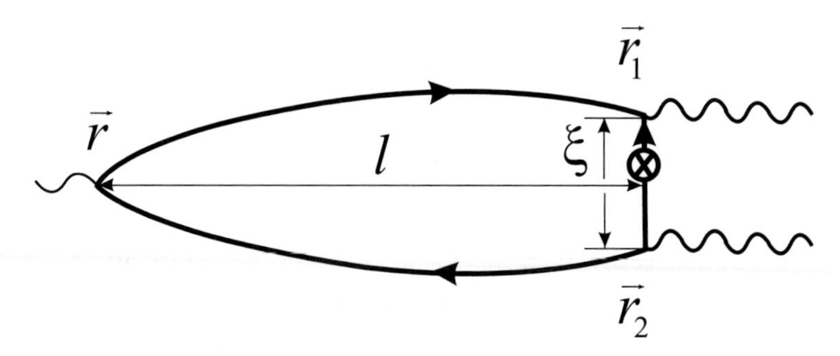

FIG. 8.9. Current vertex and its expansion in the clean limit: sign $\times$ means the expansion of the Green function.

0. Thus, we see that the block $B_{\alpha}^{R,A}(\mathbf{q},\Omega_k,\omega)$ is a nonanalytic function of the external frequency and it is not clear how to perform the analytic continuation of the corresponding contribution to the operator of electromagnetic response, and consequently, to calculate the paraconductivity.

The solution of this delicate situation was given in [109]. The difficulties mentioned above were related to non-locality of the problem, when the Green function block cannot be anymore assumed as a constant multiplied by the Cooper pair center of mass momentum.

The most transparent and convenient way to consider the AL conductivity in the clean limit is to operate in the real space [109]. The expression (8.6) can be rewritten for the paraconductivity tensor in the invariant form as[47]

$$\sigma_{\alpha\beta}^{\mathrm{AL}} = T\sum_{\Omega_k} \mathrm{tr}\left[\widehat{\Gamma}_{\alpha\gamma}\widehat{q}_{\gamma}D_{\gamma\delta}\widehat{q}_{\delta}\widehat{\Gamma}_{\delta\beta}\right]. \tag{8.56}$$

The vertex $\widehat{\Gamma}_{\alpha} = \widehat{\Gamma}_{\alpha\gamma}\widehat{q}_{\gamma}$ is given in Fig. 8.9. $D_{\gamma\delta}\left(\widehat{q},\Omega_k\right)$ is the product of two fluctuation propagators. Let us consider the vertex $\widehat{\Gamma}_{\alpha}$ in details. The distance between points $\mathbf{r}_1$ and $\mathbf{r}_2$ is of the order of ξ_{GL}, while the distances $|\mathbf{r}-\mathbf{r}_1|, |\mathbf{r}-\mathbf{r}_2| \sim l \gg \xi_{\mathrm{GL}}$. This means that we can consider $\mathbf{r}_1$ and $\mathbf{r}_2$ as the same point. In the $\mathbf{p}$ representation we can expand the Green function $G\left(\mathbf{q}-\mathbf{p}\right)$ in $\mathbf{q}$:

$$G\left(\mathbf{q}-\mathbf{p}\right) = G^{(0)}\left(-\mathbf{p}\right) - G^{(0)}\left(-\mathbf{p}\right)\frac{\mathbf{qp}}{m_e}G^{(0)}\left(-\mathbf{p}\right) + \cdots \tag{8.57}$$

This means we can represent the vertex $\widehat{\Gamma}_{\alpha}$ in the following manner Fig. 8.9:

$$\widehat{\Gamma}_{\alpha\gamma} = -\frac{e}{m_e^2}\Lambda\left\langle \mathbf{p}_{\alpha}\mathbf{p}_{\gamma}\right\rangle, \tag{8.58}$$

[47]Compare with Eq. (8.6). The electron charge e and related with the number of diagrams combinatorial coefficient 2 in the present expression are attributed to the vertex $\widehat{\Gamma}_{\alpha\gamma}$.

where Λ is some constant which we will find later. The $\langle \cdots \rangle$ means that we calculate a trace over the tensor indices and perform the summation on the Matsubara fermionic frequency.

Now let us express $\widehat{\Gamma}_\alpha$ in terms of the Drude conductivity. We know that the current density is related to the electric field by means of the electromagnetic response operator (see Eq. (8.3)):

$$\mathbf{j}_\alpha(\omega) = -Q_{\alpha\beta}(\omega)\mathbf{A}_\beta(\omega), \tag{8.59}$$

where $\mathbf{A}_\beta(\omega) = \mathbf{E}_\beta(\omega)/i\omega$ is the vector-potential of the ac electric field. In its turn $Q_{\alpha\beta}(\omega)$ can be expressed [228] in terms of the normal conductivity tensor $\sigma_n^{\alpha\beta}$:

$$Q_{\alpha\beta}(\omega) = -i\omega\sigma_n^{\alpha\beta} = \frac{e^2\mathcal{N}_e}{m_e}\delta_{\alpha\beta} - 2\frac{e^2}{m_e^2}\langle \mathbf{p}_\alpha \mathbf{p}_\beta \rangle, \tag{8.60}$$

Here $\mathcal{N}_e$ is the electron concentration in metal. The minus sign in front of the second term in Eq. (8.60) is due to the fact that the defined above operation of the trace calculation (brackets) does not take into account the sign of the fermion loop. Now we can express the vertex $\widehat{\Gamma}_{\alpha\gamma}$ defined by Eq. (8.58) through $Q_{\alpha\beta}(\omega)$ (the number 2 in front of the second term is due to spin degeneracy):

$$\widehat{\Gamma}_{\alpha\beta} = \frac{\Lambda\omega_\mathrm{p}^2}{8\pi e}\left(\delta_{\alpha\beta} + i\frac{4\pi\omega}{\omega_\mathrm{p}^2}\sigma_n^{\alpha\beta}\right). \tag{8.61}$$

Here $\omega_\mathrm{p}^2 = 4\pi e^2\mathcal{N}_e/m_e$ is the plasma frequency.

The coefficient Λ can be determined from the comparison of Eq. (8.61) with the static limit of $\widehat{\Gamma}_{\alpha\gamma}$, which, in its turn, can be related to the block B defined by Eq. (8.10): $\widehat{\Gamma}_{\alpha\alpha}(\omega = 0) = -2eB_\alpha/q_\alpha = 4e\nu\xi_{(D)}^2$ and therefore $\Lambda = 8\nu\xi_{(D)}^2 m_e/\mathcal{N}_e$.

For the free electron gas (in absence of magnetic field)

$$\sigma_{\alpha\alpha}(\omega) = \frac{\omega_\mathrm{p}^2\tau}{4\pi}\frac{1}{1-i\omega\tau} = \frac{\sigma_{\alpha\alpha}^n}{1-i\omega\tau}, \tag{8.62}$$

and therefore, Γ has the simple form

$$\Gamma_{\alpha\alpha}(\omega) = \frac{4e\nu\xi_{(D)}^2}{1-i\omega\tau}. \tag{8.63}$$

Our consideration shows that in the coordinate space current density can be represented as

$$j_\alpha(\mathbf{r}) = \int d\mathbf{r}_1 d\mathbf{r}_2 d\mathbf{r}'\Gamma_\alpha(\mathbf{r}-\mathbf{r}_1)L(\mathbf{r}_1-\mathbf{r}_2)L(\mathbf{r}_2-\mathbf{r}_1)\Gamma_\beta(\mathbf{r}_2-\mathbf{r}')E_\beta(\mathbf{r}') \tag{8.64}$$

and the current in homogeneous electric field can be expressed through the local vertices.

Using Eqs. (8.56)–(8.61) the paraconductivity in the studied nonlocal case can be written as

$$\sigma_{\alpha\alpha}^{\mathrm{AL(nl)}}(\omega) = \Gamma_{\alpha\alpha}^{2}(\omega)\, T \sum_{\Omega_k} \mathrm{tr}\left[\widehat{q}_\alpha D_{\alpha\alpha} \widehat{q}_\alpha\right] \tag{8.65}$$

since the vertex (8.61) does not depend on $\mathbf{q}$ and Ω_k. The residual sum is nothing else as, with the accuracy to the coefficient included in $\Gamma_{\alpha\alpha}^{2}$, the local expression for paraconductivity (e.g. (3.40), (3.43)), so the nonlocal paraconductivity can be expressed by means of the local one as [109]

$$\sigma_{\alpha\alpha}^{\mathrm{AL(nl)}}(\omega) = \frac{1}{(1 - i\omega\tau)^2}\sigma_{\alpha\alpha}^{AL}(\omega). \tag{8.66}$$

Hence, the paraconductivity in clean superconductor has the pole of the second order in frequency. It is different from the one-electron Drude conductivity which has the pole of the first order.

The physical sense of the obtained result is convenient to analyze, rewriting the full conductivity by means of (8.66) in the form

$$\sigma^{\mathrm{tot}} = \frac{\sigma_n}{1 - i\omega\tau - \sigma^{AL}/\sigma_n}.$$

This gives for resistivity

$$\rho^{\mathrm{tot}} = \rho_n - \rho_n^2 \sigma^{AL} - \frac{i\omega m_e}{\mathcal{N}_{\rceil} e^2}.$$

In the clean limit the spectrum of the plasma oscillations does not depend on the inter-electron interaction and therefore does contain any fluctuation correction.

8.5.3 *Discussion*

Let us discuss the results obtained. First of all it is necessary to stress the observed cancelation of the divergent parts of the DOS by the MT contribution. It is interesting that this cancelation happens up to the third order in the corresponding Laurent series in $T\tau$. Let us remember that the same cancelation of the divergent in frequency (ω^{-2}, ω^{-1} instead of τ^2, τ) terms in DOS and MT contributions (see Eqs. (8.50)–(8.51)) was found for the ballistic regime, where the similar problem aroused. Hence, one can conclude that the value of the quantum fluctuation correction in the clean limit does not depend on the order of calculation of $(\lim_{\omega\to 0,\tau\to\infty}(\sigma^{\mathrm{DOS}} + \sigma^{\mathrm{MT}}) = 0)$ and it turns negligible with respect to the quasiclassic paraconductivity.

Now let us discuss the nature of the second-order pole in the expression for paraconductivity. Its presence differs considerably the mechanisms of paraconductivity and one-electron Drude conductivity and indicates that the phenomenon of paraconductivity would be naive to imagine as the simple electric transport by particles with charge $2e$. The physical reason which reflects this fact is the following. The electric filed does not produce the direct effect on fluctuation Cooper pairs, although it immediately influences electrons. Thereby the retard in time by the value τ takes place. This means that an applied electric field firstly induces the Ohmic current of normal electrons, which, in turn, with the retard τ generates the supercurrent. Corresponding fluctuation correction to conductivity in this case depends on the order in which the limits $\omega \to 0, \tau \to \infty$ are performed: if ω is fixed even as infinitesimally small but nonzero value the paraconductivity vanishes in clean limit. In result all the effect of electric field on a superconductor in its normal phase in the limit $\tau \to \infty$ is reduced to the acceleration of electron gas and corresponding Drude conductivity with the further crossover to the Bloch oscillations in the case of perfectly pure metal. The interaction of the electric field with normal electrons without impurity scattering does not produce any effective force acting on the fluctuation condensate. Therefore, it is impossible to distinguish the motion of the electron liquid from the condensate motion in current experiments without additional scattering.

Let us stress that the nonlocal form of the Cooperon and fluctuation propagator have to be taken into account not only for the ultra-clean case but in every problem where relatively large bosonic momenta are involved: the consideration of dynamical and short wavelength fluctuations beyond the vicinity of critical temperature, the effect of relatively strong magnetic fields on fluctuations etc. Recently such an approach was developed in a number of studies [275, 278–282].

8.6 Fluctuation conductivity far from T_c

As we demonstrated above the role of fluctuations is especially pronounced in the vicinity of the critical temperature. Nevertheless for some phenomena they can be still considerable far from the transition too. In these cases the GL theory is certainly unapplicable since the high-frequency and short-wave fluctuation contributions have to be taken into account. However, this can be done in the framework of the formulated above microscopic approach.

At the beginning of this section we will discuss the effect of superconducting fluctuations on the conductivity of clean and impured superconductors far from the transition point. As it will be shown below, this problem is tightly related to the accounting for the effect of inter-electron interaction in particle-particle (Cooper) channel on conductivity of impured normal metal ($g < 0$ in Eq. (6.1)), where the role of T_c formally plays some fictitious temperature $T_0 \sim E_F \exp(\frac{1}{\nu|g|})$ and any temperature at which the metal exists satisfies the condition $T \ll T_0$. In this section we will mainly follow the results of the articles [108, 241, 271].

8.6.1 *Paraconductivity of clean superconductor for $T \gg T_c$*[48]

Let us consider the fluctuation contribution to conductivity of a clean superconductor at temperatures far from transition. As we already saw the contributions

[48] In this section we base on the results of [271].

of the DOS (8.50) and MT (8.51) type exactly canceled each other. This cancelation was demonstrated for nonzero bosonic frequency and it happens before the q-integration, so does not depend on the form of the propagator. Hence, in the case of clean superconductor, even being far from T_c, one can restrict the consideration of fluctuation contributions to conductivity only by the AL process.

Its general discussion we terminated above with the expressions (8.6), (8.7) and (8.13), so now we can continue basing on these results. We will be interested in the diagonal component of the $2D$ superconductor, so let us fix $\alpha = \beta = x$.

As it was already mentioned above the integrand function in (8.13) has breaks of analyticity at the lines $\operatorname{Im} z = 0$ and $\operatorname{Im} z = -i\omega_\nu$. Being in the local regime in the vicinity of the transition we assumed the Green functions blocks $B_\alpha(\mathbf{q}, -iz)$ to be frequency independent and have ignored their analytic properties. Here we cannot more use this approximation and the analytic structure of these blocks has to be taken into account.

In the previous section we have shown that in clean superconductor the ω-dependence of the block $B_\alpha(\mathbf{q}, \Omega_k = 0, \omega)$ has the pole structure and is responsible for the process of the indirect acceleration of Cooper pairs by electric field. The other source of the ω-dependence of the electromagnetic response operator is one of two propagators. Namely its expansion gives the linear term which determines paraconductivity. Therefore, being far from the critical temperature, we can proceed with Eq. (8.13) assuming in $B_\alpha(\mathbf{q}, \Omega_k, \omega_\nu)$ that $\omega_\nu = 0$ but carrying on its Ω_k-dependence.

Ignoring completely the electron scattering on impurities one can easily calculate the Green function block (8.7)

$$B_\alpha(\mathbf{q},\Omega_k, \omega_\nu = 0) = -\frac{\nu}{16\pi^2 T^2} \left\langle \sum_{n=0}^{\infty} \frac{v_\alpha v_\beta q_\beta}{\left[\left(n + \frac{1}{2} + \frac{|\Omega_k|}{4\pi T}\right)^2 + \left(\frac{\mathbf{vq}}{4\pi T}\right)^2\right]^{3/2}} \right\rangle_{FS}, \tag{8.67}$$

which, naturally, in the case $\mathbf{q} \to \mathbf{0}, \Omega_k = 0$ reproduces the clean limit of Eq. (8.10). The line $\operatorname{Im} z = 0$ separate the domains of the analyticity of the Green function block $B_\alpha(\mathbf{q},\Omega_k, \omega_\nu = 0)$, so the analytic functions B^R and B^A can be introduced.

As it was already done for the case of the vicinity of critical temperature the calculation of the contour integral in (8.13) can be reduced to the sum of three integrals along the contours C_1, C_2, C_3 which enclose domains of well defined analyticity of the integrand function. The integral along the large circle evidently vanishes and the contour integral is reduced to four integrals (8.14) along the cuts of the plane in Fig.8.2. Then one can shift the variable in the last integral to $z = z' - i\omega_\nu$, take into account that $i\omega_\nu$ is the period of $\coth \frac{z}{2T}$ and get an expression analytic in ω (according to the hypothesis that all $\omega-$dependence of the B_α is already extracted in the form of the Drude type pole, we ignore that one appearing in the second integral). As a result of the variable shift:

$$Q_{\alpha\alpha}^{AL}(\omega_\nu) = -\frac{2e^2}{\pi}\int\frac{d^2\mathbf{q}}{(2\pi)^2}\int_{-\infty}^{\infty}dz\coth\frac{z}{2T}\left[L^R(\mathbf{q},-iz+\omega_\nu)+\right.$$
$$\left.+L^A(\mathbf{q},-iz'-\omega_\nu)\right]\operatorname{Im}\left\{\left[B_\alpha^R(\mathbf{q},-iz)\right]^2 L^R(\mathbf{q},-iz)\right\}$$

Now one can make and analytic continuation $\omega_\nu \to -i\omega$ and to expand the propagator. In this way we get:

$$\sigma_{\alpha\alpha}^{AL}(T) = \frac{4e^2}{\pi}\int\frac{d^2\mathbf{q}}{(2\pi)^2}\int_{-\infty}^{\infty}dz\coth\frac{z}{2T}\times \tag{8.68}$$
$$\operatorname{Im}\left\{\left[B_\alpha^R(\mathbf{q},-iz)\right]^2 L^R(\mathbf{q},-iz)\right\}\frac{\partial}{\partial z}\left[\operatorname{Im}L^R(\mathbf{q},-iz)\right].$$

In the assumption of $\ln T/T_c \gg 1$ one can substitute summation in (8.53) by integration and find the explicit expression for the propagator:

$$L(\mathbf{q},\Omega_k) = -\frac{1}{\nu}\frac{1}{\ln\frac{T}{T_c}+\ln\left[\frac{1}{2}+\frac{|\Omega_k|}{4\pi T}+\sqrt{\left(\frac{1}{2}+\frac{|\Omega_k|}{4\pi T}\right)^2+\frac{v_F^2\mathbf{q}^2}{16\pi^2T^2}}\right]-\psi\left(\frac{1}{2}\right)}. \tag{8.69}$$

It is clear that the propagator is a constant for arguments $v_F\mathbf{q},\Omega_k \lesssim T^2/T_c$, while the block $B_\alpha^R(\mathbf{q},-iz)$ varies much before:

$$B_\alpha(\mathbf{q},\Omega_k,\omega_\nu=0) = -\nu\frac{v_F^2 q_\alpha}{16\pi^2T^2}\phi\left(\frac{|\Omega_k|}{4\pi T},\frac{v_F q}{4\pi T}\right) \tag{8.70}$$

with

$$\phi(t,x) = \frac{1}{\sqrt{\left(\frac{1}{2}+t\right)^2+x^2}}\frac{1}{\ln\left[\frac{1}{2}+t+\sqrt{\left(\frac{1}{2}+t\right)^2+x^2}\right]}. \tag{8.71}$$

The main term in the expansion of (8.68) over $1/\ln(T/T_c)$ appears from the contribution $\sim \operatorname{Im}B_\alpha^R(\mathbf{q},-iz)$

$$\sigma_{\alpha\alpha}^{AL}(T) = -\frac{2e^2}{\pi T}\int\frac{d^2\mathbf{q}}{(2\pi)^2}\int_{-\infty}^{\infty}dz\operatorname{Re}L^R(\mathbf{q},-iz)\times$$
$$\frac{\partial}{\partial z}\left[\operatorname{Im}L^R(\mathbf{q},-iz)\right]\operatorname{Re}B_\alpha^R(\mathbf{q},-iz)\operatorname{Im}B_\alpha^R(\mathbf{q},-iz)\coth\frac{z}{2T}$$
$$\approx\frac{2e^2}{\pi^2}\frac{1}{\ln^3\frac{T}{T_c}}\int x^3dx\int_{-\infty}^{\infty}\operatorname{Re}\phi(-it,x)\operatorname{Im}\phi(-it,x)\coth(2\pi t)\,dt.$$

Here $t=\frac{z}{4\pi T}$ and $x=\frac{v_F q}{4\pi T}$. One can see that for large arguments $\operatorname{Re}\phi(-it,x) \sim \left(t^2+x^2\right)^{-1/2}\ln^{-1}\left(t^2+x^2\right)$, while $\operatorname{Im}\phi(-it,x)\sim tx^{-1}\left(t^2+x^4\right)^{-1/2}\ln^{-1}(t^2+x^2)$. As a result the last integral is convergent and one can find

$$\sigma_{\alpha\alpha}^{AL}(T) \sim \frac{e^2}{\ln^3 \frac{T}{T_c}}. \tag{8.72}$$

8.6.2 *Fluctuation conductivity of impure superconductor for $T \gg T_c$*[49]

Let us start our analysis of the fluctuation contribution to the conductivity of superconductor far from the critical temperature from the expressions carried out above (8.6), (8.18), (8.30), (8.31), which have the general character. Let us suppose that the superconductor is dirty ($T\tau \ll 1$) and even far from T_c the local approximation for description of fluctuations in it is applicable (at the end of this section we will check what does it mean). Above we already restricted our consideration to this condition in the case of the vicinity of the critical temperature (see the discussion after the formula (6.26)) and will revise it in the following sections. Now, intending to move away from the region of the critical temperature, we will revise the expressions for all objects entering in the mentioned general expressions refraining from the assumption of $T - T_c \ll T_c$ ($\epsilon \ll 1$) and its consequences. We will carry on our discussion mainly on the example of the $2D$ system but at the end will list the results obtained for the $3D$ case.

8.6.2.1 *Propagator* One-electron Green function of the normal metal does not feel the superconducting transition point and it remains valid in the same form far from the transition. Hence, we still can use the expression (6.28) for the polarization operator, but now, supposing $\max\{\varepsilon_n, \Omega_k\} \ll \tau^{-1}$, instead of (6.26) and (6.8) one can simplify

$$\mathcal{P}(\mathbf{q}, \widetilde{\varepsilon}_{n+k}, \widetilde{\varepsilon}_{-n}) \approx \frac{2\pi\nu\theta(-\varepsilon_{n+k}\varepsilon_{-n})}{|\widetilde{\varepsilon}_{n+k} - \widetilde{\varepsilon}_{-n}|} \tag{8.73}$$

and get $\lambda(\mathbf{q}, \varepsilon_{n+k}, \varepsilon_{-n})$ in the form:

$$\lambda(\mathbf{q}, \varepsilon_{n+k}, \varepsilon_{-n}) = \frac{|\widetilde{\varepsilon}_{n+k} - \widetilde{\varepsilon}_{-n}|}{|\varepsilon_{n+k} - \varepsilon_{-n}| + \mathcal{D}q^2\theta(-\varepsilon_{n+k}\varepsilon_{-n})}. \tag{8.74}$$

Using these formula in (6.28) one can find

$$\Pi(\mathbf{q}, \Omega_k) = 2\pi\nu T \sum_{\varepsilon_n} \frac{\theta(-\varepsilon_{n+k}\varepsilon_{-n})}{|\varepsilon_{n+k} - \varepsilon_{-n}| + \mathcal{D}q^2\theta(-\varepsilon_{n+k}\varepsilon_{-n})} \tag{8.75}$$

$$\approx \nu\left[\ln\frac{\omega_D}{2\pi T} - \psi\left(\frac{1}{2} + \frac{|\Omega_k|}{4\pi T} + \frac{\mathcal{D}q^2}{4\pi T}\right)\right].$$

The corresponding expression for the propagator far from T_c can be found in the same way as it was done in the vicinity of transition: by means of summation

[49] In this section we base on the results of [241].

of the polarization operator bubbles and further solution of the Dyson equation (6.4). Let us stress that the selection of the ladder type diagrams is valid with the accuracy of $\ln \omega_D / T_c \gg 1$. Therefore the interval of temperatures where our consideration will be valid is restricted from above: $T \ll \omega_D$. Using the definition of the critical temperature (6.13) one can finally write the propagator as

$$L^{-1}(\mathbf{q}, \Omega_k) = -\nu \left[\ln \frac{T}{T_c} + \psi \left(\frac{1}{2} + \frac{|\Omega_k|}{4\pi T} + \frac{\mathcal{D}q^2}{4\pi T} \right) - \psi \left(\frac{1}{2} \right) \right]. \qquad (8.76)$$

Close to T_c and for momenta $\mathcal{D}q^2 \ll T$ it coincides with the expression (6.32).

Let us check the domain of applicability of this expression. First of all it is restricted to the limits of the applicability of the ladder approximation in the Dyson equation: $T \ll \omega_D$. Then, dealing with the Green functions in the expression for $\mathcal{P}(\mathbf{q}, \widetilde{\varepsilon}_{n+k}, \widetilde{\varepsilon}_{-n})$, we supposed the validity of the local approximation, which means that it is supposed that the momenta $q \ll l^{-1}$, while for frequencies in the expression (8.74) the restriction $|\Omega_k| \ll \tau^{-1}$ was accepted. The last conditions restrict the values of $\mathcal{D}q^2 \ll \tau^{-1}$ and $|k| \ll (T\tau)^{-1}$, but, as we see, these restrictions are not too severe: the obtained expression (8.76) is valid even for the large arguments of the digamma function and the value of $\ln(T/T_c)$ can be supposed in it as an arbitrary (with the only restriction of the validity of the BCS approximation $T \ll \omega_D$).

8.6.2.2 *DOS contribution* It turns out that far from the transition point the main contribution to the fluctuation conductivity originates from all diagrams besides the AL one, namely from the diagrams 2-10 of Fig. 8.1.

We start our discussion from the calculation of the characteristic diagram 5. Operating beyond the GL region we loose the smallness $\epsilon \ll 1$ in propagator and have no more grounds to restrict our consideration to the only frequency $\Omega_k = 0$ and small momenta. Let us remember that Eq. (8.21) was obtained in very general assumptions and we can use it in the case under consideration. Adopting $\varepsilon_n, \Omega_k, \omega_\nu \ll \tau^{-1}, q \ll l^{-1}$ one can simplify (8.21) up to the following expression

$$\begin{aligned} I_{\alpha\beta}^{(5)}(\mathbf{q} = 0, \varepsilon_n, \Omega_k, \omega_\nu) = & -2\pi\nu\tau^2 \delta_{\alpha\beta} \mathcal{D} \left\{ \theta\left(\varepsilon_n \varepsilon_{n+\nu}\right) \theta\left[\varepsilon_n\left(\varepsilon_n - \Omega_k\right)\right] \right. \\ & \left. + \theta\left(-\varepsilon_n \varepsilon_{n+\nu}\right) \theta\left[-\varepsilon_n\left(\varepsilon_n - \Omega_k\right)\right] \right\}. \qquad (8.77) \end{aligned}$$

Due to the presence of θ-functions we can rearrange its terms in such a way that the further summation over ε_n in (8.18) could be presented by two sums with finite and infinite limits:

$$\begin{aligned} \sum\nolimits_{\alpha\beta}^{(5)}(\mathbf{q}, \Omega_k, \omega_\nu) &= T \sum_{\varepsilon_n} \lambda^2(\mathbf{q}, \varepsilon_n, \Omega_k - \varepsilon_n) I_{\alpha\beta}^{(5)}(\mathbf{q} = 0, \varepsilon_n, \Omega_k, \omega_\nu) \\ &= -2\pi\nu\mathcal{D}T\delta_{\alpha\beta} \left(\sum_{n=-\infty}^{\infty} - \sum_{n=-\nu}^{-1} \right) \frac{\theta\left[\varepsilon_n\left(\varepsilon_n - \Omega_k\right)\right]}{\left(|2\varepsilon_n - \Omega_k| + \mathcal{D}q^2\right)^2}. \end{aligned}$$

The first sum is independent of the external frequency ω_ν and is consequently canceled out by analogous contributions from the remaining diagrams. Calculating the second sum, we obtain

$$\sum\nolimits_{\alpha\beta}^{(5)}(\mathbf{q},\Omega_k,\omega_\nu)=\sum\nolimits_{\alpha\beta}^{(5,1)}(\mathbf{q},\Omega_k,\omega_\nu)+\sum\nolimits_{\alpha\beta}^{(5,2)}(\mathbf{q},\Omega_k,\omega_\nu),$$

where

$$\begin{aligned}\sum\nolimits_{\alpha\beta}^{(5,1)}(\mathbf{q},\Omega_k,\omega_\nu) = {}& \theta(\Omega_k)\frac{\nu\mathcal{D}}{8\pi T}\delta_{\alpha\beta}\left[\psi'\left(\frac{1}{2}+\frac{|\Omega_k|}{4\pi T}+\frac{\omega_\nu}{2\pi T}+\frac{\mathcal{D}q^2}{4\pi T}\right)\right.\\ &\left.-\psi'\left(\frac{1}{2}+\frac{|\Omega_k|}{4\pi T}+\frac{\mathcal{D}q^2}{4\pi T}\right)\right]\end{aligned}$$

and

$$\begin{aligned}&\sum\nolimits_{\alpha\beta}^{(5,2)}(\mathbf{q},\Omega_k,\omega_\nu)\\ &=\theta(-\Omega_k)\,\theta(\Omega_k+\omega_{\nu-1})\frac{\nu\mathcal{D}}{8\pi T}\delta_{\alpha\beta}\left[\psi'\left(\frac{1}{2}-\frac{|\Omega_k|}{4\pi T}+\frac{\omega_\nu}{2\pi T}+\frac{\mathcal{D}q^2}{4\pi T}\right)\right.\\ &\quad\left.-\psi'\left(\frac{1}{2}+\frac{|\Omega_k|}{4\pi T}+\frac{\mathcal{D}q^2}{4\pi T}\right)\right].\end{aligned}\tag{8.78}$$

The function $\sum_{\alpha\beta}^{(5)}(\mathbf{q},\Omega_k,\omega_\nu)$ must now be continued analytically in the frequency ω_ν into the upper half-plane. The contributions to the electromagnetic response operator, analytic in the upper half-plane of the complex frequency and corresponding to $\sum_{\alpha\beta}^{(5,1)}(\mathbf{q},\Omega_k,\omega_\nu)$ and $\sum_{\alpha\beta}^{(5,2)}(\mathbf{q},\Omega_k,\omega_\nu)$, will be denoted by $Q_{\alpha\beta}^{(5,1)R}(\omega)$ and $Q_{\alpha\beta}^{(5,2)R}(\omega)$. Since the function $\theta(\Omega_k)$ in the expression for $\sum_{\alpha\beta}^{(5,1)}(\mathbf{q},\Omega_k,\omega_\nu)$ does not contain ω_ν, the analytic continuation is effected by the simple substitution $\omega_\nu\to -i\omega$. Putting next $\omega\ll T$, we obtain

$$Q_{\alpha\beta}^{(5,1)R}(\omega)=-\frac{i\omega e^2\mathcal{D}}{8\pi^2 T}\delta_{\alpha\beta}\sum_{k=0}\int\frac{d\mathbf{q}}{(2\pi)^D}\frac{\psi''\left(\frac{1}{2}+\frac{k}{2}+\frac{\mathcal{D}q^2}{4\pi T}\right)}{\ln\frac{T}{T_c}+\psi\left(\frac{1}{2}+\frac{k}{2}+\frac{\mathcal{D}q^2}{4\pi T}\right)-\psi\left(\frac{1}{2}\right)}.\tag{8.79}$$

Let us suppose that we are so far from the transition temperature that $\ln(T/T_c)\gg 1$. The sum over k can be replaced according to the condition $\Omega_k\tau\ll 1$ by an integral. One can easily see that it is enough to integrate over k only the numerator of the integrand, which decays as k^{-3} for $k\gg \mathcal{D}q^2/T$. As a result

$$\sigma_{\alpha\beta}^{(5,1)}(T)=-\frac{e^2\mathcal{D}}{4\pi^2 T}\delta_{\alpha\beta}\int\frac{d\mathbf{q}}{(2\pi)^D}\frac{\psi'\left(\frac{1}{2}+\frac{\mathcal{D}q^2}{4\pi T}\right)}{\ln\frac{T}{T_c}+\psi\left(\frac{1}{2}+\frac{\mathcal{D}q^2}{4\pi T}\right)-\psi\left(\frac{1}{2}\right)}.\tag{8.80}$$

When integrating over the momentum let us restrict here to $D=2$. In this case, one can recognize in the integrand function the complete derivative, so the integration becomes trivial. The only problem appears with the upper limit cut-off since the integral turns out double logarithmically divergent. It is resolved just by recalling that our local consideration is valid for $q \ll l^{-1}$, or for $x = \mathcal{D}q^2/4\pi T \lesssim 1/(T\tau)$. Finally, using the logarithmic asymptotic of $\psi(x)$ for large arguments (see Eq. (B.16)), one can find

$$\sigma_{\alpha\beta}^{(5,1)}(T) = -\frac{e^2}{4\pi^2}\delta_{\alpha\beta}\int_0^{1/(T\tau)} d\ln\left[\ln\frac{T}{T_c} + \psi\left(\frac{1}{2}+x\right) - \psi\left(\frac{1}{2}\right)\right]$$
$$= -\frac{e^2}{4\pi^2}\delta_{\alpha\beta}\left(\ln\ln\frac{1}{T_c\tau} - \ln\ln\frac{T}{T_c}\right). \tag{8.81}$$

The analytic continuation of $\sum_{\alpha\beta}^{(5,2)}(\mathbf{q},\Omega_k,\omega_\nu)$ turns out to more complicated because the θ-functions in (8.78) limit the region of summation over k to finite limits that depend on the frequency ω_ν. When calculating the sum over k in $Q_{\alpha\beta}^{(5,2)}(\omega_\nu)$ and in the subsequent analytic continuation of this function in ω_ν, we follow [108]. To this end we represent $Q_{\alpha\beta}^{(5,2)}(\omega_\nu)$ in the form:

$$Q_{\alpha\beta}^{(5,2)}(\omega_\nu) = -\frac{e^2\mathcal{D}}{4\pi}\delta_{\alpha\beta}\int\frac{d\mathbf{q}}{(2\pi)^2}\left[F(\mathbf{q},\omega_\nu)\right.$$
$$\left. -\frac{1}{2}\frac{\psi'\left(\frac{1}{2}+\frac{\omega_\nu}{2\pi T}+\frac{\mathcal{D}q^2}{4\pi T}\right) - \psi'\left(\frac{1}{2}+\frac{\mathcal{D}q^2}{4\pi T}\right)}{\ln\frac{T}{T_c}+\psi\left(\frac{1}{2}+\frac{\mathcal{D}q^2}{4\pi T}\right)-\psi\left(\frac{1}{2}\right)}\right], \tag{8.82}$$

where

$$F(\mathbf{q},\omega_\nu) = \sum_{k=0}^{\nu-1}{}' \frac{\psi'\left(\frac{1}{2}+\frac{k}{2}+\frac{\mathcal{D}q^2}{4\pi T}\right) - \psi'\left(\frac{1}{2}-\frac{k}{2}+\frac{\mathcal{D}q^2}{4\pi T}\right)}{\ln\frac{T}{T_c}+\psi\left(\frac{1}{2}+\frac{k}{2}+\frac{\mathcal{D}q^2}{4\pi T}\right)-\psi\left(\frac{1}{2}\right)}, \tag{8.83}$$

prime on the summation sign means that the term with $k = 0$ is taken with a factor 1/2.

The analytic continuation of the last term in (8.82) is carried out just as above ($\omega_\nu \to -i\omega$), after which we can expand it in terms of ω/T. To calculate $F(\mathbf{q},\omega_\nu)$ we transform the sum into a contour integral

$$F(\mathbf{q},\omega_\nu) = \sum_{k=0}^{\nu-1}{}' f(k,\mathbf{q},\omega_\nu) = \frac{1}{2}f(0,\mathbf{q},\omega_\nu) + \frac{1}{2i}\oint dz \coth(\pi z) f(-iz,\mathbf{q},\omega_\nu), \tag{8.84}$$

where the contour C is shown in Fig. 8.10. The integrals along the vertical parts of the contour tend zero when contour inflates to infinity. What concerns its horizontal parts the integral along them reduces, by change of variable, to an integral along the real axis. In result the dependence on ω_ν goes over from the integration limits into the argument of the function $f(x)$. The further analytic

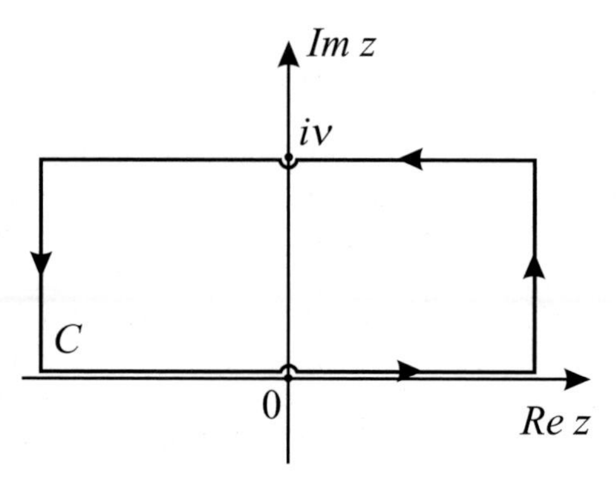

FIG. 8.10. Integration contour used in the analytic continuation of $Q^{(5,2)}(\omega_\nu)$.

continuation is carried out in standard way by the substitution $\omega_\nu \to -i\omega$, and we get

$$F^R(\mathbf{q},\omega) = \frac{1}{2i}\int_{-\infty}^{\infty} dz \coth(\pi z)\left[f(-iz,\mathbf{q},0) - f\left(-iz + \frac{i\omega}{2\pi T},\mathbf{q},0\right)\right] + \frac{1}{2}f(0,\mathbf{q},0). \tag{8.85}$$

Expanding the obtained expression in terms of the parameter ω/T and integrating by parts, we obtain

$$F^R(\mathbf{q},\omega) = \frac{i\omega}{4T}\int_{-\infty}^{\infty}\frac{dz}{\sinh^2(\pi z)}\frac{\psi'\left(\frac{1+iz}{2}+\frac{\mathcal{D}q^2}{4\pi T}\right) - \psi'\left(\frac{1-iz}{2}+\frac{\mathcal{D}q^2}{4\pi T}\right)}{\ln\frac{T}{T_c} + \psi\left(\frac{1-iz}{2}+\frac{\mathcal{D}q^2}{4\pi T}\right) - \psi\left(\frac{1}{2}\right)}. \tag{8.86}$$

Supposing $\ln(T/T_c) \gg 1$ and taking into account the fast convergence of the last integral over z, one can try to neglect z-dependence in the denominator and immediately to find that due to the oddness of the function the first term of the expansion is equal zero. The next term can be calculated by the expansion of the integrand function over z (see Appendix C):

$$F^R(\mathbf{q},\omega) = -\frac{i\omega}{24\pi T}\frac{\psi''\left(\frac{1}{2}+\frac{\mathcal{D}q^2}{4\pi T}\right)\psi'\left(\frac{1}{2}+\frac{\mathcal{D}q^2}{4\pi T}\right)}{\left[\ln\frac{T}{T_c} + \psi\left(\frac{1}{2}+\frac{\mathcal{D}q^2}{4\pi T}\right) - \psi\left(\frac{1}{2}\right)\right]^2}. \tag{8.87}$$

Finally, calculating the remaining integral with respect to q in (8.82) one gets

$$\sigma_{\alpha\beta}^{(5,2)}(T) = -\frac{e^2}{32}\delta_{\alpha\beta}\left(\frac{1}{\ln\frac{T}{T_c}} - \frac{\pi^2}{24\ln^2\frac{T}{T_c}}\right). \tag{8.88}$$

From a comparison of the expressions (8.81) and (8.88) it can be seen that the contribution of the considered diagram to the conductivity is determined by the term $\sigma_{\alpha\beta}^{(5,1)}$ and allowance for the term $\sigma_{\alpha\beta}^{(5,2)}$ together with $\sigma_{\alpha\beta}^{(5,1)}$ is an exaggeration of the accuracy.

Examination of the sixth diagram of Fig. 8.1, as it was demonstrated above, shows that its contribution to the conductivity coincides with the contribution of the fifth one: $\sigma^{(5)} = \sigma^{(6)}$. The contributions of the seventh and eighth diagrams are also equal to each other. As was shown for the dirty case in the vicinity of T_c they are smaller than $\sigma^{(5)}$ by a factor of three and are of opposite sign, i.e. $\sigma^{(7)} = \sigma^{(8)} = -\frac{1}{3}\sigma^{(5)}$. One can make sure that this statement was obtained before integration of propagator, so it has the general character. Therefore the total contribution of the four considered diagrams is $\sigma^{(5-8)} = \frac{4}{3}\sigma_1^{(5)}$.

We proceed to calculate diagrams 3–4, 9–10, containing the four-point vertex $\Gamma\ (\mathbf{q},\omega_1,\omega_2)$ (Fig. 8.1). In view of the vector character of the current vertices $e\mathbf{v}$, to obtain an nonzero result the Green functions of both sides of the vertex $\Gamma\ (\mathbf{q},\omega_1,\omega_2)$ must be expanded in powers of $\mathbf{q}\cdot\mathbf{v}$. Leaving out the intermediate calculations, which are similar to the preceding ones, we present only the expression for the conductivity due to the diagram 9:

$$\sigma_{\alpha\beta}^{(9)}(T) = -\frac{e^2}{4\pi}\frac{\mathcal{D}^2}{(2\pi T)^2}\delta_{\alpha\beta}\sum_{k=0}{}' \int \frac{q^2 d\mathbf{q}}{(2\pi)^D} \quad \frac{\psi'''\left(\frac{1}{2}+\frac{|k|}{2}+\frac{\mathcal{D}q^2}{4\pi T}\right)}{\ln\frac{T}{T_c}+\psi\left(\frac{1}{2}+\frac{|k|}{2}+\frac{\mathcal{D}q^2}{4\pi T}\right)-\psi\left(\frac{1}{2}\right)}.$$

In the $2D$ case

$$\sigma_{\alpha\beta}^{(9)}(T) = -\frac{e^2}{4\pi^2}\delta_{\alpha\beta}\left[\ln\ln\frac{1}{T_c\tau} - \ln\ln\frac{T}{T_c}\right]. \tag{8.89}$$

The tenth diagram reduces to the same expression, while the third and fourth make twice as large a contribution to the conductivity of the opposite sign, i.e. $\sigma^{(3)} = \sigma^{(4)} = -2\sigma^{(9)} = -2\sigma^{(10)}$, or $\sigma^{(3+4)}+\sigma^{(9+10)} = -2\sigma^{(9)}$. Gathering all eight diagrams together, we find that at temperatures $T \gg T_c$ the contribution made to the conductivity from diagrams 3–10 accurate to terms of order $[\ln T/T_c]^{-1}$ is of the form

$$\sigma_{\alpha\beta}^{(3-10)}(T) = \frac{e^2}{6\pi^2}\delta_{\alpha\beta}\left[\ln\ln\frac{1}{T_c\tau} - \ln\ln\frac{T}{T_c}\right]. \tag{8.90}$$

Let us attract the readers attention to the fact that at high temperatures the appearance of nonzero $\sigma^{(3)}, \sigma^{(4)}$ changed the sign of the DOS type diagrams. But this has no physical consequences: as we will see below the accounting for the regular part of the MT correction restores the correct sign. This fact indicates on the conventional character of the fluctuation contributions classification by the diagrams.

8.6.2.3 *MT contribution*[50] As we already know the second diagram from Fig. 8.1, when calculating fluctuation conductivity, contains an anomalous MT con-

[50]In this section we base on the results of [108].

tribution and in $2D$ case can turn out to be dominate one in the GL region. Let us consider its contribution to conductivity at temperatures remote from the critical one.

We can start from the general expressions (8.30) and (8.31) and use there corresponding to high temperature region expressions for the Cooperon (Eq. (8.74)) and fluctuation propagator (Eq. (8.76)). After some cumbersome exercise in application of the Cauchy theorem to the integral with twelve possible dispositions of the four poles for (8.31) instead of simplified expression (8.32) [108] one can find

$$Q^{MT}(\omega_\nu) = 2e^2\nu T \langle v_\alpha(\mathbf{p})v_\alpha(\mathbf{q}-\mathbf{p})\rangle \sum_{\Omega_k}\int \frac{d\mathbf{q}}{(2\pi)^D} L(\mathbf{q},\Omega_k) \times \left[\Sigma^{(\mathrm{reg1})} + \Sigma^{(an)} + \Sigma^{(\mathrm{reg2})}\right], \tag{8.91}$$

where

$$\Sigma^{(\mathrm{reg1})}(\Omega_k \geq 0) = \tau\frac{1}{4\omega_\nu}\left[\psi\left(\frac{1}{2}+\frac{\Omega_k+2\omega_\nu}{4\pi T}+\frac{\mathcal{D}q^2}{4\pi T}\right)-\psi\left(\frac{1}{2}+\frac{\Omega_k}{4\pi T}+\frac{\mathcal{D}q^2}{4\pi T}\right)\right]$$

and

$$\Sigma^{(an)}(\Omega_k \geq 0) + \Sigma^{(\mathrm{reg2})}(\Omega_k \geq 0) = = \tau\frac{\theta(\omega_\nu - \Omega_{k+1})}{4(\omega_\nu + \mathcal{D}q^2)}\left[\psi\left(\frac{1}{2}+\frac{-\Omega_k+2\omega_\nu}{4\pi T}+\frac{\mathcal{D}q^2}{4\pi T}\right)-\psi\left(\frac{1}{2}+\frac{\Omega_k}{4\pi T}+\frac{\mathcal{D}q^2}{4\pi T}\right)\right].$$

These results are obtained for $\Omega_k \geq 0$. In the following summation over Ω_k we will, instead of summing over all values of Ω_k, extract the term with $\Omega_k = 0$ and multiply the sum over $\Omega_k > 0$ by a factor 2. At $\Omega_k = 0$ these general expressions for Σ reduce to Eqs. (8.34) and (8.37) (in limit $\tau \to 0$).

The calculation of the regular part is trivial: one can substitute the summation by integration and then, with the logarithmic accuracy to calculate the momentum integral of the full derivative

$$\begin{aligned} Q^{MT(reg)}(\omega_\nu) &= \frac{i\omega}{4(2\pi T)^2}e^2\nu T v^2\tau \sum_{\Omega_k\geq 0}\int \frac{d\mathbf{q}}{(2\pi)^2}L(\mathbf{q},\Omega_k)\psi''\left(\frac{1}{2}+\frac{\Omega_k}{4\pi T}+\frac{\mathcal{D}q^2}{4\pi T}\right) \\ &= -\frac{i\omega e^2}{8\pi^2}\left[-2\int dx \frac{\psi'\left(\frac{1}{2}+x\right)}{\ln\frac{T}{T_c}+\psi\left(\frac{1}{2}+x\right)-\psi\left(\frac{1}{2}\right)}+\frac{1}{2}\frac{\psi'\left(\frac{1}{2}\right)}{\ln\frac{T}{T_c}}\right] \\ &= \frac{i\omega e^2}{4\pi^2}\left[\ln\frac{\ln\frac{1}{\tau T_c}}{\ln\frac{T}{T_c}}+O\left(\frac{1}{\ln\frac{T}{T_c}}\right)\right], \end{aligned} \tag{8.92}$$

or

$$\sigma^{MT(\mathrm{reg1})} = -\frac{e^2}{4\pi^2}\left[\ln\ln\frac{1}{T_{\mathrm{c}}\tau} - \ln\ln\frac{T}{T_{\mathrm{c}}}\right]. \tag{8.93}$$

Summarizing one can see that the total contribution of the DOS and the regular part of the MT diagrams consists of

$$\sigma^{DOS} + \sigma^{MT(\mathrm{reg})} = -\frac{e^2}{12\pi^2}\left[\ln\ln\frac{1}{T_{\mathrm{c}}\tau} - \ln\ln\frac{T}{T_{\mathrm{c}}}\right]. \tag{8.94}$$

We see that it is still negative, as it was in the vicinity of the critical temperature, and its magnitude very slowly decreases with temperature. By substitution $\epsilon \to \ln T/T_{\mathrm{c}}$ in Eq. (8.28) one can observe its surprisingly good match with the formula (8.90) in the region $T \sim 2T_{\mathrm{c}}$.

Let us analyze the anomalous part of the MT diagram. Using the same Eliashberg transformation of the sum in finite limits to the integral over the contour C of Fig. 8.10 one can find

$$Q^{MT(an)}(\omega) = \frac{i\omega e^2}{8\pi}\int\frac{dx}{x}\int_{-\infty}^{\infty}\frac{dt}{\sinh^2\pi t}\frac{\psi\left(\frac{1+it}{2}+x\right) - \psi\left(\frac{1-it}{2}+x\right)}{\ln\frac{T}{T_{\mathrm{c}}} + \psi\left(\frac{1-it}{2}+x\right) - \psi\left(\frac{1}{2}\right)}. \tag{8.95}$$

Taking into account the fast convergence of the integral over t one can carry it out expanding the integrand over $it/2$:

$$\sigma^{MT(an)} = \frac{e^2}{48\pi^2}\int_{\gamma_\phi}^{\infty}\frac{dx}{x}\frac{\left[\psi'\left(\frac{1}{2}+x\right)\right]^2}{\left[\ln\frac{T}{T_{\mathrm{c}}} + \psi\left(\frac{1}{2}+x\right) - \psi\left(\frac{1}{2}\right)\right]^2} = \frac{\pi^2 e^2}{192}\frac{\ln\frac{1}{\gamma_\phi}}{\ln^2\frac{T}{T_{\mathrm{c}}}}. \tag{8.96}$$

It is interesting to note that far from T_{c} the anomalous interval of summation does not contribute to the regular part of the MT contribution and as a result it changes the DOS contribution only by factor 3/2 instead of 2 as it was in the vicinity of the transition.

In conclusion of this section let us note the following circumstance. The main contribution (8.90) to the conductivity was obtained accurate to terms of order $\ln^{-1} T/T_{\mathrm{c}}$. Therefore at first glance the retention of (8.96) with it is not valid. However, there is a large logarithm $\ln\frac{1}{\delta}$, which can make the contribution (8.96) comparable to that one of (8.90).

In addition, we recall that the result (8.96) was obtained in the approximation with $\ln T/T_{\mathrm{c}} \gg 1$. However, in analogy with [283], the region of its applicability can be expanded by replacing $\pi^2/\left(12\ln^2 T/T_{\mathrm{c}}\right)$ by the function $\beta(T/T_{\mathrm{c}})$ which is tabulated in [283].

8.6.2.4 *AL contribution*[51] Let us discuss the contribution to conductivity of the AL process at temperatures far from the transition point. We have already

[51] In this section we base on the results of [108].

studied it in details both in the case of an arbitrary impurity concentration in the vicinity of transition and far from it for clean superconductor, so the structure of the diagram elements is familiar for us. In particular, we have seen above that the domains of different analyticity of the Green functions block (8.7) are separated by the lines $\operatorname{Im} z = 0$ and $\operatorname{Im} z = -i\omega_\nu$. In the case of impure superconductor under consideration the block $B_\alpha(\mathbf{q}, \Omega_k, \omega_\nu)$ can be calculated exactly without any additional assumptions.

We start from the ξ-integration of the three Green functions which (the disposition of the poles one can see in Fig. 8.3). One has to calculate the integral

$$J_\alpha^{AL}(\mathbf{q}, \varepsilon_n, \Omega_k, \omega_\nu) = \nu \left\langle v_\alpha(\mathbf{p}) \int_{-\infty}^{\infty} d\xi \frac{1}{\xi - i\widetilde{\varepsilon}_{n+\nu}} \frac{1}{\xi - i\widetilde{\varepsilon}_n} \frac{1}{\xi + i\widetilde{\varepsilon}_{n-k} - \mathbf{vq}} \right\rangle_{FS}. \tag{8.97}$$

This integration, as usual, can be performed by means of the Cauchy theorem. Four different poles dispositions contribute to J_α^{AL} :

$$\begin{aligned} &(1+2).\ \theta\left(\varepsilon_{n+\nu}\varepsilon_n\right) = 1;\quad \theta\left(\varepsilon_{n-k}\varepsilon_n\right) = 1; \\ &(3).\ \theta\left(-\varepsilon_{n+\nu}\varepsilon_n\right) = 1;\quad \theta\left(\varepsilon_{n-k}\varepsilon_n\right) = 1; \\ &(4).\ \theta\left(-\varepsilon_{n+\nu}\varepsilon_n\right) = 1;\quad \theta\left(-\varepsilon_{n-k}\varepsilon_n\right) = 1. \end{aligned} \tag{8.98}$$

The q-dependence of $B_\alpha(\mathbf{q}, \Omega_k, \omega_\nu)$ appears rather from the impurity vertices $\lambda\left(q, \varepsilon_1, \varepsilon_2\right)$, (which depends on q already for $lq \gtrsim T\tau\left(\mathcal{D}q^2 \sim T\right)$) than from Green functions (where the essential $lq \sim 1$). It is why the integrated by means of the Cauchy theorem expression for J_α^{AL} can be expanded in lq what gives

$$\begin{aligned} J_\alpha^{AL}(\mathbf{q}, \varepsilon_n, \Omega_k, \omega_\nu) = 2\pi\nu \left\langle v_\alpha \mathbf{vq} \right\rangle \Bigg\{ & \frac{\theta\left(\varepsilon_{n+\nu}\varepsilon_n\right)\theta\left(\varepsilon_{n-k}\varepsilon_n\right) \operatorname{sign}\varepsilon_n}{\omega_\nu} \\ & \times \left(\frac{1}{\left(\widetilde{\varepsilon}_{n-k} + \widetilde{\varepsilon}_n\right)^2} - \frac{1}{\left(\widetilde{\varepsilon}_{n-k} + \widetilde{\varepsilon}_{n+\nu}\right)^2} \right) \\ & - \frac{\theta\left(-\varepsilon_{n+\nu}\varepsilon_n\right)}{\omega_\nu + 1/\tau} \left(\frac{\theta\left(\varepsilon_{n-k}\varepsilon_n\right)}{\left(\widetilde{\varepsilon}_{n-k} + \widetilde{\varepsilon}_n\right)^2} + \frac{\theta\left(-\varepsilon_{n-k}\varepsilon_n\right)}{\left(\widetilde{\varepsilon}_{n-k} + \widetilde{\varepsilon}_{n+\nu}\right)^2} \right) \Bigg\}. \end{aligned} \tag{8.99}$$

One can see that the second term in the parenthesis gives contribution to $B_\alpha(\mathbf{q}, \Omega_k, \omega_\nu)$ which is small by the parameter $T\tau \ll 1$. Therefore, leaving in J_α^{AL} only the first term one can find:

$$\begin{aligned} \mathbf{B}(\mathbf{q}, \Omega_k, \omega_\nu) = -4\pi\nu\tau \left\langle \mathbf{v}(\mathbf{vq}) \right\rangle & \left(T \sum_{n=-\infty}^{-\nu-1} \theta\left(-\varepsilon_{n-k}\right) + T \sum_{n=0}^{\infty} \theta\left(\varepsilon_{n-k}\right) \right) \\ & \times \frac{1}{\left|\varepsilon_{n-k} + \varepsilon_{n+\nu}\right| + \mathcal{D}q^2} \frac{1}{\left|\varepsilon_{n-k} + \varepsilon_n\right| + \mathcal{D}q^2}. \end{aligned} \tag{8.100}$$

Summation over the fermionic frequency is performed by means of enumeration of different relations between bosonic frequencies Ω_k and ω_ν, what leads to the final result

$$B_\alpha(\mathbf{q}, \Omega_k, \omega_\nu) = -\nu \mathcal{D} q_\alpha \qquad (8.101)$$
$$\times \left(\frac{1}{\omega_\nu}\right) \left[\psi\left(\frac{1}{2} + \frac{|\Omega_k| + \omega_\nu + \mathcal{D}q^2}{4\pi T}\right) - \psi\left(\frac{1}{2} + \frac{|\Omega_k| + \mathcal{D}q^2}{4\pi T}\right)\right.$$
$$\left. + \psi\left(\frac{1}{2} + \frac{|\Omega_{k+\nu}| + \omega_\nu + \mathcal{D}q^2}{4\pi T}\right) - \psi\left(\frac{1}{2} + \frac{|\Omega_{k+\nu}| + \mathcal{D}q^2}{4\pi T}\right)\right].$$

Now in order to find the paraconductivity let us calculate the contour integral in (8.13). In the vicinity of critical temperature, where the block B_α was assumed to be frequency independent, three domains of analyticity of this integral were defined by the product of two propagators and were separated by the horizontal lines $\operatorname{Im} z = 0$ and $\operatorname{Im} z = -i\omega_\nu$. One can notice, that the block $B_\alpha(\mathbf{q}, \Omega_k, \omega_\nu)$ in the form (8.101) can be analytically continued in single-valued form in the same three domains of the complex plane. Namely, transforming the sum over bosonic frequencies in corresponding contour integral, one can introduce three functions B^{RR}, B^{RA} and B^{AA}. Each of them is chosen with the corresponding for the appropriate domain sign of the absolute values in the expression (8.101). These functions coincide with the block (8.101) for Matsubara frequencies, but they already are analytic in the corresponding domains of the plane of complex frequency. Hence we can again reduce the calculation of the contour integral in (8.13) to the sum of three integrals along the contours C_1, C_2, C_3 which enclose domains of a well-defined analyticity (see Fig. 8.2). The further procedure is known: shifting the variable in the integral along $\operatorname{Im} z = -i\omega_\nu$ to $z' = z - i\omega_\nu$, as it was done above, and taking into account that $i\omega_\nu$ is the period of $\coth \frac{z}{2T}$ one gets the expression for the electromagnetic response operator:

$$Q_{xx}^{AL}(\omega_\nu) = \frac{e^2}{2\pi i} \int \frac{d^2\mathbf{q}}{(2\pi)^2} \left\{ \int_{-\infty}^{\infty} dz \coth \frac{z}{2T} L^R(\mathbf{q}, -iz + \omega_\nu) \right.$$
$$\times \left[\left[B_x^{RR}(\mathbf{q}, -iz)\right]^2 L^R(\mathbf{q}, -iz) - \left[B_x^{RA}(\mathbf{q}, -iz)\right]^2 L^A(\mathbf{q}, -iz) \right]$$
$$+ \int_{-\infty}^{\infty} dz' \coth \frac{z'}{2T} L^A(\mathbf{q}, -iz' - \omega_\nu) \left[\left[B_x^{RA}(\mathbf{q}, -iz' - \omega_\nu)\right]^2 \right.$$
$$\left.\left. \times L^R(\mathbf{q}, -iz') - \left[B_x^{AA}(\mathbf{q}, -iz' - \omega_\nu)\right]^2 L^A(\mathbf{q}, -iz') \right] \right\}. \qquad (8.102)$$

Let us write down the explicit expressions for the analytic functions $B_x^{RR}(\mathbf{q}, \Omega_k, \omega_\nu)$, $B_x^{RA}(\mathbf{q}, \Omega_k, \omega_\nu)$, and $B_x^{AA}(\mathbf{q}, \Omega_k, \omega_\nu)$. This can be done in two steps. First let us present them in each domain of analyticity simply choosing in (8.101) the appropriate sign of the corresponding absolute value. Then the first analytic continuation $\Omega_k \to -iz$ can be easily done. For instance

$$B_x^{RA}(\mathbf{q}, \Omega_k \to -iz, \omega_\nu) = -\nu \mathcal{D} q_x \left(\frac{1}{\omega_\nu}\right) \left[\psi\left(\frac{1}{2} + \frac{iz + \omega_\nu + \mathcal{D}q^2}{4\pi T}\right)\right.$$

$$
-\psi(\frac{1}{2}+\frac{iz+\mathcal{D}q^2}{4\pi T})+\psi(\frac{1}{2}+\frac{-iz+2\omega_\nu+\mathcal{D}q^2}{4\pi T})-\psi(\frac{1}{2}+\frac{-iz+\omega_\nu+\mathcal{D}q^2}{4\pi T})\Big].
$$

Performing now the second analytic continuation $\omega_\nu \to -i\omega$ and tending $\omega \to 0$, one can expand the expressions for B in the series over ω and finally find:

$$
\begin{aligned}
\mathbf{B}^{RR}(\mathbf{q},-iz,-i\omega) &= -\frac{\nu\mathcal{D}\mathbf{q}}{2\pi T}\left[\psi'\left(\frac{1}{2}+\frac{-iz+\mathcal{D}\mathbf{q}^2}{4\pi T}\right)\right.\\
&\qquad\left.-\frac{i\omega}{4\pi T}\psi''\left(\frac{1}{2}+\frac{-iz+\mathcal{D}\mathbf{q}^2}{4\pi T}\right)\right],\\
\mathbf{B}^{RA}(\mathbf{q},-iz,-i\omega) &= -\frac{\nu\mathcal{D}\mathbf{q}}{4\pi T}\left\{\left[\psi'\left(\frac{1}{2}+\frac{iz+\mathcal{D}\mathbf{q}^2}{4\pi T}\right)-\frac{i\omega}{8\pi T}\psi''\left(\frac{1}{2}+\frac{iz+\mathcal{D}\mathbf{q}^2}{4\pi T}\right)\right]\right.\\
&\quad\left.+\left[\psi'\left(\frac{1}{2}+\frac{-iz+\mathcal{D}\mathbf{q}^2}{4\pi T}\right)-\frac{3i\omega}{8\pi T}\psi''\left(\frac{1}{2}+\frac{-iz+\mathcal{D}\mathbf{q}^2}{4\pi T}\right)\right]\right\},\\
\mathbf{B}^{RA}(\mathbf{q},-iz+i\omega,-i\omega) &= -\frac{\nu\mathcal{D}\mathbf{q}}{4\pi T}\left\{\left[\psi'\left(\frac{1}{2}+\frac{iz+\mathcal{D}\mathbf{q}^2}{4\pi T}\right)-\frac{3i\omega}{8\pi T}\psi''\left(\frac{1}{2}+\frac{iz+\mathcal{D}\mathbf{q}^2}{4\pi T}\right)\right]\right.\\
&\quad\left.+\left[\psi'\left(\frac{1}{2}+\frac{-iz+\mathcal{D}\mathbf{q}^2}{4\pi T}\right)-\frac{i\omega}{8\pi T}\psi''\left(\frac{1}{2}+\frac{-iz+\mathcal{D}\mathbf{q}^2}{4\pi T}\right)\right]\right\},\\
\mathbf{B}^{AA}(\mathbf{q},-iz+i\omega,-i\omega) &= -\frac{\nu\mathcal{D}\mathbf{q}}{2\pi T}\left[\psi'\left(\frac{1}{2}+\frac{iz+\mathcal{D}\mathbf{q}^2}{4\pi T}\right)-\frac{i\omega}{4\pi T}\psi''\left(\frac{1}{2}+\frac{iz+\mathcal{D}\mathbf{q}^2}{4\pi T}\right)\right].
\end{aligned}
$$

Let us evaluate first the contribution to $\delta Q_1^{(AL)R}(\omega)$ arising from the expansion in (8.102) of the propagator $L^R(\mathbf{q},-iz-i\omega)$ over $-i\omega$:

$$
\delta Q_1^{(AL)R}(\omega) = i\omega\frac{e^2}{\pi}\int\frac{d^2\mathbf{q}}{(2\pi)^2}\int_{-\infty}^{\infty}dz\coth\frac{z}{2T}\frac{\partial L^R(\mathbf{q},-iz)}{\partial z}\times
$$
$$
\left\{\left[B_\alpha^{RR}(\mathbf{q},-iz)\right]^2 L^R(\mathbf{q},-iz)-\left[B_\alpha^{RA}(\mathbf{q},-iz)\right]^2 L^A(\mathbf{q},-iz)\right\}.
$$

The integration in symmetric limits over z of any even function convoluted with $\coth z/2T$ cancels it out. So we have to leave in the integrand only the odd terms:

$$
\delta Q_1^{(AL)R}(\omega) = -i\omega\frac{e^2}{\pi}\int\frac{d^2\mathbf{q}}{(2\pi)^2}\int_{-\infty}^{\infty}dz\coth\frac{z}{2T}\frac{\partial\,\mathrm{Im}\,L^R}{\partial z}\cdot
$$
$$
\left\{\mathrm{Re}\,L^R\,\mathrm{Im}\left[B_\alpha^{RR}(\mathbf{q},-iz)\right]^2+\mathrm{Re}\left[B_\alpha^{RR}(\mathbf{q},-iz)\right]^2\mathrm{Im}\,L^R\right\},
$$

with $L^R = L^R(\mathbf{q},-iz)$. One can see that the first term will have a lower power of the large $\ln T/T_c \gg 1$ in the denominator than the last one, so we will be interested in it:

$$
\begin{aligned}
\delta\sigma_1^{(AL)} &= \frac{e^2}{\pi}\int\frac{d^2\mathbf{q}}{(2\pi)^2}\int_{-\infty}^{\infty}dz\coth\frac{z}{2T}\frac{\partial\,\mathrm{Im}\,L^R}{\partial z}\,\mathrm{Re}\,L^R\,\mathrm{Im}\left[B_\alpha^{RR}(\mathbf{q},-iz)\right]^2\\
&= \frac{e^2}{2\pi\nu^2}\int\frac{d^2\mathbf{q}}{(2\pi)^2}\int_{-\infty}^{\infty}dt\coth t\,\mathrm{Re}\,\psi'\left(\frac{1-it}{2}+\frac{\mathcal{D}\mathbf{q}^2}{4\pi T}\right)\mathrm{Im}\left[B_\alpha^{RR}(\mathbf{q},-iz)\right]^2
\end{aligned}
$$

$$\times \frac{\left[\ln\frac{T}{T_{\rm c}} + \operatorname{Re}\psi\left(\frac{1-it}{2} + \frac{\mathcal{D}\mathbf{q}^2}{4\pi T}\right) - \psi\left(\frac{1}{2}\right)\right]}{\left[\left[\ln\frac{T}{T_{\rm c}} + \operatorname{Re}\psi\left(\frac{1-it}{2} + \frac{\mathcal{D}\mathbf{q}^2}{4\pi T}\right) - \psi\left(\frac{1}{2}\right)\right]^2 + \operatorname{Im}^2\psi\left(\frac{1-it}{2} + \frac{\mathcal{D}\mathbf{q}^2}{4\pi T}\right)\right]^2}. \tag{8.103}$$

In spite of the importance in the further integration of the large arguments one can see that the term $\operatorname{Im}^2\psi$ in the denominator can be omitted in comparison with $\ln T/T_{\rm c}$:

$$\delta\sigma_1^{(AL)} = \frac{e^2}{\pi}\left(\frac{\mathcal{D}}{8\pi T}\right)^2 \int q_x^2 \frac{d^2\mathbf{q}}{(2\pi)^2}\int_{-\infty}^{\infty} dt \coth t$$
$$\times \frac{\operatorname{Re}^2\psi'\left(1/2 - it/2 + \frac{\mathcal{D}\mathbf{q}^2}{4\pi T}\right)\operatorname{Im}\psi'\left(1/2 - it/2 + \frac{\mathcal{D}\mathbf{q}^2}{4\pi T}\right)}{\left[\ln T/T_{\rm c} + \operatorname{Re}\psi\left(1/2 - it/2 + \frac{\mathcal{D}\mathbf{q}^2}{4\pi T}\right) - \psi\left(1/2\right)\right]^3}. \tag{8.104}$$

The singularity of $\coth t$ at small arguments is removed by $\operatorname{Im}\psi'(1/2 - it/2 + x)$. As we will see below the integral turns out to be logarithmically divergent, so we can substitute $\psi'\left(1/2 - it/2 + \frac{\mathcal{D}\mathbf{q}^2}{4\pi T}\right)$ by its asymptotic and get

$$\delta\sigma_1^{(AL)} = \frac{e^2}{8\pi^2}\int_0^{1/T\tau}\frac{dx}{x}\frac{1}{\left[\ln T/T_{\rm c} + \operatorname{Re}\psi(1/2 + x) - \psi(1/2)\right]^3} = \frac{e^2}{8\pi^2}\frac{1}{\ln^2 T/T_{\rm c}}.$$

After this exercise one can recognize that the largest contributions to AL paraconductivity occur from the ω-expansion in (8.102) of the blocks B^{RR} and B^{RA}, since the power of large logarithm in corresponding denominators is one less with respect to the expression for $\delta\sigma_1^{(AL)}$ and the final result turns out of the order

$$\delta\sigma^{(AL)} \sim O\left(\frac{1}{\ln T/T_{\rm c}}\right). \tag{8.105}$$

Let us recall, that the terms of this order we already obtained both in the DOS and MT contributions and they were found to be beyond the logarithmic accuracy of the main double logarithmic contribution (8.90).

Let us point out the considerable difference between the temperature dependencies of the paraconductivity far from $T_{\rm c}$ in clean (see Eq. (8.72)) and dirty (see Eq. (8.105)) cases. In the clean case the Green functions block B does not give any linear contribution in ω and the $\ln^{-3} T/T_{\rm c}$ dependence is direct consequence of the ω-expansion of the propagator. In the dirty case the main contribution to paraconductivity arises from the ω-expansion of the Green functions block. Moreover, the slow, logarithmic convergency of the momentum integration increases the power of the logarithm for another 1 and the result turns out $\sim \ln^{-1} T/T_{\rm c}$).

Thus, the effect of the fluctuations on the conductivity of a $2D$ dirty superconductor at temperatures far enough above critical one is determined by the expressions (8.90) and (8.96), which correspond to a fluctuation change of the DOS and to the MT process.

8.6.2.5 *Fluctuation conductivity of the 3D disordered superconductor* We discuss now the influence of the fluctuation pairing on the conductivity in $3D$ disordered superconductor far above the transition point. The corresponding corrections are determined by the same diagrams of Fig.8.1 and the same analytic expressions like (8.82), (8.92), (8.104) with the only specifics that the finale **q**-integrations must be carried out with allowance for the fact that the electronic spectrum is $3D$.

In the integration with respect to **q** the expression corresponding to $\sigma_{(3)}^{DOS}$ diverges formally at large momenta. This divergence is due to the fact that the expressions employed for the propagator and the vertices were obtained for momenta $q \ll l^{-1}$, and by taking this circumstance into account it is easy to eliminate this divergence by subtracting the corresponding quantity taken at $T = 0$. This yields

$$\Delta\left(\sigma_{(3)}^{DOS} + \sigma_{(3)}^{MT(\mathrm{reg})}\right) = -\frac{1.37e^2}{\pi^2}\left(\frac{T}{\mathcal{D}}\right)^{1/2}\frac{1}{\ln\left(T/T_{\mathrm{c}}\right)}. \tag{8.106}$$

The MT contribution has no singularities in the $3D$ case and can be obtained from expression (8.95) with account taken of the integration over the $3D$ electron momentum **q**. Since it cannot be assumed beforehand that t is small, we expand the numerator and denominator of the integrand in (8.92) in Taylor series in powers of it. Integrating the obtained series with respect to t and confining ourselves to the first nonvanishing term in the expansion in reciprocal powers of the large $\ln T/T_{\mathrm{c}}$, we obtain

$$\begin{aligned}\delta\sigma_3^{MT} &= -\frac{3^{1/2}\left(T\tau\right)^{1/2}}{16\pi^{1/2}}\int\frac{dx}{\sqrt{x}}\left[\ln T/T_{\mathrm{c}} + \psi\left(1/2+x\right) - \psi\left(1/2\right)\right]^{-2} \\ &\times\sum_{k,n=0}\left(-1\right)^{n+k}\frac{\psi^{(2n+1)}\left(1/2+x\right)\psi^{(2k+1)}\left(1/2+x\right)}{(2n+1)!\,(2k+1)!4^{n+k}}\left|B_{2(n+k+1)}\right|,\end{aligned}$$

where B_{2m} are Bernoulli numbers. An analysis of the convergence of this series shows that it suffices, with good accuracy, to retain its first term, after which we obtain

$$\Delta\sigma_3^{MT} = -\frac{3.4e^2}{\pi^2}\left(\frac{T}{\mathcal{D}}\right)^{1/2}\frac{1}{\ln^2\left(T/T_{\mathrm{c}}\right)}.$$

Calculation of the AL contribution is carried out in analogy with that for the first diagrams and leads to the result

$$\Delta\sigma_3^{AL} = -\frac{1.15e^2}{\pi^2}\left(\frac{T}{\mathcal{D}}\right)^{1/2}\frac{1}{\ln^2\left(T/T_c\right)}.$$

Thus, for a $3D$ impure superconductor far from the critical temperature, the same as in the vicinity of T_c, the contributions from the MT and AL processes turn out to be of the same order of magnitude. They are, however, smaller by the large logarithm compared with the main conductivity (8.106) correction.

8.6.3 *The effect of inter-electron interaction in Cooper channel on the conductivity of disordered metal*[52]

In contrast to the statement of classical theory of metals, requiring the saturation of the resistance at its residual value at low temperatures, it was demonstrated [284–289] that the effects of quantum interference of the non-interacting electrons and electron–electron interaction result in the appearance of a non-trivial temperature dependence of the resistance in this region. This range of problems had been extensively studied (see, e.g. [253, 243]) in the 1980s which resulted in the creation of the new beautiful chapter of condensed matter physics. In this relation special interest attracted the properties of the $2D$ disordered electron systems, where the corresponding localization phenomena side by side with the inter-electron interaction are well-pronounced. As we will see below, the accounting for dynamical and short-wave fluctuation effects made above for the case of superconducting fluctuations far from critical temperature allows almost automatically to write down the quantum corrections to conductivity related to the repulsive electron–electron interaction in Cooper channel. That is why, in order to diversify our presentation, we will continue it discussing such type of corrections on the important example of the conductivity of the disordered electron system at low temperatures.

In the first studies of the role of the electron–electron interaction on the low temperature conductivity [290–292] it was assumed to be of the screened Coulomb nature, what in the diagrammatic representation corresponds to the particle–hole, or so-called diffusive channel. Later, in the investigation of the magnetoresistance of disordered electronic systems, the importance of the electron–electron interaction in the particle-particle (Cooper) channel was revealed [283, 293]. In spite of its repulsion character it was described by means of the effective coupling constant in the complete analogy with the BCS scheme. The result of the summation of the bubble diagrams in the Dyson equation (6.4) turns out exactly in the same as (8.76)form but with some formal huge temperature $T_0 \gg E_F$ instead of the critical temperature of the superconducting transition. Such an effective propagator of the electron–electron repulsion in the Cooper channel for the disordered electron system (non superconducting metal) turns out to be of the opposite sign with respect to fluctuation propagator and

[52]In this section we base on the results of [241].

is small in the magnitude because of the presence of the large logarithm in the denominator of (8.109). Nevertheless, as it will be shown below, because of the specifics of the corresponding corrections to conductivity in some cases this type of interaction can be important for the analysis of the experimental results.

Considering the influence of the electron–electron interaction on the conductivity of the disordered electron system we can use the same formalism as in the case of fluctuation conductivity. Namely, supposing the metal being still far enough from the metal-insulator transition ($p_F l \gg 1$), one can simulate the disorder by scattering of the electrons on impurities with short-range and isotropic interaction potential and to use the one-electron Green's function in the usual form (6.2). The vertex part $\lambda(\mathbf{q}, \varepsilon_1, \varepsilon_2)$ corresponding to the accounting for impurity averaging of the Green functions product (Cooperon) and the Green functions bubble in the Cooper channel of interaction (polarization operator) from Fig. 6.1 are defined by the same expressions (6.26) and (6.28) taken in the dirty limit:

$$\lambda(\mathbf{q}, \varepsilon_1, \varepsilon_2) = \frac{|\widetilde{\varepsilon}_1 - \widetilde{\varepsilon}_2|}{|\varepsilon_1 - \varepsilon_2| + \mathcal{D}q^2\theta(-\varepsilon_1\varepsilon_2)}, \tag{8.107}$$

$$\begin{aligned}\frac{1}{\nu}\Pi(\mathbf{q}, \Omega_k) &= T\sum_{\varepsilon_n} \frac{\theta(\varepsilon_{n+k}\varepsilon_n)\left(1 - \tau\mathcal{D}q^2\right)}{|2\varepsilon_n + \Omega_k| + \mathcal{D}q^2\theta(\varepsilon_{n+k}\varepsilon_n)} \\ &= 4\pi T\nu\sum_{n=0} \frac{1}{2\varepsilon_n + \Omega_k + \mathcal{D}q^2} = \ln\frac{E_F}{2\pi T} - \psi\left(\frac{1}{2} + \frac{|\Omega_k| + \mathcal{D}q^2}{4\pi T}\right).\end{aligned}$$

Here we cut off the divergence at $\varepsilon_N \sim E_F$ in contrast to ω_D, as it was done in the case of superconducting metal. Introducing the parameter

$$T_0 = \frac{2\gamma_E E_F}{\pi}\exp\left(\frac{1}{\nu|g|}\right) \tag{8.108}$$

from the condition $\frac{1}{\nu}\Pi(0,0) = (g\nu)^{-1}$, one can solve the Dyson equation (6.4) and find the effective electron–electron interaction in the Cooper channel in form

$$\widetilde{L}(\mathbf{q}, \Omega_k) = \frac{1}{\nu}\frac{1}{\ln\frac{T_0}{T} - \psi\left(\frac{1}{2} + \frac{|\Omega_k|}{4\pi T} + \frac{\mathcal{D}q^2}{4\pi T}\right) + \psi\left(\frac{1}{2}\right)}. \tag{8.109}$$

Here it is necessary to emphasize that the found effective propagator $\widetilde{L}(\mathbf{q}, \Omega_k)$ noticeably differs from the Coulomb interaction. The fluctuation interaction takes in fact accounting for the Coulomb interaction in Cooper channel diagrams in all orders of perturbation theory in the ladder approximation. However, the exact dynamic screened Coulomb interaction here is replaced by a model interaction described by a positive coupling constant; in this way both the dominate Coulomb repulsion and the electron-phonon attraction are taken into account.

In order to calculate the corresponding to this interaction corrections to conductivity it is enough to repeat the same consideration which already was done in section 8.3 for the fluctuation corrections i.e. to calculate ten diagrams from the Fig. 8.1 without the assumptions of the importance of small momenta and frequencies only. Namely this program was accomplished in the previous section and we can use the corresponding results with the minimal modifications. Substituting $\ln T/T_c \to -\ln T_0/T$ one can rewrite the final expressions (8.94), (8.96) and (8.105) as

$$\sigma_2^{DOS} + \sigma_2^{MT(\text{reg})} = \frac{3e^2}{4\pi^2}\left[\ln\frac{\ln\frac{T_0}{T}}{\ln T_0\tau}\right], \tag{8.110}$$

$$\sigma_2^{MT(an)} = \frac{\pi^2 e^2}{192}\frac{\ln\frac{1}{\gamma_\phi}}{\ln^2\frac{T_0}{T}}, \tag{8.111}$$

$$\delta\sigma_2^{(AL)} \sim -O\left(\frac{1}{\ln T_0/T}\right). \tag{8.112}$$

We see that the main contribution (8.110), related to the one electron DOS renormalization, changes its sign with respect to the analogous contribution in superconductor: in the case of non-superconducting dirty film this quantum correction is positive. The sign of the anomalous MT contribution does not depend on the sign of the inter-electron interaction: it is always positive. Finally, in the case of non-superconducting metal the paraconducting contribution turns out negative, but this does not matter since it lies beyond the accuracy of the calculations of the main contributions.

In the case of $3D$ dirty metal, in order to eliminate the unphysically divergent terms, we can again introduce the difference $\Delta\sigma(T) = \sigma(T) - \sigma(0)$ and to find

$$\Delta\left(\sigma_{(3)}^{DOS} + \sigma_{(3)}^{MT(\text{reg})}\right) = \frac{1.37e^2}{\pi^2}\left(\frac{T}{\mathcal{D}}\right)^{1/2}\frac{1}{\ln(T_0/T)}, \tag{8.113}$$

$$\Delta\sigma_3^{MT} = -\frac{3.4e^2}{\pi^2}\left(\frac{T}{\mathcal{D}}\right)^{1/2}\frac{1}{\ln^2(T_0/T)}, \tag{8.114}$$

$$\Delta\sigma_3^{AL} = -\frac{1.15e^2}{\pi^2}\left(\frac{T}{\mathcal{D}}\right)^{1/2}\frac{1}{\ln^2(T/T_c)}, \tag{8.115}$$

In conclusion, let us compare the results obtained with other temperature-dependent corrections to the conductivity and discuss the situation as a whole.

The foregoing analysis of the influence of the fluctuation interaction on the conductivity of $2D$ and $3D$ disordered electron systems has shown that the decisive corrections are those connected with the change of the DOS under the influence of the electron–electron interaction in the Cooper channel, σ^{DOS}. In the $2D$ case with a weak pair-breaking the MT contribution σ_2^{MT} may also turn out to be significant.

It must be noted that summation of all the ladder diagrams is most significant in the considered problem. In [294] were calculated the analogous corrections to the conductivity in first order perturbation theory in the coupling constant. As seen from (8.94) and (8.96), we can confine ourselves to first order perturbation theory only in the case of extremely weak coupling constant $g \ll [\nu \ln T/T_c]^{-1}$. In the opposite case, as shown in [241], allowance for all the diagrams in the Cooper channel leads to a considerable change in the form of the singularity of the temperature-dependent corrections to the conductivity: in place of $\ln(T\tau)$, as in [294], the contribution σ^{DOS} turns out according to Eq. (8.94) to be proportional to $\ln\ln T/T_c$. In addition, in first order in g it is impossible in principle to calculate the MT contribution.

It is of interest to compare the conductivity corrections due to the fluctuation interaction with the contribution made to the conductivity by the interaction in the diffusion channel, $\delta\sigma^D$. As shown in [295], it is convenient to present the expression for $\delta\sigma_2^D$ as a sum of contributions from the interaction with a total spin of the electron and hole equal to 0 and 1:

$$\delta\sigma_2^D = \delta\sigma_2^{j=0} + \delta\sigma_2^{j=1}, \tag{8.116}$$

$$\delta\sigma_2^{j=0} = \frac{e^2}{2\pi^2}\ln(T\tau), \tag{8.117}$$

$$\delta\sigma_2^{j=1} = \frac{3\lambda_\sigma^{j=1}e^2}{4\pi^2}\ln(T\tau), \tag{8.118}$$

where $\lambda_\sigma^{j=1}$ is a constant that depends on the magnitude and the character of the electron–electron interaction. It can be seen from (8.117) that the correction $\delta\sigma_2^{j=0}$ is larger and increases more rapidly with temperature than $\delta\sigma_2^{DOS}$(at $\ln T_0/T \gg 1$). We note, however, the following circumstances:

- The contributions $\delta\sigma_2^{j=0}$ and $\delta\sigma_2^{j=1}$ can cancel one another to a considerable degree.
- The value of $\delta\sigma_2^D$ in the presence strong spin-orbit scattering is universal: $\delta\sigma_2^D = \delta\sigma_2^{j=0}$.
- The quantity $\lambda_\sigma^{j=1}$ can sometimes be determined independently.
- The corrections to the conductivity on accounting for the interaction in the Cooper channel and in the diffusion channel depend differently on the magnetic field H: the contribution $\delta\sigma_2^{DOS}$ ceases to depend on the temperature already at

$$H \gtrsim H_{\rm int} = \frac{\pi T}{2\mathcal{D}e}.$$

The corresponding contribution to the magnetoresistance is considered in [293], and the corrections to the density of states, proportional to $\ln\ln(eH/mcT_0)$, were obtained in [296]. At the time, $\delta\sigma_2^{j=0}$ does not depend at all on H in the region of classically weak magnetic fields, while in $\delta\sigma_2^{j=1}$ the temperature dependence is suppressed [297] at $\gamma\mu_B H \sim T$ (γ is the gyromagnetic ratio and μ_B is the Bohr magneton). Since usually $H_{\text{int}} \ll T/\gamma\mu_B$ a study of the temperature dependence of the conductivity in different magnetic fields makes it possible to separate the contributions $\delta\sigma_2^{j=0}$, $\delta\sigma_2^{j=1}$, and $\delta\sigma_2^{DOS}$.

For all the reasons above, the contribution $\delta\sigma_2^{DOS} + \delta\sigma_2^{MT}$ of the interaction in the Cooper channel should be taken into account even at $\ln T_0/T \gg 1$, which is a sufficient condition for the validity of expressions (8.94), (8.96) and (8.105). The same arguments remain in force also in the $3D$ case, where the temperature dependence of σ_3^D differs from σ_3^{DOS} (see Eq. (8.115)) in that the former does not contain the factor $\ln^{-1} T_0/T$.

The qualitative form of the dependence of the resistance on the temperature for normal and superconducting disordered metals is shown in Fig. 8.11. The downward sequence of the designations of the corrections to the resistance corresponds to the hierarchy of these quantities in this temperature region. In experiments on thin aluminum films [41] it was observed that the resistance depends on temperature in just this manner, and there was also a quantitative agreement with the results cited above.

8.7 Nonlinear fluctuation effects[53]

As we have already seen in the temperature region $Gi \ll \epsilon \ll 1$ the thermodynamic fluctuations of the order parameter Ψ can be considered to be Gaussian. Nevertheless in transport phenomena the nonlinear effects, related to the interaction of fluctuations (higher order corrections) can manifest themselves much earlier. It has been found [276], that nonlinear fluctuation phenomena restrict the Gaussian region in the fluctuation conductivity of a superconducting film to a new temperature scale: $\sqrt{Gi_{(2d)}} \ll \epsilon \ll 1$ (see also [251,252,257,260,277,299]).

Why do nonlinear fluctuation effects manifest themselves in transport coefficients so much earlier than in thermodynamic characteristics? We have already seen above that fluctuation induced phase-breaking time $\tau_\varphi^{(\text{fl})}$ decreases when temperature tends to T_c. This leads to the saturation of the MT contribution in the region $\epsilon \ll \sqrt{Gi_{(2d)}}$. In this section we will show that the AL contribution, in contrast to MT one, does not saturate at $\epsilon \ll \sqrt{Gi_{(2d)}}$ and even can diverge strongly than in the GL region. This is due to the facts that: (a). AL contribution is not phase sensitive; (b). in this temperature region the TDGL theory coefficient $\gamma^{(\text{fl})}(\epsilon) = a(\epsilon)\,\gamma_{\text{GL}}$ growths.

Let us recall that TDGL theory was derived as the expansion over small frequencies of the microscopic expression for fluctuation propagator, which gave $\gamma_{\text{GL}} = \pi\alpha/8$. The possibility of such expansion supposed that the life time of the fluctuation Cooper pair τ_{GL} is the largest characteristic time in the system. Nevertheless, at least in conventional, low critical temperature, superconductors some other slow decay processes exist and they can result in the increase of $\gamma^{(\text{fl})}$ just as chemical reactions can increase viscosity of liquid (the Mandelstam–Leontovich effect). One of such mechanisms is the electron phase relaxation,

[53] In this section we base on the results of [75].

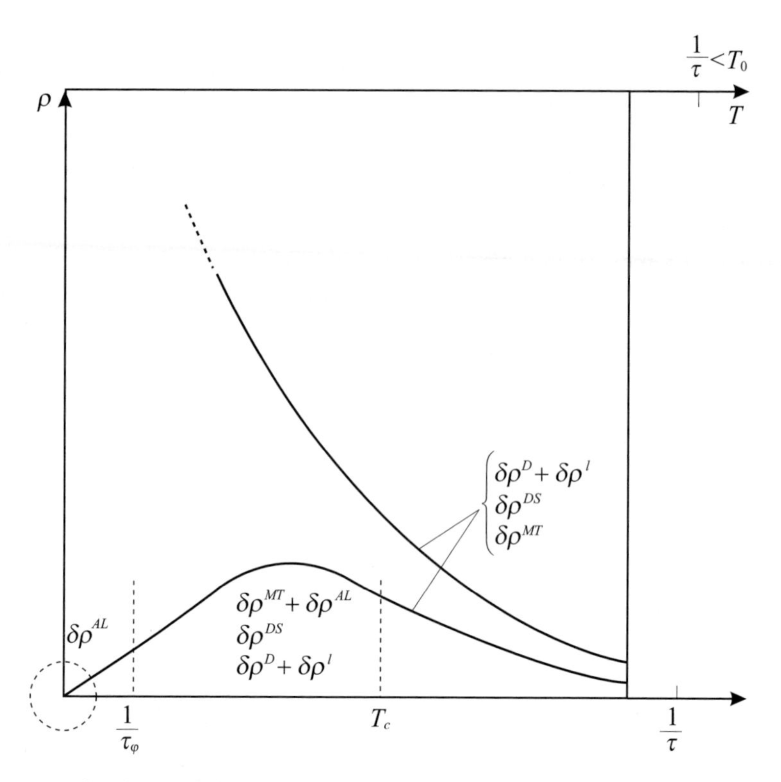

FIG. 8.11. Qualitative form of the temperature dependence of the resistance of a disordered superconductor (lower curve) and of a normal metal (upper curve). $\delta\rho^D$ is contribution made to the resistance by the interaction in the diffusion channel [290, 291]; $\delta\rho^L$ is correction to the resistance from the localization effect [298]; $\delta\rho^{DOS} = \delta\sigma^{DOS}\sigma_n^{-2}$ is the resistance contribution obtained above and due to the change of the density of the single-electron states as a result of the electron–electron interaction in the Cooper channel; $\delta\rho^{MT}$and $\delta\rho^{AL}$-MT [40, 41, 108] and AL [36, 108] contributions. The dashed circle separates the critical region. We note that in the $2D$ case $\delta\rho^D \sim \delta\rho^L$, and in the $3D$ case $\delta\rho^D \gg \delta\rho^L$. For a superconductor in the temperature region $\tau_\varphi^{-1} \ll T - T_c \ll T_c$ in the $2D$ case, the predominant contribution to the resistance is $\delta\rho^{MT}$ ($\left|\delta\rho^{MT}\right| \gg \left|\delta\rho^{AL}\right|\}$, whereas in the $3D$ case these contributions turn out to be of the same order of magnitude.

another is its energy relaxation. The characteristic time of the latter τ_ε is determined by the electron inelastic scattering and in the impure $2D$ case within the accuracy of large logarithm is given by Eq. (8.46). When $\epsilon \ll \sqrt{Gi_{(2d)}}$ the main contribution to $\gamma^{(\mathrm{fl})}(\epsilon)$ give the electrons with energies of the order of already formed fluctuation pseudogap Δ_{pg}. They are too slow to follow in their relaxation the fast variation of the fluctuation order parameter (characteristic time τ_{GL}). As result $\gamma^{(\mathrm{fl})}$ growths with respect to γ_{GL} as temperature tends to T_{c}, what leads, in accordance with Eq. (3.26), to the additional growth of the paraconductivity.

Let us obtain expression for the paraconductivity in the temperature region $Gi_{(2d)} < \epsilon < \sqrt{Gi_{(2d)}}$, where both the perturbation theory works well and the nonlinear fluctuation effects are already important. In this region fluctuations still weakly affect on the thermodynamic parameters of the GL functional but the dynamical coefficient $\gamma^{(\mathrm{fl})}(\epsilon) = a\gamma_{\mathrm{GL}}$ undergoes strong changes since it is sensitive to the form of the excitation spectrum. We will restrict our consideration to the simplest case of dirty superconductor with s-pairing and relatively low critical temperature $T_{\mathrm{c}} \ll \omega_D$.[54]

Let us start from the correlator (6.32) which can be expressed by means of the $Gi_{(2d)}$ number:

$$\langle \Psi_{\mathbf{k}}^* \Psi_{\mathbf{k}} \rangle = \frac{T}{\nu} \frac{1}{\epsilon + \frac{\pi \mathcal{D}}{8T} \mathbf{k}^2} = \frac{32\pi^3}{7\zeta(3)} Gi_{(2d)} \frac{T^2}{k^2 + \frac{8T\epsilon}{\pi \mathcal{D}}}. \tag{8.119}$$

The long-wavelength fluctuations with $k^2 < k_{\min}^2 = 8T\epsilon/\pi\mathcal{D}$ can be considered as a local condensate. They lead to the formation of the pseudogap

$$\Delta_{pg} = \left[\int_{k^2 \lesssim k_{\min}^2} \frac{d^2 k}{(2\pi)^2} \langle \Psi_{\mathbf{k}}^* \Psi_{\mathbf{k}} \rangle \right]^{1/2} \simeq T \sqrt{Gi_{(2d)}} \tag{8.120}$$

in the single-particle excitations spectrum.

Not very close to the transition ($\epsilon > \sqrt{Gi_{(2d)}}$) only excitations with energies $E > \Delta_{pg}$ are important. The pseudogap does not play any role for them. Thus, in this region of temperatures it is sufficient to consider fluctuations in the linear approximation only (see [36, 40, 41]). However, in the temperature region $\epsilon < \sqrt{Gi_{(2d)}}$ the nonlinear fluctuation contribution of the excitations with energies $E < \Delta_{pg}$ becomes essential.

To take into account the spatial dependence of the order parameter we will use the results obtained in [300]. It was shown there that the spatial variations of Δ_{pg} act on single-particle excitations in the same way as magnetic impurities do (the analogy between the effect of fluctuations and magnetic impurities was observed in many papers, see e.g. [301]). In this case, the total pair-breaking rate Γ can be written as a sum of the pair-breaking rate due to the magnetic impurities and the fluctuation term. Thus, the self-consistent equation for Γ can be written as [300]:

$$\Gamma = \int \frac{d^2 k}{(2\pi)^2} \frac{\langle \Psi_k^* \Psi_k \rangle}{E + \frac{1}{2}\mathcal{D}k^2 + \Gamma} + \frac{1}{\tau_s}. \tag{8.121}$$

[54] If one of these conditions is violated, the other, mechanisms of the $\gamma^{(\mathrm{fl})}(\epsilon)$ energy dependence different from fluctuation can become important. For instance in the case of HTS this can be the specifics of the phonon or paramagnon spectrum.

In the region $E \lesssim \Gamma$, $\Gamma \gg T\epsilon$ we obtain from Eqs.(8.121), (8.119):

$$\Gamma \sim T\left(Gi_{(2d)}\right)^{1/2} \simeq \Delta_{pg}, \tag{8.122}$$

which coincides with the results obtained in [251, 257].

Let us note, that the pair-breaking rate Γ was found to be of the order of the pseudogap Δ_{pg}. Thus, a wide maximum appears in the DOS at $E \sim \Delta_{pg}$. As we already saw (8.44), in purely $2D$ case the MT correction to the conductivity saturates for $T\epsilon < \Gamma$ (where $\Gamma = 8T\gamma_\varphi/\pi$) and takes the form [75]:

$$\frac{\delta\sigma^{MT}}{\sigma_n} \sim \frac{T}{\Gamma} Gi_{(2d)} \ln \frac{\pi\Gamma}{8T\epsilon}. \tag{8.123}$$

As it can be seen from Eq. (8.122) such a saturation takes place when $\epsilon < \sqrt{Gi_{(2d)}}$. Similar results have been obtained in [251, 252, 260], with slightly different numerical coefficients.[55] However, its exact value is not very important since in the region $T\epsilon < \Gamma$ the MT correction is less singular than the AL one and can be neglected. The latter does not saturate when T tends to T_c but becomes more singular.

In the presence of the pseudogap the fluctuating Cooper pair lifetime increases with respect to the GL one: $\gamma^{(\mathrm{fl})} = a\gamma_{\mathrm{GL}}$ $(a > 1)$. Recall, that analogous changes in the coefficient a in the TDGL equations appear below the transition temperature (see e.g. [98, 104, 174, 302, 303]). The growth of the coefficient a and, consequently, the increase of the lifetime of fluctuation pairs occurs because the quasiparticles require more time to attain thermal equilibrium (the corresponding time we denote as τ_e). A rough estimate gives $a \sim \Delta_{pg}\tau_e$. In the case of a weak energy relaxation, τ_e has to be determined from the diffusion equation taking account for the pseudogap (see [174, 303, 304]). Note, that in this complicated case the coefficient a becomes a non-local operator. Rough estimates give the following value for the thermal equilibrium transition time $\tau_e \sim (\mathcal{D}k_{\min}^2)^{-1} \sim (T\epsilon)^{-1}$. Taking into account (8.120) we obtain from (1.9) for the paraconductivity contribution in the discussed limit of a weak energy relaxation [75] :

$$\frac{\delta\sigma}{\sigma_n} \sim \frac{Gi_{(2d)}^{3/2}}{\epsilon^2}. \tag{8.124}$$

Let us discuss now the role of the energy relaxation processes, characterized by a quasiparticle lifetime τ_ε. Nonelastic electron scattering by phonons and other possible collective excitations can decrease τ_e significantly. These processes together with additional pair-breaking processes (due to magnetic impurities or a magnetic field) lead to a decrease of the nonlinear effects. In view of these

[55] Note that the numerical coefficient in Eq. (8.123) depends on the definition of $Gi_{(2d)}$ and how the summation of higher order diagrams is made.

processes, one can write the following interpolation formula for the nonlinear fluctuation conductivity [75]:

$$\frac{\delta\sigma}{\sigma_n} = \frac{Gi_{(2d)}}{\epsilon}\left[1 + \frac{Gi_{(2d)}}{\left(\epsilon + \frac{1}{T\tau_\varepsilon}\right)} \operatorname{Im} \frac{1}{\left(\epsilon + \frac{i}{T\tau_\varphi}\right)}\right]. \tag{8.125}$$

One can see that in the region of temperatures $\epsilon \gg \sqrt{Gi_{(2d)}}$ the nonlinear correction is still small, while closer to transition $\epsilon \lesssim \sqrt{Gi_{(2d)}}$ it can exceed the linear AL contribution. Indeed, as we demonstrated above (see Eq. (8.49)) in this interval of temperatures $1/T\tau_\varphi \sim \sqrt{Gi_{(2d)}}$. The problem of the energy relaxation time τ_ε is studied even less than that one for τ_φ. It can be assumed to be the same as in normal metal: $\tau_\varepsilon^{-1} \sim TGi_{(2d)}$ [243] in the belief that fluctuations are almost static. The analysis of Eq. (8.125) in this assumption demonstrates that in the interval of temperatures $Gi_{(2d)} \lesssim \epsilon \lesssim \sqrt{Gi_{(2d)}}$ nonlinear correction $\delta\sigma$ exceeds the linear fluctuation correction. Yet it is enough for energy relaxation time to coincide with phase-breaking time $\tau_\varepsilon = \tau_\varphi \sim 1/T\sqrt{Gi_{(2d)}}$ and contribution of the nonlinear effects will be of the same order as of the linear ones.

PART III

MANIFESTATION OF FLUCTUATIONS IN OBSERVABLES

9

FLUCTUATIONS IN MAGNETIC FIELD

Experimental investigations of the fluctuation magnetoconductivity are of special interest first because this physical value weakly depends on the normal state properties of a superconductor and second due to its special sensitivity to temperature and magnetic field. The role of the AL contribution for both the in-plane and out-of-plane magnetoconductivities was studied above in the framework of the phenomenological approach. The microscopic calculations of the other fluctuation corrections to the in-plane magnetoconductivity show that the MT contribution has the same positive sign and temperature singularity as the AL one. In the case of weak pair-breaking it can even considerably exceed the latter. The negative DOS contribution, like in the case of the zero-field conductivity, turns out to be considerably less singular and many authors (see e.g. [305–316]) successfully explained the in-plane magnetoresistance data in HTS using the AL and MT contributions only [274, 281, 317, 318].

Turning to the out-of-plane magnetoconductivity of a layered superconductor one can find quite a different situation. Both the AL and MT contributions turn out here to be of the second order in the interlayer transparency and this circumstance makes the less singular DOS contribution (which remains, however, of first order in transparency) competitive with the main terms [319]. The large number of microscopic characteristics involved in this competition, such as the Fermi velocity, interlayer transparency, phase-breaking and elastic relaxation times, gives rise to the possibility of the occurrence of different scenarios for various compounds. The c-axis magnetoresistance of a set of HTS materials shows a very characteristic behavior above T_{c0}. In contrast to the in-plane magnetoresistance which is positive at all temperatures, the out-of-plane magnetoresistance has been found in many HTS compounds (BSSCO [320–324], LSSCO [325], YBCO [326] and TlBCCO [327]) to have a negative sign not too close to T_{c0} and turn positive at lower temperatures. We will show how this behavior find its explanation within the fluctuation theory.

9.1 Magnetoconductivity in the vicinity of transition

We consider here the effect of a magnetic field parallel to the c-axis. In this case both quasiparticles and Cooper pairs move along Landau orbits within the layers. The c-axis dispersion remains unchanged from the zero-field form. In the chosen geometry one can generalize the zero-field results reported in the previous section to finite field strengths simply by the replacement of the $2D$ integration over $\mathbf{q}$ by a summation over the Landau levels

$$\int \frac{d^2\mathbf{q}}{(2\pi)^2} \to \frac{H}{\Phi_0}\sum_n = \frac{h}{2\pi\eta_{(2)}}\sum_n$$

(let us recall that $\eta_{(2)} = \xi_{xy}^2$). So the general expressions for all fluctuation corrections to the c-axis conductivity in a magnetic field can be simply written in the form [240]:

$$\sigma_{zz}^{AL} = \frac{e^2 s r^2 h}{64\xi_{xy}^2}\sum_{n=0}^{\infty}\frac{1}{\{[\epsilon + h(2n+1)][r+\epsilon+h(2n+1)]\}^{3/2}} \tag{9.1}$$

$$\sigma_{zz}^{DOS} = -\frac{e^2 s r\kappa h}{8\xi_{xy}^2}\sum_{n=0}^{1/h}\frac{1}{\{[\epsilon + h(2n+1)][r+\epsilon+h(2n+1)]\}^{1/2}} \tag{9.2}$$

$$\sigma_{zz}^{\mathrm{MT(reg)}} = -\frac{e^2 s\tilde{\kappa} h}{4\xi_{xy}^2}\sum_{n=0}^{\infty}\left(\frac{\epsilon + h(2n+1) + r/2}{\{[\epsilon + h(2n+1)][r+\epsilon+h(2n+1)]\}^{1/2}} - 1\right) \tag{9.3}$$

$$\sigma_{zz}^{\mathrm{MT(an)}} = \frac{e^2 s h}{8\xi_{xy}^2(\epsilon-\gamma_\varphi)}\sum_{n=0}^{\infty}\left(\frac{\gamma_\varphi + h(2n+1) + r/2}{\{[(\gamma_\varphi + h(2n+1)][\gamma_\varphi + h(2n+1)+r)]\}^{1/2}}\right.$$
$$\left. - \frac{\epsilon + h(2n+1) + r/2}{\{[\epsilon + h(2n+1)][r+\epsilon+h(2n+1)]\}^{1/2}}\right). \tag{9.4}$$

For the in-plane component of the fluctuation conductivity tensor the only additional problem appears in the AL diagram, where the matrix elements of the harmonic oscillator type, originating from the $B_{\|}$ $(q_{\|})$ blocks, have to be calculated. How to do this was demonstrated in detail in section 3.3. The other contributions are essentially analogous to their out-of-plane counterparts:

$$\sigma_{xx}^{AL} = \frac{e^2}{4s}\sum_{n=0}^{\infty}(n+1)\left(\frac{1}{\{[\epsilon + h(2n+1)][r+\epsilon+h(2n+1)]\}^{1/2}}\right.$$
$$-\frac{2}{\{[\epsilon + h(2n+2)][r+\epsilon+h(2n+2)]\}^{1/2}}$$
$$\left. + \frac{1}{\{[\epsilon + h(2n+3)][r+\epsilon+h(2n+3)]\}^{1/2}}\right), \tag{9.5}$$

$$\sigma_{xx}^{DOS} + \sigma_{xx}^{\mathrm{MT(reg)}} = -\frac{e^2 h(\kappa+\tilde{\kappa})}{2s}\sum_{n=0}^{1/h}\frac{1}{\{[\epsilon + h(2n+1)][r+\epsilon+h(2n+1)]\}^{1/2}}, \tag{9.6}$$

and

$$\sigma_{xx}^{\mathrm{MT(an)}} = \frac{e^2 h}{4s(\epsilon-\gamma_\varphi)}\sum_{n=0}^{\infty}\left(\frac{1}{\{[(\gamma_\varphi + h(2n+1)][\gamma_\varphi + h(2n+1)+r)]\}^{1/2}}\right.$$

$$- \frac{1}{\{[\epsilon + h(2n+1)][r + \epsilon + h(2n+1)]\}^{1/2}} \Bigg) . \tag{9.7}$$

These results can in principle already be used for numerical evaluations and fitting of the experimental data which was indeed successfully done in a series of papers [326–328].

Looking attentively the above expressions one can find some general relation between anomalous and regular parts of the MT contributions, which evidently is valid for $H = 0$, too:

$$\sigma_{\alpha\beta}^{\mathrm{MT(an)}}(\epsilon, h) = \frac{1}{2\tilde{\kappa}(\epsilon - \gamma_\varphi)} \left[\sigma_{\alpha\beta}^{\mathrm{MT(reg)}}(\epsilon, h) - \sigma_{\alpha\beta}^{\mathrm{MT(reg)}}(\gamma_\varphi, h) \right] . \tag{9.8}$$

Contributions to in-plane magnetoconductivity in the $2D$ limit can be easily summed up exactly in terms of ψ-functions just putting $r = 0$:

$$\sigma_{xx}^{AL}(\epsilon, h) = \frac{e^2}{8hs} \left[\frac{\epsilon}{h} \left[\psi(1/2 + \epsilon/2h) - \psi(1 + \epsilon/2h) \right] + 1 \right] , \tag{9.9}$$

(what exactly coincides with the obtained above phenomenological result (3.38)), and also

$$\sigma_{xx}^{DOS}(\epsilon, h) + \sigma_{xx}^{MT(reg)}(\epsilon, h) = - \frac{e^2(\kappa + \tilde{\kappa})}{4s} \left[\ln \frac{1}{h} - \psi(1/2 + \epsilon/2h) \right] , \tag{9.10}$$

$$\sigma_{xx}^{MT(an)}(\epsilon, h) = \frac{e^2}{8s(\epsilon - \gamma_\varphi)} \left[\psi(1/2 + \epsilon/2h) - \psi(1/2 + \gamma_\varphi/2h) \right] . \tag{9.11}$$

Out-of-plane magnetoconductivity for the $2D$ case is evidently absent, but one can find how it disappears when $r \to 0$:

$$\sigma_{zz}^{AL}(\epsilon, h) = - \frac{e^2 s r^2}{2^{10} \xi_{xy}^2 h} \psi'' \left(\frac{1}{2} + \frac{\epsilon}{2h} \right) , \tag{9.12}$$

$$\sigma_{zz}^{DOS}(\epsilon, h) = - \frac{e^2 s r \kappa}{16 \xi_{xy}^2} \left[\ln \frac{1}{h} - \psi \left(\frac{1}{2} + \frac{\epsilon}{2h} \right) \right] , \tag{9.13}$$

$$\sigma_{zz}^{\mathrm{MT(reg)}}(\epsilon, h) = - \frac{e^2 s \tilde{\kappa} r^2}{128 \xi_{xy}^2 h} \psi' \left(\frac{1}{2} + \frac{\epsilon}{2h} \right) , \tag{9.14}$$

$$\sigma_{zz}^{\mathrm{MT(an)}}(\epsilon, h) = \frac{e^2 s r^2}{2^8 \xi_{xy}^2 (\epsilon - \gamma_\varphi) h} \left[\psi' \left(\frac{1}{2} + \frac{\gamma_\varphi}{2h} \right) - \psi' \left(\frac{1}{2} + \frac{\epsilon}{2h} \right) \right] . \tag{9.15}$$

The detailed comparison of the cited results with the experimental data [326, 329], especially in strong fields, raised the problem of regularization of the DOS contribution. If in the absence of the magnetic field its ultraviolet divergence was successfully cut off at $q \sim \xi^{-1}$, in the case under consideration the cut-off parameter depends on the magnetic field and makes the fitting procedure ambiguous (in the above formulae (9.10)-(9.13) we cut the ultraviolet divergence with the logarithmic accuracy at $n_c \sim 1/h$). The solution to this problem was proposed in [106], where the authors calculated the difference $\Delta\sigma_{zz}^{DOS} = \sigma_{zz}^{DOS}(h,\epsilon) - \sigma_{zz}^{DOS}(0,\epsilon)$ applying to formulae (9.2) and (9.6) the same trick which was used in section 2.3) for the regularization of the free energy in magnetic field (Eq. (2.51)). The corresponding asymptotics for all out-of-plane fluctuation contributions are presented in Table 9.1.

Table 9.1 *The asymptotics of the various fluctuation contributions to the out--of-plane conductivity of the layered superconductor in different magnetic field domains*

	$h \ll \epsilon$	$\epsilon \ll h \ll r \quad (3D)$	$\max\{\epsilon, r\} \ll h \quad (2D)$
$\Delta\sigma_{zz}^{DOS}$	$\frac{e^2 s\kappa}{3\,2^5 \xi_{xy}^2}\frac{r(\epsilon+r/2)}{[\epsilon(\epsilon+r)]^{3/2}}h^2$	$0.428\frac{e^2 s\kappa}{16\xi_{xy}^2}r\sqrt{\frac{h}{r}}$	$\frac{e^2 s\kappa}{8\xi_{xy}^2}r\ln\frac{\sqrt{h}}{\sqrt{\epsilon}+\sqrt{\epsilon+r}}$
$\Delta\sigma_{zz}^{\mathrm{MT(reg)}}$	$\frac{e^2 s\tilde{\kappa}}{3\,2^6 \xi_{xy}^2}\frac{r^2}{[\epsilon(\epsilon+r)]^{3/2}}h^2$	$0.428\frac{e^2 s\tilde{\kappa}}{8\xi_{xy}^2}r\sqrt{\frac{h}{r}}$	$-\sigma_{zz}^{\mathrm{MT(reg)}}(0,\epsilon)-$ $\frac{\pi^2 e^2 s\tilde{\kappa}}{2^8\xi_{xy}^2}\frac{r^2}{h}$
$-\Delta\sigma_{zz}^{AL}$	$\frac{e^2 s}{2^8 \xi_{xy}^2}\frac{r^2(\epsilon+r/2)}{[\epsilon(\epsilon+r)]^{5/2}}h^2$	$\sigma_{zz}^{AL}(0,\epsilon)-$ $\frac{3.24e^2 s}{\xi_{xy}^2}\sqrt{\frac{r}{h}}$	$\sigma_{zz}^{AL}(0,\epsilon)-$ $\frac{7\zeta(3)e^2 s}{2^9\xi_{xy}^2}\frac{r^2}{h^2}$
$-\Delta\sigma_{zz}^{\mathrm{MT(an)}}$ $\min\{\epsilon,r\} \ll \gamma_\varphi$	$\frac{e^2 s}{3\,2^7 \xi_{xy}^2}\frac{r^2}{[\epsilon(\epsilon+r)]^2}h^2$	$\sigma_{zz}^{\mathrm{MT(an)}}(0,\epsilon)-$ $\frac{e^2 s}{32\xi_{xy}^2}\sqrt{\frac{r}{\gamma_\varphi}}$	$\sigma_{zz}^{\mathrm{MT(an)}}(0,\epsilon)-$ $\frac{3\pi^2 e^2 s}{2^8\xi_{xy}^2}\frac{\max\{r,\gamma_\varphi\}}{h}$
$-\Delta\sigma_{zz}^{\mathrm{MT(an)}}$ $\gamma_\varphi \ll \min\{\epsilon,r\}$	$\frac{e^2 s}{3\,2^7 \xi_{xy}^2}\frac{\sqrt{r}}{\epsilon\gamma_\varphi^{3/2}}h^2$	$\sigma_{zz}^{\mathrm{MT(an)}}(0)-$ $\frac{3.24e^2 s}{64\xi_{xy}^2}\sqrt{\frac{r}{h}}$	$\sigma_{zz}^{\mathrm{MT(an)}}(0,\epsilon)-$ $\frac{3\pi^2 e^2 s}{2^8\xi_{xy}^2}\frac{(r+\epsilon)}{h}$

The procedure described gives an excellent fitting up to very high fields [330] which is shown in Fig. 9.1.

Let us start the analysis from the $2D$ case ($r \ll \epsilon$). One can see that here the positive DOS contribution to magnetoconductivity turns out to be dominant. It grows as H^2 up to $H_{c2}(\epsilon)$ and then crosses to a slow logarithmic asymptote. At $H \sim H_{c2}(0)$ the value of $\Delta\sigma_{zz}^{DOS}(h \sim 1, \epsilon) = -\sigma_{zz}^{DOS}(0,\epsilon)$, which means the total suppression of the fluctuation correction in such a strong field. The regular part of the MT contribution does not manifest itself in this case while the AL term can compete with the DOS one in the immediate vicinity of T_c, where the small anisotropy factor r can be compensated for the additional ϵ^3 in the denominator. The anomalous MT contribution can contribute in the case of small pair-breaking only, which is opposite to what is expected in HTS.

In the $3D$ case ($\epsilon \ll r$) the behavior of the magnetoconductivity is more complex. In weak and intermediate fields the main, negative, contribution to the magnetoconductivity occurs from the AL and MT terms. At $H \sim H_{c2}(\epsilon)(h \sim \epsilon)$ the paraconductivity is already considerably suppressed by the magnetic field and the h^2-dependence of the magnetoconductivity changes through the $\sqrt{\frac{r}{h}}$ tendency to the high field asymptote $-\sigma_{zz}^{\mathrm{fl}}(0,\epsilon)$. In this intermediate region of fields ($\epsilon \ll h \ll r$), side by side with the decrease ($\sim \sqrt{\frac{r}{h}}$) of the main AL and MT contributions, the growth of the still relatively small DOS term takes

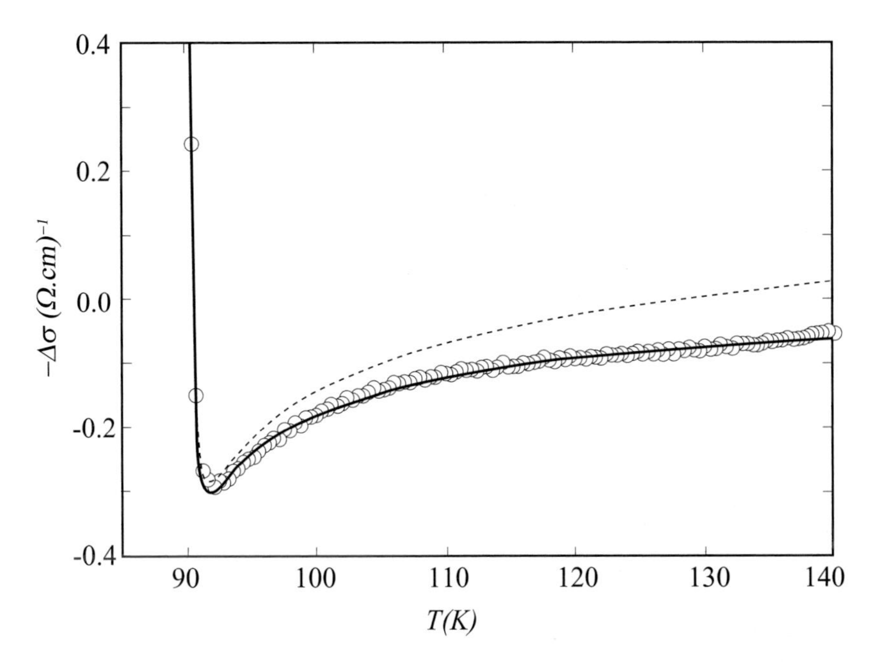

FIG. 9.1. Magnetoconductivity versus temperature at 27 T for an underdoped Bi-2212 single crystal. The solid line represents the theoretical calculation. The symbols are the experimental magnetoconductivity $\Delta\sigma_{zz}(B \parallel c \parallel I)$ [330].

place. At the upper limit of this region ($h \sim r$) its positive contribution is of the same order as the AL one and at high fields ($r \ll h \ll 1$) the DOS contribution determines the slow logarithmic decay of the fluctuation correction to the conductivity which is completely suppressed only at $H \sim H_{c2}(0)$. The regular part of the Maki- Thompson contribution is not of special importance in the 3D case. It remains comparable to the DOS contribution in the dirty case at fields $h \lesssim r$, but decreases rapidly ($\sim \frac{r}{h}$) at strong fields ($h \gtrsim r$), in the only region where the robust $\Delta\sigma_{zz}^{DOS}(h,\epsilon) \sim \ln \frac{h}{r}$ shows up surviving up to $h \sim 1$.

The temperature dependence of the different fluctuation contributions to the magnetoconductivity calculated for an underdoped Bi-2212 single crystal at the magnetic field 27 T is presented in Fig. 9.2.

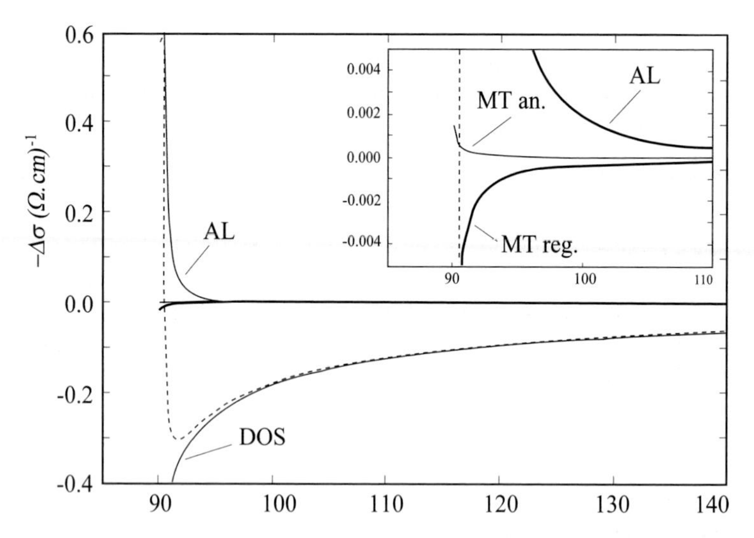

FIG. 9.2. Decomposition of the calculation of total theoretical magnetoconductivity for an underdoped Bi-2212 single crystal at 27 T. The inset shows the regular and anomalous parts of the MT contribution which are too small to be presented in the same scale as the AL and DOS contributions [330].

The formulae for the fluctuation corrections to the in-plane magnetoconductivity are presented in the Table 9.2.

Table 9.2 *The asymptotics of the various fluctuation contributions to the in-plane conductivity of the layered superconductor in different magnetic field domains*

	$h \ll \epsilon$	$\epsilon \ll h \ll r$; $\max\{\epsilon, r\} \ll h$
$\Delta\sigma_{xx}^{AL}$	$-\frac{e^2}{2^8 s}\frac{[8\epsilon(\epsilon+r)+3r^2]}{[\epsilon(\epsilon+r)]^{5/2}}h^2$	$-\sigma_{xx}^{AL}(0,\epsilon)+\frac{e^2}{4s}\frac{1}{\sqrt{2hr}}$; $-\sigma_{xx}^{AL}(0,\epsilon)+\frac{e^2}{8s}\frac{1}{h}$
$\Delta\sigma_{xx}^{\mathrm{MT(an)}}$ $(\min\{\epsilon, r\} \ll \gamma_\varphi)$	$-\frac{e^2}{3\ 2^5 s}\frac{(\epsilon+r/2)}{[\epsilon(\epsilon+r)]^{3/2}}h^2$	$-\sigma_{xx}^{MT}(0,\epsilon)+\frac{e^2}{8s}\frac{1}{\gamma_\varphi}\ln\frac{\sqrt{\gamma_\varphi}}{\sqrt{2h}+\sqrt{2h+r}}$
$\Delta\sigma_{xx}^{\mathrm{MT(an)}}$ $(\gamma_\varphi \ll \min\{\epsilon, r\})$	$-\frac{e^2}{3\ 2^5 s}\frac{1}{\epsilon\gamma_\varphi^{3/2}r^{1/2}}h^2$	$-\sigma_{xx}^{MT}(0,\epsilon)+\frac{0.2e^2}{s}\frac{1}{\sqrt{hr}}$; $-\sigma_{xx}^{MT}(0,\epsilon)+\frac{3\pi^2e^2}{32s}\frac{1}{h}$
$\Delta(\sigma_{xx}^{DOS}+\sigma_{xx}^{\mathrm{MT(reg)}})$	$\frac{e^2(\kappa+\tilde{\kappa})}{3\ 2^7 s}\frac{(\epsilon+r/2)}{[\epsilon(\epsilon+r)]^{3/2}}h^2$	$0.428\frac{e^2(\kappa+\tilde{\kappa})}{2^6 s}\sqrt{\frac{h}{r}}$; $\frac{e^2(\kappa+\tilde{\kappa})}{32s}\ln\frac{\sqrt{h}}{(\sqrt{\epsilon}+\sqrt{\epsilon+r})}$

Analyzing it one can see that in almost all regions the negative AL and MT contributions govern the behavior of in-plane magnetoconductivity. Nevertheless, similar to the c-axis case, the high field behavior is again determined by the positive logarithmic $\Delta(\sigma_{xx}^{DOS}+\sigma_{xx}^{\mathrm{MT(reg)}})$ contribution, which is the only one to survive in strong field. It is important to stress that the suppression of the DOS contribution by a magnetic field takes place very slowly. Such robustness with respect to the magnetic field is of the same physical origin as the slow logarithmic dependence of the DOS-type corrections on temperature.

9.1.1 *Self-consistent treatment of the critical temperature shift in magnetic field*

Another important problem which appears in the fitting of the resistive transition shape in relatively strong fields with the fluctuation theory is the much larger broadening of the transition than was predicted by the Abrikosov-Gor'kov theory [331]. In [219, 103] the self-consistent Hartree approach was proposed for the extension of fluctuation theory beyond the Gaussian approximation. It results in the considerable shift of $T_c(H)$ toward low temperatures with a corresponding broadening of the transition. Let us discuss it following the paper by Ullah and Dorsey [103].

We already saw how in the framework of the renormalization group approach the presence of the fourth order term in the GL free energy results in the shift of the critical temperature (see Eq. (2.121)). One could capture these results also in much more simple way using the Hartree approximation. Indeed, one can substitute $b|\Psi|^4$ in (2.4)) by $2b|\Psi|^2\left\langle|\Psi|^2\right\rangle$ and to use the explicit expression for the average value of the fluctuation order parameter:

$$< |\Psi_{(D)}|^2 >= \int \frac{d\mathbf{k}}{(2\pi)^D} \frac{1}{\alpha(\epsilon+\xi^2\mathbf{k}^2)} = \begin{cases} \frac{1}{4\pi\alpha\xi^2}\ln\frac{1}{\epsilon}, & D=2, \\ \frac{1}{\pi^2\alpha\xi^3}\left(1-\sqrt{\epsilon}\right), & D=3. \end{cases} \tag{9.16}$$

As a result the first coefficient of the GL functional is renormalized and the requirement of its vanishing gives the definition of the critical temperature $\widetilde{\epsilon}$ reduced by fluctuations:

$$\widetilde{a}\left(\widetilde{\epsilon}\right) = \left[a\left(\widetilde{\epsilon}\right) + b\left\langle|\Psi_{(D)}\left(\widetilde{\epsilon}\right)|^2\right\rangle\right] = 0. \tag{9.17}$$

Using the relation (2.6) between α and b one easily reproduces Eqs. (2.120)-(2.121).

Analyzing it one can see that in almost all regions the negative AL and MT contributions govern the behavior of in-plane magnetoconductivity. Nevertheless, similar to the c-axis case, the high field behavior is again determined by the positive logarithmic $\Delta(\sigma_{xx}^{DOS}+\sigma_{xx}^{\mathrm{MT(reg)}})$ contribution, which is the only one to survive in strong field. It is important to stress that the suppression of the DOS contribution by a magnetic field takes place very slowly. Such robustness with respect to the magnetic field is of the same physical origin as the slow logarithmic dependence of the DOS-type corrections on temperature.

The same procedure can be repeated in the presence of a magnetic field, but instead of integration over the momenta in plane perpendicular to the direction of a magnetic field one must sum over the Landau states. As a result the renormalized reduced temperature $\widetilde{\epsilon}_H = \widetilde{\epsilon} + h$ is determined in the self-consistent way [103]. For a layered superconductor:

$$\epsilon_H = \widetilde{\epsilon}_H - 4\ Gi_{(2)}\ h \sum_{n=0}^{n_c \sim 1/h} \frac{1}{[(\widetilde{\epsilon}_H + 2hn)(\widetilde{\epsilon}_H + 2hn + r)]^{1/2}}, \qquad (9.18)$$

with $\epsilon_H = \epsilon + h$. The cut-off, $n_c = L_H^2/\xi_{xy}^2 \sim 1/h$, reflects the fact that the GL theory is not valid on length scales less than the zero temperature coherence length ξ_{xy}.

In the $2D$ case the summation in Eq. (9.18) can be performed exactly:

$$\epsilon_H = \widetilde{\epsilon}_H - 2\ Gi_{(2)} \left[\ln\left(n_c + \widetilde{\epsilon}_H/2h\right) - \psi\left(\widetilde{\epsilon}_H/2h\right)\right]. \qquad (9.19)$$

In the weak field limit one can use the asymptotic expression of $\psi(x)$ and reproduce Eq. (2.120). In the strong field case ($\widetilde{\epsilon}_H \ll 2h$) the main contribution is determined by the pole of the $\psi(z)$-function:

$$\epsilon_H = \widetilde{\epsilon}_H - 4\ Gi_{(2)} \frac{h}{\widetilde{\epsilon}_H}. \qquad (9.20)$$

In this region of strong fields and weak fluctuations ($\epsilon_H \gg Gi_{(2)}(H) = \sqrt{2Gi_{(2)}h}$) $\widetilde{\epsilon}_H$ is close to ϵ_H ($\widetilde{\epsilon}_H - \epsilon_H \sim Gi_{(2)}^2(H)/\epsilon$) and no renormalization is necessary.

Let us discuss the strong fluctuations regime. In Part I, being in the region of weak fields, we described strong fluctuations by the scaling and RG methods. Here, being in the region of strong magnetic fields, in order to describe strong fluctuations we use the Hartree approximation. It turns out as good as the interpolation formula, providing results are correct by the order of value. The calculation of the fluctuation heat capacity in a strong magnetic field in Borel–Pade approximation performed up to the 11-th order of the perturbation theory [69] turns out to be very similar to the exact solution for the $0D$ case (see Fig. 2.1). Nevertheless, the Hartree approximation is unable to detect the phase transition of fluctuating normal phase to superconductor in vortex state. Indeed, the phase transition corresponds to the value $\widetilde{\epsilon}_H = 0$ while Eq. (9.20) does not have such a solution at any nonzero temperature. The absence of the nonzero solution for $\widetilde{\epsilon}_H$ could be interpreted as the lack of a phase transition at any temperature [103]. Indeed, the reduction of the effective dimensionality of the system by the magnetic field suppresses the effective fluctuation dimensionality by two, since the order parameter correlations transverse to the magnetic field have length scale set by magnetic length $L_H = (\Phi_0/H)^{1/2}$, which is always larger than ξ_{xy}. Hence, fluctuations in a $2D$ system placed in a perpendicular magnetic field are effectively $0D$ and as we have seen in section 2.2, they should destroy the phase transition widely smearing it.

Nevertheless the phase transition in such a system exists. This transition turns out to be of the first order and it corresponds to the melting of the vortex lattice, already discussed in section 5.4.

The authors of [332], following the procedure proposed by Dorsey and Ullah [103], modified expressions (9.1)-(9.7) by taking account of (9.18). As a result they succeeded in fitting quantitatively both the in-plane resistivity transition and the transverse resistivity peak for BSCCO films strongly broadened by the applied magnetic field.

9.2 Magnetoconductivity far from transition[56]

Let us discuss the conductivity of the $2D$ electron system with impurities in a magnetic field at low temperatures. Even in the absence of the field the effects of quantum interference of the noninteracting electrons in their scattering on elastic impurities already results in the appearance of a nontrivial temperature dependence of the resistance. This result contradicts the statement of the classical theory of metals requiring the saturation of the resistance at its residual value at low temperatures. In a superconductor above the critical temperature this, the so-called weak-localization (WL), effect is amplified by the Andreev reflection of electrons on the fluctuation Cooper pair leading to appearance of the MT correction to the conductivity. The characteristic feature of both the MT and WL corrections is their extreme sensitivity to the dephasing time τ_φ and to weak magnetic fields. It turns [283] out that the fluctuation MT contribution to magnetoconductivity has the same magnetic field dependence as the WL contribution [292], but with opposite sign and with a coefficient called $\beta(T)$ which diverges at T_c.

Let us start from the discussion of the MT contribution for the wide range of temperatures $T_c \lesssim T \ll \tau^{-1}$ and magnetic fields $H \ll H_{c2}(0)$. Beyond the GL region ($T \gtrsim T_c$) the MT correction is determined by the same diagram 2 of Fig. 8.1 but now, besides the dynamic and short wavelength ($q \sim \xi^{-1}$) fluctuation modes, the presence of magnetic field has to be taken into account. In this limit of a dirty metal ($T\tau \ll 1$) $|\widetilde{\epsilon}_{n+\mu} - \widetilde{\epsilon}_{-n}| \approx \tau^{-1}$ and the Cooperon (8.74) can be approximated in the form:

$$\lambda(\mathbf{q}, \varepsilon_1, \varepsilon_2) = \frac{\tau^{-1}\theta(-\varepsilon_1\varepsilon_2)}{|\varepsilon_1 - \varepsilon_2| + \widehat{\mathcal{D}}q^2}, \tag{9.21}$$

while fluctuation propagator is determined by Eq. (8.76). The prominent characteristic of these expressions is that they are valid even far from the critical temperature (for temperatures $T \ll \min\{\tau^{-1}, \omega_D\}$) and for $|\mathbf{q}| \ll l^{-1}, |\Omega_k| \ll \omega_D$.

Let us assume that a magnetic field is applied perpendicularly to the film. In this case one can rewrite both Eqs. (8.76) and (9.21) in the Landau representation simply replacing $\widehat{\mathcal{D}}q^2 \Rightarrow \omega_c(n+1/2)$. The parameter ω_c was introduced in the Part I (see (2.52))and it is related to the electron diffusion coefficient $\omega_c = 4\mathcal{D}eH$ [333].[57] As a result:

$$\lambda_n(\varepsilon_1, \varepsilon_2) = \frac{\tau^{-1}\theta(-\varepsilon_1\varepsilon_2)}{|\varepsilon_1 - \varepsilon_2| + \omega_c(n+1/2) + \tau_\varphi^{-1}}, \tag{9.22}$$

$$L_n^{-1}(\Omega_k) = -\nu\left[\ln\frac{T}{T_c} + \psi\left(\frac{1}{2} + \frac{|\Omega_k|}{4\pi T} + \frac{\omega_c(n+1/2)}{4\pi T}\right) - \psi\left(\frac{1}{2}\right)\right]. \tag{9.23}$$

[56] In this section we base on the results of [283] and its further extension [285].

[57] Comparison of the expressions (2.52), (3.28) and (9.33) relates the diffusion coefficient with the phenomenological GL constants $\mathcal{D} = 2/(\pi m\alpha) = 8T\xi^2/\pi$.

In contrast to the previous consideration, where τ_φ^{-1} was used as the infrared cut-off parameter of the momentum (energy) integrations, we introduced it here directly in the Cooperon. Moreover, one has to remember that the presence of pair-breaking processes reduced the critical temperature: $T_c = T_{c0} - \pi/(8\tau_\varphi)$.

Now one can write down the general expression for MT contribution to the electromagnetic response operator (8.30)-(8.31) in magnetic field as

$$Q^{MT}(\omega_\nu) = 4\pi\sigma_n T^2 \sum_{\substack{-\omega_\nu < \varepsilon_l < 0 \\ -\omega_\nu < \varepsilon_{l'} < 0}} \frac{eH}{\pi d} \sum_{n=0}^{\infty} L_n(\varepsilon_l + \varepsilon_{l'} + \omega_\nu)$$
$$\times \frac{1}{|\varepsilon_l - \varepsilon_{l'} + \omega_\nu| + \omega_c(n+1/2) + \tau_\varphi^{-1}} \frac{1}{|\varepsilon_{l'} - \varepsilon_l + \omega_\nu| + \omega_c(n+1/2) + \tau_\varphi^{-1}}. \tag{9.24}$$

Here, following [285], in order to simplify the further summations, we have parameterized the internal frequencies in a different way with respect to (8.30)-(8.31), using instead of the bosonic frequency the second fermionic one $\varepsilon_{l'}$: $\Omega_k \to \varepsilon_l + \varepsilon_{l'} + \omega_\nu$.

The frequency summation in Eq. (9.24) can be performed by contour integration (see details in [283, 285]) and after cumbersome algebra one can find

$$\sigma^{MT}(T,H) = -\frac{e^2}{2\pi^2 d}\beta(T,\tau_\varphi)\left[Y(4\mathcal{D}eH\tau_\varphi) - Y\left(\frac{\pi\mathcal{D}eH}{2T\ln T/T_c}\right)\right]. \tag{9.25}$$

The function $\beta(T,\tau_\varphi)$ is determined as

$$\beta(T,\tau_\varphi) = -\frac{\pi^2}{4}\sum_m (-1)^m \Gamma(|m|) - \sum_{n\geq 0}\Gamma(2n+1), \tag{9.26}$$

where m is an integer $m = 0, \pm1, \pm2, \ldots$ and

$$\Gamma(|m|) = \left[\ln\frac{T}{T_c} + \psi\left(\frac{1}{2} + \frac{|m|}{2}\right) - \psi\left(\frac{1}{2}\right) - \psi'\left(\frac{1}{2} + \frac{|m|}{2}\right)\frac{1}{4\pi T\tau_\varphi}\right]^{-1}. \tag{9.27}$$

The function

$$Y(x) = \ln x + \psi\left(\frac{1}{2} + \frac{1}{x}\right) = \begin{cases} \frac{x^2}{24}, & x \ll 1, \\ \ln x, & x \gg 1. \end{cases} \tag{9.28}$$

In the vicinity of transition, when $\ln T/T_c = \epsilon \ll 1$,

$$\beta(T,\tau_\varphi) = \frac{\pi^2}{4}\frac{1}{\epsilon - \gamma_\varphi}, \tag{9.29}$$

and (9.33) reduces to the already studied MT correction in the vicinity of critical temperature described by Eq. (9.6).

In the limit

$$\ln \frac{T}{T_c} \gg \min \left\{ \gamma_\varphi, \frac{4\mathcal{D}eH}{T} \right\}, \tag{9.30}$$

Eq. (9.33) can be simplified:

$$\sigma^{MT}(T, H) = -\frac{e^2}{2\pi^2 d} \beta(T) Y(4\mathcal{D}eH\tau_\varphi), \tag{9.31}$$

where

$$\beta(T) = \beta(T, \tau_\varphi \to \infty) \tag{9.32}$$

and it was tabulated in [283].

One can see that although the magnetic field dependence of the MT correction (9.31) is exactly the same as was found for the WL contribution:

$$\delta\sigma_{WL+MT} = \frac{e^2}{2\pi^2} [\alpha - \beta(T)] Y(4\mathcal{D}eH\tau_\varphi), \tag{9.33}$$

but their magnitudes depend on different microscopic characteristics of the sample. Namely, the first term in this formula corresponds to the WL contribution [292] and depends on the constant of the spin–orbit interaction $\alpha(\alpha = 1$ if the spin–orbit interaction of the electrons with the impurities is small while in the opposite limiting case $\alpha = -1/2)$. The second, describing the MT contribution, does not depend on spin–orbit interaction but contains phase-breaking time τ_φ.

The function $\beta\left(\ln \frac{T}{T_c}\right)$ at $T \to T_c$ has asymptotic form $\beta(x) = 1/x$ and (9.33) reduces to the already studied MT correction in the vicinity of critical temperature. For $T \gg T_c$ $\beta(x) = 1/x^2$ and the MT contribution gives a logarithmically small correction to the WL result. Its zero-field value, being proportional to $\ln^{-2}(T/T_c)$, decreases with the growth of the temperature faster than both the AL contribution (in the dirty case $\delta\sigma_{AL} \sim 1/\ln(T/T_c)$) and the especially slow DOS contribution $\left(\delta\sigma^{DOS} \sim \ln\ln(1/T_c\tau) - \ln\ln(T/T_c)\right)$ (see [108, 241]).

Presented results has been verified by several experimental groups [286, 334–338] and yielded direct measurements of the electron inelastic scattering time τ_φ near critical temperature.

It worth mentioning that for the region of temperatures $T \gg T_c$, analogous to Eq. (9.56)-(9.57), the result (9.33) can be applied both to superconducting and normal metals $(g > 0)$, if in place of the critical temperature the formal value $T_c \sim E_F \exp(1/\nu g)$ is undermined. The interplay of the localization and fluctuation corrections was extensively studied (see [285–289]).

9.3 Effect of fluctuations on the Hall conductivity

9.3.1 *AL contribution to the Hall conductivity*[58]

Let us start with a discussion of the physical meaning of the Hall resistivity ρ_{xy}. In the case of only one type of carriers it depends on their concentration n and

[58] In this section we base on the results [109].

turns out to be independent of the electron diffusion coefficient: $\rho_{xy} = H/(en)$. The fluctuation processes of the MT and DOS types contribute to the diffusion coefficient, so their expected contribution to the Hall resistivity is zero. For the Hall conductivity in a weak field one can write

$$\sigma_{xy} = \rho_{xy}\sigma_{xx}^2 = \rho_{xy}\sigma_{xx}^{(n)2} + 2\rho_{xy}\sigma_{xx}^{(n)}\delta\sigma_{xx} = \sigma_{xy}^{(n)}\left(1 + 2\frac{\delta\sigma_{xx}}{\sigma_{xx}}\right) \tag{9.34}$$

so, evidently, the relative fluctuation correction to Hall conductivity is twice as large as the fluctuation correction to the diagonal component. This qualitative consideration is confirmed by the direct calculation of the MT type diagram [339].

The AL process corresponds to an independent charge transfer which cannot be reduced to a renormalization of the diffusion coefficient. It contributes weakly to the Hall effect, and this contribution is related to the Cooper pair particle–hole asymmetry. This effect was investigated in a set of papers: [103,109,339–343]. Let us recall that the proper expressions describing the paraconductivity contribution to the Hall conductivity (in the TDGL theory limits) were already carried out above in the phenomenological approach. The microscopic consideration of this value can be done in the spirit of the calculation of σ_{xx}^{AL} (see (8.13)) and after the analytic continuation results in

$$\sigma_{xy}^{AL} = \left(\frac{2h\nu(0)}{\pi}\right)^2 \sum_{n=0}^{\infty}(n+1)\int_{-\pi/s}^{\pi/s}\frac{dk_z}{2\pi}\int_{-\infty}^{\infty}\coth\frac{z}{2T}dz \times$$
$$\times\left[\operatorname{Im} L_n^R(z)\frac{\partial}{\partial z}\operatorname{Re} L_{n+1}^R(z) - \operatorname{Im} L_{n+1}^R(z)\frac{\partial}{\partial z}\operatorname{Re} L_n^R(z)\right]. \tag{9.35}$$

The phenomenological expression (3.23) can be obtained from this formula by carrying out the frequency integration in the same way as was done in the calculation of (8.13) (the essential region of integration is $z \ll T$).

One can see from (9.35) that if $\operatorname{Im} L_n^R(-z) = -\operatorname{Im} L_n^R(z)$ and $\operatorname{Re} L_n^R(-z) = \operatorname{Re} L_n^R(z)$ the Hall conductivity is equal to zero, or, in terms of the phenomenological parametrization, the reality of γ_{GL} results in a zero Hall effect. Physically it is possible to say that this zero is the direct consequence of electron–hole symmetry. However, from the formula (6.5) one can see that an energy dependence of the DOS or the electron interaction constant g immediately results in the appearance of an imaginary part of γ_{GL}. In the weak-interaction approximation

$$\operatorname{Im}\gamma_{\mathrm{GL}} = -\frac{\nu(0)}{2}\left(\frac{\partial \ln T_{\mathrm{c}}}{\partial E}\right)_{E=E_F}. \tag{9.36}$$

Usually this value is small in comparison with $\operatorname{Re}\gamma_{\mathrm{GL}}$ by a ratio of the order of T_{c}/E_F. Taking into account the terms of the order of $\operatorname{Im}\gamma_{\mathrm{GL}}$ in (9.35) and using the explicit form of the fluctuation propagator for layered superconductor (6.39) one can find

$$\sigma_{xy}^{AL} = e^2 T \frac{\operatorname{Im}\gamma_{\mathrm{GL}}}{2\nu(0)} h \int_{-\pi/s}^{\pi/s} \frac{dk_z}{2\pi} \frac{1}{\left[\epsilon + r\sin^2(k_z s/2))\right]^2} F_H\left(\frac{\epsilon + r\sin^2(k_z s/2))}{2h}\right),$$

where the function $F_H(x)$ was introduced above (see Eq. (3.50)), in the process of consideration of the fluctuation Cooper pairs contribution to the Hall conductivity.

For $H \to 0$ the expression for the fluctuation Hall paraconductivity takes the form

$$\sigma_{xy}^{AL} = \frac{e^2 T}{6\pi s}\left(\frac{\operatorname{Im}\gamma_{\mathrm{GL}}}{\nu(0)}\right)\frac{\epsilon + r/2}{\left[\epsilon(\epsilon + r)\right]^{3/2}} h = \frac{e^2}{48 s}\left(\frac{\operatorname{Im}\gamma_{\mathrm{GL}}}{\operatorname{Re}\gamma_{\mathrm{GL}}}\right)\frac{\epsilon + r/2}{\left[\epsilon(\epsilon + r)\right]^{3/2}} h. \tag{9.37}$$

One can see that in the $2D$ case the temperature dependence of the AL fluctuation correction to the Hall conductivity

$$\sigma_{xy}^{AL} \approx \frac{e^2}{12\pi s}\left(\frac{T_{\mathrm{c}}}{E_F}\right)\frac{h}{\epsilon^2}. \tag{9.38}$$

turns out to be more singular than that one originating from the MT contribution to the diagonal component of conductivity (see Eq. (9.34)). Nevertheless the absolute value of the former contains the additional smallness $\operatorname{Im}\gamma_{\mathrm{GL}}$.

9.3.2 *Effect of topological singularity on fluctuation Hall effect*[59]

We have seen that superconducting fluctuations induce a characteristic deviation from the normal state temperature dependence of the Hall conductivity above T_{c}. In particular, it has been shown that a Hall sign reversal takes place below T_{c} [103, 345]. Although admittedly σ_{xy}^{AL} arises as a result of an electron–hole asymmetry in the band structure [339], it turns out that this is not the unique source of the appearance of $\operatorname{Im}\gamma_{\mathrm{GL}}$. Recently it was demonstrated [344] that the sign and value of $\operatorname{Im}\gamma_{\mathrm{GL}}$ strongly depend also on the topology of Fermi surface. Evidence for a universal behavior of the Hall conductivity as a function of doping has been reported in the cuprate superconductors [346].

An electronic topological transition (ETT) consists of a change of topology of the Fermi surface, and may be induced by doping, as well as by changing the impurity concentration, or applying pressure or anisotropic stress [347–349]. In all such cases, one may introduce a critical parameter z, measuring the proximity to the ETT occurring at $z = 0$. In the most interesting case of quasi-$2D$ materials, such as the cuprates, the electronic band is locally characterized by a hyperbolic-like dispersion relation. Therefore, one is particularly interested in the study of an ETT of the 'neck disruption' kind, according to the original classification of Lifshitz [347].

[59]In this section we base on the results of [344].

Concerning the normal state transport properties of a superconductor, the effect of the proximity to an ETT has been studied for the thermoelectric power in a quasi–$2D$ metal [350], and for the Nernst and the weak-field Hall effects for both $3D$ and quasi-$2D$ metals [351].

Below we will follow [344], where $\operatorname{Im}\gamma$ was studied for a quasi-2D superconductor close to an ETT as a function of the ETT parameter z and temperature T, with a realistic band dispersion typical of the high-T_c cuprate compounds. Close to the ETT, $\operatorname{Im}\gamma$ is characterized by a steep inflection point, surrounded by a minimum and a maximum, whose height increases with decreasing temperature. In the presence of electron–hole symmetry, we will show that $\operatorname{Im}\gamma$ is an odd function of the ETT parameter z, and that $\operatorname{Im}\gamma$ vanishes and rapidly changes sign at the ETT point. The parameter $\operatorname{Im}\gamma$ can be extracted by comparison with experimental data for the excess Hall effect [309,321,324,352,353]. Table 9.3 lists values of $\operatorname{Im}\gamma$ for several layered cuprate superconductors and HTS superlattices. One can immediately observe that $\operatorname{Im}\gamma$ shows a direct correlation with T_c, i.e. $|\beta|$ increases as T_c increases.

Table 9.3 *Electron–hole asymmetry parameter $\beta \propto \operatorname{Im}\gamma$ and critical temperature T_c for several layered cuprates and cuprate superlattices. The values of β listed here have been obtained from a fit of the AL+MT corrections to conductivity and Hall conductivity, against data for excess Hall effect.*

compound	T_c [K]	β	Ref.
YBCO/PBCO (36 Å/96 Å)	68.68	−0.0003	[353]
YBCO/PBCO (120 Å/96 Å)	86.33	−0.075	[353]
YBCO	88.55	−0.17	[352]
Bi-2223	105.00	−0.38	[309]
(Bi,Pb)-2223	109.00	−1.00	[352]

9.3.2.1 *Evaluation of* $\operatorname{Im}\gamma$ *in the presence of an ETT* Several models of electron spectrum with topological singularity are available [348]. Here having in mind the further description of the highly anisotropic HTS cuprates., we use specifically the $2D$ tight-binding dispersion relation:

$$\xi_{\mathbf{k}} = -2t(\cos k_x + \cos k_y) + 4t' \cos k_x \cos k_y - \mu, \tag{9.39}$$

with t, t' being hopping parameters between nearest and next-nearest neighbors of a square lattice, respectively. For $\mu = \mu_c = -4t'$, the Fermi surface defined by $\xi_{\mathbf{k}} = 0$ has a critical form and undergoes an ETT (see [354, 355]). Below, we will make use of the parameters $z = (\mu - \mu_c)/4t$, measuring the distance from the ETT ($z = 0$), and of the hopping ratio $r = t'/t$ ($0 < r < \frac{1}{2}$) [356]. A nonzero value of r implies a breaking of electron–hole symmetry, with the electron subband width decreasing, and the hole subband width increasing of an equal amount $8rt$ [355]. The DOS associated to Eq. (9.39) is characterized by a logarithmic singularity at $z = 0$. Such a logarithmic cusp becomes weakly asymmetric around $z = 0$ in the case $r \neq 0$.

From a microscopic point of view, the TDGL relaxation rate γ is related to the static limit of the frequency derivative of the retarded polarization operator as:

$$\gamma = i \lim_{\Omega \to 0} \frac{\partial \Pi^R}{\partial \Omega}. \tag{9.40}$$

The polarization operator (6.11) is calculated with the Green function for free electrons with dispersion relation Eq. (9.39), and the outer sum is performed over the N wave vectors $\mathbf{q}$ in the first Brillouin zone (1BZ). With the help of standard methods [228], the sum over electronic Matsubara frequencies is readily evaluated, and after analytic continuation to the upper complex plane one obtains:

$$\Pi^R(\Omega, z, T) = -\lim_{\delta \to 0} \frac{1}{2N} \sum \frac{1}{\Omega + 2\xi_{\mathbf{k}} + i\delta} \left[\tanh \frac{2\xi_{\mathbf{k}} + \Omega}{2T} + \tanh \frac{2\xi_{\mathbf{k}}}{2T} \right]. \tag{9.41}$$

Performing the frequency derivative and passing to the static limit, as required by Eq. (9.40), one has:

$$\gamma = \frac{i}{8T} \frac{1}{N} \sum_{\mathbf{k}} \frac{1}{\xi_{\mathbf{k}} + i\delta} F\left(\frac{\xi_{\mathbf{k}}}{2T} \right), \tag{9.42}$$

where

$$F(y) = \frac{1}{y} \tanh y - \frac{1}{\cosh^2 y}. \tag{9.43}$$

In the electron–hole symmetric case ($r = 0$), $\operatorname{Im} \gamma(z)$ is an odd function of the ETT parameter, vanishing at $z = 0$, i.e. at the ETT, for all temperatures. Close to the ETT point, $\operatorname{Im} \gamma$ rapidly changes sign, with two symmetric peaks occurring very close to the ETT point. The height of these peaks decreases with increasing temperature (Fig. 9.3), and eventually diverges as $T \to 0$ (see Eq. (9.46)). Such a behavior, in particular, implies a sign-changing Hall effect as a function of doping, and a large Hall effect close to the ETT. Moreover, the result $\operatorname{Im} \gamma(z = 0) = 0$ is consistent with the absence of electron–hole asymmetry [339]. A similar z-dependence have been demonstrated also for the thermoelectric power in the proximity of an ETT [350].

On the other hand, in the electron–hole asymmetric case ($r \neq 0$), one in general has $\operatorname{Im} \gamma(z) \neq -\operatorname{Im} \gamma(-z)$. However, one still recovers a sign-changing $\operatorname{Im} \gamma(z)$, with $\operatorname{Im} \gamma(z)$ vanishing very close to the ETT. Moreover, the two peaks around the ETT have increasing heights with increasing hopping ratio r. Given that a non-zero value of the hopping ratio r can be associated with structural distortions in the ab plane of the cuprates [355], one may conclude that in-plane anisotropy enhances the fluctuation effects associated to a nonzero value of $\operatorname{Im} \gamma(z)$. Moreover, on the basis of the direct correlation existing between

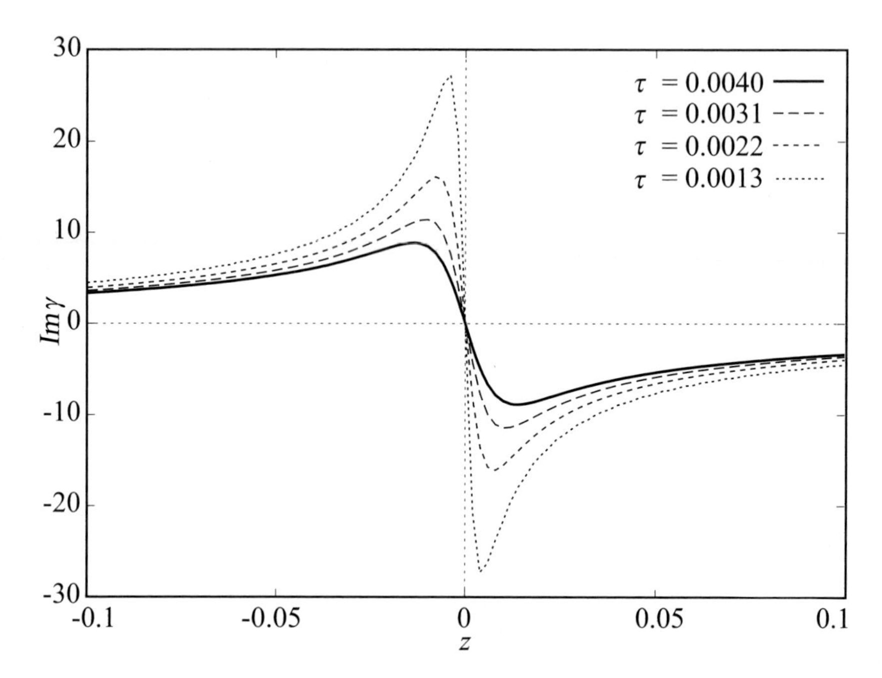

FIG. 9.3. $\operatorname{Im}\gamma(z)$ in the electron–hole symmetric case ($r = 0$), for decreasing temperatures τ (in units such that $4t = 1$).

$T_{c,\max}$ and the hopping ratio r [356], it follows that the heights of the peaks in $\operatorname{Im}\gamma(z)$ around the ETT increase with increasing $T_{c,\max}$ across different classes of cuprates. Such a result is in agreement with the data listed in Table 9.3 for the excess Hall parameter $\beta \propto \operatorname{Im}\gamma(z)$.

A further justification of the above numerical results can be drawn from an analysis of the continuum limit ($\mathcal{N}^{-1}\sum_{\mathbf{k}} \mapsto \int \frac{d^2\mathbf{k}}{(2\pi)^2}$) of Eq. (9.42). Making use of the DOS $\nu(z)$ corresponding to the dispersion relation (9.39) for the imaginary part of the relaxation rate one obtains:

$$16t^2 \operatorname{Im}\gamma(z) = \frac{1}{8\tau} \int_{-1+2r}^{1+2r} \frac{\nu(x)}{x-z} F\left(2t\frac{x-z}{T}\right) dx, \tag{9.44}$$

Eq. (9.44) confirms that $\operatorname{Im}\gamma(z)$ is an odd function of the ETT parameter z in the electron–hole symmetric case, a source of asymmetry being provided by a nonzero value of the hopping ratio r, both through a change of the integration limits, and through a change in the DOS. Equation (9.44) can be integrated analytically, yielding the cumbersome expression [344] which qualitatively recovers the z-dependence shown in Fig. 9.3 for $\operatorname{Im}\gamma(z)$, with $\operatorname{Im}\gamma(z)$ being an odd function of z at any given temperature T. In particular, $\operatorname{Im}\gamma$ vanishes at $z = 0$, where it behaves like

$$\operatorname{Im}\gamma \simeq -\frac{8t^3}{\pi^2}\frac{z}{T^3}, \tag{9.45}$$

for $|z| \ll 1$, $\tau \ll 1$. $\mathrm{Im}\,\gamma(z)$ is also characterized by two antisymmetric peaks occurring at $z_\pm \simeq \pm 2s\tau$. In the symmetric case ($r = 0$), the height of such peaks diverge in the limit $T \to 0$ as

$$|\,\mathrm{Im}\,\gamma_{\mathrm{peak}}| \approx \frac{\ln 2}{2\pi^2}\frac{1}{T^2}. \tag{9.46}$$

The singular behavior of $\mathrm{Im}\,\gamma(z)$ as a function of the ETT critical parameter z in the limit $T \to 0$ is a fingerprint of quantum criticality [354,357]. In the case $r \neq 0$ this singular behavior becomes asymmetric on the two sides of the ETT, thus showing that particle–hole asymmetry enhances the singular behavior of $\mathrm{Im}\,\gamma$ close to the ETT at low temperature.

Far away from the ETT, in the limit $|\mu - \mu_c|/T \gg 1$:

$$\mathrm{Im}\,\gamma \approx \frac{t\nu_0}{T(\mu - \mu_c)} \ln \frac{2T}{|\mu - \mu_c|}, \tag{9.47}$$

where $\nu_0 = (4t\pi^2\sqrt{1 - 4r^2})^{-1}$ is the DOS in the isotropic limit. Such a result again confirms that $\mathrm{Im}\,\gamma$ is a sign-changing function of doping, with $\mathrm{Im}\,\gamma(z) < 0$ in the hole-like doping range ($z > 0$). As expected, we find that $\mathrm{Im}\,\gamma(z)$ increases with decreasing temperature, with a jump-like structure at the ETT whose height diverges as $T \to 0$, and increases with increasing in-plane anisotropy, given by a non-zero hopping ratio r.

9.4 Fluctuation magnetic susceptibility far from transition[60]

The qualitative estimations given above (2.40)-(2.46) for the fluctuation diamagnetic susceptibility, based on the Langevin formula, demonstrate that even at high temperatures $T \gg T_c$ it turns to be of the order of χ_P for clean $3D$ superconductors and exceeds noticeably this value for $2D$ systems. In order to develop the microscopic theory [52,333,358] let us start from the same one-loop approximation for the fluctuation part of the free energy which can be written in the most general form as [57]

$$F_1(H,T) = -T\sum_{\Omega_k} \mathrm{tr}\left[\ln\ (1 - g\,\Pi(\Omega_k, \mathbf{r}, \mathbf{r}'))\right], \tag{9.48}$$

where $tr\,(\cdots)$ is calculated over the space variables of the polarization operator (7.22) (compare with Eq. (2.51)). Evidently it corresponds to the ladder approximation for the fluctuation propagator (see (6.4)). Nevertheless, being interested here in calculation of fluctuation magnetic susceptibility far from transition we have to carry out the momentum integration over all momenta, accounting for the contributions of short wavelength fluctuations, dominating far from T_c.

For this purpose let us express diamagnetic susceptibility in terms of fluctuation propagator, differentiating expression (9.48). We will be interested in the

[60] In this section we base on the results of [52].

case of weak fields, so let us work in momentum representation, including magnetic field by means of gauge invariant momentum $q_\alpha - 2e e_{\alpha\beta\gamma}\widehat{x}_\beta H_\gamma$ (here $e_{\alpha\beta\gamma}$ is unitary antisymmetric tensor) in polarization operator:

$$\chi_{\mu\nu} = \frac{\partial^2}{\partial H_\mu \partial H_\nu} T \sum_{\Omega_k} \int \frac{d\mathbf{q}}{(2\pi)^3} \left[\ln \left(1 - g\,\Pi \left(\Omega_k, q_\alpha - 2e e_{\alpha\beta\gamma}\widehat{x}_\beta H_\gamma \right) \right) \right]$$
$$= -4e^2 e_{\alpha\gamma\mu}\, e_{\beta\kappa\nu} T \sum_{\Omega_k} \int \frac{d\mathbf{q}}{(2\pi)^3} \widehat{x}_\gamma \widehat{x}_\kappa L^2 \Pi'_\alpha \Pi'_\beta. \tag{9.49}$$

The coordinate operators in momentum representation are reduced to the corresponding derivatives and one can find:

$$\chi_{\mu\nu} = -4e^2 e_{\alpha\gamma\mu}\, e_{\beta\kappa\nu} T \sum_{\Omega_k} \int \frac{d\mathbf{q}}{(2\pi)^3}$$
$$\times \left[6L^4 \Pi'_\alpha \Pi'_\beta \Pi'_\gamma \Pi'_\kappa - 2L^3 \Pi'_\gamma \left(\Pi''_{\alpha\kappa} \Pi'_\beta - \Pi'_\alpha \Pi''_{\beta\kappa} \right) + L^2 \frac{\partial^2}{\partial q_\gamma \partial q_\kappa} \Pi'_\alpha \Pi'_\beta \right]. \tag{9.50}$$

Averaging for isotropic case can be performed using the expression for partial convolution of the unitary antisymmetric tensors:

$$e_{\alpha\gamma\mu}\, e_{\beta\kappa\mu} = \det \begin{pmatrix} \delta_{\alpha\beta} & \delta_{\alpha\kappa} \\ \delta_{\gamma\beta} & \delta_{\gamma\kappa} \end{pmatrix} = \delta_{\alpha\beta}\delta_{\gamma\kappa} - \delta_{\gamma\beta}\delta_{\alpha\kappa}. \tag{9.51}$$

As a result, due to the symmetry of subscripts, the first and last terms in (9.50) are averaged out and one gets:

$$\overline{\chi} = \frac{1}{3}\chi_{\mu\mu} = \frac{8}{3} e^2 \left(\delta_{\alpha\beta}\delta_{\gamma\kappa} - \delta_{\gamma\beta}\delta_{\alpha\kappa} \right) T \sum_{\Omega_k} \int \frac{d\mathbf{q}}{(2\pi)^3} L^3 \Pi'_\gamma \left(\Pi''_{\alpha\kappa} \Pi'_\beta - \Pi'_\alpha \Pi''_{\beta\kappa} \right). \tag{9.52}$$

The further summation over the subscripts finally gives:

$$\overline{\chi} = -\frac{16}{3} e^2 T \sum_{\Omega_k} \int \frac{d\mathbf{q}}{(2\pi)^3} \Pi'_x L^3 \left[\Pi'_x \Pi''_{yy} - \Pi'_y \Pi''_{xy} \right]. \tag{9.53}$$

Now one can use expressions found above (8.73) and (8.76) for propagator and polarization operator of dirty superconductor, valid for wide range of temperatures $T_c \ll T \ll \tau^{-1}$, and analogously section 8.6 find the expressions for the diamagnetic susceptibility for $2D$ and $3D$ cases. In both cases Eq. (9.53) turns out to be divergent at the upper limit of the momentum integration. In $2D$ case this divergence is double logarithmic [333, 358]:

$$\chi_{\mathrm{fl}}^{(2)}(T) = \frac{2\pi^2}{3}\chi_P \left(\ln\ln\frac{1}{T_c\tau} - \ln\ln\frac{T}{T_c} \right), \tag{9.54}$$

and one can see that the temperature dependence of the diamagnetic susceptibility coincides with that one of the DOS part of conductivity. The same analogy takes place in the $3D$ case. Here the ultraviolet divergence is stronger and, as it was done in section 8.6, it can be easily eliminated by subtracting the corresponding quantity taken at $T \sim T_c$. This yields

$$\Delta\chi_{\mathrm{fl}}^{(3)}(T) = -\frac{2}{3}\chi_P \frac{\sqrt{T\tau}}{\ln(T/T_c)}. \tag{9.55}$$

In the clean case $\tau^{-1} \ll T \ll \omega_D$ one can use the propagator in the form (8.69) and the final expressions for the fluctuation diamagnetic susceptibility can be written as:

$$\chi_{\mathrm{fl}}^{(3)}(T) = \frac{0.05\chi_P}{3}[\ln^{-2}(\omega_D/T_c) - \ln^{-2}(T/T_c)], \tag{9.56}$$

$$\chi_{\mathrm{fl}}^{(2)}(T) = -0.05\chi_P \left(\frac{E_F}{T}\right) \frac{1}{\ln^3(T/T_c)}. \tag{9.57}$$

Let us stress that these results are valid for the fluctuation diamagnetism of a normal metal with $g > 0$ too, if by T_c one uses the formal value $T_c \sim E_F \exp(\frac{1}{\nu g})$.

9.5 Fluctuations in magnetic fields above $H_{c2}(0)$[61]

As was mentioned above the role of fluctuations is especially pronounced in the vicinity of the critical temperature. Nevertheless for some phenomena they can be still considerable far from the transition too. In these cases the GL theory is certainly unapplicable since the short-wave and dynamical fluctuation contributions have to be taken into account. It can be done in the microscopic approach which we will demonstrate in several examples.

9.5.1 *Conductivity*

As one can see from (9.5)-(9.7), in the vicinity of the upper critical field $H_{c2}(T)$ the fluctuation corrections diverge as ϵ_h^{-1} for the $2D$ case and as $\epsilon_h^{-1/2}$ for the $3D$ case[62] (it is enough to keep just the terms with $n = 0$ in these formulae). This behavior is preserved in strong magnetic fields too, but the coefficients undergo changes. A case of special interest is $T \ll T_c$ (which means $H \to H_{c2}(0)$) which represents an example of a quantum phase transition [76]. Microscopic analysis of

[61]The systematic theory of low temperature fluctuations was proposed firstly in paper [359]. In this section we base on the results of [76].

[62]ϵ_h is the renormalized by the magnetic field reduced temperature $\epsilon_h = \epsilon + h$.

the magnetoconductivity allows us to study the effect of fluctuations in magnetic fields of the order of $H_{c2}(0)$, where the GL functional approach is inapplicable.

We restrict our consideration to the case of a dirty metal ($T\tau \ll 1$). In this limit, as it was demonstrated above, the fluctuation propagator takes form (8.76). This, very general expression, being calculated for temperatures $T \ll \min\{\tau^{-1}, \omega_D\}$) and momenta $|\mathbf{q}| \ll l^{-1}, |\Omega_k| \ll \omega_D$, can already be used for the account of quantum fluctuations. In the case of a superconductor placed in magnetic field $\mathbf{H}$ (applied along the c-axis) this expression can be rewritten to Landau representation by the simple substitution $\left(\widehat{\mathcal{D}}q^2\right)_{\|} \Rightarrow \omega_c(n+1/2)$:

$$L_n^{-1}(q_z, \Omega_k) = -\nu \left[\ln \frac{T}{T_c} + \psi \left(\frac{1}{2} + \frac{|\Omega_k|}{4\pi T} + \frac{\omega_c(n+1/2) + 4\tau J^2 \sin^2(q_z s/2)}{4\pi T} \right) - \psi \left(\frac{1}{2} \right) \right]. \tag{9.58}$$

Introduced in the Part I parameter ω_c (see (2.52)) presents itself the effective cyclotron frequency of the Cooper pair in its motion in magnetic field and it is related to the electron diffusion coefficient $\omega_c = 4\mathcal{D}eH$ [333]. The latter in our case plays the role of the inverse mass of the diffusive motion of the Cooper pair.[63]

In the case of arbitrary temperatures and magnetic fields the expression for the AL contribution to the electromagnetic response operator takes the form:

$$Q_{xx}^{AL}(\omega_\nu) = -4e^2 T \sum_{\Omega_k} \sum_{\{n,m\}=0}^{\infty} B_{n,m}(\Omega_k + \omega_\nu, \Omega_k) L_m(\Omega_k) \times B_{m,n}(\Omega_k, \Omega_k + \omega_\nu) L_n(\Omega_k + \omega_\nu). \tag{9.59}$$

Here we have restricted our consideration by the $2D$ case. Summation over the eigen-states of the Cooper pair in magnetic field is performed with the corresponding density of states H/Φ_0.

The expression for the block $B_{n,m}(\Omega_k, \omega_\nu)$ in the Landau representation can be rewritten as

$$B_{n,m}(\Omega_k + \omega_\nu, \Omega_k) = -4\pi\nu\tau^2 \mathcal{D}_{(2)} \left[\sqrt{eH(n+1)}\delta_{m,n+1} + \sqrt{eHn}\delta_{m,n-1} \right] T \sum_{\varepsilon_i} \lambda_n(\varepsilon_i + \omega_\nu, \Omega_k - \varepsilon_i) \lambda_m(\varepsilon_i, \Omega_k - \varepsilon_i) \tag{9.60}$$

with the Cooperon

[63] Comparison of the expressions (2.52), (3.28) and (9.33) relates the diffusion coefficient with the phenomenological GL constants $\mathcal{D} = 2/(\pi m\alpha) = 8T\xi^2/\pi$.

$$\lambda_m(\varepsilon_1,\varepsilon_2) = \frac{1}{\tau}\frac{\theta(-\varepsilon_1\varepsilon_2)}{|\varepsilon_1-\varepsilon_2|+\omega_c(m+1/2)} \tag{9.61}$$

written down in the same representation.

The critical field $H_{c2}(T)$ is determined by the equation $L_0^{-1}(q_z = 0, \Omega_k = 0) = 0$. That is why in the vicinity of $H_{c2}(T)$ the singular contribution to (9.59) originates only from the terms with $L_0(0,\Omega_k)$. The frequency dependencies of the functions $B_{n,m}(\Omega_k+\omega_\nu,\Omega_k)$ and $L_1(\Omega_k)$ are weak although we cannot omit them to get non-vanishing answer. It is enough to restrict ourselves to the linear approximation in their frequency dependencies. If the temperature $T \ll T_{c0}$ the sum over frequencies in (9.60) can be approximated by an integral. Transforming the boson frequency Ω_k summation to a contour integration as was done above and making the analytic continuation in the external frequency ω_ν one can get an explicit expression for the d.c. paraconductivity. In the same spirit the contributions of all other diagrams from Fig. 8.1 which contribute to fluctuation conductivity in the case under discussion are calculated side by side with the AL one. The final answer can be presented in the form:

$$\delta\sigma_{tot} = \frac{2e^2}{3\pi^2}\left\{-\ln\frac{\pi}{2\gamma_E t}+\frac{3\gamma_E t}{\widetilde{h}}+\psi\left(\frac{\widetilde{h}}{2\gamma_E t}\right)+4\left(\frac{1}{2\widetilde{h}\gamma_E t}\psi'\left[\frac{\widetilde{h}}{2\gamma_E t}\right]-1\right)\right\},$$

where γ_E is the Euler constant, $t = T/T_{c0}$ and $\widetilde{h} = (H - H_{c2}(T))/H_{c2}(T)$.

Let us consider some limiting cases. If the temperature is relatively high $t \gg \widetilde{h}$, we obtain the following formula for the fluctuation conductivity:

$$\delta\sigma = \frac{2\gamma_E e^2}{\pi^2}\frac{t}{\widetilde{h}}. \tag{9.62}$$

If $H < H_{c2}(0)$, we can introduce $T_c(H)$ and rewrite Eq. (9.62) in the usual way

$$\delta\sigma = \frac{3e^2}{2\gamma_E\pi^2}\frac{T_{c0}}{T-T_c(H)}. \tag{9.63}$$

If $H > H_{c2}(0)$, in the low-temperature limit $t \ll \widetilde{h}$ we have

$$\delta\sigma = -\frac{2e^2}{3\pi^2}\ln\frac{1}{\widetilde{h}}. \tag{9.64}$$

One can see, that even at zero temperature a logarithmic singularity remains and the corresponding correction is negative. It results from all three fluctuation contributions, although the DOS one exceeds the others by numerical factor. Let us recall that in the case of the out-of-plane conductivity of a layered superconductor, or in granular superconductors above T_c, the DOS contribution exceeds the MT and AL ones parametrically [360].

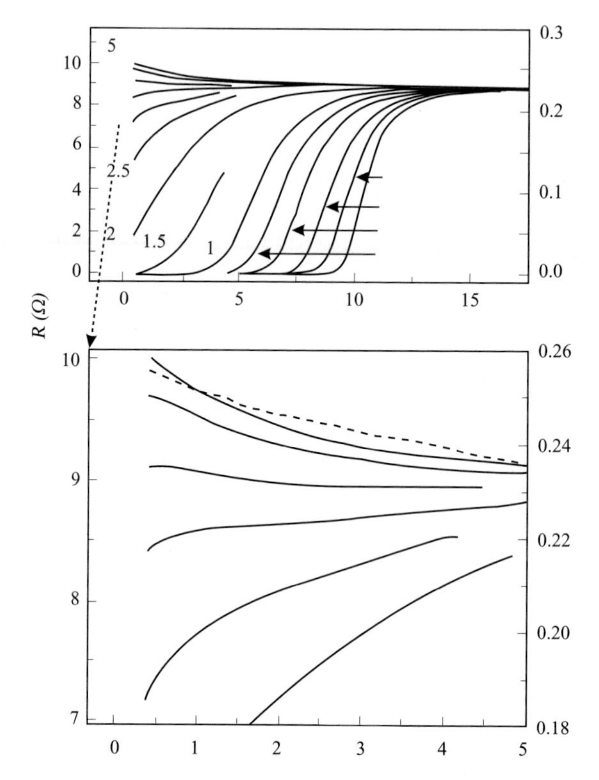

FIG. 9.4. Low temperature resistivity data for NdCeCuO film. The enlarged designated area of panel (a) is shown on panel (b). Curve at 7 T (dashed line) is crossing the other ones manifesting the negative magnetoresistance below 1 K.

It is interesting to compare Eq. (9.64) with the WL correction to conductivity in strong magnetic fields:

$$\delta\sigma_{wl} = -\frac{e^2}{2\pi^2}\ln\left(\omega_c\tau_\varphi\right). \tag{9.65}$$

Both contributions have the same sign but the former depends on $H - H_{c2}(T)$ and is dominant near the line $H_{c2}(T)$.

The central result of this section is the existence of the logarithmic correction to the conductivity which persists down to zero temperature. This correction is shown to be negative in the dirty case. The minus sign comes from the DOS diagrams as well as from the MT term. The AL contribution is positive but numerically smaller. Let us note, that similar results (negative fluctuation correction to the conductivity) exist for the granular and layered superconductors [360, 107]. In these cases, the AL and MT contributions are parametrically small compared to the DOS term.

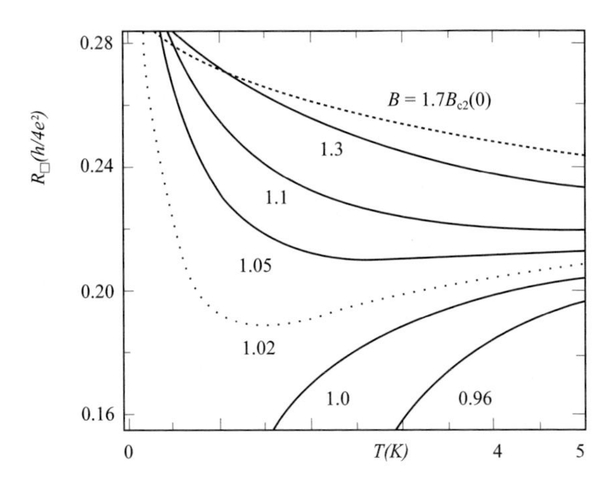

FIG. 9.5. Functions R(T) at different B calculated basing on theory of Galitsky and Larkin [76]. The curves are labeled by reduced field values. The curve that shows the negative magnetoresistance is marked by dashed line. The dotted curve should not be compared to experiment.

Let us note, that the singular behavior of the transport and thermodynamic quantities at low temperature is due to the low dimensionality of the system. In the $3D$ case the leading correction to the conductivity is not singular $\delta\sigma_{3\mathrm{D}} \propto \sqrt{\widetilde{h}}$.

Finally, let us mention some recent experiments of Gantmakher *et al.* [361, 362] In these experiments the magnetic-field-tuned quantum phase transition has been studied in dirty In–O films at low temperatures. It was found that in the vicinity of the transition, the magnetoresistance reaches a maximum. It is possible that the theory presented above can give an explanation for the observed effects. Moreover, in paper [363] the data on magneto-transport of superconducting $\mathrm{Nd_{2-x}Ce_xCuO_{4+y}}$ films obtained in high fields and at low temperatures are compared with the results presented in this section. The coincidence of the complex nonmonotonic dependencies is impressive (see Figs. 9.4 and Fig. 9.5).

The clean case may be relevant to HTS [364] and, probably, to $2D$ organic superconductors. Let us note, that our results assume s-pairing and isotropic Fermi surface which is not true for HTS. However, it can be shown, that the logarithmic singularity remains for any pairing type. It is worth mentioning, that in the overdoped HTS the Ginzburg-Levanyuk parameter is small and, thus, the fluctuations are negligible. In the underdoped superconductors the fluctuations are extremely large and they lead to the formation of a large pseudogap which makes the conventional Fermi liquid theory inapplicable. Hence, optimally doped superconductors should be used to check the results obtained.

9.5.2 *Magnetization*[64]

9.5.2.1 *One-loop approximation* Considering thermodynamic properties, we can calculate the free energy directly. In the one-loop approximation, the fluctuation part of the free energy is given by Eq. (9.48). The polarization operator $\Pi(\Omega_k, \mathbf{r}, \mathbf{r}')$ is determined by the correlator of two Green functions (see Eq. (7.22)) but in the case of an applied magnetic field the homogeneity of the system is lost and $\Pi(\Omega_k, \mathbf{r}, \mathbf{r}')$ depends not on the space variable difference $\mathbf{r}-\mathbf{r}'$ but on each separately. Using Eqs. (9.48),(9.58) and (9.61), one can easily obtain the magnetization

$$M_1(H,T) = -\frac{1}{V}\frac{\partial F_1(H,T)}{\partial H} = \frac{1}{\gamma_E}\left(\frac{T_{c0}}{\Phi_0 d}\right) I\left(\widetilde{h}, t\right), \tag{9.66}$$

where d is the thickness of the film or the interlayer distance and $I(\widetilde{h}, t)$ is defined as

$$I(\widetilde{h}, t) = \ln\frac{1}{2\gamma_E t} - \frac{\gamma_E t}{\widetilde{h}} - \psi\left(\frac{1}{2\gamma_E}\frac{\widetilde{h}}{t}\right). \tag{9.67}$$

Using the asymptotics of the ψ-function one can write

$$I\left(\widetilde{h}, t\right) = \begin{cases} -\ln\widetilde{h}, & t \ll \widetilde{h}, \\ \ln\frac{1}{2\gamma_E t} + \frac{\gamma_E t}{\widetilde{h}}, & \widetilde{h} \ll t \end{cases} \tag{9.68}$$

and the fluctuation diamagnetic susceptibility takes form:

$$\chi_1(H,T) = -\frac{\partial M_1(H,T)}{\partial H} = 12\chi_L\left(\frac{l}{d}\right)\begin{cases} \frac{1}{\widetilde{h}}, & t \ll \widetilde{h}, \\ \frac{\gamma_E t}{\widetilde{h}^2}, & \widetilde{h} \ll t, \end{cases} \tag{9.69}$$

where $\chi_L = e^2 v_F/12\pi^2$ is the value of Landau diamagnetic susceptibility. We see that the fluctuating magnetization exceeds conventional Landau diamagnetism for a very large range of fields. It is shown to be logarithmically divergent as well at $T \to 0$.

9.5.2.2 *Two-loop approximation* In the previous sections we found the fluctuation corrections to the transport and thermodynamic properties of a superconductor in a magnetic field in the first (one-loop) approximation. The purpose of the given section is to find the order of the subleading corrections. This will determine the area of applicability of the results obtained. We shall calculate the magnetization in the two-loop approximation for a dirty superconductor. This correction can be easily found in view of the simplifications described above.

[64] In this section we base on the results of [76].

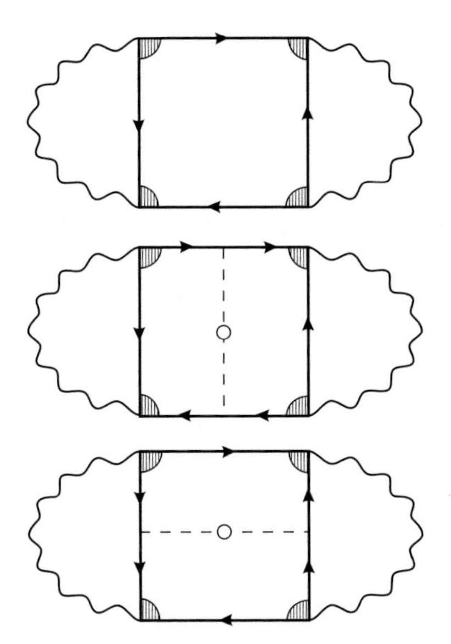

FIG. 9.6. Diagrams contributing to the free energy in the two-loop approximation. Similar diagrams appear in the derivation of the GL equations from the microscopic theory.

In the two-loop approximation, we have to deal with diagrams presented in Fig. 9.6. The corresponding contribution can be written in the coordinate representation in the following way

$$F_2 = T^3 \sum_{\varepsilon,\Omega,\Omega'} \int d^2r_1 d^2r_2 d^2r_3 d^2r_4 K_\varepsilon(\mathbf{r}_1,\mathbf{r}_2;\mathbf{r}_3,\mathbf{r}_4)\mathcal{L}_\Omega(\mathbf{r}_1,\mathbf{r}_2)\mathcal{L}_{\Omega'}(\mathbf{r}_3,\mathbf{r}_4), \tag{9.70}$$

where K_ε is the operator corresponding to the square blocks in the diagrams presented on Fig. 9.6. This operator is familiar from the usual BCS theory. It has been calculated by Maki [365] and Caroli *et al.* [366] and has the form.

$$\begin{aligned} K_{\varepsilon_n}(\mathbf{r}_1,\mathbf{r}_2;\mathbf{r}_3,\mathbf{r}_4) &= \frac{\pi\nu(0)}{2} \\ &\times\delta(\mathbf{r}_1-\mathbf{r}_2)\delta(\mathbf{r}_1-\mathbf{r}_3)\delta(\mathbf{r}_1-\mathbf{r}_4)\left\{\prod_{k=1}^{4}\frac{1}{|\varepsilon_n|+\frac{1}{2}\mathcal{D}\partial^2_{(k)}}\right\} \\ &\times\left[|\varepsilon_n|+\frac{1}{8}\mathcal{D}\left(\left[\partial_{(1)}-\partial_{(3)}\right]^2+\left[\partial_{(2)}-\partial_{(4)}\right]^2\right)\right], \end{aligned} \tag{9.71}$$

where we make use of the Maki's notations:

$$\partial_{(k)} = -i\nabla - 2e(-1)^k\mathbf{A}(\mathbf{r}).$$

In the coordinate representation, the fluctuation propagator can be expanded on the basis of the eigenfunctions in the magnetic field and has the form (see [367]):

$$\mathcal{L}_\Omega(\mathbf{r},\mathbf{r}') = \int\limits_{-\infty}^{+\infty} \frac{dp_y}{2\pi} \sum_{n=0}^{\infty} \mathcal{L}_n(\Omega) \psi^*_{np_y}(\mathbf{r}) \psi_{np_y}(\mathbf{r}'), \tag{9.72}$$

where $\mathcal{L}_n(\Omega)$ are matrix elements of the fluctuation propagator in the magnetic field (see Eq. (9.58)), $\psi_{np_y}(\mathbf{r})$ is the eigenfunction for an electron in a magnetic field in the Landau gauge and p_y is the y-component of the momentum, which determines the orbit's center. Again, in the vicinity of the transition line we keep the $n = 0$ term only in Eq. (9.72). From Eqs. (9.70)-(9.72), we obtain the free energy per unit volume

$$\frac{F_2}{V} = \frac{\nu(0)\,(eH)^2}{2\pi d^2} T^3 \left(\sum_\Omega L_0(\Omega) \right)^2 \sum_{\varepsilon_n} \frac{1}{\left(|\varepsilon_n| + \omega_c/4\right)^3}. \tag{9.73}$$

Thus, the magnetization takes the form:

$$M_2 = \frac{2}{\pi \gamma_E p_F l} \left(\frac{T_{c0}}{\Phi_0 d} \right) \frac{\partial I^2(\widetilde{h}, t)}{\partial \widetilde{h}^2}. \tag{9.74}$$

At low temperatures $t \ll \widetilde{h}$ we have

$$M_2 = -\frac{4}{\pi \gamma_E p_F l} \left(\frac{T_{c0}}{\Phi_0 d} \right) \frac{1}{\widetilde{h}} \ln \frac{1}{\widetilde{h}}. \tag{9.75}$$

We see, that the second order correction is negative.

From Eqs. (9.66) and (9.74) we obtain the ratio

$$\frac{M_2}{M_1} = \frac{2}{\pi p_F l} \left[2\gamma_E \frac{t}{\widetilde{h}^2} - \frac{1}{\gamma_E t} \psi' \left(\frac{1}{2\gamma_E} \frac{\widetilde{h}}{t} \right) \right] = \tag{9.76}$$

$$= \frac{1}{7\zeta(3)} Gi_{(2d)} \left[2\gamma_E \frac{t}{\widetilde{h}^2} - \frac{1}{\gamma_E t} \psi' \left(\frac{1}{2\gamma_E} \frac{\widetilde{h}}{t} \right) \right],$$

where we used the definition (2.87) of the Ginzburg parameter.

The one-loop approximation is valid unless this ratio becomes of the order of unity. Equation (9.76) yields the following conditions:[65]

$$\widetilde{h} \gg 2Gi_{(2d)} \begin{cases} 1, & t \ll \widetilde{h}, \\ \sqrt{\frac{t}{Gi_{(2d)}}}, & t \gg \widetilde{h}. \end{cases} \tag{9.77}$$

[65] Let us underline that this estimation of the width of fluctuation region, obtained from comparison of the results of two- and one-loop approximations coincides with that one accomplished in Part I and based on comparison between the result of the one-loop approximation and the magnetization jump in type-II superconductor at zero temperature.

Equation (9.77) indicates that at large enough temperatures the fluctuation region becomes wider (see Fig. 2.4).

These results stand for the kinetic coefficients as well. In the clean case the formula (9.77) is valid with $Gi^{(2c)}(T_c, H = 0) \sim T_{c0}/\varepsilon_F$ (see Table 2.1). However, the explicit calculations are more complicated due to the nonlocal structure of the K-operator.

Let us note that at an exponentially low temperature some other effects may reveal themselves. In the dirty case, the quenched disorder may be important [368, 369, 435] (see below sections 12.2 and 12.6).

10

DOS AND TUNNELING

10.1 Density of states of superconductor in fluctuation regime[66]

The appearance of nonequilibrium Cooper pairing above T_c leads to a redistribution of the one-electron states around the Fermi level. A semi-phenomenological study of the fluctuation effects on the density of states (DOS) of a dirty superconducting material was first carried out while analyzing the tunneling experiments of granular Al in the fluctuation regime just above T_c [370]. The second metallic electrode was in the superconducting regime and its well-developed gap gave a bias voltage around which a structure, associated with the superconducting fluctuations of Al, appeared. The measured DOS energy dependence has a dip at the Fermi level,[67] reaches its normal value at some energy $E_0(T)$, show a maximum at an energy value equal to several times E_0, finally decreases towards its normal value at higher energies. The characteristic energy E_0 was found to be of the order of the inverse of the GL relaxation time τ_{GL} introduced above. The presence of a depression at $E = 0$ and of a peak at $E \sim (1/\tau_{GL})$ in the DOS above T_c are precursor effects of the appearance of the superconducting gap in the quasiparticle spectrum at temperatures below T_c. The microscopic calculation of the fluctuation contribution to the one-electron DOS can be carried out within the diagrammatic technique.

Let us start from the discussion of a clean superconductor. The one-electron DOS is determined by the imaginary part of the retarded Green function integrated over momentum. This definition allows us to express the appropriate fluctuation correction in terms of the fluctuation propagator:

$$\delta\nu(E,\epsilon) = -\frac{1}{\pi}\operatorname{Im}\int \frac{d^D\mathbf{p}}{(2\pi)^D}\delta G^R(\mathbf{p},E) = -\frac{1}{\pi}\operatorname{Im} R^R(E) \tag{10.1}$$

where $R^R(E)$ is the retarded analytic continuation of the expression corresponding to the diagram of Fig. 10.1:

$$R(\varepsilon_n) = T\sum_{\Omega_k}\int \frac{d^D\mathbf{q}}{(2\pi)^D} L(\mathbf{q},\Omega_k)\lambda^2(q,\varepsilon_n,\Omega_k-\varepsilon_n)\int \frac{d^D\mathbf{p}}{(2\pi)^D} G^2(\mathbf{p},\varepsilon_n)G(\mathbf{q}-\mathbf{p},\Omega_k-\varepsilon_n). \tag{10.2}$$

[66] In this section we base on the results of [169, 237].

[67] Here we refer the energy E to the Fermi level, where we assume $E = 0$.

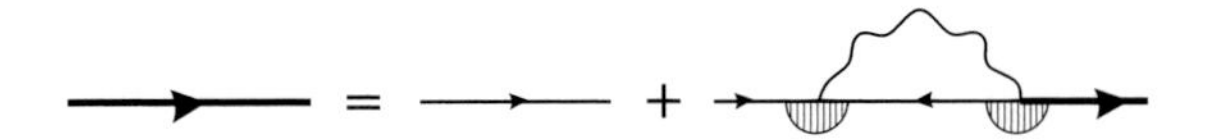

FIG. 10.1. The one-electron Green function with the first-order fluctuation correction.

The result of the integration of the last expression depends strongly of the electron spectrum dimensionality: for the two important cases of isotropic $3D$ and $2D$ electron spectra one finds [237]

$$\frac{\delta\nu_{(3)}^{(c)}(E,\epsilon)}{\nu_{(3)}} = -\frac{(4\pi)^{3/2}}{7\zeta(3)}\sqrt{Gi_{(3,c)}}\,\mathrm{Re}\,\frac{\sqrt{T_{\mathrm{c}}}}{\sqrt{\tau_{\mathrm{GL}}^{-1}-2iE+\varkappa_3^2T_{\mathrm{c}}}} \times\left\{\frac{1}{\tau_{\mathrm{GL}}^{-1}-iE+\tau_{\mathrm{GL}}^{-1/2}\left[\tau_{\mathrm{GL}}^{-1}-2iE+\varkappa_3^2T_{\mathrm{c}}\right]^{1/2}}\right\}, \tag{10.3}$$

$$\frac{\delta\nu_{(2)}^{(c)}(E,\epsilon)}{\nu_{(2)}} = -\frac{(4\pi)^{2}}{7\zeta(3)}Gi_{(2,c)}\frac{T_{\mathrm{c}}^2}{\left[E^2+\varkappa_2^2T_{\mathrm{c}}\tau_{\mathrm{GL}}^{-1}\right]} \times\left\{1-\frac{E}{\sqrt{E^2+\varkappa_2^2T_{\mathrm{c}}\tau_{\mathrm{GL}}^{-1}}}\ln\frac{E+\sqrt{E^2+\varkappa_2^2T_{\mathrm{c}}\tau_{\mathrm{GL}}^{-1}}}{\varkappa_2\sqrt{T_{\mathrm{c}}\tau_{\mathrm{GL}}^{-1}}}\right\}, \tag{10.4}$$

where $\varkappa_D=\pi\sqrt{\pi D/7\zeta(3)}$.

In a dirty superconductor the calculations may be carried out in a similar way with the only difference that the impurity renormalization of the Cooper vertices has to be taken into account [169]. The value of the fluctuation dip at the Fermi level can be written in the form:

$$\frac{\delta\nu_{(D)}^{(d)}(0,\epsilon)}{\nu_{(D)}} \sim -\begin{cases}\sqrt{Gi_{(3,d)}}\epsilon^{-3/2}, & D=3,\\ Gi_{(2,d)}\epsilon^{\ 2}, & D=2.\end{cases} \tag{10.5}$$

At large energies $E\gg\tau_{\mathrm{GL}}^{-1}$ the DOS recovers its normal value, according to the same laws (10.5) but with the substitution $\epsilon\to E/T_{\mathrm{c}}$. It is interesting that the critical exponents of the fluctuation correction of the DOS change when moving from a dirty to a clean superconductor [237]: the analysis of (10.3)-(10.4) gives

$$\frac{\delta\nu_{(D)}^{(c)}(0,\epsilon)}{\nu_{(D)}} \sim -\begin{cases}\sqrt{Gi_{(3,c)}}\epsilon^{-1/2}, & D=3,\\ Gi_{(2,c)}\epsilon^{-1}, & D=2.\end{cases} \tag{10.6}$$

Another important aspect in which the character of the DOS renormalization is substantially different for the clean and dirty cases is the energy scale at which this renormalization occurs. In the dirty case this energy turns out to

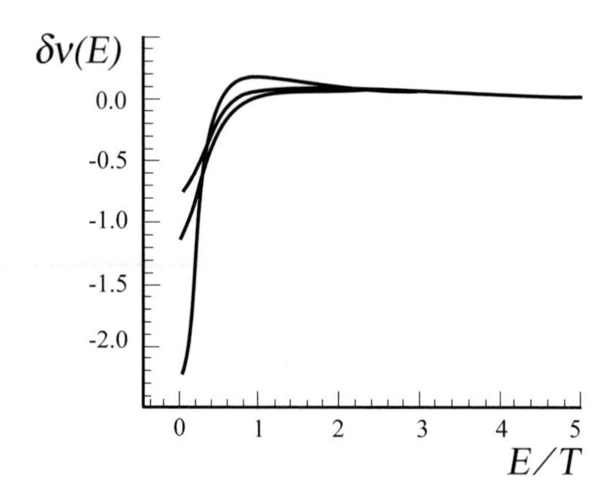

FIG. 10.2. The theoretical curve of the energy dependence for the normalized correction to the single-particle DOS versus energy for a clean $2D$ superconductor above T_c.

be [169] $E_0^{(d)} \sim T - T_c \sim \tau_{GL}^{-1}$, while in the clean case $E_0^{(c)} \sim \sqrt{T_c(T-T_c)}$ [237]. To understand this important difference one has to study the character of the electron motion in both cases [237]. The relevant energy scale in the dirty case is the inverse of the time necessary for the electron to diffuse over a distance equal to the coherence length $\xi(T)$. This energy scale coincides with the inverse relaxation time: $t_\xi^{-1} = \mathcal{D}\xi^{-2}(T) \sim \tau_{GL}^{-1} \sim T - T_c$. In the clean case, the ballistic motion of the electrons gives rise to a different characteristic energy scale $t_\xi^{-1} \sim v_F\xi^{-1}(T) \sim (T_c\tau_{GL}^{-1})^{1/2} \sim \sqrt{T_c(T-T_c)}$ (see Fig. 10.2)

One can check that the integration of (10.3)-(10.4) over all positive energies gives zero:

$$\int_0^\infty \delta\nu(E)\, dE = 0 \tag{10.7}$$

This "sum rule" is a consequence of a conservation law: the number of quasiparticles is determined by the number of cells in the crystal and cannot be changed by the interaction. So the only effect which can be produced by the inter-electron interaction is a redistribution of the energy levels near the Fermi energy. The sum rule (10.7) plays an important role in the understanding of the manifestation of the fluctuation DOS renormalization in the observable phenomena.

10.2 Fluctuation phenomena in tunnel junction above T_c[68]

10.2.1 *Preliminaries*

In this section we will show that the generated by the DOS density of states renormalization zero-bias singularity of the tunneling current turns out to be

[68] In this section we base on the results of [277, 372, 373].

much weaker than that in the DOS itself ($\ln \epsilon$ instead of ϵ^{-1} or ϵ^{-2}, see (10.5)-(10.6)). A similar smearing of the DOS singularity occurs in the opening of the pseudo-gap in the out-of-plane a.c. conductivity, in the NMR relaxation rate etc. These features are due to the fact that we must always form the convolution of the DOS with some slowly varying function: e.g. a difference of Fermi functions in the case of the tunnel current. The sum rule then leads to an almost perfect cancelation of the principal singularity at low energies. The main nonzero contribution then comes from the high energy region where the DOS correction has its 'tail'. Another important consequence of the conservation law (10.7) is the considerable increase of the characteristic energy scale of the fluctuation pseudo-gap opening with respect to E_0: this is $eV_0 = \pi T$ for tunneling and $\omega \sim \tau^{-1}$ for the c-axis a.c. conductivity.

10.2.1.1 *Ambegaokar–Baratoff formula for tunneling current* It is quite evident that the renormalization of the density of states near the Fermi level, even of only one of the electrodes, will lead to the appearance of anomalies in the voltage–current characteristics of a tunnel junction. The quasiparticle current flowing through the barrier may be written as a convolution of the densities of states with the difference of the electron Fermi distributions in each electrode (L and R) [374, 375].

In order to derive the corresponding explicit expression let us start from the tunneling Hamiltonian:

$$\widehat{\mathcal{H}} = \widehat{\mathcal{H}}_L + \widehat{\mathcal{H}}_R + \widehat{\mathcal{H}}_T, \tag{10.8}$$

where $\widehat{\mathcal{H}}_L$ and $\widehat{\mathcal{H}}_R$ are the Hamiltonians of the left and the right electrodes while the part describing the process of electron tunneling is

$$\widehat{\mathcal{H}}_T = \sum_{\mathbf{p},\mathbf{k}} \left(T_{\mathbf{p},\mathbf{k}} \widehat{a}_{\mathbf{p}}^{+} \widehat{b}_{\mathbf{k}} + T_{\mathbf{k},\mathbf{p}}^{*} \widehat{a}_{\mathbf{p}} \widehat{b}_{\mathbf{k}}^{+} \right). \tag{10.9}$$

Here $T_{\mathbf{p},\mathbf{k}}$ is the tunneling matrix element and $\widehat{a}_{\mathbf{p}}^{+}, \widehat{b}_{\mathbf{k}}$ are the creation and annihilation operators in the first and the second electrode, respectively.

Tunneling current can be determined as the average of the time derivative of the particle number operator taken in the Heisenberg representation [372, 374]:

$$I_T = e \left\langle \frac{d\widehat{\mathcal{N}}_L(t)}{dt} \right\rangle, \tag{10.10}$$

where

$$\widehat{\mathcal{N}}_L = \left\langle \exp\left(i \int_{-\infty}^{t} \widehat{\mathcal{H}} dt_1 \right) \sum_{\mathbf{p}} \widehat{a}_{\mathbf{p}}^{+} \widehat{a}_{\mathbf{p}} \exp\left(-i \int_{-\infty}^{t} \widehat{\mathcal{H}} dt_1 \right) \right\rangle, \tag{10.11}$$

$\langle ... \rangle$ means the averaging over the Gibbs ensemble of the electrons of both electrodes. In the second order of the perturbation theory over the tunneling Hamiltonian one finds

$$I_T = e\left\langle \int_{-\infty}^{t} \left[\widehat{\mathcal{H}}_T\left(t_1\right), \left[\widehat{\mathcal{N}}_L\left(t\right), \widehat{\mathcal{H}}_T\left(t\right)\right]\right] dt_1 \right\rangle. \tag{10.12}$$

Supposing the electrodes to be in superconducting state one can present the quasiparticle branch of the tunnel current as the trace of product of the Green functions [372]

$$I_{qp}\left(t\right) = -4e\,\mathrm{Re}\sum_{\mathbf{p},\mathbf{k}} \left|T_{\mathbf{p},\mathbf{k}}\right|^2 \int_{-\infty}^{t} \left[G_R\left(\mathbf{p},t_1,t\right)G_L\left(\mathbf{k},t,t_1\right)\right.$$
$$\left. -G_R^*\left(\mathbf{p},t,t_1\right)G_L^*\left(\mathbf{k},t_1,t\right)\right]dt_1 \tag{10.13}$$

with

$$G_{L,R}\left(\mathbf{p},t_1,t\right) = \exp\left[-ie\varphi_{L,R}\left(t-t_1\right)\right]\int \frac{d\omega}{2\pi}G_{L,R}\left(\omega\right)\exp\left[-i\omega\left(t-t_1\right)\right]. \tag{10.14}$$

The Green function $G_{R,L}\left(\omega\right)$ can be related to the corresponding retarded one $G^R\left(x\right)$ by means of the Leman representation [228]:

$$G\left(\omega\right) = \frac{1}{2\pi}\int_{-\infty}^{\infty} dx\,\mathrm{Im}\,G^R\left(x\right)\left[\frac{1-\tanh x/2T}{x-\omega+i\delta} + \frac{1+\tanh x/2T}{x-\omega-i\delta}\right]. \tag{10.15}$$

In the typical assumption of the weak dependence of the matrix elements $T_{\mathbf{p},\mathbf{k}}$ on energy in the vicinity of the Fermi level one can pass from the momenta summation to the energy integration

$$e\sum_{\mathbf{p},\mathbf{k}}\left|T_{\mathbf{p},\mathbf{k}}\right|^2\left(\cdots\right) = \frac{1}{4\pi eR_n}\int d\xi_{\mathbf{p}}\int d\xi_{\mathbf{k}}\left(\ldots\right), \tag{10.16}$$

for the junction of unit area. Substituting in Eq. (10.13) the expressions (10.14)-(10.16) and recalling the relation between the $\mathrm{Im}\,G^R\left(x\right)$ and DOS (10.1) one can reproduce so-called Ambegaokar–Baratoff formula [388] for the quasiparticle current flowing through the junction of unit area:

$$I_{qp}\left(V\right) = \frac{1}{eR_n\nu_L\nu_R}\int_{-\infty}^{\infty}\left(\tanh\frac{E+eV}{2T} - \tanh\frac{E}{2T}\right)\nu_R(E)\nu_L(E+eV)dE, \tag{10.17}$$

where R_n is the Ohmic resistance per unit area of the junction, ν_L, ν_R are the DOS at the Fermi levels in each of electrodes in the absence of tunneling.

One can see that for low temperatures and voltages the expression in parenthesis is a sharp function of energy near the Fermi level. Nevertheless, depending on the properties of the DOS functions, the convolution (10.17) may exhibit different properties. Let us consider the properties of the asymmetric junction, which one electrode has constant DOS while in the second it varies. If the energy scale of the $\nu(E)$ is much larger than T, the expression in parenthesis in (10.17) acts as a delta-function and the zero-bias anomaly in the tunnel conductivity strictly reproduces the anomaly of the DOS around the Fermi level:

$$\frac{\delta G(V)}{G_n(0)} = \frac{\delta \nu(eV)}{\nu(0)}, \tag{10.18}$$

where $G(V)$ is the differential tunnel conductance and $G_n(0)$ is the background value of the Ohmic conductance supposed to be bias independent, $\delta G(V) = G(V) - G_n(0)$. This situation, for instance, occurs in a junction with one amorphous electrode [290], where the dynamically screened Coulomb interaction is strongly retarded, which leads to a considerable suppression of the DOS in the vicinity of the Fermi level, within $\tau^{-1} \gg T$.

It is worth stressing that the proportionality between the tunneling current and the electron DOS of the electrodes is widely accepted as an axiom, but generally speaking this is not always so. As one can see from the previous subsection, the opposite situation occurs in the case of the DOS renormalization due to the electron–electron interaction in the Cooper channel: in this case the DOS correction varies strongly already in the scale of $E_0 \sim E_{\text{ker}} \ll T$ and the convolution in (10.17) with the DOS (10.4) has to be carried out without the simplifying approximations assumed to obtain (10.18). We will show that the fluctuation induced pseudo gap-like structure in the tunnel conductance differs drastically from the anomaly of the DOS (10.4), both in its temperature singularity near T_c and in the energy range of its manifestation.

10.2.1.2 *Diagrammatic representation of the tunneling current* For our further purposes will be convenient to present the tunneling current (10.13) in the form of the imaginary part of analytically continued correlator of the exact Matsubara Green functions of both electrodes [277]:

$$I_T(V) = -e \operatorname{Im} K^R(\omega_\nu \to -ieV), \tag{10.19}$$

where

$$K(\omega_\nu) = 4T \sum_{\varepsilon_n} \sum_{\mathbf{p},\mathbf{k}} |T_{\mathbf{p},\mathbf{k}}|^2 G_R(\mathbf{p}, \varepsilon_n + \omega_\nu) G_L(\mathbf{k}, \varepsilon_n). \tag{10.20}$$

The corresponding diagrammatic representation of the $K(\omega_\nu)$ as the Green functions loop is demonstrated in Fig. 10.3.

The external bosonic frequency ω_ν here accounts for the potential difference between the electrodes, factor 4 reflects the result of the spin indices summation.

Indeed, one can make sure in the validity of such presentation of the tunneling current substituting the integrated over energy Matsubara Green's functions of each electrode in their spectral form [228]

$$g(\varepsilon_n) = \int d\xi_{\mathbf{k}} G(\mathbf{k}, \varepsilon_n) = \frac{1}{\nu} \int \frac{d\mathbf{k}}{(2\pi)^D} G(\mathbf{k}, \varepsilon_n) = \int \frac{\nu(E)\, dE}{E - i\varepsilon_n} \tag{10.21}$$

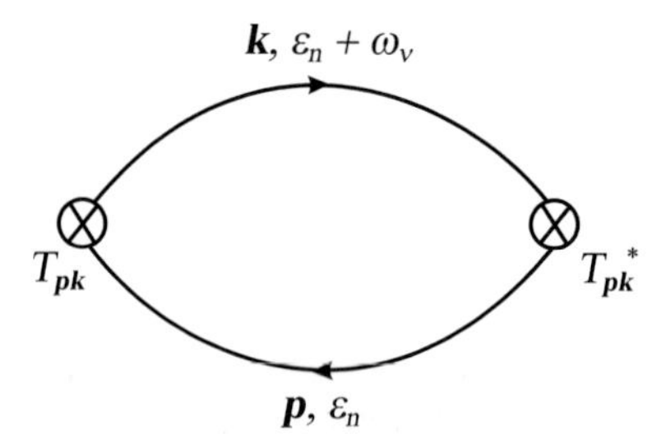

FIG. 10.3. Green's functions correlator which determines the tunneling current.

to Eq. (10.20). The further presentation of the product of energy denominators in the form of simple fractions gives

$$K(\omega_\nu) = \frac{1}{\pi e R_n \nu_L \nu_R} \int\int \frac{\nu_L(E_L)\,\nu_R(E_R)\,dE_L dE_R}{E_R - E_L - i\omega_\nu} \times T\sum_{\varepsilon_n}\left(\frac{1}{i\varepsilon_{n+\nu} - E_R} - \frac{1}{i\varepsilon_n - E_L}\right). \tag{10.22}$$

Following summation over fermionic frequencies and integration over one of energies restores the Ambegaokar–Baratoff result (10.17) what confirms the validity of Eq. (10.19).

10.2.2 *The effect of fluctuations on the tunnel current above T_c*

Let us start our discussion from the effect of the fluctuation suppression of the DOS on the properties of a tunnel junction between a normal metal N and a superconductor in fluctuation regime (i.e. above it critical temperature T_c). The effect under discussion turns out to be most pronounced in the case of the fluctuating electrode performed in the form of thin superconducting film $(d \ll \xi(T))$.

In the framework of the Matsubara temperature diagrammatic technique the accounting for fluctuations in such a junction is straightforward. It is enough to insert fluctuation propagator in the Green function loop in Fig. 10.4.

One can see that usually principal AL (diagram 6) and MT (diagram 7) processes here manifest themselves only in the second order $\left(\sim |T_{\mathbf{p},\mathbf{k}}|^4\right)$ in the barrier transparency. That is why the main contribution originates from the diagram 1, which corresponds to the effect of the DOS renormalization in the electrode being in fluctuation regime:

$$K^{(1)}(\omega_\nu) = 4T\sum_{\varepsilon_n}\sum_{\mathbf{p},\mathbf{k}} |T_{\mathbf{p},\mathbf{k}}|^2 G_R(\mathbf{p}, \varepsilon_n + \omega_\nu)\, G_L^2(\mathbf{k}, \varepsilon_n) \times T\sum_{\Omega_k}\int \frac{d\mathbf{q}}{(2\pi)^D}\lambda^2(\mathbf{q}, \varepsilon_n, \Omega_k - \varepsilon_n)\, L(\mathbf{q}, \Omega_k)\, G_L(\mathbf{q}-\mathbf{k}, \Omega_k - \varepsilon_n). \tag{10.23}$$

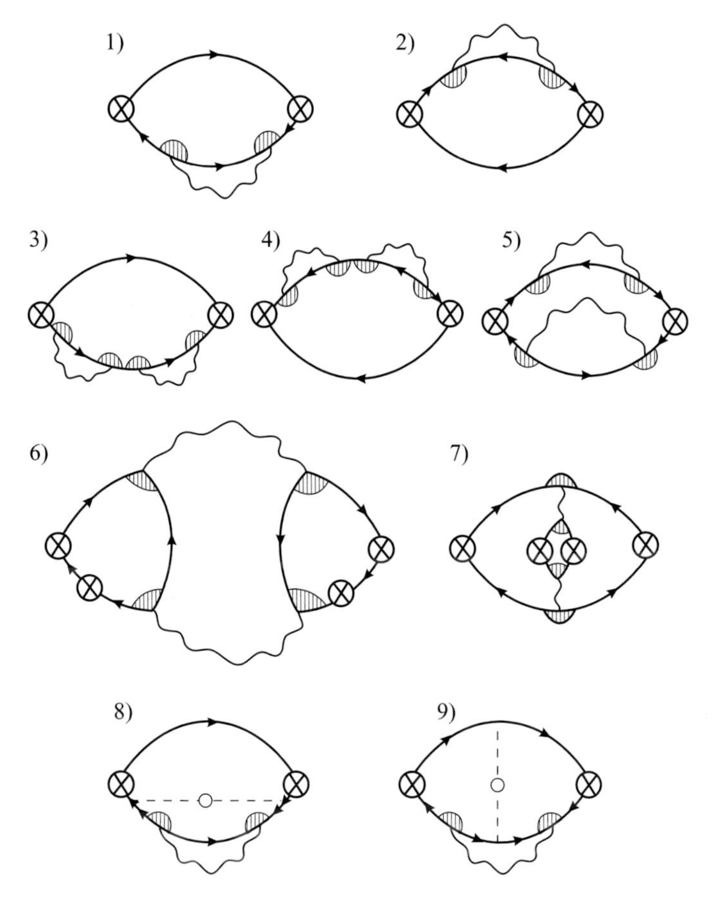

FIG. 10.4. Fluctuation corrections to the tunneling current of the N–I–N(S) junction in different orders in transparency and fluctuations in the system.

In the case of a dirty film electrode the simple integrations reduce Eq. (10.23) to

$$K^{(1)}(\omega_\nu) = -\frac{\pi}{e^2 R_n} T \sum_{\varepsilon_n} \operatorname{sign}(\varepsilon_n) \operatorname{sign}(\varepsilon_n + \omega_\nu) T \sum_{\Omega_k} \theta(\varepsilon_n(\varepsilon_n - \Omega_k)) \times \int \frac{d\mathbf{q}}{(2\pi)^2} \frac{L(\mathbf{q}, \Omega_k)}{[|2\varepsilon_n - \Omega_k| + \mathcal{D}q^2]^2}. \tag{10.24}$$

The following transformation of the Ω_k-summation in the contour integral and further summation over ε_n and momentum integration result in

$$I_{qp}^{(1)}(V) = -\frac{3T}{\pi e R_n} \frac{1}{p_F^2 l d} \left(\ln \frac{1}{\epsilon}\right) \operatorname{Im} \psi' \left(\frac{1}{2} - \frac{ieV}{2\pi T}\right), \tag{10.25}$$

or

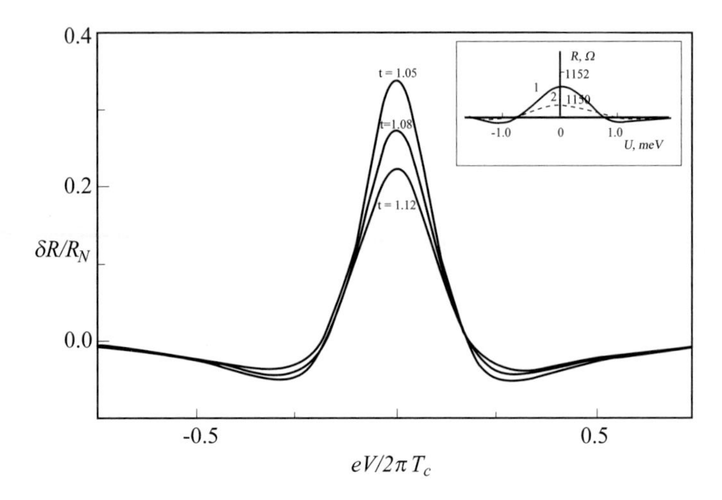

FIG. 10.5. The theoretical prediction for the fluctuation-induced zero-bias anomaly in the tunnel-junction resistance as a function of voltage for reduced temperatures $\epsilon = 0.05$ (top curve), $\epsilon = 0.08$ (middle curve) and $\epsilon = 0.12$ (bottom curve). The insert shows the experimentally observed differential resistance as a function of voltage in an Al–I–Sn junction just above the transition temperature.

$$G_{qp}^{(1)}(V) = \frac{Gi_{(2d)}}{7\zeta(3)\,R_n}\left(\ln\frac{1}{\epsilon}\right)\operatorname{Re}\psi''\left(\frac{1}{2}-\frac{ieV}{2\pi T}\right) \tag{10.26}$$

for the tunneling conductance.

It is important to emphasize several nontrivial features of the result obtained. First, the sharp decrease ($\epsilon^{-2(1)}$) of the density of electron states in the immediate vicinity of the Fermi level generated by fluctuations surprisingly results in a much more moderate growth of the tunnel resistance at zero voltage ($\ln 1/\epsilon$). Second, in spite of the manifestation of the DOS renormalization at the characteristic scales $E_0^{(d)} \sim T - T_c$ or $E_0^{(cl)} \sim \sqrt{T_c(T-T_c)}$, the energy scale of the anomaly developed in the I-V characteristic is much larger: $eV = \pi T \gg E_0$ (see Fig. 10.5).

In the inset of Fig. 10.5 the result of measurements of the differential resistance of the tunnel junction Al-I-Sn at temperatures slightly above the critical temperature of Sn electrode is presented. This experiment was done [376] with the purpose of checking the theory proposed [277]. The nonlinear differential resistance was precisely measured at low voltages which allowed the observation of the fine structure of the zero-bias anomaly. The reader can compare the shape of the measured fluctuation part of the differential resistance (the inset in Fig. 10.5) with the theoretical prediction. It is worth mentioning that the experimentally measured positions of the minima are $eV \approx \pm 3T_c$, while the theoretical prediction following from (10.26) is $eV = \pm\pi T_c$. Recently similar results on an aluminum film with two regions of different superconducting transition temperatures were reported [377]. The observations of the pseudogap anomalies in tunneling experiments at temperatures above T_c obtained by a variety of experimental techniques were reported in [378–382].

We will now consider the case of a symmetric junction between two superconducting electrodes at temperatures above T_c. In this case, evidently, the correction (10.26) has to be multiplied by a factor of "two" because of the possibility of fluctuation pairing in both electrodes. Furthermore, in view of the extraordinarily weak ($\sim \ln 1/\epsilon$) temperature dependence of the first order correction, different types of high order corrections may manifest themselves on the energy scale $eV \sim T - T_c$ or $\sqrt{T_c(T - T_c)}$. Among them are the familiar AL and MT corrections which take place in the first order of Gi but in the second order of the barrier transparency. Another type of higher order correction appears in the first order of barrier transparency but in the second of fluctuation strength ($\sim Gi^2$) (see diagram 5 in Fig. 10.4) [277]. Such corrections are generated by the interaction of fluctuations through the barrier and they can be evaluated directly from (10.17) applied to a symmetric junction. The second-order correction in Gi can be written as [277]:

$$\delta G_{\text{fl}}^{(2d)}(0, \epsilon) \sim \int_{-\infty}^{\infty} \frac{dE}{\cosh^2\left(\frac{E}{2T}\right)} \left[\delta \nu_{(2)}^{(d)}(E, \epsilon)\right]^2 \sim \frac{Gi_{(2d)}^2}{\epsilon^3}. \tag{10.27}$$

This nonlinear fluctuation correction turns out to be small by Gi^2 but its strong singularity in temperature and opposite sign with respect to $\delta G_{\text{fl}}^{(1)}$ make it interesting. Apparently it leads to the appearance of a sharp maximum at zero voltage in $G(V)$ with a characteristic width $eV \sim T - T_c$ in the immediate vicinity of T_c (one can call this peak as the hyperfine structure). This result was confirmed in [383] but to our knowledge such corrections were never observed in tunneling experiments.

As we have seen above the fluctuation correction to the DOS $\delta \nu_{(2)}^{(d)}$ at temperatures $\epsilon \sim \sqrt{Gi_{(2d)}}$ and energies $E \sim T\sqrt{Gi_{(2d)}}$ reaches the order of its Fermi liquid value. Evaluating the integral in Eq. (10.27) as the product of the integrand function (~ 1) and the characteristic domain of integration ($\sqrt{Gi_{(2d)}}$) one can find that $\delta G_{\text{fl}}^{(2d)}\left(0, \sqrt{Gi_{(2d)}}\right) \sim \sqrt{Gi_{(2d)}}$, hence can exceed the first order correction.

10.3 Fluctuation phenomena in N(S)-I-S junction

In the most second-order phase transitions the order parameter can be directly coupled to an external field by means of some susceptibility function [373]. In the case of paramagnetic–ferromagnetic transition this is the magnetic susceptibility, which connects the magnetization and external magnetic field. For superconductor the role of magnetization plays the order parameter Δ , which is off diagonal in electron number space and presents itself the condensate wave function. So its coupling with some classical field and the possibility of the direct observation of the corresponding susceptibility $\langle \Delta_q \Delta_q^* \rangle$ (see Eq. (6.6)) in principle is questionable. As we saw above, usually corresponding fluctuation propagator is involved in the expressions for observable physical values in the form of convolution. Nevertheless in the paper [384] was suggested the original method and in [373] it was concretized how the fluctuation propagator could be measured by frequency dependent conductivity of the tunnel junction in which one electrode is near its transition point T_{c1} while the second is deeply in superconducting state (its critical temperature $T_{c2} \gg T_{c1}$).

10.3.1 *Quasiparticle current in N(S)-I-S junction*

Let us calculate the quasiparticle current in the described tunneling junction. It is determined by the correlator (10.20), where one Green's function corresponds to the electron state in normal metal and other to the electron state in superconductor (for the further use we do not suppose the space homogeneity):

$$K(\omega_\nu) = T \sum_{\varepsilon_n} \sum_{\mathbf{p}_1, \mathbf{k}_1} \sum_{\mathbf{p}_2, \mathbf{k}_2} T_{\mathbf{k}_1 \mathbf{p}_1} G_n(\mathbf{p}_1, \mathbf{p}_2, \varepsilon_n + \omega_\nu) T^*_{\mathbf{p}_2 \mathbf{k}_2} G_s(\mathbf{k}_2, \mathbf{k}_1, \varepsilon_n). \tag{10.28}$$

Substitution in this expression of the electron Green functions of the normal metal and superconductor results in the nonlinear gap I-V characteristics of the N-I-S junction (see [385]):

$$I_{qp}(V) = \frac{2C_{NN}}{e} \Delta_2 \sum_{n=1}^{\infty} (-1)^{n+1} K_1\left(\frac{n\Delta_2}{T}\right) \sinh\left(\frac{neV}{T}\right), \tag{10.29}$$

where $K_1(x)$ is the elliptic integral. When $T \ll \Delta_2$ this current is exponentially small. At such temperatures there are only few quasiparticles in the superconducting electrode and Cooper pair tunneling from the Bose condensate to the normal electrode becomes significant. Penetrating into the normal electrode these pairs acquire a fluctuation nature. Such a tunneling current appears only in the fourth order in the tunneling matrix elements $T_{\mathbf{kp}}$ (or, which is the same, in the second order in tunneling probability) analogously to the out-of-plane AL paraconductivity of layer superconductor σ_{zz} (see section 8.3). Nevertheless, it does not contain the exponential smallness at $T \ll T_{c2}$. This current turns out to be proportional to the fluctuation propagator and especially increases close to T_{c1}.

Phenomenological theory of this phenomenon was developed in papers [373, 384, 386]. Below we will propose its microscopic analysis which will allow us to follow connection between the fluctuation tunneling conductivity at $T \to T_{c1}$ and Andreev conductance [387] at $T \gg T_{c1}$.

10.3.2 *Effective boundary Hamiltonian*

Above we demonstrated how the tunneling Hamiltonian defined by Eqs. (10.8)-(10.9) can be applied to study of the problem of quasiparticle tunneling current.

Below we will be interested in the Cooper pair charge transfer in $N(S)$-I-S junction. In order to do this we will derive the analogous Hamiltonian for the tunneling of Cooper pairs.

The quasiparticle spectrum of superconducting electrode S at low temperatures $T \ll \Delta_2$ has a large gap Δ_2 and one particle tunneling from the N(S) electrode at small voltages ($V \ll \Delta_2/e$) is forbidden by the energy conservation law. Nevertheless there remains the possibility of pair tunneling. The quantum mechanical amplitude of two electrons (with momenta $\mathbf{p}$ and $\mathbf{p}'$) tunneling from the normal electrode to the superconducting one appears in the second order of the perturbation theory in the tunneling matrix element. Since the momentum in tunneling process is not conserved, any two electrons, being already in the superconducting electrode, with the opposite momenta $(\mathbf{k}, -\mathbf{k})$ ($\mathbf{k}$ is arbitrary) can form a Cooper pair. Hence the total amplitude of such process can be written in accordance with the laws of quantum mechanics

$$A(\mathbf{p}, \mathbf{p}') = \sum_{\mathbf{k}} T_{\mathbf{kp}} T_{-\mathbf{kp}'} u_{\mathbf{k}} v_{-\mathbf{k}} \left(\frac{1}{\varepsilon_{\mathbf{p}} - E_{\mathbf{k}}} + \frac{1}{\varepsilon_{\mathbf{p}'} - E_{\mathbf{k}}} \right). \tag{10.30}$$

The energy denominators in Eq. (10.30) are equal to the difference of energy of the initial state $\varepsilon_{\mathbf{p}} + \varepsilon_{\mathbf{p}'}$ and the virtual excited state with one quasiparticle in superconducting electrode (with energy $E_{\mathbf{k}} = \left(\xi_{\mathbf{k}}^2 + |\Delta_2|^2\right)^{1/2}$) and one electron (with energy $\varepsilon_{\mathbf{p}}$ or $\varepsilon_{\mathbf{p}'}$) in normal one. $u_{\mathbf{k}} v_{-\mathbf{k}}$ in Eq. (10.30) is the product of Bogolyubov's coherence factors [32]:

$$u_{\mathbf{k}} v_{-\mathbf{k}} = \frac{\Delta_2}{2E_{\mathbf{k}}}. \tag{10.31}$$

Below we will be interested in the initial states with energies $\varepsilon_{\mathbf{p}}, \varepsilon_{\mathbf{p}'} \sim T \ll \Delta_2$. It is why one can rewrite the amplitude Eq. (10.30) as

$$A(\mathbf{p}, \mathbf{p}') = -\Delta_2 \sum_{\mathbf{k}} \frac{T_{\mathbf{kp}} T_{-\mathbf{kp}'}}{E_{\mathbf{k}}^2}. \tag{10.32}$$

Now one can write the effective Hamiltonian in a weak superconducting or normal electrode (N(S)) as:

$$\begin{aligned} \widehat{\mathcal{H}}_b &= \frac{1}{2} \sum_{\mathbf{p}.\mathbf{p}'} \left[A^*(\mathbf{p}, \mathbf{p}') \widehat{a}_{\mathbf{p}\uparrow} \widehat{a}_{\mathbf{p}'\downarrow} + A(\mathbf{p}, \mathbf{p}') \widehat{a}^+_{\mathbf{p}\uparrow} \widehat{a}^+_{\mathbf{p}'\downarrow} \right] \\ &= -\frac{1}{2} \sum_{\mathbf{k}} \sum_{\mathbf{p}} \sum_{\mathbf{q}} \frac{1}{E_{\mathbf{k}}^2} \left[T^*_{\mathbf{k},\mathbf{p}+\mathbf{q}/2} T^*_{-\mathbf{k},\mathbf{p}-\mathbf{q}/2} \Delta_2^*(\mathbf{q}) \widehat{a}_{\mathbf{p}+\mathbf{q}/2\uparrow} \widehat{a}_{\mathbf{p}-\mathbf{q}/2\downarrow} \right. \\ &\left. + \, T_{\mathbf{k},\mathbf{p}+\mathbf{q}/2} T_{-\mathbf{k},\mathbf{p}-\mathbf{q}/2} \Delta_2(\mathbf{q}) \widehat{a}^+_{\mathbf{p}+\mathbf{q}/2\uparrow} \widehat{a}^+_{\mathbf{p}-\mathbf{q}/2\downarrow} \right]. \end{aligned} \tag{10.33}$$

Presenting $\Delta_2 = \Delta_2(\mathbf{q})$ in the assumption $|\Delta_2|^2 = const$ we leave the possibility of the order parameter phase variation along the barrier. One can pass in

Eq. (10.33) to integration instead of summation and in assumption of a weak dependence of the matrix elements on momentum $\mathbf{q}$, analogously to Eq. (10.16) get

$$\widehat{\mathcal{H}}_b = \frac{1}{4\pi e^2 R_n} \int \frac{d\xi_k}{2\left(\xi_{\mathbf{k}}^2 + |\Delta_2|^2\right)} \int d\xi_p \times \sum_{\mathbf{q}} \left(\Delta_2^*(\mathbf{q}) \widehat{a}_{\mathbf{p}+\mathbf{q}/2\uparrow} \widehat{a}_{\mathbf{p}-\mathbf{q}/2\downarrow} + \Delta_2(\mathbf{q}) \widehat{a}^+_{\mathbf{p}+\mathbf{q}/2\uparrow} \widehat{a}^+_{\mathbf{p}-\mathbf{q}/2\downarrow} \right). \quad (10.34)$$

The integral over ξ_k can be easily carried. Then $\Delta_2(\mathbf{q})$ can be presented in the canonical form of the Fourier integral. Defining the order parameter phase $\varphi(\mathbf{r})$ as $\exp\left[i\varphi(\mathbf{r})\right] = \Delta_2(\mathbf{r})/|\Delta_2|$, one can find

$$\widehat{\mathcal{H}}_b = \frac{1}{8e^2 R_n} \int d\xi_p \int \exp(i\mathbf{q}\mathbf{r}) d^2\mathbf{r} \times \sum_{\mathbf{q}} \left(e^{-i\varphi(\mathbf{r})} \widehat{a}_{\mathbf{p}+\mathbf{q}/2\uparrow} \widehat{a}_{\mathbf{p}-\mathbf{q}/2\downarrow} + e^{i\varphi(\mathbf{r})} \widehat{a}^+_{\mathbf{p}+\mathbf{q}/2\uparrow} \widehat{a}^+_{\mathbf{p}-\mathbf{q}/2\downarrow} \right). \quad (10.35)$$

Let us apply to the junction voltage V. The order parameter phase is related with it by the equation

$$\frac{\partial \varphi}{\partial t} = 2eV. \quad (10.36)$$

When in addition magnetic field is applied along the junction current flows along the confine in the direction perpendicular to magnetic field and the phase φ turns out to be space dependent:

$$\nabla \varphi = 2eH\lambda_L, \quad (10.37)$$

where λ_L is the London penetration depth for the superconducting electrode. Correspondingly in the sum of Eq. (10.35) remains the only term with

$$q = 2eA = 2eH\lambda_L \quad (10.38)$$

The electron moving from the fluctuating electrode to that one being in deeply superconducting state ($|\Delta_2|$ is large enough) can penetrate in it only virtually, for times $t \lesssim |\Delta_2|^{-1}$.[69] It is why the effective boundary Hamiltonian equation (10.33) can be applied only to the problems where this time is negligible. For example the Josephson energy (per unit area) of the junction between two different superconductors with $\Delta_2 \gg \Delta_1$ is equal to the average value of $\widehat{\mathcal{H}}_b$.

[69] Let us remember that the use of the tunneling Hamiltonian is possible only when the time of quasiparticle under barrier motion is small in comparison with the characteristic time of the problem.

Taking into account that $\langle \widehat{a}_{\mathbf{p}\uparrow}\widehat{a}_{\mathbf{p}\downarrow}\rangle = \left\langle \widehat{a}^{+}_{\mathbf{p}\uparrow}\widehat{a}^{+}_{\mathbf{p}\downarrow}\right\rangle = \Delta_1/\sqrt{\xi_p^2+|\Delta_1|^2}$ one can write

$$E_J = \frac{|\Delta_1|\cos\varphi}{4e^2R_n}\int\frac{d\xi_p}{\sqrt{\xi_p^2+|\Delta_1|^2}} = \frac{|\Delta_1|\cos\varphi}{2e^2R_n}\ln\left(\frac{4\Delta_2}{\Delta_1}\right). \tag{10.39}$$

The energy integration cut-off is done at $\xi_p \sim \Delta_2$[70] since only for energies $\xi_p \ll \Delta_2$ the boundary Hamiltonian is applicable. This expression coincides with the exact result of Ambegaokar and Baratoff [388].

10.3.3 *General formula for normal tunneling current*

At temperatures $T > T_{c1}$ the N(S) electrode stays in the normal state and with the exponential accuracy there is no normal current in the first order over R_n^{-1}. Such current appears as the response to $\widehat{\mathcal{H}}_b$, i.e. in the second order over transparency ($\sim R_n^{-1}$).[71] Indeed, as we have seen above, the boundary Hamiltonian describes the processes of the Cooper pair tunneling from the condensate of the superconducting electrode to the normal one and vice versa: of the pairs of normal electrons to the condensate.

The general expression for the normal current can be obtained in the complete analogy with Eq. (10.12) substituting $\widehat{\mathcal{H}}_T$ by $\widehat{\mathcal{H}}_b$:

$$I_b = e\int_{-\infty}^{0}\left\langle\left[\widehat{\mathcal{H}}_b(t),\left[\sum_{\mathbf{p},\sigma=\pm 1/2}\widehat{a}^{+}_{\mathbf{p}\sigma}\widehat{a}_{\mathbf{p}\sigma},\widehat{\mathcal{H}}_b(0)\right]\right]dt\right\rangle. \tag{10.40}$$

Substituting here $\widehat{\mathcal{H}}_b$ in the form Eq. (10.35) one can find

$$I_b = \frac{4e}{\left(8\nu e^2R_n\right)^2}\,\mathrm{Im}\int_{-\infty}^{0}\exp(2ieVt)\left\langle\left[\widehat{\mathcal{A}}^{+}(t),\widehat{\mathcal{A}}(0)\right]dt\right\rangle, \tag{10.41}$$

where

$$\widehat{\mathcal{A}}(t) = \sum_{\mathbf{p}}\widehat{a}_{\mathbf{p}+\mathbf{q}/2\uparrow}\widehat{a}_{\mathbf{p}-\mathbf{q}/2\downarrow}. \tag{10.42}$$

Let us mention that the operators $\widehat{a}_{\mathbf{p}\uparrow}$ and $\widehat{a}^{+}_{\mathbf{p}\downarrow}$ are nothing else as the Fourier transforms of the electron field operators $\widetilde{\psi}(\mathbf{r})$ in Heisenberg representation which define the two-particle Green function in Eq. (6.3). Hence

[70] The numerical coefficient in cut-off we choose in a way to reproduce exactly the logarithm in the exact Ambegaokar–Baratoff result.

[71] The formalism of the tunneling Hamiltonian strictly speaking does not guarantee the correctness of the coefficient for the result obtained in the second order over transparency. The point is that the overfull basis of electron wave functions of both electrodes is used in it, and the lack of their orthogonality manifests itself already in the second order over transparency. One of the ways to avoid this problem was proposed in [389] where the exact electron Green function in Keldysh representation for tunnel junction was used.

$$I_b(V,H) = -\frac{e}{(4\nu e^2 R_n)^2} \operatorname{Im} \sum_{\mathbf{p},\mathbf{p}'} \mathcal{L}^R(\mathbf{p},\mathbf{p}',\mathbf{q},2eV). \tag{10.43}$$

In its turn the two-particle Green function $\mathcal{L}^R(\mathbf{p},\mathbf{p}',\mathbf{q},2eV)$ summed over momenta $\mathbf{p},\mathbf{p}'$ can be expressed in terms of the vertex part of the interaction in the Cooper channel (i.e. fluctuation propagator $L^R(\mathbf{q},\Omega_k)$, see Eqs. (6.4), (6.6), (6.7)) with its further analytic continuation from Matsubara frequencies by the rule: $\omega_\nu \to -2ieV$. For our purposes will be convenient to present the propagator in terms of the polarization operator

$$L^R(\mathbf{q},2eV) = \sum_{\mathbf{p},\mathbf{p}'} \mathcal{L}^R(\mathbf{p},\mathbf{p}',\mathbf{q},2eV) \propto \frac{\Pi^R}{1-g\Pi^R} = \Pi^R + \Pi^R \frac{1}{1/g - \Pi^R} \Pi^R. \tag{10.44}$$

The substitution of this expression to Eq. (10.43) one gets two terms

$$\begin{aligned} I_b(V,H) = &-\frac{e}{(4\nu e^2 R_n)^2} \operatorname{Im} \sum_{\mathbf{p},\mathbf{p}'} \Pi^R(\mathbf{p},\mathbf{p}',\mathbf{q},2eV) \\ &-\frac{e}{(4\nu e^2 R_n)^2} \operatorname{Im} \sum_{\mathbf{p},\mathbf{p}'} \Pi^R \frac{1}{1/g - \Pi^R} \Pi^R. \end{aligned} \tag{10.45}$$

The first one (originating from Π) corresponds to the Andreev scattering of the electrons of normal electrode at the boundary with superconducting electrode (without their interaction). The second term describes the current appearing due to the tunneling of Cooper pairs from the condensate of the superconducting electrode to the normal one with their further transformation there to fluctuation Cooper pairs.

10.3.4 *Andreev conductance*

The general formula for Andreev conductance between superconductor and dirty normal metal was found in [390]. Equation (10.45) represents its limit in the case of a weak tunneling (expansion over R_n^{-1}, see Fig. 10.6) but even in this simplified case Andreev conductance depends on the sample geometry, temperature, voltage and applied magnetic field.

Let us use for the polarization operator Eq. (6.35). One has write:

$$\operatorname{Im} \Pi(2eV) = -\nu \operatorname{Im} \psi \left(\frac{1}{2} + \frac{-2ieV + \mathcal{D}q^2}{4\pi T} \right). \tag{10.46}$$

Let us consider a thin film separated by the tunnel barrier from superconductor. When the linear size exceed the thermal length l_T ($S \gg l_T^2 = \mathcal{D}/T$) and magnetic field along the junction is absent $q = 0$ (see Eq. (10.38)) one can write:

$$I(V) = \frac{S}{16 d\nu e^3 R_n^2} \operatorname{Im} \psi \left(\frac{1}{2} - \frac{ieV}{2\pi T} \right) = \frac{\pi S}{32 d\nu e^3 R_n^2} \tanh \left(\frac{eV}{2T} \right). \tag{10.47}$$

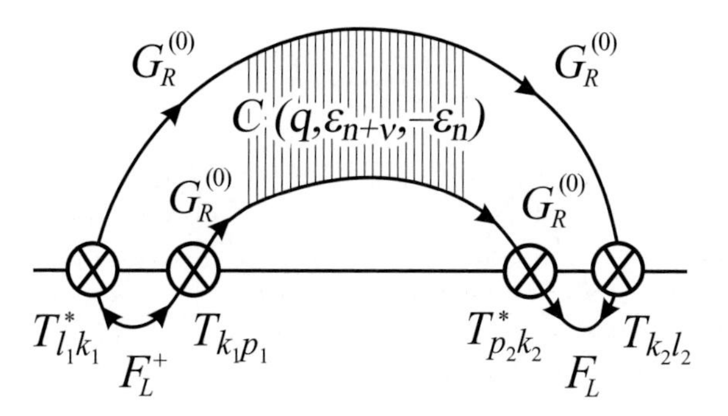

FIG. 10.6. Diagram for the Andreev reflection contribution to the tunneling current in S-I-$N(S)$ junction.

In the case when the superconducting electrode is small ($S \ll l_T^2$) while the normal film is still large ($S_n \gg l_T^2$), i.e. we consider the point contact, one finds

$$I(V) = \frac{S^2}{16d\nu e^3 R_n^2} \int_{q^2 S \lesssim 1} d^2\mathbf{q} \,\mathrm{Im}\,\psi\left(\frac{1}{2} + \frac{-2ieV + \mathcal{D}q^2}{4\pi T}\right). \tag{10.48}$$

Separating the contribution from large q and using the relation $\sigma_n = 2\nu e^2 \mathcal{D}$[72] one can find with the logarithmic accuracy

$$I(V) = \frac{G_T^2}{4\pi\sigma_n d}\left[V \ln\frac{\mathcal{D}}{TS} + \frac{2\pi T}{e}\,\mathrm{Im}\ln\Gamma\left(\frac{1}{2} - \frac{ieV}{2\pi T}\right)\right]. \tag{10.49}$$

In the case when there is nonzero magnetic field applied perpendicularly to the junction plane the integral over momenta in Eq. (10.48) has to be substituted by the corresponding sum. This results in additional suppression of the Cooperon in diagram (see Fig. 10.6), and with the logarithmic accuracy one gets

$$G_A(V) = \frac{G_T^2}{4\pi\sigma_n d} \ln\frac{\mathcal{D}/S}{\max\{T, eV, e\mathcal{D}H\}}. \tag{10.50}$$

Such geometry was considered in [391]. Let us mention that film resistance between concentric contacts with radii R and r is equal to

$$R_N = \frac{1}{2\pi\sigma_n d} \ln\frac{R}{r}, \tag{10.51}$$

hence

$$G_A = G_T^2 R_N. \tag{10.52}$$

This result, obtained as the expansion over G_T, is valid until $G_T R_N \ll 1$.

[72] Let us stress that σ_n is the normal conductivity of the electrode material, while R_n is the unit area resistance of the tunnel barrier.

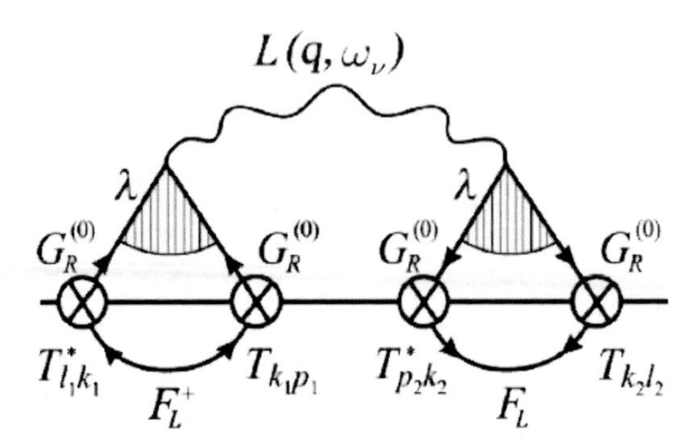

FIG. 10.7. Diagram for the fluctuation contribution to the tunneling current in S-I-$N(S)$ junction.

10.3.5 *Pair-field susceptibility of a superconductor*

Now we are ready to investigate the manifestation of the fluctuation Cooper pairing in the normal electrode. Formally the corresponding contribution to the quasiparticle junction is represented by the diagram drawn in Fig. 10.7. It is very similar to the Andreev reflection but instead of the Cooperon it contains the ladder of electron–electron interactions, i.e. fluctuation propagator.

In contrast to impurity lines the interaction lines can transfer energy and as a result the fermionic frequency ε_{n2} passing through the exiting Green function lines of the fluctuation propagator is not the same as ε_{n1} at the entrance to it.

Let us rewrite the second term in Eq. (10.44) in the form

$$L_b = \Pi_b L \Pi_b. \tag{10.53}$$

Let us stress that the polarization operators Π_b , standing at left and right from L, are slightly different from Π standing in Eq. (6.4) for L. The difference consists in the cut-off of the summation over ε_n: usually it is cut off at ω_D, while the summation in Π_b has to be restricted to Δ_2. Indeed, at large ε_n the description with help of the boundary Hamiltonian fails and the cutting off must be done at frequencies $\varepsilon_n \sim \Delta \ll \omega_D$:

$$\Pi_b = \nu \ln \frac{\Delta}{T} \tag{10.54}$$

(the momentum and frequency dependencies of the polarization operator within the logarithmic accuracy can be omitted).

Let us consider again the wide contact between normal metal and superconducting film. When a magnetic field is applied along the junction plane the momentum q is determined by Eq. (10.38) and

$$I(\omega) = -\frac{S}{16e^3 R_n^2 d} \ln^2 \frac{\Delta_2}{T} \operatorname{Im} L^R(q,\omega). \tag{10.55}$$

Substituting here the explicit expression Eq. (6.17) for $L^R(q,\omega)$ one obtains

$$I_{\text{fl}}(q,\omega) = \frac{S}{16e^3 R_n^2 d\epsilon} \ln^2 \frac{\Delta_2}{T} \frac{\omega\tau_{\text{GL}}}{(1+\xi^2 q^2)^2 + \omega^2 \tau_{\text{GL}}^2}, \qquad (10.56)$$

where $\omega = 2eV$ and $q = 2eH\lambda_L$. The similar expression was obtained firstly by Scalapino [373] in a phenomenological way. Let us stress that in spite of the fact that this fluctuation correction to the tunnel current is obtained in the second order in the barrier transparency ($\sim R_n^{-2}$) it can noticeably exceed the first order quasiparticle contribution since at temperatures $T \ll \Delta_2$ it is exponentially small.

10.3.6 *Goldman's group experiments*[73]

Soon after appearance of the papers [373, 384, 386] the pair field susceptibility of a superconductor was successfully investigated experimentally [392, 394, 395]. The authors measured *I-V* characteristics of aluminum–aluminum oxide–lead tunneling junctions biased from a constant-current source. Then the background current $I_{qp}(V)$, computed according to Eq. (10.29), was subtracted and the obtained excess current–voltage characteristics identified with Eq. (10.56) for $I_{\text{fl}}(q,\omega)$. Such fitting procedure was carried out over a wide enough range of voltages and temperatures. In Fig. 10.8 are shown the results of such procedure carried out at several temperatures and in absence of magnetic field.

In Fig. 10.9 is shown the variation of the reciprocal of the square root of the dynamic conductance as the function of temperature. The curve turns to be linear function of temperature over almost 500 mK, what means $\approx 25\%$ of T_c.

One can see that the experimental findings are in a good agreement with the described above theory, but what concerns of the magnitude of the excess current the agreement was found to be only qualitative. Nevertheless, the variation of the peak voltage with temperature was found to be in an excellent agreement with theory [373, 386], except in the immediate vicinity of T_c. This constitutes a determination of the GL Cooper pair lifetime $\tau_{\text{GL}}(\epsilon)$, a quantity which depends only on transition temperature, T and fundamental constants.

The next step in investigations of the excess current–voltage characteristics of an asymmetric tunnel junction between two superconductors was undertaken by the same group of researchers: they, in a continuous way, passed the lower transition temperature and moved to study fluctuations below T_{c1} [393, 396, 397]. This is due to the fact, that the microscopic description of fluctuations becomes much more complicated at temperatures below T_c. In contrast to the simple fluctuation picture of the normal phase, where nonequilibrium pairs appear and decay, below T_c the consequent construction of the fluctuation theory, starting from the one-particle representation (see Part II), requires us to go beyond the simple BCS description of superconductivity. Indeed, fluctuations below T_c, where the coexistence of, thermodynamically, in equilibrium Cooper pair condensate and

[73] In this section we base on the results of [392, 393].

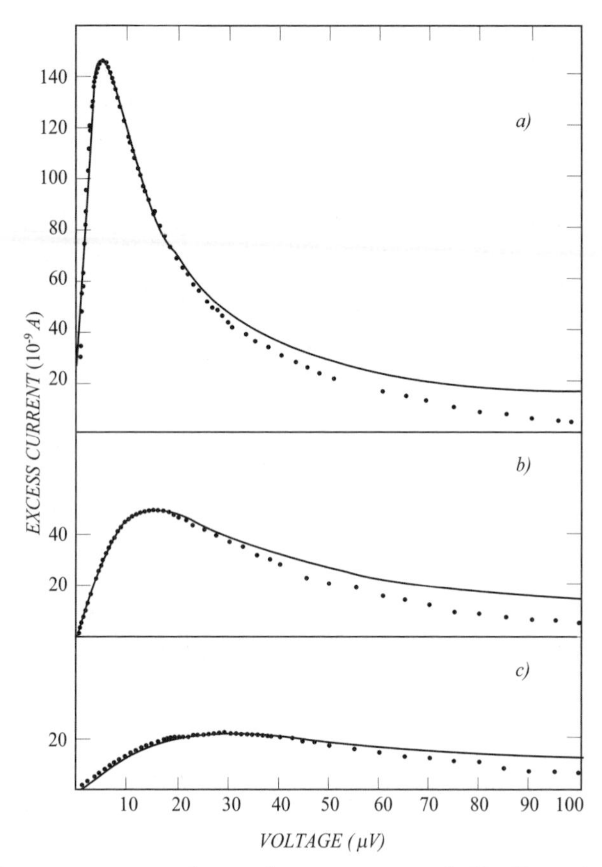

FIG. 10.8. Excess current-voltage characteristics of *Al*–*Al* oxide–*Pb* junction [392]. Curves (a)-(c) corresponds to $T = 1.84726, 1.93276$ and $2.09892\ K$, respectively. The solid line is the theoretical excess current obtained by fitting the theory to the experimental values of the peak current I_{p} and the peak voltage V_{p}.

one-particle excitations takes place, present a variety of options: the appearance and decay of nonequilibrium Cooper pairs, different types of quasiparticle scattering processes in diffusion (screened Coulomb interaction) and Cooper channels involving the condensate. Formally, the matrix elements beyond the usual $< N|...|N >$ and $< N|...|N+2 >$ have to be taken into account [398,399]. As a result, instead of the fluctuation propagator (6.3), the matrix propagator 4×4 has to be introduced. Formally it is determined by the same Dyson equation (6.4), but the polarization operator has to be calculated with the matrix Green functions of the superconducting state and the bare interaction $\widehat{g}$ becomes a matrix [398,399,92,301]. Being in the framework of the phenomenological approach one can say that at temperatures above T_{c} the order parameter fluctuations are well described by the diffusive TDGL equation. Indeed, the pair-field susceptibility obtained experimentally is quite consistent with such a picture of the order parameter dynamics. Below T_c a diffusive type equation cannot be used to describe the dynamics of order parameter.

The same excess current–voltage characteristics as above were experimentally studied at temperatures below T_{c1}. Three structures were observed [397]. There was a lateral, low voltage peak, or sometimes a shoulder, which can be related to longitudinal fluctuations or fluctuations in the amplitude of the order parameter [398–401]. The main peak, which develops continuously from the peak observed above T_c, can be associated with the fluctuations in the phase. Finally, a third peak occurred at a voltage $eV = \Delta(T)$, where $\Delta(T)$ is the order parameter in the aluminium.

This peak, which may be a tunneling anomaly associated with the creation of real pairs by the a.c. Josephson current, provides a convenient reference to the magnitude of the order parameter at the particular temperature and magnetic field.

In accordance with Eq. (10.56) the excess current $I_{fl}(q,\omega)$ determines the imaginary part of the fluctuation propagator which, in its turn can be related to the so-called structure factor by means of the fluctuation–dissipation theorem:

$$S(\omega,q) = -\frac{T}{\omega}\operatorname{Im} L^R(q,\omega). \tag{10.57}$$

This function is nothing other than the Fourier transform of the order-parameter – order-parameter correlation function and its frequency dependence for different temperatures above and below T_c is presented in Fig. 10.10. On the frequency scale used, the peak at aluminum gap observed in the excess current is not on the curve.

Above T_c the structure factor is a Lorentzian centered at $\omega = 0$. Below and in the immediate vicinity of T_c, $S(\omega,q)$ has a peak at nonzero frequency ω_φ in addition to the usual one at the origin. As the temperature is reduced below T_c this peak broadens and disappears into the tail of the central peak.

Usually a peak in a dynamical structure factor at nonzero frequency implies the existence of the propagating mode. In the present case the mode was identified with the fluctuations of the phase of the order parameter [400, 401]. The range of temperatures over which the mode is well defined corresponds to that over which the main peak was found to be distorted from the form exhibited above T_c.

The observed nontrivial manifestation of the phase fluctuations indicates the necessity to include into the scheme the theoretical description of fluctuations below T_c the Coulomb interaction. Indeed, above T_c the order parameter fluctuations, or electron–electron interaction in the Cooper (particle–particle) channel, and the dynamical screening of the Coulomb interaction (particle–hole channel) could be considered independently. Below T_c, fluctuations of scalar potential cannot be more separated from the fluctuations of the modulus and phase of

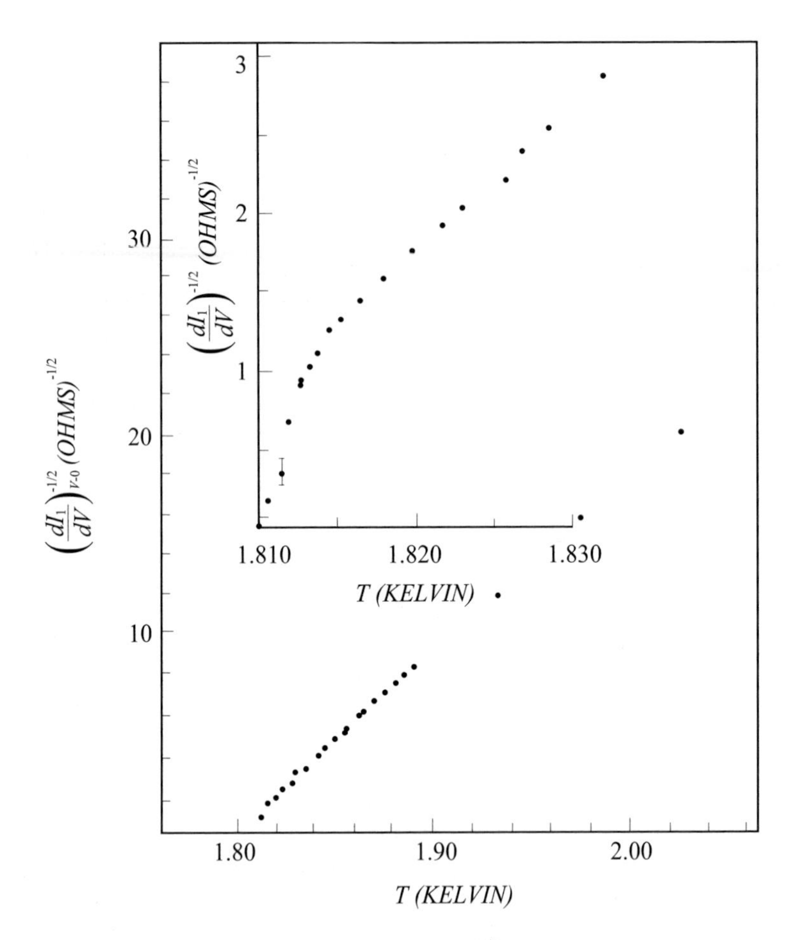

FIG. 10.9. The main figure shows $(dI/dV)_{V=0}^{-1/2}$ versus T for Al–Al oxide–Pb junction over the entire temperature range covered [392]. The insert shows in detail the departure close to $T_c = 1.801 \pm 0.001\ K$ of the curve from a linear temperature dependence.

the order parameter. This is due to the fact, that above T_c the screening of the charge fluctuation occurs only by the redistribution of electrons in space. Below critical temperature charge transfer by means of supercurrent flow, i.e. by spatial change of the order parameter phase, is also possible. One can see that this means the linking of scalar potential and phase fluctuations, i.e. the appearance of off-diagonal elements in the matrix of propagators. The found by Carlson and Goldman mode in the spectrum of collective excitations of superconductor below T_c, corresponds to the relative motion of the superconducting and quasiparticle subsystems.

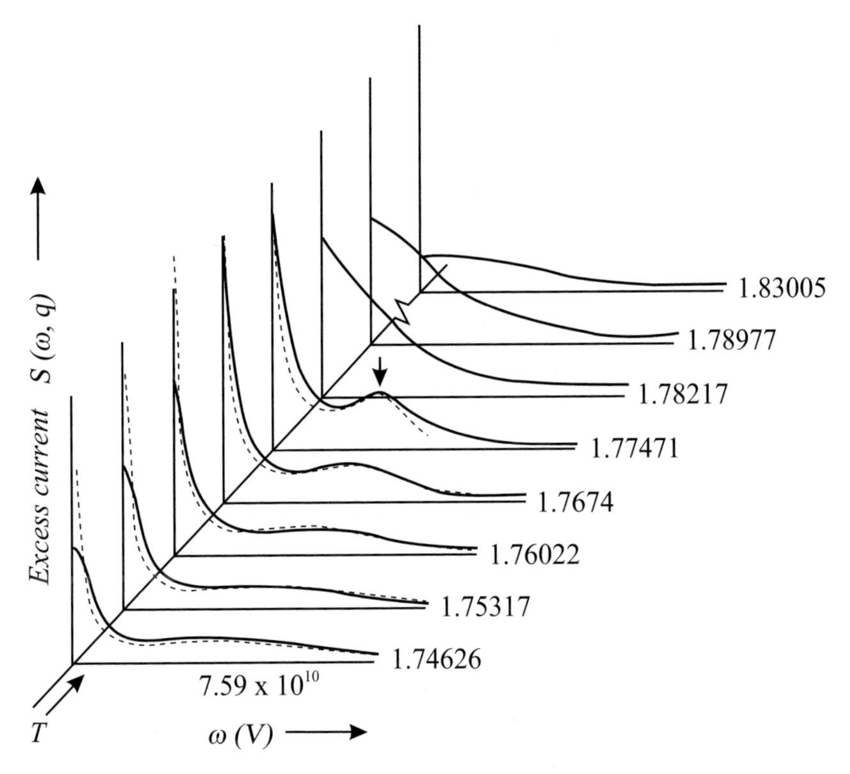

FIG. 10.10. The structure factor in arbitrary units plotted as a function of frequency and temperature [397]. A magnetic field of 125 G is applied parallel to the plane of junction reducing the transition temperature to 1.7795 K from 1.786 K in zero field. It should be noted that the curves are symmetric about $\omega = 0$, although only positive ω is shown. The high frequency peak is out of the range of the graphs.

10.4 Fluctuation tunneling anomaly in a superconductor above the paramagnetic limit[74]

As we already have seen the magnetic field due to the break of time-reversal symmetry, suppresses both superconductivity and superconducting fluctuations. This effect can be separated into two mechanisms: the effect of the magnetic field on the orbital motion coupling associated with the Aharonov–Bohm phase and the Zeeman splitting of the states with the same spatial wave functions but opposite spin directions.

In the bulk systems the suppression of superconductivity usually is related to the first mechanism in the fields of the order of $H_{c2}(0)$ (see (2.52)). The effect of magnetic field on the spin structure of the Cooper pair determines the so-called

[74]In this section we base on the results of [402, 403].

Clogston limit:

$$g_L \mu_B H_c^{spin} \simeq \Delta, \tag{10.58}$$

where g_L is the Landé g-factor, μ_B is the Bohr magneton. Comparison of the Clogston (often it is called as Zeeman) critical field H_c^{spin} with $H_{c2}(0)$ demonstrates that for a normal metal the former is far in excess of the latter:

$$\frac{H_c^{spin}}{H_{c2}(0)} \simeq E_F \tau \gg 1, \tag{10.59}$$

and Clogston limit is practically unaccessible.

The situation may change in the restricted geometries. For example, in the superconducting film of the thickness d placed in the magnetic field parallel to the plane of the film, the Cooper pairs are restricted in the transversal direction to the film thickness. As a result, the geometrical area covered by the pairs rotation can be estimated as $d\xi$ than as ξ^2. Therefore, the evaluation of the corresponding second critical field should be changed to

$$H_{c2}^{\parallel} \xi d \simeq \Phi_0 \Rightarrow H_{c2}^{\parallel} \simeq H_{c2}(0) \left(\frac{\xi}{d} \right). \tag{10.60}$$

On the other hand, Zeeman splitting $E_Z = g_L \mu_B H$ is not affected by geometrical restriction. Accordingly, instead of Eq. (10.59) the ratio of the two scales of magnetic field is given by

$$\frac{H_c^{spin}}{H_{c2}^{\parallel}} \simeq E_F \tau \left(\frac{\xi}{d} \right), \tag{10.61}$$

Thus, for sufficiently thin films, $d \ll \xi / E_F \tau$, the spin effects become dominant. One can easily check that the same estimate (10.61) holds for other restricted systems, i.e. superconducting grains or wires. In these cases, d is the size of the grain or the diameter of the wire respectively. Quite generally, d is determined by the minimal size of the sample in the plane perpendicular to the magnetic field.

The transition from superconductor to paramagnet is of the first order: the superconducting state is the only stable state at $E_Z \leq \Delta$; while at $E_Z \geq 2\Delta$ the normal state is the only stable state. Both phases are locally stable in the interval of magnetic fields where $\Delta \leq E_Z \leq 2\Delta$. The normal state becomes lowest in energy and thus globally stable at $E_Z \geq \sqrt{2}\Delta$. From now on, we will assume that this condition is fulfilled.

One of the most fundamental manifestations of the superconductivity is the gap in the tunneling DOS around zero bias. One can expect that after the paramagnetic transition not only BCS order parameter vanishes but also the energy dependence of tunneling DOS becomes similar to those in superconductors above critical temperature T_c.

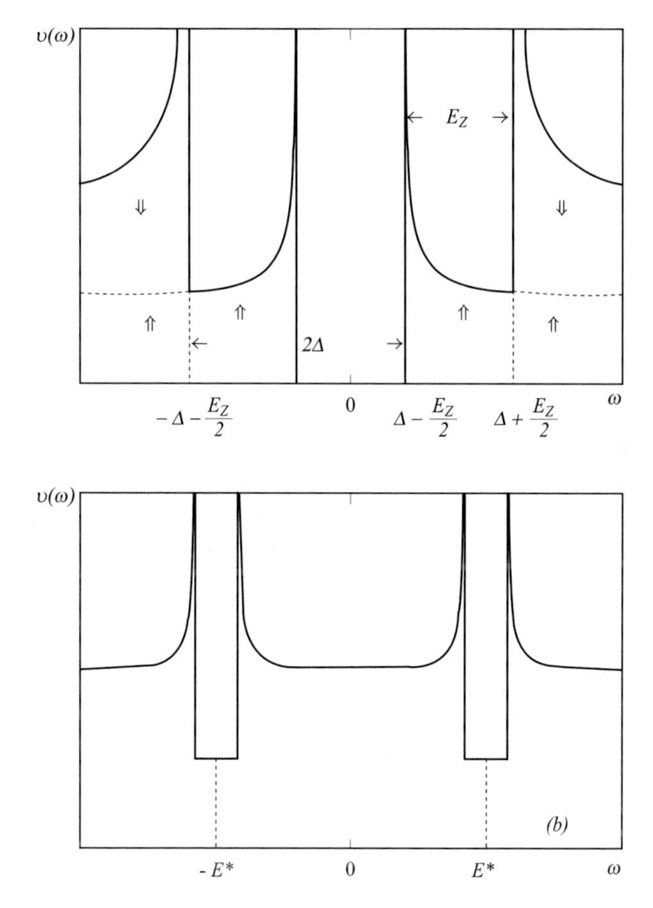

FIG. 10.11. Evolution of the tunneling DOS with the Zeeman splitting E_Z for (a) the superconducting state, $E_Z \leq \sqrt{2}\Delta$, and (b) for the paramagnetic state $E_Z \geq \sqrt{2}\Delta$. The usual zero bias anomaly in the paramagnetic state (b) is not shown for simplicity. Shape of the singularity at E^* corresponds to the 0-dimensional case, [403].

Below will be demonstrated that, on the contrary, there are clear observable superconducting effects in the normal state even far from the transition region. We will show that at the transition point there appears a dip in the DOS (schematic evaluation of the DOS with the magnetic field is shown in Fig. 10.11).

The shape and the width of this dip depend on dimensionality of the system. However, its position turns out to be remarkably universal:

$$E^* = \frac{E_Z + \sqrt{E_Z^2 - \Delta^2}}{2} \tag{10.62}$$

for OD (grain), $1D$ (strip), and $2D$ (film) cases.

Let us consider an isolated disordered superconducting grain which is small so that the Zeeman splitting dominates over the orbital magnetic field effect (see, e.g. [404, 405] for recent experiments on such grains). We assume that the size of the grain exceeds the electronic mean free path l, and, at the same time, it is much smaller than the superconducting coherence length ξ. We also assume that $p_F l \gg 1$. This results in the large dimensionless conductance of the grain $g (g \sim p_F^2 l d)$. Finally, we assume that the grain is already driven into the paramagnetic state by the Zeeman splitting. Our goal is to find effects of the superconducting fluctuations on the DOS of the system.

The Hamiltonian $\mathcal{H}$ of the system consists of noninteracting part $\mathcal{H}_0$ and interacting one $\mathcal{H}_{\text{int}}$. Using the basis of the exact eigenstates of $\mathcal{H}_0$ labeled by integers i and j, one can rewrite the Hamiltonian (6.1) as:

$$\mathcal{H} = \sum_{i,\sigma} E_{i\sigma} a_{i\sigma}^{+} a_{i\sigma} - g\delta \sum_{i,j} a_{i\uparrow}^{+} a_{i\downarrow}^{+} a_{j\uparrow} a_{j\downarrow}. \tag{10.63}$$

Here operator $a_{i\sigma}^{+}$ $(a_{i\sigma})$ creates (annihilates) an electron in a state i with spin $\sigma = \uparrow, \downarrow$ and energy $E_{i\uparrow(\downarrow)} = E_i \mp E_Z/2$, where E_i is the orbital energy of i-th state. $g \ll 1$ is the dimensionless interaction constant, and δ is the average level spacing:

$$\langle E_{i+1} - E_i \rangle = \delta, \tag{10.64}$$

which was defined in the Part I (see Eq.(2.26)).

To evaluate the effect of superconducting fluctuations on the DOS of electrons in the paramagnetic state $\nu_\sigma(\omega)$, we will use the same diagrammatic expression (10.1) as above, but undermining as $G_{i\sigma}^{R}(E)$ the one-electron Green function for grain in state i and with spin σ and summing over all eigenstates instead of momentum integration:

$$\nu_\sigma(E) = -\frac{1}{\pi} \operatorname{Im} \sum_i G_{i\sigma}^{R}(E). \tag{10.65}$$

Here

$$G_{i\sigma}^{-1}(\varepsilon_n) = G_{i\sigma}^{(0)-1}(\varepsilon_n) - \Sigma_{i\sigma}(\varepsilon_n), \tag{10.66}$$

while the Green function for the corresponding non-interacting electron system

$$G_{i\uparrow(\downarrow)}^{(0)-1}(\varepsilon_n) = i\varepsilon_n - E_i \pm E_Z/2, \tag{10.67}$$

and $\Sigma_{i\sigma}$ is the one particle self-energy; leading contribution to it is shown in Fig. 10.1.

The corresponding to the problem under discussion fluctuation propagator can be obtained as usually by summing the electron loops in the Cooper channel shown in Fig. 6.1 but calculated with the Green functions (10.67). The polarization operator is given by

$$\Pi(\Omega_k) = T\sum_{\varepsilon_n}\sum_i G_{i\uparrow}^{(0)}(\varepsilon_n+\Omega_k)\,G_{i\downarrow}^{(0)}(-\varepsilon_n) = 2\pi\frac{T}{\delta}\sum_{\omega_n}\frac{\theta(\varepsilon_n(\varepsilon_n+\Omega_k))\,\mathrm{sign}\varepsilon_n}{2\varepsilon_n+\Omega_k+iE_Z}$$
$$= \frac{1}{2\delta}\left[2\ln\frac{\omega_D}{2\pi T} - \psi\left(\frac{1}{2}+\frac{\Omega_k}{4\pi T}+\frac{iE_Z}{4\pi T}\right) - \psi\left(\frac{1}{2}+\frac{\Omega_k}{4\pi T}-\frac{iE_Z}{4\pi T}\right)\right], \tag{10.68}$$

where ω_D is the high-energy cut-off of the BCS theory. Here we passed from summation over the electron states in the grain to the energy integration using the notion of the average spacing introduced above.

Solving the Dyson equation (6.4), we obtain the propagator in the form:

$$L(\Omega_k) = -\frac{g\delta}{1-g\delta\Pi(\Omega_k)}$$
$$= -\frac{2\delta}{\psi\left(\frac{1}{2}+\frac{\Omega_k}{4\pi T}+\frac{iE_Z}{4\pi T}\right)+\psi\left(\frac{1}{2}+\frac{\Omega_k}{4\pi T}-\frac{iE_Z}{4\pi T}\right)-2\ln\frac{\Delta}{4\pi T}}, \tag{10.69}$$

where $\Delta = 2\omega_D\exp(-1/\nu g)$ is the BCS gap. Taking into account that $T \ll E_Z$, one can substitute $\psi(x)$ by its asymptotic (B.16) and for the retarded polarization operator and propagator get

$$\Pi^R(\Omega) = \frac{1}{2\delta}\ln\left(\frac{4\omega_D^2}{E_Z^2-\Omega^2}\right), \tag{10.70}$$

$$L^R(\Omega) = -\frac{2\delta}{\ln\left(\frac{E_Z^2-\Omega^2}{\Delta^2}\right)}. \tag{10.71}$$

The propagator (10.71) has the pole at $\Omega = \pm\Omega^*$,

$$\Omega^* = \sqrt{E_Z^2-\Delta^2}. \tag{10.72}$$

This pole can be interpreted as the bound state of two quasiparticles with energy Ω^*.

Now we can write the analytic expression for the self-energy. For this reason, following Eliashberg [244], let us transform the sum over bosonic frequencies into the contour integral

$$\Sigma_{\uparrow\downarrow}(\varepsilon_n) = T\sum_{\Omega_k} L(\Omega_k)G_{i\uparrow}^{(0)}(\Omega_k-\varepsilon_n)$$
$$= \frac{1}{4\pi i}\oint_C \coth\left(\frac{z}{2T}\right)L(-iz)\,G_{i\uparrow}^{(0)}(-iz-\varepsilon_n)dz$$
$$= \frac{1}{4\pi i}\int_{-\infty+i\varepsilon_n}^{\infty+i\varepsilon_n} dz\coth\left(\frac{z}{2T}\right)L^R(-iz)\left[G_{i\uparrow}^{(0)R}(-iz-\varepsilon_n)-G_{i\uparrow}^{(0)A}(-iz-\varepsilon_n)\right]$$

$$+\frac{1}{2\pi}\int_{-\infty}^{\infty} dz \coth\left(\frac{z}{2T}\right) G_{i\uparrow}^{(0)A}(-iz-\varepsilon_n)\,\mathrm{Im}\,L^R(-iz)\,. \tag{10.73}$$

We see that there are two contributions to the self-energy: one comes from the pole of L and the other is due to the branch-cut of propagator. The pole contribution gives a singularity of the self-energy at certain ε_n and ε_i while the contribution of the branch-cut is smooth. To find the singularity in the DOS, only the pole contribution to Σ may be retained. Let us shift in it the variable $z \to z' + i\varepsilon_n$ and take into account that the shift of the argument of coth on ε_n transforms it in tanh:

$$\Sigma_{\uparrow\downarrow}(\varepsilon_n) \approx \frac{1}{2\pi}\int_{-\infty}^{\infty} \tanh\left(\frac{z'}{2T}\right) L^R(-iz'+\varepsilon_n)\left[G_{i\uparrow}^{(0)R}(-iz') - G_{i\uparrow}^{(0)A}(-iz')\right] dz'. \tag{10.74}$$

Taking into account that $T \ll E_Z$ one can substitute $\tanh\frac{z'}{2T} \to \mathrm{sign} z$, and make the analytic continuation $\varepsilon_n \to -i\varepsilon$. Both poles of the propagator (10.71) are located in the lower half-plane of z. To find the singularity in the DOS only the pole contribution to $\Sigma_{\uparrow\downarrow}^R(\varepsilon)$ may be retained:

$$\Sigma_{\uparrow\downarrow}^R(E) = \frac{\delta\Delta^2}{2\Omega^*}\frac{1}{E + E_i - E_Z/2 + \Omega^*\mathrm{sign}(E_i - E_Z/2)}. \tag{10.75}$$

At certain ε the pole of the self-energy coincides with the pole of $G^{(0)}$. This causes the singularity in the DOS. One can check that singularities of Eqs. (10.67) and (10.75) coincide provided $E_i = \Omega^*/2$ (what means $\mathrm{sign}(E_i - E_Z/2) = -1$) and

$$E = \frac{E_Z + \Omega^*}{2} = \frac{E_Z + \sqrt{E_Z^2 - \Delta^2.}}{2} \equiv E^*. \tag{10.76}$$

Substituting Eqs. (10.75) into (10.66) we obtain the Green function for the down-spin electron with the energy close to E^*:

$$G_{i\downarrow}^R(E) = \frac{E + E_i - E_Z/2 - \Omega}{(E - E_i - E_Z/2)(E + E_i - E_Z/2 - \Omega) - W_0^2}, \tag{10.77}$$

where the energy scale of the singularity is given by

$$W_0 = \sqrt{\frac{\delta\Delta^2}{\Omega^*}}. \tag{10.78}$$

Since E_Z , $\Delta \gg \delta$, one can neglect the fine structure of the DOS on the scale of δ and substitute the summation over i by the integration over ε_i:

$$\sum_i G_{i\downarrow}^R(E) = \nu_0 \int d\varepsilon_i \frac{E + E_i - E_Z/2 - \Omega^*/2}{-E_i^2 + (E - E_Z/2 - \Omega^*/2)^2 - W_0^2}$$

$$= -i\pi\nu_0 \frac{E - E^*}{\sqrt{(E - E^*)^2 - W_0^2}}. \tag{10.79}$$

Analogously, the Green function for the up-spin electron can be obtained by changing the signs of E_Z and Ω^*, so that the singularity occurs at $\varepsilon = -E^*$.

Substituting Eqs. (10.79) into (10.65), we obtain the final expression for the tunneling DOS in the ultra-small grains:

$$\nu_{\uparrow(\downarrow)}(E) = \nu_0 F_0 \left(\frac{E \pm E^*}{W_0} \right), \tag{10.80}$$

where the function

$$F_0(x) = \mathrm{Re} \left(\frac{x^2}{x^2 - 1} \right)^{1/2}, \tag{10.81}$$

ν_0 is the bare DOS per one spin, energy E^* is defined by Eq. (10.62) and width of the singularity W_0 is given by Eq. (10.78). Equations (10.80) and (10.78) are the main results of our discussion. They predict the hard gap in the spin resolved DOS: $\nu_\sigma(E)$ vanishes at $|E + \sigma E^*| < W_0$ (see Fig. 10.11). Overall DOS $\nu_\uparrow + \nu_\downarrow$ is suppressed by a factor of two near the singularity.

In this calculation we neglected higher order corrections to the self-energy, e.g. those shown in Fig. 10.12 (a,b). In order to justify this approximation, we have to compare contributions shown in Figs. 10.12(a), b with the reducible diagram shown in Fig. 10.12(c) included in Eqs. (10.75) and (10.77). Singular contribution originates from the pole of L. It means that L carries frequency Ω^*. The singularity in the DOS at $E = E^*$ appears when the pole of the self-energy and the pole of G^0 coincide. This happens when Green function G^0 for up-spin carries energy $\Omega^* - E$. In Fig. 10.12(a-c) the intermediate G^0 for down-spin should carry energy E to give singularity to the DOS at E^*. This condition cannot be satisfied for the diagrams Fig. 10.12(a,b). As a result, after the integration over the intermediate frequency, these higher order corrections turns out to be smaller than the reducible contribution (c) by a small factor $W_0/\Delta \simeq \sqrt{\delta/\Delta} \ll 1$.

10.5 The magnetoresistance of a granular superconducting metal[75]

We have already demonstrated above that the common belief that the fluctuation corrections consist only of the AL and MT terms often turns to be wrong. Namely, both of these corrections can be suppressed by the tunneling probability, as in the case of c-axis transport in layered superconductors (see section 8.3), or the tunneling probability be of the same order as the DOS and regular MT corrections as it takes place far from T_c or, vice versa, at very low temperatures, when quantum fluctuations are of the first importance (see section 9.5). Now we will appeal to the system where both mentioned reasons result in the interplay of all, positive and negative, fluctuation contributions which allows to explain existing nontrivial experimental findings.

[75]In this section we base on the results of [360].

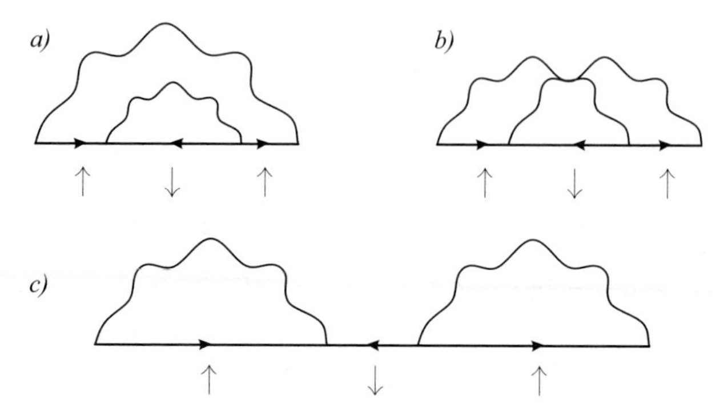

FIG. 10.12. Higher order correction to the self-energy (a) and (b), which were neglected in comparison with reducible diagram (c).

In a recent experiments [406–408], transport properties of the superconducting grain systems in a strong magnetic field were studied. The samples were quite homogeneous and the grains formed a $3D$ array. As usual, sufficiently strong magnetic fields destroyed the superconductivity in the samples and a finite resistivity could be seen above a critical magnetic field. The applied magnetic fields reached tens of Tesla, which was more than sufficient to destroy even the superconducting gap in each grain. At extremely strong fields the resistivity was almost independent of it, while it *increased* when decreasing the magnetic field. At sufficiently weak magnetic fields the resistivity started to decrease and finally the samples displayed superconducting properties.

We consider below the system of superconducting grains in a strong magnetic field at low temperatures and assume that the system is macroscopically a good metal. This corresponds to a sufficiently large tunneling energy t

$$t \gg \delta, \tag{10.82}$$

where δ is the average level spacing in a single granule (see Eq. (2.26)). In other words, we assume the validity of the limit of large conductance

$$J_g \equiv (\pi^2/4)(t/\delta)^2 \gg 1. \tag{10.83}$$

All effects of the WL and the charging effects have to be small for $J_g \gg 1$, which would imply that the resistivity could not considerably depend on the magnetic field. Nevertheless, we demonstrate that the magnetoresistance of a such granulated metal in a strong magnetic field and at low temperature *must be negative.* A strong magnetic field destroys the superconducting gap in each granule. However, even at magnetic fields H exceeding the critical field H_c virtual Cooper pairs can still be formed. It turns out that the influence of these pairs on the macroscopical transport is drastically different from that near T_c. The existence of the virtual pairs leads to a reduction of the DOS but, in the limit $T \to 0$, these pairs cannot travel from one granule to another. As a result, the conductivity σ can be at $H > H_\mathrm{c}$ considerably lower than conductivity σ_0 of the normal metal without an electron–electron interaction. It approaches the value σ_0 only in the limit $H \gg H_\mathrm{c}$, when all the superconducting fluctuations are completely suppressed by the magnetic field.

As it was already mentioned above the superconducting pairing inside the grains is destroyed by both the orbital mechanism and the Zeeman splitting of the electron states forming a Cooper pair. The critical magnetic field $H_\mathrm{c}^{\mathrm{or}}$ corresponding to the first effect can be estimated from the relation $H_\mathrm{c}^{\mathrm{or}} R\xi \approx \Phi_0$, where Φ_0 is a flux quantum, R is a radius of single grain and $\xi = \sqrt{\xi_0 l}$ is the superconducting coherence length. The Zeeman, or Clogston, critical magnetic field H_c^{spin} we already introduced in the previous section and it is determined by the Eq. (10.58). We notice here that H_c^{spin} is independent of the size of the grain whereas for $H_\mathrm{c}^{\mathrm{or}}$ the size of the grain is important. The ratio of this two fields can be written in the form

$$H_\mathrm{c}^{\mathrm{or}} / H_\mathrm{c}^{spin} \approx R_\mathrm{c}/R, \tag{10.84}$$

where $R_\mathrm{c} = \xi (p_0 l)^{-1}$. We can see from Eq. (10.84) that for $R > R_\mathrm{c}$ the orbital critical magnetic field is smaller than the Zeeman critical magnetic field $H_\mathrm{c}^{\mathrm{or}} < H_\mathrm{c}^{spin}$ and the suppression of superconductivity is due to the orbital mechanism. This condition is well-satisfied in grains with $R \sim 100\mathring{A}$ studied in [406].

10.5.1 *Choice of the model*

We consider a $3D$ array of superconducting grains coupled to each other. The grains are not perfect and there can be impurities inside the grains as well as on the surface. We assume that electrons can hop from grain to grain and can interact with phonons.

The Hamiltonian $\hat{H}$ of the system can be written as

$$\hat{H} = \hat{H}_0 + \hat{H}_T, \tag{10.85}$$

where $\hat{H}_0$ is a conventional Hamiltonian for a single grain with an electron-phonon interaction in the presence of a strong magnetic field

$$\hat{H}_0 = \sum_{i,k} E_{i,k} a^\dagger_{i,k} a_{i,k} - g \sum_{i,k,k'} a^\dagger_{i,k} a^\dagger_{i,-k} a_{i,-k'} a_{i,k'} + \hat{H}_{\mathrm{imp}}, \tag{10.86}$$

where i stands for the numbers of the grains, $k \equiv (\mathbf{k}, \uparrow)$, $-k \equiv (-\mathbf{k}, \downarrow)$; g is an interaction constant, and $\hat{H}_{\mathrm{imp}}$ describes elastic interaction of the electrons with impurities. The interaction in Eq. (10.86) contains diagonal matrix elements only. This form of the interaction can be used provided the superconducting gap Δ_0 is not very large

$$\Delta_0 \ll E_T \tag{10.87}$$

where E_T is the Thouless energy of the single granule and Δ_0 is the BCS gap at $T = 0$ in the absence of a magnetic field. Equation (10.87) is equivalent to the condition $R \ll \xi_0$, where R is the radius of the grain and ξ_0 is the superconducting coherence length. In this limit, superconducting fluctuations in a single grain are $0D$.

The term $\hat{H}_T$ in Eq. (10.85) describes tunneling from grain to grain and has the form (see, e.g. [374])

$$\hat{H}_T = \sum_{i,j,p,q} t_{ijpq} a_{ip}^{\dagger} a_{jq} \exp(i\frac{e}{c}\mathbf{A}\mathbf{d}_{ij}) + \text{h.c.}, \tag{10.88}$$

where $\mathbf{A}$ is the external vector-potential; $\mathbf{d}_{ij}$ are the vectors connecting centers of two neighboring grains i and j ($|\mathbf{d}_{ij}| = 2R$); $a_{ip}^{\dagger}$ (a_{ip}) are the creation (annihilation) operators for an electron the grain i and state p. We carry out the calculation of the conductivity making expansion both in fluctuation modes and in the tunneling term H_T. This implies that the tunneling energy t is not very large. Proper conditions will be written later but now we mention only that the tunneling energy t will be everywhere much smaller than the energy E_T.

General formula (10.17), written for the symmetric junction, allows to express the tunneling conductivity in the form

$$\sigma^{DOS} = \sigma_0 (4T)^{-1} \int_{-\infty}^{+\infty} [\nu(E)/\nu(0)]^2 \cosh^{-2}(\frac{E}{2T})dE. \tag{10.89}$$

In Eq. (10.89), $\sigma_0 = 2\pi e^2 R^{-1} (t/\delta)^2$ is the classical conductivity of the granular metal. It can be rewritten in terms of the dimensionless conductance of the junction J_g as

$$\sigma_0 = \frac{8J_g e^2}{\pi R}. \tag{10.90}$$

Without the electron–electron interaction $\nu(E) = \nu(0)$ that gives $\sigma^{DOS} = \sigma_0$. Taking into account the electron–electron attraction we can write the contribution σ_{DOS} to the classical conductivity as

$$\sigma^{DOS} = \sigma_0 + \delta\sigma_{DOS}(J_g, T, H) \tag{10.91}$$

The correction $\delta\sigma_{DOS}(t, T, H)$ depends on temperature T and magnetic field H. Using Eq. (10.89) the correction to the conductivity $\delta\sigma^{DOS}$ at low temperatures can be written in terms of the correction to the DOS at zero energy $\delta\nu(0)$ as

$$\delta\sigma_{DOS}(J_g, H)/\sigma_0 = 2(\delta\nu(0)/\nu_0). \tag{10.92}$$

10.5.2 *Cooperon and propagator for granular superconductor*

Starting to deal with a such new object as the granular superconductor it is necessary to define in explicit form the averaged one-particle Green functions, the impurity vertices $\lambda(\mathbf{q},\varepsilon_1,\varepsilon_2)$ and the propagator of the superconducting fluctuations. The functions $\lambda(\mathbf{q},\varepsilon_1,\varepsilon_2)$ and $L(\mathbf{q},\Omega_k)$ depend on the coordinates and time slower than the averaged one-particle Green functions because the characteristic scale for both the impurity vertices and the propagator of superconducting fluctuations is the coherence length ξ which is much larger than l. As a result, the magnetic field affects only the vertex λ and the propagator L, whereas the Green functions should be assumed in the form (6.2).

For the impurity vertex $\lambda(\mathbf{q},\varepsilon_1,\varepsilon_2)$ above was obtained the general expression (6.26). One can easily include in it the effect of magnetic field recognizing that in its presence the momentum $\mathbf{q}$ must be substituted by $\mathbf{q} - 2e\mathbf{A}(\mathbf{r})$. If the shape of the grain is close to spherical, the vector-potential is expressed through the magnetic field $\mathbf{H}$ as $\mathbf{A}(\mathbf{r}) = [\mathbf{H}\times\mathbf{r}]/2$ (we choose it in the London gauge). The momentum $\mathbf{q}$ enters in Eq. (6.26) in the combination $\mathcal{D}\mathbf{q}^2$. In the case of superconducting grain with $R \ll \xi$, the Cooper pairs are localized in confines of the grains and the kinetic energy of the diffusion motion has to be substituted by the energy of space quantization. All relevant energies in the problem are assumed to be much smaller than the energy of the first harmonics $E_T = \mathcal{D}_{(3)}\pi^2/R^2$ playing the role of the Thouless energy of a single grain and this allows to keep in (6.26) only the zero harmonics. One can find the eigenvalue $\mathcal{E}_0(H)$ of this harmonics as:

$$\mathcal{D}\mathbf{q}^2 \to \mathcal{D}(2e\mathbf{A}(\mathbf{r}))^2 \to 4e^2\mathcal{D}_{(3)} < \mathbf{A}^2 > \to \mathcal{E}_0(H), \tag{10.93}$$

where $< \cdots >$ stands for the averaging over the volume of the grain. Such procedure we already performed in section 2.3 and for the grain of a nearly spherical form the relation $< \mathbf{A}^2 > = \frac{1}{10}H^2R^2$ was found

$$\mathcal{E}_0(H) = \frac{2}{5}\left(\frac{eHR}{c}\right)^2\mathcal{D}_{(3)} = \frac{2}{5\pi^2}\left(\frac{\Phi}{\Phi_0}\right)^2 E_T. \tag{10.94}$$

Within the zero-harmonics approximation, the function λ does not depend on coordinates and equals

$$\lambda_0(\varepsilon_n,\Omega_k-\varepsilon_n,H) = \frac{1}{\tau}\frac{\theta(-\varepsilon_1\varepsilon_2)}{|2\varepsilon_n-\Omega_k|+\mathcal{E}_0(H)}. \tag{10.95}$$

To calculate the fluctuation propagator for the granular superconductor L one should sum the sequence of the same ladder diagrams as it was done in Part II. Yet, as it was already mentioned the characteristic energies of the propagator L are low and therefore, when calculating the function L, one should take into account the tunneling processes from grain to grains described by the tunneling Hamiltonian H_T.

We will demonstrate here the alternative way: namely we will decouple the electron–electron interaction in Eq. (10.86) by a Gaussian integration over an auxiliary field Δ (Hubbard–Stratonovich transformation). Then, we will perform averaging over the electron quantum states, thus reducing the calculation to computation of a functional integral over the field Δ. In principle, one obtains within such a scheme a complicated free energy functional and the integral cannot be calculated exactly. The situation simplifies if the fluctuations are not very strong. Then, one can expand the free energy functional in Δ and come to Gaussian integrals that can be treated without difficulties. For the problem considered the propagator L is proportional to the average of the square of the field $\langle\, |\Delta|^2 \rangle$. In terms of the functional integral this quantity is written as

$$\langle |\Delta|^2 \rangle = \frac{\int |\Delta|^2 \exp(-\mathcal{S}_{\text{eff}}[\Delta, \Delta^*])\mathfrak{D}\Delta\mathfrak{D}\Delta^*}{\int \exp(-\mathcal{S}_{\text{eff}}[\Delta, \Delta^*])\mathfrak{D}\Delta\mathfrak{D}\Delta^*}, \tag{10.96}$$

where $S_{\text{eff}}[\Delta, \Delta^*]$ is the effective action functional. We have chosen the parameters in such a way that the grains are zero-dimensional. Therefore, it is sufficient to integrate over the zero space harmonics only, which means that the field Δ in the integral in Eq. (10.96) does not depend on coordinates. In the quadratic approximation in the field Δ the action functional $\mathcal{S}_{\text{eff}}$ includes two different terms

$$\mathcal{S}_{\text{eff}} = \mathcal{S}_{\text{eff}}^{(1)} + \mathcal{S}_{\text{eff}}^{(2)} \tag{10.97}$$

where $\mathcal{S}_{\text{eff}}^{(1)}$ describes the superconducting fluctuations in an isolated grain and $\mathcal{S}_{\text{eff}}^{(2)}$ takes into account tunneling from grain to grain. For the first term we obtain

$$\mathcal{S}_{\text{eff}}^{(1)}[\Delta, \Delta^*] = VT \sum_{\Omega_k} \left(1/g - 4\pi T \nu \tau \sum_{2\varepsilon_n > |\Omega_k|} \lambda_0(\varepsilon_n, \Omega_k - \varepsilon_n, H) \right) |\Delta(\Omega_k)|^2. \tag{10.98}$$

This formula evidently is reduced to the expansion of Eq. (7.16) in classical limit $\Omega_k = 0$. In the limit of low temperatures $T \ll \mathcal{E}_0(H)$ the sum over the frequencies ε_n in Eq. (10.98) can be replaced by the integral and we reduce the functional $\mathcal{S}_{\text{eff}}^{(1)}[\Delta, \Delta^*]$ to the form

$$\mathcal{S}_{\text{eff}}^{(1)}[\Delta, \Delta^*] = \frac{1}{\delta} T \sum_{\Omega_k} \left(\ln \left(\frac{\mathcal{E}_0(H) + |\Omega_k|}{\Delta_0} \right) \right) |\Delta(\Omega_k)|^2. \tag{10.99}$$

Close to the critical magnetic field H_c destroying the superconducting gap in a single grain the energy of the first harmonics $\mathcal{E}_0(H)$ is equal to the BCS gap at zero temperature Δ_0. This means that Eq. (10.99) can be written in this case in the region $T \ll \Delta_0$. Near H_c small frequencies Ω_k are most important and one can expand Eq. (10.99) in powers of the small parameter Ω_k/Δ_0. Then, we obtain

$$\mathcal{S}_{\text{eff}}^{(1)}[\Delta, \Delta^*] = \frac{1}{\delta} \sum_{\Omega_k} \left(\ln \left(\frac{\mathcal{E}_0(H)}{\Delta_0} \right) + \frac{|\Omega_k|}{\Delta_0} \right) |\Delta(\Omega_k)|^2. \tag{10.100}$$

At strong magnetic fields $H \gg H_c$, one should use the more general formula, Eq. (10.99).

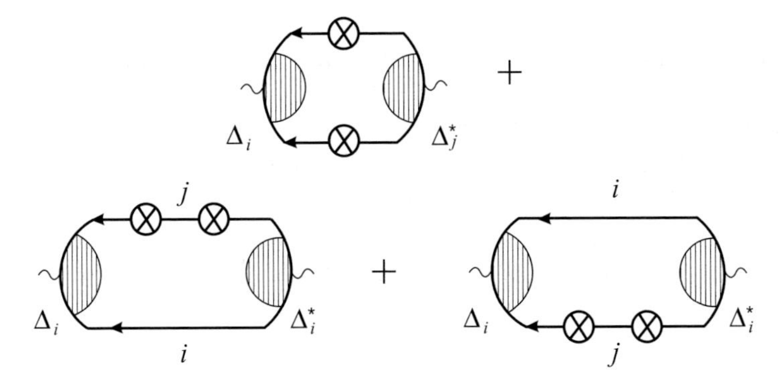

FIG. 10.13. Diagrams describing $F_{\text{eff}}^{(2)}$.

The term $\mathcal{S}_{\text{eff}}^{(2)}[\Delta, \Delta^*]$ describing the tunneling includes three different contributions represented in Fig. 10.13.

The analytic expression $F_{\text{eff}}^{(21)}[\Delta, \Delta^*]$ corresponding to the first diagram Fig. 10.13 can be written as

$$\mathcal{S}_{\text{eff}}^{(21)}[\Delta, \Delta^*] = -t^2/(2\pi\nu_0\tau)^2 T \sum_{i,j} (\Delta_i \Delta_j^* + c.c.) V^2 \int \frac{d^3\mathbf{p}_i d^3\mathbf{p}_j}{(2\pi)^6}$$
$$T \sum_{\varepsilon_n} G(i\varepsilon_n \mathbf{p}_i) G(-i\varepsilon_n \mathbf{p}_i) G(i\varepsilon_n \mathbf{p}_j) G(-i\varepsilon_n \mathbf{p}_j) C^2(i\varepsilon_n) \tag{10.101}$$

Writing Eq. (10.101) we put $\Omega_k = 0$ in the expression for the vertex λ and in the Green functions. This is justified because the action $\mathcal{S}_{\text{eff}}^{(2)}[\Delta, \Delta^*]$ is already small because it includes the parameter $J_g(\delta/\mathcal{E}_0(H))$ that is assumed to be small, where J_g is the dimensionless conductance of the system specified by Eq. (10.83). Next terms of the expansion are of the order of $J_g(\delta/\mathcal{E}_0(H))(\Omega_k/\mathcal{E}_0(H))$ and can be neglected for small Ω_k. The second and third diagrams in Fig. 10.13 are equal to each other and have the opposite sign with respect to the first diagram. For simplicity we assume that the granules are packed into a cubic lattice. Using the momentum representation with respect to the coordinates of the grains and taking into account all diagrams in Fig. 10.13 we reduce the action functional $\mathcal{S}_{\text{eff}}^{(2)}[\Delta, \Delta^*]$ to the form

$$\mathcal{S}_{\text{eff}}^{(2)}[\Delta, \Delta^*] = \frac{8}{3\pi\delta} \sum_{i=1}^{3} J_g(\delta/\mathcal{E}_0(H)) \left(1 - \cos q_i d\right) |\Delta|^2, \tag{10.102}$$

where $\mathbf{q}$ is the quasi-momentum and $d = 2R$. Equation (10.102) is written in the limit

$$J_g \ll \mathcal{E}_0(H)/\delta. \tag{10.103}$$

The inequality (10.103) is compatible with the inequality (10.83) provided the inequality

$$\mathcal{E}_0(H) \gg \delta \tag{10.104}$$

is fulfilled. If $\mathcal{E}_0(H) \sim \Delta_0$, the condition, Eq. (10.104), is at the same time the condition for the existence of the superconducting gap in the single granule. The inequality (10.103) allows us also to neglect influence of the tunneling on the form of the λ, so we use for calculations Eq. (10.95).

Writing the previous equations for $\mathcal{S}_{\text{eff}}^{(2)}[\Delta, \Delta^*]$ we neglected the influence of magnetic field on the phase of the order parameter Δ. In other words we omitted the phase factor $\exp(2ie \int \mathbf{A}(\mathbf{r})d\mathbf{r})$.

Although the final result for the correction to the DOS can be written for arbitrary temperatures T and magnetic fields H, let us concentrate on the most interesting case $T \ll T_c$, $H > H_c$. Using Eqs. (8.76), (10.99), (10.102) we obtain for the propagator of the superconducting fluctuations $L(\Omega_{k,}\mathbf{q})$

$$L(\Omega_k, \mathbf{q}) = -\nu_0^{-1} \left(\ln \left(\frac{\mathcal{E}_0(H) + |\Omega_k|}{\Delta_0} \right) + \eta(\mathbf{q}) \right)^{-1},$$

$$\eta(\mathbf{q}) \equiv \frac{8}{3\pi} \sum_{i=1}^{3} J_g\left(\delta/\mathcal{E}_0(H)\right)\left(1 - \cos q_i d\right), \tag{10.105}$$

The pole of the propagator $L(\Omega_k, \mathbf{q})$ at $\mathbf{q} = 0$, $\Omega_k = 0$ determines the field H_c, at which the BCS gap disappears in a single grain. From the form of Eq. (10.105) we find

$$\mathcal{E}_0(H_c) = \Delta_0. \tag{10.106}$$

The result for H_c, Eqs. (10.94), (10.106), agrees with the one obtained long ago by another method [234]. We can see from Eqs. (10.105), (10.106) that the term $\eta(\mathbf{q})$ describing tunneling is very important if H is close to H_c.

10.5.3 *Suppression of the conductivity due to DOS fluctuations*

Equations (10.95), (10.105) give the explicit formulae for the functions λ and L and allow us to calculate the correction $\delta\nu$ to the DOS which is determined by Eqs. (10.1), (10.2). Equation (10.1) contains integration over the momentum in the single grain $\mathbf{p}$ and the quasi-momentum $\mathbf{q}$. First, we integrate over the momentum $\mathbf{p}$ and reduce it for $\varepsilon_n > 0$ to the form

$$\delta\nu(i\varepsilon_n) = 4i\pi\nu_0\tau^2 T \sum_{\Omega_k < \varepsilon_n} \int L(\Omega_k, \mathbf{q})\lambda^2\left(\varepsilon_n, \Omega_k - \varepsilon_n\right) \frac{d^3\mathbf{q}}{(2\pi)^3} \tag{10.107}$$

After calculation of the sum over Ω_k in Eq. (10.107), one should make the analytic continuation $i\varepsilon_n \to \varepsilon$. At low temperatures, it is sufficient to find the correction to the DOS at zero energy, $\delta\nu \equiv \delta\nu(0)$.

Remarkably, Eqs. (10.105)–(10.107) do not contain explicitly the mean free time τ (the product $\tau\lambda_0$ according to Eq. (10.95) is τ-independent). This is a consequence of the zero-harmonics approximation, which is equivalent to using the random matrix theory (RMT) [409]. (The parameter τ enters only Eq. (10.94) giving the standard combination $\mathcal{E}_0(H)$ describing in RMT the crossover from the orthogonal to the unitary ensemble). This justifies the claim that the results can be used also for clean grains with a shape providing a chaotic electron motion.

Using Eqs. (6.27), (10.105)–(10.107) one can easily obtain an explicit expression for σ_{DOS} for $H - H_c \ll H_c$. In this limit, one expands the logarithm in the denominator of Eq. (10.105) and neglects the dependence of λ_0 on ε_n and Ω_k because the main contribution in the sum over Ω_k comes from $\Omega_k \sim \mathcal{E}_0(H) - \mathcal{E}_0(H_c) \ll \Delta_0$. Using Eq. (10.92) the result for $\delta\sigma_{DOS} = \sigma_0 - \sigma_{DOS}$ can be written as

$$\frac{\delta\sigma_{DOS}}{\sigma_0} = -\frac{2\delta}{\Delta_0}\begin{cases} -\pi^{-1} < \ln\tilde{\eta}(\mathbf{q}) >_q, & T/\Delta_0 \ll \tilde{\eta}, \\ \frac{2T}{\Delta_0} < \tilde{\eta}^{-1}(\mathbf{q}) >_q, & \tilde{\eta} \ll T/\Delta_0 \ll 1, \end{cases} \tag{10.108}$$

where $\tilde{\eta}(\mathbf{q}) = \eta(\mathbf{q}) + 2\widetilde{h}$,

$$< \cdots >_q \equiv V \int_0^{2\pi/d} (\cdots)\, d\mathbf{q}/(2\pi)^3,$$

$\widetilde{h} = (H - H_c)/H_c$ and V is a volume of the single grain. We see that the correction to the conductivity is negative and its absolute value decreases when the magnetic field increases. The correction reaches its maximum at $H \to H_c$. At zero temperature and close to the critical field H_c such that $J_g(\delta/\Delta_0) \gg \widetilde{h}$, the maximum value of $\delta\sigma_{DOS}/\sigma_0$ from Eq. (10.108) is

$$\left|\frac{\delta\sigma_{DOS}}{\sigma_0}\right| = \frac{2}{\pi}\frac{\delta}{\Delta_0}\left\langle \ln\left(\frac{1}{\eta(\mathbf{q})}\right)\right\rangle_q = \frac{1}{3}\frac{\delta}{\Delta_0}\ln\left(\frac{\Delta_0}{J_g\delta}\right). \tag{10.109}$$

In the limit $J_g(\delta/\Delta_0) \ll \widetilde{h} \lesssim 1$, one can expand the logarithm in Eq. (10.108). Then, taking $\widetilde{h} \sim 1$ the correction to the conductivity at zero temperature can be estimated as

$$\left|\frac{\delta\sigma_{DOS}}{\sigma_0}\right| \sim J_g\left(\frac{\delta}{\Delta_0}\right)^2. \tag{10.110}$$

Schematically, the suppression of the DOS due to the superconducting fluctuations is shown in Fig. 10.14.

As temperature grows, the correction to the conductivity due to the reduction of the DOS can become larger and reach for $T \sim \Delta_0$ and $J_g(\delta/\Delta_0) \gg \widetilde{h}$ the order of magnitude of J_g^{-1}.

$$\frac{|\delta\sigma_{DOS}|}{\sigma_0} = 4(\delta T/\Delta_0^2)\left\langle\frac{1}{h + \eta(\mathbf{q})}\right\rangle_q \sim \frac{1}{J_g}. \tag{10.111}$$

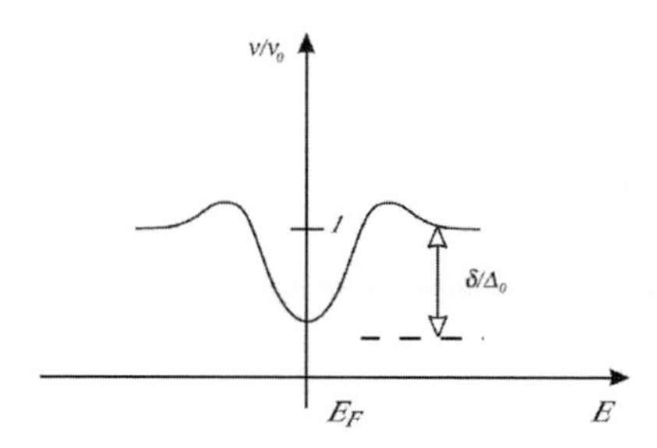

FIG. 10.14. Suppression of DOS due to superconducting fluctuations.

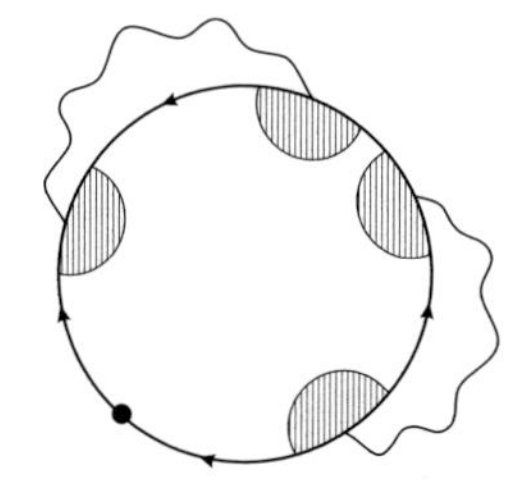

FIG. 10.15. An example of high order corrections to DOS.

In the limit $\widetilde{h} \approx 1 \gg J_g(\delta/\Delta_0)$ and at temperature $T \sim \Delta_0$ this correction can be estimated as

$$\frac{\delta\sigma_{DOS}}{\sigma_0} \approx \frac{\delta}{\Delta_0}. \tag{10.112}$$

We see from Eqs. (10.109)–(10.112) that the corrections to the conductivity are smaller than unity provided we work in the regime of a good metal (see Eqs. (10.82), (10.83)), so the diagrammatic expansion we use is justified. Indeed, we can neglect the corrections of higher orders. For example, the diagram shown in Fig. 10.15 has the additional small factor of $(\delta/\Delta_0)\ln\tilde{\eta}$ at $T/\Delta_0 \ll \tilde{\eta}$ and $\delta/(\Delta_0\tilde{\eta})$ at $\tilde{\eta} \ll T/\Delta_0$.

The correction to the conductivity calculated in this section could become comparable to σ_0 when $J_g \sim 1$. However, such values of J_g mean that we would be in this case not far from the metal–insulator transition. Then, we would have to take into account all localization effects. For values of $J_g \sim 1$ one can use Eq. (10.108) for rough estimates only. Apparently, the parameters of the samples of [406] correspond to the region $J_g \sim 1$, $\delta/\Delta_0 \sim 1/3$.

In the limit of strong magnetic fields $H \gg H_c$ the correction to σ_0 can still be noticeable. In this case we can use Eq. (10.99) as before but, with a logarithmic accuracy, we can neglect the dependence of the superconducting propagator, Eq. (10.105), on Ω_k and on the tunneling term. Then we obtain finally

$$\delta\sigma_{DOS}/\sigma_0 = -(1/3)\left(\delta/\mathcal{E}_0(H)\right)\ln^{-1}\left(\mathcal{E}_0(H)/\Delta_0\right) \tag{10.113}$$

Equation (10.113) shows that in the region $H \gg H_c$ the correction to the conductivity decays essentially as $\delta\sigma_{DOS} \sim H^{-2}$.

Let us emphasize that the correction to the conductivity coming from the DOS remains finite in the limit $T \to 0$, thus indicating the existence of the virtual Cooper pairs even at $T = 0$.

In the region of not very small $\widetilde{h} \gg J_g\delta/\Delta_0$, we can neglect the tunneling term $F_{\text{eff}}^{(2)}[\Delta, \Delta^*]$ in the free energy functional $F_{\text{eff}}[\Delta, \Delta^*]$. Then, we can write the correction to the conductivity in a rather general form. The superconducting propagator K can be written in this case as

$$L(\Omega_k) = -\nu_0^{-1}\ln^{-1}\left(\frac{\mathcal{E}_0(H) + |\Omega_k|}{\Delta_0}\right). \tag{10.114}$$

Using Eq. (10.107) for the correction to the DOS and Eq. (10.92) we obtain for the correction to the conductivity at zero temperature

$$\frac{\delta\sigma_{DOS}}{\sigma_0} = -\frac{1}{3}\left(\frac{\delta}{\Delta_0}\right)\int_a^\infty \frac{\exp(-x)}{x}dx = -\frac{1}{3}\left(\frac{\delta}{\Delta_0}\right)Ei\left[2\ln(1+h)\right]. \tag{10.115}$$

In the limit $\widetilde{h} \ll 1$, we reproduce with logarithmic accuracy Eq. (10.109), whereas in the opposite limiting case $\widetilde{h} \gg 1$ we come to Eq. (10.113).

In order to calculate the entire conductivity, one must investigate the AL and MT contributions. This cumbersome calculations were performed in [360] and it was demonstrated that in the granular materials both the AL and MT contributions, being proportional to T^2, *vanish* in the limit $T \to 0$ at all $H > H_c$ and thus, the correction to the conductivity comes from the DOS only. So, at low temperatures, estimating the total correction to the classical conductivity σ_0, one can use the above results.

At not-very-low temperatures, and not far from the critical field the AL correction to the conductivity is the most important. This means that approaching the transition in this region the resistivity decreases, which is in contrast to the behavior at very low temperature where the correction to the resistivity is determined entirely by the contribution to the DOS and is positive. The temperature and magnetic field dependence of σ_{AL} and σ_{MT} is rather complicated but they are definitely positive. The competition between these corrections and σ_{DOS} determines the sign of the magnetoresistance. As we already have seen at $T \to 0$ the magnetoresistance is negative for all H. In contrast, at $T \sim T_c$ and close to H_c, the AL and MT corrections can become larger than σ_{DOS} resulting in a positive magnetoresistance in this region. Far from H_c the magnetoresistance is negative again.

10.5.4 *Zeeman splitting*

In our previous consideration we neglected interaction between the magnetic field and spins of the electrons. This approximation is justified if the size of the grains is not very small. Then, the critical field H_c^{or} destroying the superconducting gap is smaller than the paramagnetic limit H_c^{spin} and the orbital mechanism dominates the magnetic field effect on the superconductivity. However, the Zeeman splitting leading to the destruction of the superconducting pairs can become important if one further decreases the size of the grains.

Let us discuss now the effect of Zeeman splitting. We can rewrite Eq. (10.84) for the ratio of orbital H_c^{or} magnetic field to the Zeeman magnetic field in the following form

$$\frac{H_c^{or}}{H_c^{spin}} = \left(\frac{\delta}{\delta_c}\right)^{1/3}, \tag{10.116}$$

where $\delta_c \approx 1/(\nu R_c^3)$, $R_c = \xi(p_0 l)^{-1}$ and $\xi = \sqrt{\xi_0 l}$. To understand whether the Zeeman splitting is important for an experiment we can estimate the ratio δ_c/Δ_0 and compare it with the proper experimental result. We find easily

$$\frac{\delta_c}{\Delta_0} = (p_0 l)\sqrt{\frac{l}{\xi_0}} \tag{10.117}$$

which shows δ_c/Δ_0 can be both smaller than $(\delta/\Delta_0)_{\text{exp}}$ and larger depending on the values of l and ξ_0. Using the result for δ_c, Eq. (10.117), we can rewrite Eq. (10.116) as

$$\frac{H_c^{or}}{H_c^{spin}} = \left(\frac{\delta}{\Delta_0}\frac{R}{l}\right)^{1/2} \tag{10.118}$$

Below, we consider the corrections to the DOS and conductivity due to the Zeeman mechanism. We will see that at temperature $T = 0$ these corrections can be of the same order of magnitude as the correction due to orbital mechanism.

Let us calculate first the critical magnetic field H_0 destroying the superconducting gap in a single grain taking into account both the orbital and Zeeman mechanisms of the destruction. The Green function for the noninteracting electrons in this case is determined by Eq. (10.67) Including the interaction between the magnetic field and electron spins we obtain the following form of the impurity vertex

$$\lambda_0(\varepsilon_n, -\varepsilon_n) = \frac{1}{\tau}\frac{1}{|2\varepsilon_n| - iE_Z \text{sign}\varepsilon_n + \mathcal{E}_0(H)} \tag{10.119}$$

Repeating the calculations of previous section with the modified vertex, Eq. (10.119), we find at $T \ll T_c$ the new critical magnetic field H_0

$$\mathcal{E}_0^2(H_0) + E_Z^2(H_0) = \Delta_0^2. \tag{10.120}$$

In the limit of a very weak Zeeman splitting, when the orbital mechanism is more important, Eq. (10.120) reproduces the previous result for the critical magnetic field Eq. (10.106). In the opposite limiting case, when the Zeeman mechanism plays the major role, we obtain $E_Z = \Delta_0$. In general case, when the both mechanisms of the destruction of the conductivity are important, one should solve Eq. (10.120) and the result reads

$$H_0 = H_c \left(-\frac{1}{2} \left(\frac{g\mu_B H_c}{\Delta_0} \right)^2 + \sqrt{\frac{1}{4} \left(\frac{g\mu_B H_c}{\Delta_0} \right)^4 + 1} \right)^{1/2} . \tag{10.121}$$

Using Eq. (10.106) for a critical field H_c we can estimate the ratio $\mu_B H_c/\Delta_0 = (\xi_0/R)(p_0 R)^{-1}(p_0 l)^{-1} \ll 1$. If this parameter is small (the orbital mechanism is more important) the critical field H_0 is close to the field H_c

$$H_0 = H_c \left(1 - \frac{1}{4} \left(\frac{g\mu_B H_c}{\Delta_0} \right)^2 \right) . \tag{10.122}$$

When Zeeman splitting is more important then orbital mechanism then from Eq. (10.121) we obtain $g\mu_B H_0 = \Delta_0$. This is the point of the absolute instability of the paramagnetic state. At $g\mu_B H < \Delta_0$, the superconducting state is the only stable one.

Let us calculate the correction to the DOS taking into account both the Zeeman splitting and orbital mechanism of the suppression of superconductivity. Using Eqs. (10.67), (10.119) we obtain for the superconducting propagator

$$L(\Omega_k) = -\nu_0^{-1} \left(\frac{1}{2} \ln \left(\frac{(\mathcal{E}_0(H) + |\Omega_k|)^2 + E_Z^2}{\Delta_0^2} \right) + \eta(\mathbf{q}) \right)^{-1} . \tag{10.123}$$

In the region $\tilde{h} = (H - H_0)/H_0 \ll 1$, where H_0 is given in Eq. (10.121), expanding the logarithm in the superconducting propagator we obtain

$$L(\Omega_k) = -\nu_0^{-1} \left(2\tilde{h} + \frac{|\Omega_k|}{\Delta_0} + \eta(\mathbf{q}) \right)^{-1} . \tag{10.124}$$

Using Eqs. (10.107), (10.92) for the correction to the conductivity we obtain the same result as before, Eq. (10.108), but with the new h_0. Equations (10.109)–(10.112) are correct for in general case.

In the limit of a strong magnetic field $\tilde{h} \gg 1$ we can neglect with logarithmic accuracy the Ω-dependence of superconducting propagator in Eq. (10.123). Using Eqs. (10.92), (10.107), (10.119) we obtain for the correction to the conductivity at strong magnetic fields

$$\frac{\delta\sigma_{DOS}}{\sigma_0} = -\frac{2}{3} \left(\frac{\delta}{E_Z(H)} \right) \arctan \left(\frac{E_Z(H)}{\mathcal{E}_0(H)} \right) \ln^{-1} \left(\frac{\mathcal{E}_0^2(H) + E_Z^2(H)}{\Delta_0^2} \right) . \tag{10.125}$$

From Eq. (10.125) we can see that if orbital mechanism is more important than the Zeeman one, that is if $E_Z/\mathcal{E}_0 \ll 1$, we reproduce the previous result for the

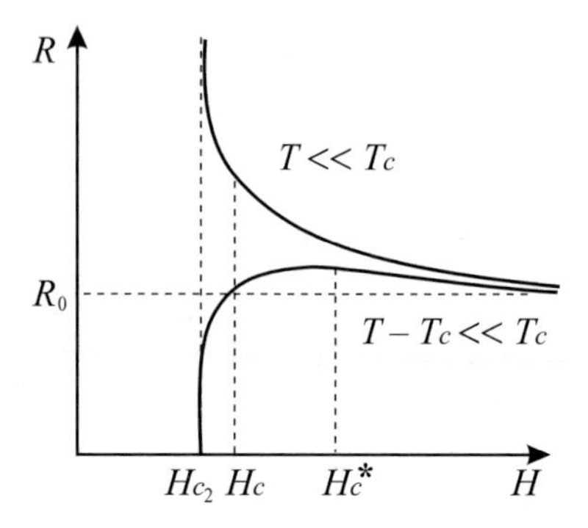

FIG. 10.16. The typical curves of resistivity for low temperatures $T \ll T_c$ and temperatures close to T_c.

correction to the conductivity at strong magnetic field, Eq. (10.113). If the Zeeman mechanism is more important $E_Z/\mathcal{E}_0 \gg 1$, then we obtain for the correction to the conductivity

$$\frac{\delta\sigma_{DOS}}{\sigma_0} = -\frac{\pi}{6}\left(\frac{\delta}{E_Z(H)}\right)\ln^{-1}\left(\frac{E_Z(H)}{\Delta_0}\right). \tag{10.126}$$

Equation (10.126) has the same structure as Eq. (10.113) but the function $\mathcal{E}_0(H)$ is replaced by $E_Z(H)$. This changes the asymptotic behavior at strong magnetic fields because $E_Z(H) \sim H$ in contrast to $\mathcal{E}_0(H) \sim H^2$. So, we conclude that $\delta\sigma_{DOS}/\sigma_0 \sim H^{-1}$.

10.5.5 *Conclusions*

It has been demonstrated that even if the superconducting gap in each granule is destroyed by the magnetic field the virtual Cooper pairs can persist up to extremely strong magnetic fields. However, the contribution of the Cooper pairs to transport is proportional at low temperatures to T^2 and vanishes in the limit $T \to 0$. In contrast, they reduce the one-particle DOS in the grains even at $T = 0$, thus diminishing the macroscopic conductivity. The conductivity can reach its classical value only in extremely strong magnetic fields when all the virtual Cooper pairs do not exist anymore. This leads to negative magnetoresistance. Moreover, both the orbital and Zeeman mechanisms of the destruction of superconductivity were analyzed, as well as the limits of low temperatures and temperatures close to the critical temperature T_c. The results demonstrate that, at low temperatures $T \ll T_c$, there must be a pronounced peak in the dependence of the resistivity on the magnetic field. This peak should be much smaller in the region of temperatures $T \ll T_c$ because, in this region the superconducting fluctuations can contribute to transport, thus diminishing the role of the reduction of the DOS.

Qualitatively, the results are summarized in Fig. 10.16, where typical curves for the low temperatures $T \ll T_c$ and temperatures close to T_c are represented.

Both the functions reach asymptotically the value of the classical resistivity R_0 only at extremely strong magnetic fields. The resistivity R at low temperatures grows monotonously on decreasing the magnetic field. The function does not have any singularity at the magnetic field H_c destroying the superconducting gap Δ in a single grain. The real transition into the superconducting state occurs at a lower field H_{c2}. This field, in the region of parameters involved, is close to the field H_c. The resistivity $R(H)$ remains finite as $H \to H_{c2}$ but its derivative diverges resulting in the infinite slope at $H = H_{c2}$. The dependence of the resistivity R on the magnetic field at temperatures near T_c is more complicated. Already far from the field H_c the superconducting fluctuations start contributing to transport and the resistivity goes down on decreasing the magnetic field. A negative magnetoresistance is possible in this region only at magnetic fields H^*, that can be much larger than the field H_c.

11

SPIN SUSCEPTIBILITY AND NMR

11.1 Preliminaries

In this section, we discuss the contribution of superconducting fluctuations to the spin susceptibility and the NMR relaxation rate. For both these effects the interplay of different fluctuation contributions is unusual with respect to the case of the conductivity. As in the case of the a.c. conductivity, the fluctuation contributions to the spin susceptibility and the NMR relaxation rate can manifest themselves as the opening of a pseudogap even in the normal phase, a phenomenon which is characteristic of HTS compounds.

We begin with the dynamic spin susceptibility $\chi_{\pm}^{(R)}(\mathbf{k},\omega) = \chi_{\pm}(\mathbf{k}, i\omega_\nu \to \omega + i0^+)$ where

$$\chi_{\pm}(\mathbf{k},\omega_\nu) = \int_0^{1/T} d\tau e^{i\omega_\nu \tau} \left\langle\left\langle \hat{T}_\tau \left(\hat{S}_+(\mathbf{k},\tau)\hat{S}_-(-\mathbf{k},0)\right)\right\rangle\right\rangle. \tag{11.1}$$

Here $\hat{S}_{\pm}$ are the spin raising and lowering operators, $\hat{T}_\tau$ is the time ordering operator, and the brackets denote thermal and impurity averaging in the usual way. The uniform, static spin susceptibility is given by $\chi_s = \chi_{\pm}^{(R)}(\mathbf{k} \to 0, \omega = 0)$ while the dynamic NMR relaxation rate is denoted as $1/T_1$ and is determined by

$$\frac{1}{T_1 T} = \lim_{\omega \to 0} \frac{A}{\omega} \int \frac{d^3\mathbf{k}}{(2\pi)^3} \operatorname{Im} \chi_{\pm}^{(R)}(\mathbf{k},\omega) \tag{11.2}$$

where A is a positive constant involving the gyromagnetic ratio.

For noninteracting electrons $\chi_{\pm}^{0}(\mathbf{k},\omega_\nu)$ is determined by the usual loop diagram. Simple calculations lead to the well known results for $T \ll E_F$: $\chi_s^0 = \nu$ (Pauli susceptibility) and $(1/T_1T)^0 = A\pi\nu^2$ (Korringa relaxation). We will present the fluctuation contributions in a dimensionless form by normalizing to the above results.

To leading order in Gi the fluctuation contributions to $\chi_{\pm}$ can be discussed with the help of the same diagrams drawn for the conductivity in Fig. 8.1. It is important to note that the role of the external vertices (electron interaction with the external field) is now played by the $\hat{S}_{\pm}(\mathbf{k},\tau)$ operators. This means that the two fermion lines attached to the external vertex must have opposite spin labels (up and down). Consequently, the AL diagram for $\chi_{\pm}$ does not exist since one cannot consistently assign a spin label to the central fermion for spin–singlet pairing.

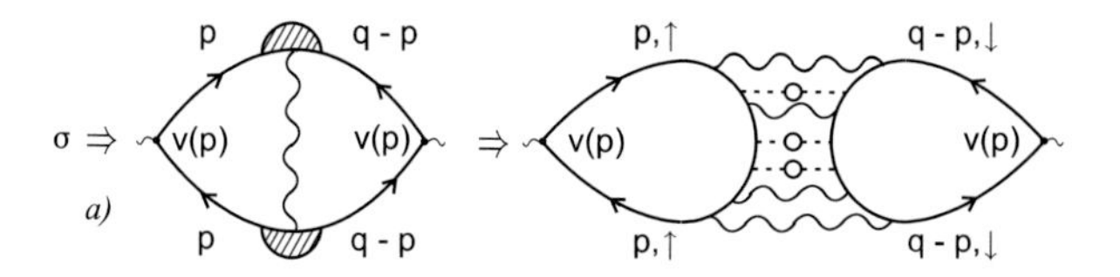

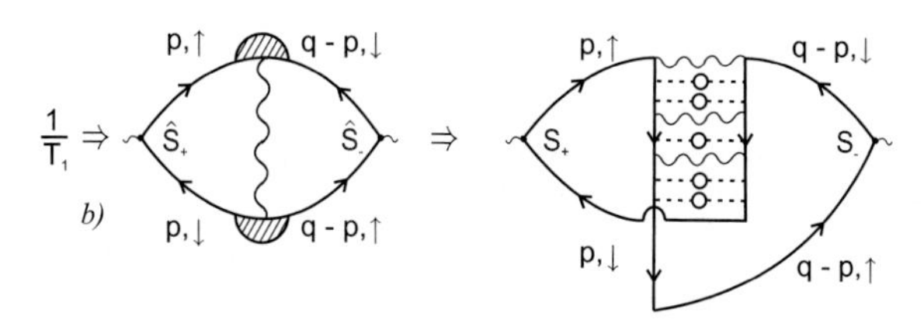

FIG. 11.1. MT diagrams for (a) conductivity; (b) spin susceptibility in representation of the propagator in the explicit form as a ladder of electron lines connected by attractive interactions. The bold wavy lines correspond to the fluctuation propagator, while the thin wavy lines correspond to bare inter–electron interaction.

The next set of diagrams to consider corresponds to the MT contribution. While the MT diagrams for $\chi_{\pm}$ appear to be identical to those for the conductivity, there is an important difference in topology which arises from their spin structure. Drawing the fluctuation propagator explicitly as a ladder of attractive interaction lines, one can seem that the MT diagram is a nonplanar graph with a single fermion loop. In contrast the MT graph for the conductivity is planar and has two fermion loops. (see Fig. 11.1). The number of loops, according to the rules of the diagrammatic technique [228], affects the sign of the contribution. In the case of spin susceptibility, which is under consideration, the topological sign of the MT diagram turns opposite to that one drawn for the conductivity.

The diagrams 5 and 6 represent the effect of fluctuations on the single-particle self-energy, leading to a decrease in the DOS at Fermi level. The DOS diagrams 7 and 8 include impurity vertex corrections (note that these have only a single impurity scattering line as additional impurity scattering in the form of a ladder has a vanishing effect). Finally 9 and 10 are the DOS diagrams with the Cooperon impurity corrections.

11.2 Spin susceptibility[76]

We note that, when the external frequency and momentum can be set to zero at the outset, as is the case for χ_s, there is no anomalous MT piece (which as we shall see below is the most singular contribution to $1/T_1$). The MT diagram 2 then yields a result which is identical to the sum of the DOS diagrams 5 and

[76] In this section we base on the results of [278, 410].

6. Let us remember that the same relation between the absolute values of the DOS and regular MT contributions was found in the case of heat conductivity, but they were of the opposite signs (the number of loops in the MT diagrams for heat conductivity and spin susceptibility are different) and compensated one another (see section 4.6).

In the clean limit ($T_c\tau \gg 1$) the fluctuation contribution is given by $\chi_s^{\text{fl}} = \chi_{s2} + \chi_{s5} + \chi_{s6}$; all other diagrams turn out to be negligible. In the dirty case ($T_c\tau \ll 1$), the DOS diagrams 5 and 6, together with the regular part of the MT diagram (2), yield the same result as in the clean limit (of the order $\mathcal{O}(T_c/E_F)$). One can see, that this contribution is negligible in comparison with the expected dominant one for the dirty case of order $\mathcal{O}(1/E_F\tau)$. A thorough study of all diagrams shows that the important graphs in the dirty case are those with the Cooperon impurity corrections MT 3 and 4, and the DOS ones 9 and 10. This is the unique example known to us where the Cooperons, which play a central role in the WL theory, give the leading order result in the study of superconducting fluctuations. Diagrams 3 and 4 give one half of the final result given below; diagrams 9 and 10 provide the other half. The total fluctuation susceptibility is $\chi_s^{\text{fl}} = \chi_{s3} + \chi_{s4} + \chi_{s9} + \chi_{s10}$. Interestingly, in both the clean and dirty cases $\chi_s^{\text{fl}}/{\chi_s}^{(0)}$ can be expressed by the same formula if one expresses the coefficient in terms of the GL number $Gi_{(2)}$ (6.36):

$$\frac{\chi_s^{\text{fl}}}{{\chi_s}^{(0)}} = -2Gi_{(2)} \ln\left(\frac{2}{\sqrt{\epsilon} + \sqrt{\epsilon + r}}\right). \tag{11.3}$$

It is tempting to explain the negative sign of the fluctuation contribution to the spin susceptibility in Eq. (11.3) as arising from a suppression of the DOS at the Fermi level. But one must keep in mind that only the contribution of diagrams 5 and 6 can strictly be interpreted in this manner; the MT graphs and the coherent impurity scattering described by the Cooperons do not allow such a simple interpretation.

11.3 Relaxation rate[77]

The calculation of the fluctuation contribution to $1/T_1$ requires rather more care than χ_s because of the subtleties of analytic continuation. Let us define the local susceptibility

$$K(\omega_\nu) = \int (d\mathbf{k}) \chi_{+-}(\mathbf{k}, \omega_\nu).$$

In order to write down the fluctuation contribution to $1/T_1$ for the case of an arbitrary impurity concentration including the ultra-clean case let us start from the anomalous MT contribution and evaluate it using the standard contour integration techniques

[77] In this section we base on results of [278, 411–413].

$$\lim_{\omega\to 0}\frac{1}{\omega}\operatorname{Im}K^{(an)R}(\omega) = -\frac{\pi\nu^2}{8}\int(d\mathbf{q})L(\mathbf{q},0)\mathcal{K}(\mathbf{q}), \tag{11.4}$$

$$\mathcal{K}(\mathbf{q}) = 2\tau\int_{-\infty}^{\infty}\frac{dz}{\cosh^2(z/4T\tau)}\frac{1}{\left(\sqrt{l^2q^2-(z-i)^2}-1\right)}$$
$$\times\frac{1}{\left(\sqrt{l^2q^2-(z+i)^2}-1\right)}. \tag{11.5}$$

We have used the impurity vertices in the general form (8.52). The first simple limiting case for (11.5) is $lq \ll 1$, when the square roots in the denominator can be expanded and $\mathcal{K}(\mathbf{q}) = 2\pi/\mathcal{D}q^2$. As we already know from section 8.5, this corresponds to the usual local approximation and covers the domain $T_c\tau \ll 1/\sqrt{\epsilon}$. Introducing the pair-breaking rate γ_φ as an infrared cut-off one can find:

$$\frac{\delta\left(1/T_1\right)^{\mathrm{MT(an)}}}{\left(1/T_1\right)^0} = \frac{28\zeta(3)}{\pi^4}Gi_{(2,d)}\frac{1}{\epsilon-\gamma_\varphi}\ln(\epsilon/\gamma_\varphi). \tag{11.6}$$

The other limiting case is the "ultra-clean limit" when the characteristic q-values satisfy $lq \gg 1$. This is obtained when $T\tau \gg 1/\sqrt{\epsilon} \gg 1$. From (11.5) we then find $\mathcal{K}(\mathbf{q}) = 4\ln(lq)/vq$, which leads to

$$\frac{\delta(1/T_1)^{\mathrm{MT(an)}}}{(1/T_1)^0} = \frac{\pi^3}{\sqrt{14\zeta(3)}}Gi_{(2,cl)}\frac{1}{\sqrt{\varepsilon}}\ln(T\tau\sqrt{\epsilon}). \tag{11.7}$$

We note that in all cases the anomalous MT contribution leads to an *enhancement* of the NMR relaxation rate over the normal state Korringa value. In particular, the superconducting fluctuations above T_c have the *opposite* sign to the effect for $T \ll T_c$ (where $1/T_1$ drops exponentially with T). One might argue that the enhancement of $1/T_1$ is a precursor to the coherence peak just below T_c. Although the physics of the Hebel–Slichter peak (pile-up of the DOS just above gap edge and coherence factors) appears to be quite different from that embodied in the MT process, we note that both effects are suppressed by strong inelastic scattering.

Basing on proposed above understanding of the physics of the MT contribution to conductivity, one can give the similar interpretation in terms of the self-intersecting trajectories also for the MT process in NMR relaxation rate. Indeed, let us consider a self-intersecting trajectory and the motion of the electron along it with fixed spin orientation (let us say "spin up"). Let the nuclei is placed in the point of trajectory self-intersection. If, after passing a full turn, the electron interacts with the nucleus and changes its spin state and momentum to the opposite value it can pass again along the previous trajectory moving in the opposite direction. Its motion between impurities is very fast, the electron moves with the Fermi velocity. Due to the retarded character of the Cooper interaction

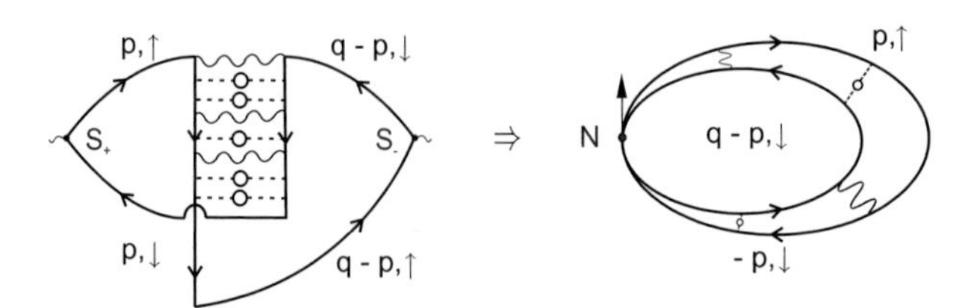

FIG. 11.2. The interpretation of the MT contribution to the relaxation rate in terms of the electron states pairing at the same self-intersecting trajectory.

one can imagine formation of some particular fluctuation pair, consisting of the only electron, which stays in the mixed state. Namely, the electron being in quantum states corresponding to the forward motion with "spin up" and backward motion with "spin down" along the self-intersecting trajectory can be paired with itself (see Fig. 11.2). This purely quantum process opens a new mechanism of spin relaxation, and so contributes positively to the relaxation rate $1/T_1$.

We now discuss the DOS and the regular MT contributions which are important when strong dephasing suppresses the anomalous MT contribution discussed above. The local susceptibility arising from diagrams 5 and 6 can be easily evaluated. The other remaining contribution is from the regular part of the MT diagram. It can be shown that this regular contribution doubles DOS-like contributions from diagrams 5 and 6, exactly as in the case of fluctuation conductivity. All other diagrams either vanish (as is the case for graphs 7 and 8) or contribute at higher order in $1/E_F\tau$ (this applies to the graphs with the Cooperon corrections). The final results can be presented in a unique way for the clean (but not ultra-clean) and dirty cases by means of the $Gi_{(2)}$ number:

$$\frac{\delta(1/T_1)^{DOS}}{(1/T_1)^0} = -16Gi_{(2)} \ln\left(\frac{2}{\sqrt{\epsilon}+\sqrt{\epsilon+r}}\right). \tag{11.8}$$

The negative sign of the DOS contribution to the NMR relaxation rate has a clear physical origin and directly follows from the Korringa law. It appears due to the fluctuation suppression of the one-electron density of states at the Fermi level.

In the case of the nuclear magnetic relaxation rate calculations, the electron interaction causing nuclear spin flip is considered. If one would try to imagine an AL process of this type he would be in trouble, because the electron–nuclei scattering with spin-flip evidently transforms the initial singlet state of the fluctuation Cooper pair into a triplet one, which is forbidden in the scheme discussed. So the formally discovered absence of the AL contribution to the relaxation rate is evident enough.

It is worth mentioning that the cancelation of the MT and DOS contributions to conductivity found in section 8.5 is crucial for the fluctuation contributions to the NMR relaxation rate. In fact, the MT and DOS contributions here have the same structure as in the conductivity while the AL contribution is absent. So the full fluctuation correction to the NMR relaxation rate in clean superconductor simply disappears.

11.4 Discussion

The main results of this section, valid for $\epsilon \ll 1$, can be summarized as follows:

- Fluctuations lead to a suppression of the spin susceptibility χ_s, due to the combined effect of the reduction of the single-particle DOS arising from the self energy contributions, and of the regular part of the MT process.
- 'Cooperon" impurity interference terms, involving impurity ladders in the particle–particle channel, are crucial for the χ_s suppression in the dirty limit.
- The hierarchy of fluctuation contributions to spin susceptibility turns out to be is unusual with respect to conductivity and many other effects. Namely, the AL and the anomalous MT terms, which usually dominate, in this case are absent.
- For weak pair-breaking ($1/\tau_\varphi \ll T_c$), an enhancement of $1/T_1T$, coming from the positive anomalous MT term, takes place [278, 412, 413].
- Strong dephasing suppresses the anomalous MT contribution, and $1/T_1$ is then dominated by the less singular DOS and the regular MT terms. Being negative, these contributions lead to a suppression of spectral weight and a decrease in $1/T_1$.

An intensive controversy took place in recent years in relation to the magnetic field dependence of the fluctuation contribution to $1/T_1$. The situation here resembles much the situation with the magnetoconductivity: a positive MT contribution is suppressed by the magnetic field while the magnetic field dependent part of the DOS contribution increases with the growth of the field. But in contrast to the magnetoconductivity, which can be measured extremely precisely, the NMR relaxation rate measurements are much more sophisticated. The result of this delicate competition, depending on many parameters (r, γ_φ, τ), was found in HTS materials to be qualitatively different in experiments of various groups. The absence of a strong positive AL contribution, possible d pairing, killing the MT contribution [412], small magnitude of the sum of MT and DOS effects even in the case of s-pairing, lack of the precise values of r, γ_φ, τ, leading to contradictive theoretical predictions [249,278,279,414], the dispersion in the quality of samples and experimental methods were the reason of this discussion [67, 249, 415–418]

Recently the effect of amplitude fluctuations in clean case but taking into account the nonzero frequency of the *a.c.* field was considered in [419]. The results obtained in the limit of zero frequency correspond to that of [278] in the limit of clean superconductor. Moreover, the authors of [419] took into account the effect of BKT vortex–antivortex fluctuations on the relaxation rate.

PART IV

FLUCTUATIONS IN NANOSTRUCTURES AND UNCONVENTIONAL SUPERCONDUCTING SYSTEMS

12

FLUCTUATIONS IN NANOGRAINS, NANODROPS AND GRANULAR SUPERCONDUCTORS

In this chapter the terms, nanodrop, nanograin and granule, appear side by side with each other. In the modern scientific literature they often are considered almost as synonyms. Nevertheless, we will distinguish them and will start this section from the definitions.

Using the term "superconducting nanograin" we will refer to the superconducting object with dimensions smaller than the Cooper pair size ξ. In spite of their small size such objects can still retain some superconducting properties, for example the gap in the excitations spectrum. The macroscopic system of such nanograins we will call a granular superconductor. Its properties depend strongly on the character of junctions between granules (grains). We already discussed above (see section 10.5) the magnetoresistance of such a granular system with a weak electron intergrain tunneling probability, and recognized the important role of the quantum fluctuations in them. In the opposite case, when this probability is high enough, properties of such system in the vicinity of the critical temperature can be described by the GL functional with some effective coefficients.

When the properties of various grains are different the GL or BCS parameters become some random functions of the space coordinates. We already met with such situation in chapter 5 discussing the effect of the quenched disorder on the properties of the vortex system. The granular superconductor, with randomly varying in space superconducting parameters, presents another example of the system with quenched disorder. Below we will consider one more manifestation of the quenched disorder which can be observed even in the absence of a magnetic field. When $T > T_c$ (or $H > \overline{H}_{c2}$) the most of the granular superconductor volume remains in the normal state but due to the quenched disorder the superconducting drops are formed in those spots where the grains with high superconducting parameters are by chance concentrated. We will demonstrate that many properties of the system of such drops resemble strongly properties of the system of superconducting grains.

12.1 Ultrasmall superconducting grains

The recent progress in nanotechnologies allowed the fabrication of superconducting grains of the nanometer scale [420,421]. Let us remember that in the chapter 1 we already studied the thermodynamics of small grains based on the GL phenomenology. The natural question arises: how small is the grain size to which one can apply the GL approach?

It turns out that the properties of superconducting grains change dramatically when their size shrinks to a few nanometers. This phenomenon has recently generated a lot of interest [422]. In particular, it has been established both theoretically and experimentally that the superconductivity cannot be observed if the quantum level spacing in a nanoparticle exceeds the superconducting gap Δ. This answer is contained in the value of the Gi number which can be rewritten in terms of the level spacing for the grain $\delta = 1/\nu V$ (let us remember that the DOS is the number of states per unit energy interval and unit volume) and superconducting gap $\Delta(0) \sim T_c$:

$$Gi_{(0)} = \frac{\sqrt{7\zeta(3)}}{2\pi} \frac{1}{\sqrt{\nu T_c V}} \sim \sqrt{\frac{\delta}{\Delta(0)}}. \quad (12.1)$$

If this value is small thermal fluctuations smear out the phase transition only in the narrow vicinity of the mean-field critical temperature. Beyond this region MFA gives the correct result for thermodynamic characteristics. What happens when $Gi_{(0)} \sim 1$, i.e. level spacing δ becomes comparable to $\Delta(0)$?

The positions of impurities, the shape of the surface and hence, the level spacing are the random values. That is why the answers on the proposed questions depend on the way of averaging over the ensembles of different random grains. Fortunately recently the new experimental technique of the Coulomb blockade was worked out [420,421], which allows one to measure the change in the energy of the grain as one electron is added. Important that nor shape nor impurities' locations are changed in such measurements. The small finite levels spacing and superconducting pairing effects can be observed on the background of the large but smooth Coulomb energy.

The energy of superconducting pairing in the MFA is determined by the self-consistency equation:

$$1 = \widehat{g} \sum_{\mathbf{k}} \frac{1}{2\sqrt{(E_{\mathbf{k}} - \mu)^2 + \Delta^2}}, \quad (12.2)$$

where $E_{\mathbf{k}}$ are the one-electron levels without accounting for the pairing interaction. If $\Delta \gg \delta$ one can substitute the summation by integration and to get the usual equation of the BCS theory. The sum has a logarithmic character: ultraviolet divergence is cut off at ω_D while the infrared divergence is cut off by Δ_{BCS}. When the spacing δ in the nanodrop becomes comparable to Δ_{BCS} and the chemical potential does not coincide with any level $E_{\mathbf{k}}$ the role of the cut-off parameter Δ_{BCS} plays $\delta + \Delta$ and with the further growth of the spacing the gap Δ tends zero.

It is worth noting that Eq. (12.2) treats the superconductivity in small grains within the self-consistent mean field approximation for the superconducting order parameter. Although this approximation works well for large systems, one should expect the quantum fluctuations of the order parameter to grow when the level spacing δ reaches Δ. Below will be presented a theory of superconductivity in ultrasmall grains which includes the effects of quantum fluctuations of the order parameter. We show that the corrections to the mean field results which are small in large grains, $\delta \ll \Delta$, become important in the opposite limit, $\delta \gg \Delta$.

Accounting for quantum fluctuations can be done by studying some physically observable value. The parameter $\Delta \sim \langle a_{\mathbf{k}} a_{-\mathbf{k}} \rangle$ has sense only in the limit $V \to \infty$. In the system of finite number of particles such an anomalous average is equal to zero due to phase fluctuations.

The superconducting gap Δ in Eq. (12.2) is not well-defined in the presence of quantum fluctuations. Therefore, we must first identify an *observable* physical quantity which characterizes the superconducting properties of small grains. The most convenient such quantity for our purposes is the ground state energy of the grain E_N as a function of the number of electrons N. More precisely, we study the so-called *parity effect* in ultrasmall grains, which is described quantitatively by parameter

$$\Delta_P = E_{2l+1} - \frac{E_{2l} + E_{2l+2}}{2}. \tag{12.3}$$

Here E_l is the ground state energy for a system with l electrons. Such physical value first was studied in the nuclear physics [423, 424]. In the ground state of a large superconducting grain with an odd number of electrons, one electron is unpaired and carries an additional energy $\Delta_P = \Delta$. This result is known in nuclear physics and was recently discussed in connection to superconducting grains in [425, 426]. The parity effect was demonstrated experimentally in [427, 428], where the Coulomb blockade phenomenon [429] in a superconducting grain was studied. In such an experiment the intervals between Coulomb blockade peaks in which the grain charge is odd shrink by an amount proportional to Δ_P.

We describe the grain by the following Hamiltonian:

$$\hat{H} = \sum_{k\sigma} \varepsilon_k a^\dagger_{k\sigma} a_{k\sigma} - \widehat{g} \sum_{kk'} a^\dagger_{k\uparrow} a^\dagger_{k\downarrow} a_{k'\downarrow} a_{k'\uparrow}. \tag{12.4}$$

Here k is an integer numbering the single particle energy levels ε_k, the average level spacing $\langle \varepsilon_{k+1} - \varepsilon_k \rangle = \delta$, operator $a_{k\sigma}$ annihilates an electron in state k with spin σ, and $\widehat{g}$ is the interaction constant. In Eq. (12.4) we assume zero magnetic field, so that the electron states can be chosen to be invariant under the time reversal transformation [236]. We include in Eq. (12.4) only the matrix elements of the interaction Hamiltonian responsible for the superconductivity; the contributions of the other terms are negligible in the weak coupling regime $\widehat{g}/\delta\varepsilon \ll 1$ we consider. Finally, we did not include in Eq. (12.4) the charging energy responsible for the Coulomb blockade, as its contribution to the ground-state energy is trivial.

In the absence of interactions, $\widehat{g} = 0$, the parity parameter Δ_P can be easily calculated. Indeed, the ground state energy E_N is found by summing up N lowest single-particle energy levels, twice degenerated in the absence of magnetic field.

This results in $E_{2l+1} = E_{2l} + \varepsilon_{l+1}$ and $E_{2l+2} = E_{2l} + 2\varepsilon_{l+1}$, since the last two electrons were placed at the same level. Substituting this into Eq. (12.3), we find that without the interactions $\Delta_P = 0$.

For weak interactions one can start with the first-order perturbation theory in $\widehat{g}$. In this approximation an electron in state k interacts only with an electron with the opposite spin in the same orbital state k. Thus when the "odd" $(2l+1)$-st electron is added to the grain, it is the only electron in the state $l+1$ and does not contribute to the interaction energy, $\delta E_{2l+1} = \delta E_{2l}$. The next, $(2l+2)$-st electron goes to the same orbital state and interacts with it: $\delta E_{2l+2} = \delta E_{2l+1} - \widehat{g}$. From Eq. (12.3) we now find

$$\Delta_P = \frac{\widehat{g}}{2}, \quad \text{at } \widehat{g} \to 0. \tag{12.5}$$

One should note that the result (12.5) is not quite satisfactory even in the weak coupling case $\widehat{g}/\delta \ll 1$. Indeed, the low-energy properties of a superconductor are usually completely described by the gap Δ. The interaction constant $\widehat{g}$ is related to the gap Δ in a way which depends on a particular microscopic model, so the result (12.5) cannot be directly compared with experiments.

This problem can be resolved by considering corrections of higher orders in $\widehat{g}$, which are known [228] to give rise to logarithmic renormalization of the interaction constant. In the leading-logarithm approximation the renormalized interaction constant is found [228] as

$$\tilde{g} = \frac{\widehat{g}}{1 - \frac{\widehat{g}}{\delta} \ln \frac{D_0}{D}}. \tag{12.6}$$

Here D_0 is the high-energy cut-off of our model, which has the physical meaning of Debye frequency, and $D \ll D_0$ is the low-energy cut-off. At zero temperature, $D \sim \delta$. Taking into account the relation between the gap in a large grain Δ and microscopic interaction constant $\widehat{g}$:

$$\Delta \sim D_0 e^{-\delta/\widehat{g}}, \tag{12.7}$$

we can exclude the constant D from Eq. (12.6) and within logarithmic accuracy find $\tilde{g} = \delta/\ln(\delta/\Delta)$. Finally, substituting the renormalized interaction constant into Eq. (12.5), we get

$$\Delta_P = \frac{\delta}{2 \ln \frac{\delta}{\Delta}}, \quad \Delta \ll \delta. \tag{12.8}$$

Unlike the first-order result (12.5), Δ_P is now expressed in terms of experimentally observable parameters Δ and δ rather than model-dependent interaction constant $\widehat{g}$.

Let us stress, that in a very small grain with $\delta \gg \Delta$, the mean field gap vanishes, and no parity effect is expected. On the contrary, our result (12.8) predicts that in small grains the parity effect is stronger than in the large ones. This behavior is due to the strong quantum fluctuations of the order parameter which persist even when its mean field value vanishes. The physics of the fluctuations of the order parameter is hidden in the renormalization procedure leading to Eq. (12.6).

The case $\delta \ll \Delta$ has been discussed in [426,430,431], where it was found that $\Delta_P = \Delta - \delta/2$. By comparing these two limits Matveev and Larkin concluded that a minimum should appear in Δ_P when the level spacing is of the order of Δ. The numerical analysis of the case $\delta \sim \Delta_{\mathrm{BCS}}$ [432,433] demonstrated that Δ_P is the smooth function of δ, what indicates on the absence of the phase transition at which Δ would turn zero.

12.2 Superconducting drops in system with quenched disorder

We will study below the phase transition in the macroscopic system of large number of granules. The situation turns analogous in superconductor with the quenched disorder. This is due to the fact, that even weak quenched disorder can change the character of the phase transition. Let us mention that such disorder is static by nature and manifests itself as the fluctuations of the GL functional parameters a, b, c.

Quenched disorder can be of two kinds. First of all it can be generated by already familiar us impurities of vacations randomly distributed over crystalline lattice. Such type of disorder is called universal. It is this disorder we undermined in the above discussion of the properties of impured superconductors. Another kind of quenched disorder can be caused by nonuniversal, or strongly correlated structure inhomogeneities of the initial crystalline lattice such as dislocations, accumulations of impurities, separation of the other phase grains. Granular superconductor is one more example of such nonuniversal disordered system.

Usually structure fluctuations are small and they weakly affect on properties of superconductor. It was mentioned above that such quenched disorder changes qualitatively the properties of the vortex structure breaking down Galilean invariance and resulting in the appearance of dry friction (collective pinning). Below, using the so-called method of optimal fluctuation, we will demonstrate that in the case when quenched disorder is stronger than the thermal or quantum fluctuations it results in the considerable smearing of the phase transition.

12.2.1 *The optimal fluctuations in the vicinity of T_c*[78]

Let us consider the superconducting system with quenched disorder which slowly varies the local critical temperature. It can be described by the GL functional with the coefficients being some random functions of the coordinates. The effect of such quenched disorder on the vortex pinning was already discussed for temperatures below T_c in section 5.2. Here we will suppose $T > \overline{T_c}$, i.e. $\overline{a} > 0$.

The realization of the situation with $a = \overline{a} + \delta a(\mathbf{r}) < 0$ in some large enough domain means that the superconducting drop can be formed there.

[78]In this section we base on the results of [434].

"Large enough" signifies that the domain must be so large that the proximity effect will not be able to suppress the superconductivity in it. That is why the probability to find such a domain is small. The problem of calculation of such probability (i.e. of the density of superconducting drops) is the particular case of the general problem of the optimal fluctuation.

Supposing the characteristic scale of the disorder much less than ξ, one can assume the distribution functions of such random values as the Gaussian ones. For instance for $a = \overline{a} + \delta a(\mathbf{r})$

$$P[\delta a(\mathbf{r})] = C \exp\left[-\frac{1}{W}\int [\delta a(\mathbf{r})]^2 \, d\mathbf{r}\right], \tag{12.9}$$

where W is the phenomenological parameter which can be determined experimentally measuring the critical current for $T < T_c$.

Let us define as the *superconducting drop* the domain with $a < 0$ and of so large size L that the superconducting gap Δ in it exceeds the level spacing δ (hence the modulus fluctuations $|\Delta|$ in it can be neglected). As a result the value $|\Psi|$ can be found from the GL equation with the depending on $\mathbf{r}$ coefficient $a(\mathbf{r})$ and then to calculate the drops density, i.e. the probability to find the drop at the point $\mathbf{r}$.

Let us first resolve the problem qualitatively. Due to the proximity effect the order parameter in the drop of the size L can be evaluated as: [79]

$$|\Psi|^2 = \frac{1}{b}\left(-a - \frac{1}{mL^2}\right), \tag{12.10}$$

so, according to (12.9), the probability to find such drop is

$$P(a, L) = C \exp\left[-\frac{(a-\overline{a})^2 L^D}{W}\right] \tag{12.11}$$

with $D = 3$. The same formula can be used for the drop formation in the superconducting film ($D = 2$) or wire ($D = 1$). Integrating over all drop sizes one can find

$$P(\Psi) = \int_0^\infty P(a, L) \, dL = \int C \exp\left[-\frac{\left(b|\Psi|^2 + \overline{a} + \frac{1}{mL^2}\right)^2 L^D}{W}\right] dL. \tag{12.12}$$

The extreme value of L for any dimensionality is the same:

[79] Here the coefficient 1 in the second term is chosen for the sake of convenience. One has to recognize that by our evaluation of $|\Delta|^2$ we cannot guarantee the coefficient since we suppose it to be invariant in all volume of a drop.

$$L \sim \frac{1}{\sqrt{m\left(b\left|\Psi\right|^{2}+\overline{a}\right)}}. \tag{12.13}$$

This gives for the probability distribution

$$P\left(\Psi\right) \sim \exp\left[-\frac{\left(\overline{a}+b\left|\Psi\right|^{2}\right)^{2-D/2}}{W}\right], \tag{12.14}$$

while for the drops density

$$\rho \sim \exp\left[-C\frac{\left(T-\overline{T_{c}}\right)^{2-D/2}}{W}\alpha^{2-D/2}\left(m\right)^{-D/2}\right]. \tag{12.15}$$

The mean value $\left\langle\left|\Psi\right|^{2}\right\rangle$ in the drop is equal to

$$\left\langle\left|\Psi\right|^{2}\right\rangle \sim \frac{W\left[\alpha\left(T-\overline{T_{c}}\right)\right]^{2-D/2}}{b}. \tag{12.16}$$

In the strong magnetic field the transverse drop size is determined by the magnetic length $L_H = \sqrt{\Phi_0/H}$. That is why in the formula (12.14)–(12.16) one has to substitute $D \to D-2$ and $W \to WL_H^2$.

Let us stress that in the obtained formula we can guarantee the functional dependencies only. The numerical factor C in the exponent in the demonstrated approach unfortunately remains unknown. In order to find the density of superconducting drops more precisely one has to optimize not only the drop size L, but the dependence $a\left(\mathbf{r}\right)$ itself too. This program was accomplished in [434] where was found that $C = 37/2^D$. Below the result obtained will be used in order to evaluate the temperature of the establishment of the global superconductivity overall the granular superconductor.

12.2.2 *Formation of the superconducting drops in magnetic fields* $H > \overline{H_{c2}}\left(0\right)$[80]

The applied magnetic field does not suppress the mechanism of superconducting drops formation in the medium with the randomly distributed interaction constant g described above. But in its presence appears another mechanism, which can prevail on the first one. This is due to the fact, that the upper critical field $H_{c2}(0)$ depends on the electron mean free path l_{tr}, i.e. impurity concentration, i.e. on the diffusion coefficient (see Eq. (A.5). At some small domain where occasionally the impurity concentration turns out to be higher than the average one, the local $H_{c2}(0) > \overline{H_{c2}(0)}$ and the formation of superconducting drop is possible.

[80] In this section we base on the results of [435].

There are two methods to accounting for the randomness of the $H_{c2}(0)$ and we will describe both of them. The first one is the usual microscopic Abrikosov–Gor'kov diagrammatic technique [228], but the impurity averaging has to be applied directly to calculation of the drop formation probability, not to the propagator L (like it was done in section 6.4). The second method is semi-phenomenological. The averaging is performed here in two stages. In the first one equation for the propagator L in dirty superconductor with the diffusion coefficient $\mathcal{D}$ as the random function of position is derived. At the second stage the averaging over the random $\delta\mathcal{D}(\mathbf{r})$ is performed. Close to the transition in the framework of the GL formalism this means accounting for the randomness in the coefficient $1/4m$ of the gradient term.

Two different methods correspond to different physical situations. The first one describes the universal fluctuations analogous to the mesoscopic fluctuations in resistance. Besides l_{tr} it does not involve new parameters: the disorder weaker than universal does not exist. The second approach is more general and it can describe the stronger disorder. Such disorder determines the critical current in the collective pinning phenomenon (see section 5.2).

One can consider a generic disordered system with a random diffusion coefficient:

$$\mathcal{D}(\mathbf{r}) = \overline{\mathcal{D}} + \delta\mathcal{D}(\mathbf{r}), \tag{12.17}$$

where a short-scale disorder characterized by the Gaussian white noise is introduced

$$\overline{\delta\mathcal{D}(\mathbf{r})\delta\mathcal{D}(\mathbf{r}')} = \overline{\mathcal{D}}^2 d^2 \delta\left(\mathbf{r}-\mathbf{r}'\right). \tag{12.18}$$

The randomness can be connected with localized and extended defects present in a superconductor. It characterizes the strength of disorder which can be connected with dislocation clusters, grain boundaries in polycrystal samples etc. Deep into the superconducting state, this randomness leads to collective pinning effects. Thus, the phenomenological constant d is directly connected with the pinning properties of a superconductor [73] and can be extracted by the order of value independently from experiments. For example, the critical current of a superconducting film [179] in the collective pinning regime is $j_c/j_{c0} \approx [H_{c2}(0)/H]\, d^2/L^2_{H_{c2}}$, where j_{c0} is the depairing current in zero field. Let us also note that a possible randomness in the BCS interaction constant g would lead to the same effects on the upper critical field. The proposed model can be realized in a system of superconducting grains.

Let us consider the superconductor in the vicinity of the BCS upper critical field $\overline{H_{c2}}(0)$ at zero temperature. The transition is controlled by the dimensionless parameter $\tilde{h} = \left[H - \overline{H_{c2}}(0)\right]/\overline{H_{c2}}(0)$. The region of strong fluctuations in an homogeneous superconductor is determined by the condition $\tilde{h} < Gi$. We suppose that the external magnetic field is such that $\tilde{h} < 1$ but lies outside the fluctuation region $\tilde{h} > Gi$.

Due to quenched disorder (structure fluctuations) superconducting drops appear even above $\overline{H_{c2}}(0)$. In this section we are interested by the $2D$ case so drops will be called below as “islands”. To find the distribution of such islands, we need to have the BCS type theory in the form not yet averaged over the quenched disorder. It can be written by means of decoupling the interaction term in the BCS Hamiltonian via a Hubbard–Stratonovich field Δ (see section 7.1). Then, the fermionic degrees of freedom can be integrated out and one gets an effective action for the superconducting order parameter. In the vicinity of the transition, an expansion on the order parameter is possible and we obtain the following action:

$$-S_\Delta = \int \Delta^*(x_1)\mathcal{L}^{-1}(x_1,x_2)\Delta(x_2)dx_1dx_2 \\ +\frac{1}{2}\int \Delta^*(x_1)\Delta^*(x_2)B(\{x_i\})\Delta(x_3)\Delta(x_4)\prod_{i=1}^{4} dx_i, \tag{12.19}$$

where we use $x = (\mathbf{r},t)$ and $dx = d^2\mathbf{r}\,dt$ for brevity. In Eq. (12.19), operator $\hat{L}$ is the fluctuation propagator: $\hat{L}_\omega = \left[-g^{-1}+\hat{\Pi}_\omega\right]^{-1}$, g is the BCS interaction constant, operator $\hat{\Pi}$ in the coordinate representation has the form $\Pi_\omega(\mathbf{r},\mathbf{r}') = T\sum_\varepsilon \Pi_\omega(\mathbf{r},\mathbf{r}';\varepsilon)$ where $\Pi_\omega(\mathbf{r},\mathbf{r}';\varepsilon) = G_\varepsilon(\mathbf{r},\mathbf{r}')G_{\omega-\varepsilon}(\mathbf{r},\mathbf{r}')$ and G_ε is the Matsubara Green's function. Let us emphasize that operator $\hat{\Pi} = \overline{\hat{\Pi}} + \delta\hat{\Pi}$ consists of a mean part and a random part $\delta\hat{\Pi}$ which is responsible for the effects under consideration. In the vicinity of the transition we can neglect the randomness in the Δ^4 term. The nonlinear operator B in Eq. (12.19) corresponds to the diagrams calculated close to $H_{c2}(0)$ explicitly by Maki [365] and Caroli *et al.* [78].

The saddle point approximation $\delta S/\delta\Delta(\mathbf{r},t) = 0$ results in the TDGL equation for a gapless superconductor [104]. When considering the spatial distribution of the islands we can disregard dynamic effects and consider the static form of the GL equation. Below the critical field $\overline{H_{c2}}(0)$ its nontrivial solutions describe the superconducting state.

If we neglect the randomness of the polarization operator there are no nontrivial solutions for the corresponding mean-field equation above the BCS upper critical field ($\tilde{h} > 0$). However, the random part of integral operator $\delta\hat{\Pi}$ possesses the positive and negative eigenvalues $\delta\epsilon$ and for that negative ones which are greater in absolute value than h the nontrivial solutions appear. They correspond to the appearance of local superconducting islands.

To find the distribution of the islands, one should find the distribution function of the eigenvalues for the random operator $\hat{\Pi}$:

$$\frac{1}{\nu}\int \Pi(\mathbf{r},\mathbf{r}')\,\psi(\mathbf{r}')d^2\mathbf{r}' = (\bar{\epsilon}+\delta\epsilon)\,\psi(\mathbf{r}), \tag{12.20}$$

where $\epsilon = \bar{\epsilon}+\delta\epsilon$ is the dimensionless eigenvalue of the polarization operator and ν ($\nu = m/2\pi$ for isotropic $2D$ metal) is the DOS per spin at the Fermi line. In the absence of a randomness in $\hat{\Pi}$, its spectrum is discrete and is parameterized by the Landau level indexes. The random part smears out the eigenvalues.

The "DOS" (density of eigenstates of $\hat{\Pi}$) can be introduced as

$$\rho(\epsilon) = \int \mathfrak{D}\{\delta\Pi\}\,\delta\left(\epsilon - \epsilon\left[\Pi\right]\right) w\left[\delta\Pi\right], \tag{12.21}$$

where $w\left[\delta\Pi\right]$ is the distribution function for the polarization operator which is supposed to be Gaussian with correlator $\overline{\delta\Pi(\mathbf{r_1},\mathbf{r_2})\delta\Pi^*(\mathbf{r_3},\mathbf{r_4})}$.

To find the DOS $\rho(\epsilon)$, we use the optimal fluctuation method [436]. This means that we evaluate the functional integral (12.21) in the saddle-point approximation. Let us note that the problem of finding $\rho(\epsilon)$ is equivalent to the problem of DOS of a particle in a random potential [437]. In the presence of an external magnetic field in $2D$, the problem is simplified, since the coordinate dependence of the wave functions is dictated by the magnetic field [434].

In the case of large enough $\delta\epsilon$, exceeding noticeably the smearing of the levels ($\overline{(\delta\epsilon)^2} \ll (\delta\epsilon)^2$), the solution has a form of rare islands. In the vicinity of a circularly symmetric island located at a point $\mathbf{r}_i$, the "wave function" can be taken in the following form:

$$\psi_i(\mathbf{r}) = \frac{1}{\sqrt{2\pi}L_H}\exp\left\{-\frac{(\mathbf{r}-\mathbf{r}_i)^2}{4L_H^2}\right\}. \tag{12.22}$$

When $(\delta\epsilon)^2 \ll 1$ the LLL approximation can be used and in the first order of the perturbation theory one can obtain:

$$\delta\epsilon = \frac{1}{\nu}\delta\Pi_{00} \equiv \frac{1}{\nu}\int \psi_i(\mathbf{r_1})\delta\Pi(\mathbf{r}_1,\mathbf{r}_2)\psi_i(\mathbf{r_2})d^2\mathbf{r}_1 d^2\mathbf{r}_2.$$

The distribution function reads:

$$\rho(\epsilon) \propto \exp\left[-\frac{(\delta\epsilon)^2}{2I}\right], \tag{12.23}$$

where $I = \overline{\delta\Pi_{00}^2}/\nu^2$.

In the case of a dirty metal, correlator I can be calculated with the help of the conventional cross diagram technique [369]. This yields the following estimate for the correlator:

$$I \sim I_1 \sim Gi_{(2)}^2 \tag{12.24}$$

and index "1" refers to the first model we consider (weak mesoscopic fluctuations in a dirty metal).

In the system with a short-scale randomness in the diffusion coefficient (12.18), we can calculate the correlator using the differential equation for the polarization operator which in the presence of an external magnetic filed has the form:

$$[\partial\,\mathcal{D}(\mathbf{r})\,\partial + i\varepsilon]\,\Pi(\mathbf{r},\mathbf{r}';\varepsilon) = (2\pi\nu)\;\delta(\mathbf{r}-\mathbf{r}'), \tag{12.25}$$

where $\partial = -i\nabla - 2e\mathbf{A}(\mathbf{r})$. One can solve Eq. (12.25) using a simple perturbation theory with respect to $\delta\mathcal{D}$. With the help of Eq. (12.18), we get the correlator and find the distribution function which can be written in the form (12.23) with

$$I = I_2 = \frac{1}{8\pi\,(L_H/d)^2}. \tag{12.26}$$

The modulus of the order parameter in a superconducting island is random and parameterized by random variable ϵ (see Eqs. (12.20) and (12.23)). The coordinate dependence of the order parameter is described by $\Delta_i(\mathbf{r}) = \Delta_0\psi_i(\mathbf{r})$, where $\psi_i(\mathbf{r})$ is defined in Eq. (12.22). Let us note that the typical size of a superconducting island is L_H. Using the explicit expression (see [366]) for the nonlinear operator B in Eq. (12.19), one can get the following "mean-field" value of the order parameter modulus for a circularly symmetric island:

$$|\Delta_0| = \sqrt{4\pi}\,\frac{\overline{\mathcal{D}}}{L_H^2}\sqrt{\delta\epsilon - \tilde{h}}. \tag{12.27}$$

Substituting $\delta\epsilon$ from Eqs. (12.27) to (12.23) one can find

$$\rho(\Delta_0) \propto \exp\left\{-\frac{1}{2I}\left[\tilde{h} + \frac{|\Delta_0|^2}{4\pi}\left(\frac{L_H^2}{\overline{\mathcal{D}}}\right)^2\right]^2\right\}. \tag{12.28}$$

The typical distance between the islands, being proportional to $\rho^{-1/2}$, is exponentially large $R \sim L_H \exp\left[\tilde{h}^2/4I\right]$. In the case of the universal disorder $I \sim Gi^2$ and in the region $Gi < \tilde{h}$ the density of such superconducting drops is exponentially small. Below we will us the results obtained here in order to evaluate that critical field when the global superconductivity establishes overall the granular superconductor at $T = 0$.

12.3 Exponential DOS tail in superconductor with quenched disorder[81]

Now let us discuss the effect opposite to the one discussed in the previous section: appearance of the exponential tail under the gap in superconductor with quenched disorder. In homogeneous superconductor in absence of magnetic field Δ plays not only the role of order parameter but characterizes the gap in the quasiparticle spectrum too. The presence in the system of the quenched disorder results in the possibility of appearance of the noticeable domains with superconductivity weaker than in average.

Let us start from the case when the random, fluctuating in space, physical value is the effective constant of the electron–electron interaction g [300]

[81] In this section we base on the results of [300].

$$\frac{1}{g} = \left\langle \frac{1}{g} \right\rangle + g_1, \tag{12.29}$$

where $g_1 \ll 1$. Namely g_1 is the random value distributed by the Gauss law:

$$P\left[g_1\left(\mathbf{r}\right)\right] = \exp\left(-\frac{1}{W}\int g_1^2\left(\mathbf{r}\right) d^D\mathbf{r}\right). \tag{12.30}$$

Close to critical temperature side by side with fluctuations of g_1 changes in space the local T_c, or, what is the same, the GL parameter a, like it was considered in section 12.2. But here we are interested in the case of $T = 0$, where the GL formalism is unapplicable and one has to operate in the framework of the Green functions method. In the case of a dirty superconductor the most convenient formulation of the Green function formalism is the so-called Uzadel equations [438], written on Green functions already integrated over the momenta $\widehat{g}^R\left(\omega\right)$. We are interested in the DOS which is expressed in terms of the normal Green function $g^R\left(\omega\right) = \nu(0)\sinh[\theta(\omega)]$:

$$\nu\left(\omega\right) = -\frac{1}{\pi}\operatorname{Im}\int \frac{d^D p}{\left(2\pi\right)^D} G^R\left(p,\omega\right) = -\frac{1}{\pi}\operatorname{Im} g^R\left(\omega\right), \tag{12.31}$$

where $\theta(\omega)$ is solution of the Uzadel equation

$$\Delta\sinh\theta - \omega\cosh\theta + \mathcal{D}\nabla^2\theta = 0. \tag{12.32}$$

In homogeneous superconductor, when $\Delta = const$,

$$\nu\left(\omega\right) = \frac{\theta\left(\omega - \Delta\right)\omega}{\sqrt{\omega^2 - \Delta^2}}. \tag{12.33}$$

In the case of nonhomogeneous superconductor one can present $\Delta = \overline{\Delta} + \Delta_1$, where Δ_1 is small random value $\Delta_1 = \overline{\Delta}g_1$. The expansion of Eq. (12.32) over Δ_1 gives the equation for the averaged over volume function $\overline{\theta(\omega)}$:

$$\overline{\Delta}\sinh\overline{\theta} - \omega\cosh\overline{\theta} - \Gamma\sinh\overline{\theta}\cosh\overline{\theta} = 0, \tag{12.34}$$

with

$$\Gamma = \int \frac{\left\langle \Delta_1^2 \right\rangle_k}{\mathcal{D}k^2} d^D k. \tag{12.35}$$

In superconductor with paramagnetic impurities the Uzadel equation for $\overline{\theta}$ has the same form (12.34) but with $\Gamma = 1/\tau_s$.

If $\Gamma \ll \overline{\Delta}$, the gap in the spectrum decreases to

$$\omega_e = \Delta \left[1 - \frac{3}{2} \left(\frac{\Gamma}{\Delta} \right)^{2/3} \right]. \tag{12.36}$$

Below this threshold, due to the formation in the domains where g_1 and Δ_1 are negative of the drops with the superconducting properties suppressed, the tail in DOS appears. The method of the optimal fluctuation gives the expression for the density of such drops and therefore, for the quasiparticle DOS [300]:

$$\rho(\omega) \approx \exp\left[-\frac{48\pi}{5W} \left(\sqrt{\frac{2}{3}\frac{\mathcal{D}}{\Delta}} \right)^{3/2} \left(\frac{\omega_e - \omega}{\Delta} \right)^{5/4} \right]. \tag{12.37}$$

In complete analogy with the case of the drop formation above T_c the numerical factor in the exponent is unusually large and it makes the tail almost unobservable.

As in the case of the formation of superconducting drops the tails in the DOS can appear in the result of the universal fluctuations of the electron wave functions. These fluctuations are important in some special cases when the gap in superconductor spectrum depends on the degree of disorder. Such a situation is realized in a superconductor with magnetic impurities, in Nb structures, in dirty $2D$ superconductors. In these cases tails appear in the DOS due to the universal fluctuations of the wave functions in random systems. Such fluctuations is convenient to account by the supersymmetry method [409]. Calculating the probability to find the drop with the small gap one has to optimize not only its size but the Green function parameters responsible for the randomness of the wave functions. That is why the DOS calculated by the supersymmetry method decreases with the energy increase more slowly than in the case of nonuniversal fluctuations. In this case DOS does not depend more on the interaction constant fluctuations but only on the value of the average conductance [439–442].

12.4 Josephson coupled superconducting grains and drops

As was already mentioned above the phase transition does not exist in the single grain. Even if the modulus of the order parameter fluctuates there weakly its phase undergoes the strong fluctuations. Nevertheless the superconducting grains (drops) array can undergo the phase transition due to the Josephson connection between them.

Let us consider the system of superconducting grains in the insulator matrix.[82] The interaction energy of neighboring superconducting grains is determined by the Josephson effect and is equal to

$$E(\phi_i - \phi_j) = E_J \cos(\phi_i - \phi_j). \tag{12.38}$$

Here [388]

[82]Such system may be obtained by the weak oxidation of the grain surface.

$$E_J = \frac{|\Delta|}{4e^2 R_N} \tanh \frac{|\Delta|}{2T}, \tag{12.39}$$

where ϕ_i is the phase of the condensate wave function in the i-th drop and R_N is the tunnel resistance between them in the normal state.

In the case when superconducting grains are settled in the normal metal matrix, or on the surface of the normal metal film, or drops appear in the superconducting medium above T_c as a result of the optimal fluctuation, the interaction between them appears due to the proximity effect. This interaction $E_J(R_{ij}) \sim \langle \Delta_1(0) \Delta_2(\mathbf{R}_{ij}) \rangle \sim \int L(\mathbf{k}, 0) \exp(i\mathbf{k}\mathbf{R}_{ij}) d\mathbf{k}$. At large distances it is determined by the logarithmic singularity of the integral and is equal to

$$E_J(R_{ij}) = \nu D |\Delta_0|^2 2\pi^2 \frac{\xi^2}{R_{ij}^D} \exp\left(-\frac{R_{ij}}{\xi(T)}\right). \tag{12.40}$$

For the case $D = 1$ this expression was firstly obtained in [443]. As it was demonstrated there, the accounting for repulsion in normal metal affects this result weakly.

It is interesting to discuss the dependence $E_J(R_{ij})$ in the magnetic field. As one can see from Eq. (9.58) E_J decreases exponentially at the distances $R_{ij} > L_H = \sqrt{\frac{\Phi_0}{H}}$. Nevertheless, this statement is correct only for the value $\overline{E_J}$ averaged over the impurity positions. The value $\overline{E_J^2}$ decays with distance only by the power law [369]:

$$\overline{E_J^2} \propto R^{-4}. \tag{12.41}$$

Such slow decrease is related to the fact that the square of Josephson energy in its diagrammatic presentation is a block of four one-electron Green functions. The average of a such product contains two Cooperons C and two diffusons D and if the former describe the transfer of the charge $2e$ the latter are neutral and are not affected by the magnetic field. Physically Eq. (12.41) means that before the averaging procedure $E_J \sim R^{-2}$, but it contains some quickly oscillating factor.

For each realization of disorder, there is a fixed spatial distribution of the superconducting islands. The interaction Hamiltonian for such a system can be obtained from Eq. (12.19) and has the standard form

$$\mathcal{H}_{\text{int}} = \sum_{ij} J_{ij} \cos(\phi_i - \phi_j + A_{ij}), \tag{12.42}$$

where J_{ij} is the Josephson energy of the interaction between islands i and j and A_{ij} is the phase-shift due to the magnetic field. The average value of the Josephson energy is $\overline{J(R)} \propto R^{-2} \exp[-R/L_H]$. When the typical distance between the islands is exponentially large compared to L_H, the average Josephson energy is negligible, while the dispersion of the Josephson energy decays as a power law only (see Eq. (12.41)).

12.5 Classical phase transition in granular superconductors

12.5.1 *XY-model for granular superconductor*

Let us consider now the peculiarities of phase transition in the system of grains connected by the interaction E_J. In the case when the transition temperature is high enough (the precise condition will be formulated below) this transition is described by the partition function

$$Z = \int \prod d\phi_i \exp\left[-\frac{E_J}{T}\left[1-\cos\left(\phi_i-\phi_j\right)\right]\right]. \tag{12.43}$$

It is interesting that this is the partition function of the classical XY model since

$$\cos\left(\phi_i-\phi_j\right) = \cos\phi_i\cos\phi_j + \sin\phi_i\sin\phi_j = n_{ix}n_{jx} + n_{iy}n_{jy}, \tag{12.44}$$

where $\mathbf{n}$ is unit $2D$ vector.

If the superconducting grains are separated by insulator the interaction of only nearest neighbors takes place. If the Josephson energy E_J is the same for all nearest neighbors the transition in the system happens at $T_c \sim E_J$.

In the $3D$ system below the transition temperature appears nonzero order parameter $\Psi \sim \overline{\mathbf{n}}$. In the MFA it is determined from the equation:

$$\overline{n} = \frac{\int \cos\phi \exp\left(\frac{E_J}{T}\overline{n}\cos\phi\right) d\phi}{\int \exp\left(\frac{E_J}{T}\overline{n}\cos\phi\right) d\phi}. \tag{12.45}$$

At the transition point $\overline{n} \to 0$, i.e.

$$T_c = E_J\overline{\cos^2\phi} = \frac{1}{2}E_J. \tag{12.46}$$

12.5.2 *GL description of the granular superconductor*

In the above consideration we took into account only the phase fluctuations, ignoring those of the modulus. Such an approximation can be justified only when $T_c \ll T_{c0}$, where T_{c0} is the critical temperature of the bulk superconducting material of which the grain is fabricated. In this case the MFA is applicable only for evaluations by the order of value and since the XY-model does not contain any small parameters $Gi \sim 1$.

In the opposite case, when $E_J(0) \gg T_{c0}$, the transition takes place in the vicinity of T_{c0} and it can be described in the framework of the GL functional. For example, in order to evaluate the shift of critical temperature and to describe phenomenologically fluctuations in this case one can consider the regular lattice of grains with the distance a between them. The full free energy of the system consists of the sum of free energy of each grain and energies of their Josephson interactions. The grain free energy and corresponding GL functional parameters we already calculated at the beginning of the book but the gradient term in those approximation was omitted. The GL parameters determined above for the isolated grain are still valid for the granular superconductor. Here we concentrate our attention on the Josephson part of the free energy which will permit to find the gradient coefficient C of the GL functional for the grain system.

The Josephson energy $E_{ij}(T)$ between the grains i and j in accordance with (12.39) in the vicinity of T_c can be written in the form

$$E_{ij}(T) = -\frac{\langle |\Delta_i^*||\Delta_j| \cos(\varphi_j - \varphi_i) \rangle}{8e^2 R_N T} = -\frac{\mathrm{Re} \langle \Delta_i^* \Delta_j \rangle}{8e^2 R_N T}. \tag{12.47}$$

When $E_J(0) \gg T_{c0}$ the values of the order parameter in the neighboring grains differ weakly and one can write

$$\Delta_j(a) = \Delta_j(0) + \frac{\partial \Delta}{\partial x} a. \tag{12.48}$$

Let us substitute Eqs.(12.47) to (12.48) and perform the averaging. In result

$$E_J(T) = \frac{|\Delta|^2}{8e^2 R_N T} + \frac{a^2}{8e^2 R_N T} \left(\frac{\partial \Delta}{\partial x} \right)^2. \tag{12.49}$$

The first term gives the small correction ($\sim Gi_{(2g)}$) to the main value of the grain free energy. Normalizing $E_J(T)$ per unit volume one can write the gradient coefficient of the GL functional (A.9) as:

$$\xi_g^2 = \frac{C}{\nu} = \frac{a^{2-D}}{8\nu e^2 R_N T}, \tag{12.50}$$

where D is the dimensionality of the grain lattice and ξ_g is the effective coherence length of granular superconductor. Equation (12.50) replaces Eqs. (A.2), (A.3) and (A.4) valid for homogeneous superconductor.

Now one can substitute Eqs. (A.2) to (2.87) and obtain $Gi_{(g)}$, while Eqs. (2.147) and (2.148) define the corresponding shift of T_c. Let us mention that for the $2D$ granular system $R_N = R_\square$. As a result the value of $Gi_{(2g)}$ number of the granular superconductor takes form of Eqs. (2.88) and (2.89).

12.5.3 *The broadening of superconducting transition by the quenched disorder*[83]

12.5.3.1 *Quenched disorder below T_c* In this section we consider how the random space dependence of the GL parameter $\delta a(\mathbf{r})$ smears the transition below the mean-field critical temperature. This smearing turns out to be much stronger than that one above transition [444]. As it follows from the GL equation the fluctuations of the coefficient a below T_c ($\overline{a} < 0$) result in the appearance of the random addition to the order parameter $\Psi = \overline{\Psi} + \delta\Psi$:

$$\delta\Psi(\mathbf{r}) = \overline{\Psi} \int \frac{d\mathbf{k}}{(2\pi)^3} \int \frac{\delta a(\mathbf{r}_1) \exp[i\mathbf{k}(\mathbf{r} - \mathbf{r}_1)]}{2|\overline{a}| + \mathbf{k}^2/4m} d\mathbf{r}_1. \tag{12.51}$$

[83] In this section we base on the results of [444].

Substituting in the GL equation (12.51) and averaging over δa with the distribution function (12.9) one can find the equation for the $\langle \Psi^2 \rangle$:

$$\left(\alpha(T_{c0} - T) - W \int \frac{d\mathbf{k}}{(2\pi)^3} \frac{1}{2\,|\overline{a}| + \mathbf{k}^2/4m} \right)$$
$$= b\overline{\Psi}^2 \left(1 + 3W \int \frac{d\mathbf{k}}{(2\pi)^3} \frac{1}{\left[2\,|\overline{a}| + \mathbf{k}^2/4m\right]^2} \right).$$

Our analysis demonstrated that due to inhomogeneities the critical temperature increases as

$$\delta T_c = T_c - T_{c0} = \frac{W}{\alpha} \int \frac{d\mathbf{k}}{(2\pi)^3} \frac{1}{2\,|\overline{a}| + \mathbf{k}^2/4m}. \tag{12.52}$$

Close to the new T_c the order parameter

$$\overline{\Psi}^2 = \alpha \frac{|T_c - T|}{b} \left(1 - \frac{7}{8\pi} \frac{W\,(4m)^{3/2}}{\sqrt{2\alpha\,|T_c - T|}} \right). \tag{12.53}$$

The heat capacity below the transition

$$C_s - C_N = (\Delta C)_0 \left(1 - \frac{3}{8\pi} \frac{W\,(4m)^{3/2}}{\sqrt{2\alpha\,|T_c - T|}} \right). \tag{12.54}$$

Thus the presence of the inhomogeneities results in the smearing of the transition in the range of temperatures

$$\frac{\delta T}{T_c} = Gi_{(\text{disorder})} \sim \frac{W^2 T^3}{\mathcal{D}^3}. \tag{12.55}$$

If this value exceeds Gi, hence the quenched disorder is more important than the thermal fluctuations. In the opposite case the quenched disorder is noticeable only in the narrow region in the vicinity of T_c, where nevertheless it changes the shape of the scaling singularity of the transition [445–447].

It is worth mentioning that the correction to heat capacity induced by inhomogeneities has the same temperature singularity as the correction generated by thermal fluctuations. Nevertheless these corrections have the opposite signs and if $W > (\nu T)^{-1}$ the inhomogeneities contribution dominates.

Let us stress that the considered granular system with the random distribution of the critical temperature presents the example of the quenched disorder. The case when $E_J(0) \gg T_{c0}$ is described by the GL functional and was already discussed above.

12.5.3.2 *Percolation superconductivity in granular and drop system*[84] The opposite to the previous case $E_J(0) \ll T_{c0}$ can be reduced to the XY-model with the random interaction $E_J(0)$ and we will shortly discuss it now. In Eqs. (12.40) and (12.39) the Josephson energy was supposed to be equal for all junctions. In reality E_J depends on the insulator thickness exponentially and if this value is random the Josephson energy will be random too and with very wide distribution function.

The Hamiltonian (12.42) describes a "frustrated" $2D$ XY-model with random bonds. The frustration comes both from the Josephson energy which is random and from the phase difference due to the magnetic field. At zero temperature such a system should show a glassy behavior if there are no effects capable of destroying phase coherence between the grains. In this case one can estimate the value of critical temperature using percolation theory [448].

The granules with $E_J \ll T$ one can assume as noninteracting, while those one for which $E_J \gg T$ is possible to treat as the unique granule with the homogeneous phase ϕ. The superconducting transition in such macroscopic system takes place when such strongly interacting granules form the infinite cluster. The distribution function of R and hence E_J in tunnel junctions are poorly known. Nevertheless it is the same percolation picture, that determines the average macroscopic resistance of the sample in the normal state [448]. As a result [434], one can find for T_c

$$T_c \sim E_J(R) \tag{12.56}$$

where now R has to be understood as the average resistance on the square of the granular film in $2D$ case or the resistance of the layer of thickness $L \sim d\left(\frac{d}{a}\right)^{0.9}$(this is the characteristic size of the typical cluster). Here d is the granule diameter while a is the interatomic distance. The analogous percolation problems were studied also in papers [449, 450] in relation respectively to conventional and HTS inhomogeneous systems.

Similar picture takes place in the dirty metallic superconductors, where the drops exist due to the optimal fluctuations. Nevertheless due to the large numerical factor in (12.15) the drops density is extremely small (in absence of magnetic field) and they give some small contribution to the increase of T_c.

Under the effect of strong magnetic field the transition is smeared not only due to the randomness of T_c (i.e. coefficient a) but also due to fluctuations of the diffusion coefficient. In order to demonstrate this let us consider the system of grains of the random size d with the same critical temperature separated by insulator barriers or placed on the $2D$ electron gas ($2DEG$) layer. The second critical field of each such grain depends on its size (see Eqs. (10.60) and (10.118)) [77, 234]. It is why in strong fields superconductivity survives in the smallest grains only.

[84] In this section we base on the results of [434]

The Josephson interaction between such grains is strongly frustrated $\overline{E_J^2} \gg \overline{E_J}^2$ (see Eqs. (12.40) and (12.41)) and the transition has the glassy character. The value of critical temperature of such system depends on the method of its experimental determination. Nevertheless as a good estimation one can accept that $T_c \sim \left(\overline{E_J^2}\right)^{1/2} \sim r^{-2}(H)$, where $r(H)$ is the average distance between grains which are still superconducting in field H. The dependence $T_c(H)$ is determined by the distribution function of the grains by their sizes. If the part of small grains is small the tail in $T_c(H)$ appears in high fields.

In accordance with this classical picture [368, 369] the global superconductivity at $T = 0$ has to remain up to the Clogston limit H_c^{spin}, when magnetic field acting on the electron spins destroys superconductivity due to the Zeeman effect $\left(g\mu_B H_c^{spin} = \Delta_0\right)$. Below we will demonstrate that quantum fluctuations at zero temperature can destroy the global superconductivity already at fields considerably lower than H_c^{spin}.

12.6 Quantum phase transition in granular superconductors

12.6.1 *Coulomb suppression of superconductivity in the array of tunnel coupled granules*[85]

In the case when the Josephson energy E_J is small the transition temperature of the array decreases and the quantum effects become of the first importance. The first consequence of it is the appearance of the quantum nature of the phase ϕ, which must be treated as the operator. This operator does not commute with the operator of the particles number corresponding to one granule. It turns out that instead of the Hamiltonian it is more convenient to deal with the action $\mathcal{S}[\phi((\tau))]$. The Coulomb interaction tends to fix the number of the particles at each granule, i.e. it increases the phase fluctuations. The Coulomb energy of the granules system is defined as

$$E_{\text{Coul}} = \frac{1}{2}\sum_{i,j} C_{ij} V_i V_j\,, \tag{12.57}$$

where C_{ij} is the capacity matrix. Due to the Josephson relation $V_i = 2e\partial\phi_i/\partial\tau$ and the action of the granules system is equal to

$$\mathcal{S} = \int_0^\beta d\tau \left(2e^2\sum_{i,j} C_{ij}\dot{\phi}_i\dot{\phi}_j + E_J\cos(\phi_i - \phi_j)\right)\,. \tag{12.58}$$

Here, as was done in the case of the classical transition, we assume the Josephson connection between the nearest neighbors only. This formula was derived microscopically by Efetov [451].

[85]In this section we base on the results of [451].

For the sake of simplicity let us restrict our consideration to the case with the diagonal capacity matrix (this case is realized when the plane granules lie at the metallic substrate). In this case

$$\mathcal{S} = \mathcal{S}_C + \mathcal{S}_J = \int_0^\beta d\tau \left(2e^2 \sum_i \frac{\dot{\phi}_i^2}{E_C} + E_J \cos(\phi_i - \phi_j) \right), \tag{12.59}$$

where $E_C = 2e^2/C$. When $T > E_C$ the first term does not play any important role and $T_c \sim E_J$. For the contact of the special shape the value of $E_J(H)$ can be suppressed applying magnetic filed. The corresponding T_c decreases and at some value of $E_J(H) \sim E_C$ it turns to zero. One can find the values of T_c and E_J in the MFA, as it was done above for the classical case:

$$E_J^{-1} = \mathcal{C}(\beta) = \int_0^\beta d\tau C(\tau), \tag{12.60}$$

where

$$C(\tau) = \langle \cos\phi(0) \cos\phi(\tau) \rangle = \frac{\int \exp(-\mathcal{S}_C[\phi(\tau)]) \cos\phi(0) \cos\phi(\tau) \mathfrak{D}\phi(\tau)}{\int \exp(-\mathcal{S}_C[\phi(\tau)]) \mathfrak{D}\phi(\tau)}.$$

The average $C(\tau)$ is calculated with the zero action $\mathcal{S}_0 = \mathcal{S}(E_J = 0)$. At $T = 0$

$$C(\tau) = \exp(-E_C \tau), \tag{12.61}$$

and Eq. (12.60) gives for the transition value $E_J^{(c)} \sim E_C$. Thus at zero temperature the critical value of the Josephson energy $E_J^{(c)}$ exists (it turns to be of the order of the Coulomb energy) such, that below it the global superconductivity cannot appear more. In the region of energies less than this critical one the considered array of so weakly coupled grains becomes the Mott insulator [451].

12.6.2 *Superconducting grains in the normal metal matrix*[86]

In this section, we will consider the quantum phase transition in the system of superconducting grains separated by the normal metal. For instance, the granules can be placed on the surface containing $2DEG$. In contrast to the system of grains in the insulator matrix (section 15.4) here $E_J \sim b^{-D}$ (b is the distance between grains) is determined by the proximity effect (12.40) instead of the Josephson tunneling (12.39). Another very interesting and important difference between the grains in metal with respect to insulator matrix is the weakening of the charge quantization effect. This is due to the fact, that the grain charge can flow through the conducting metal system, so it is not quantized. Nevertheless at low temperatures the traces of this quantization still remain. That is why some

[86] In this section we base on the results of [452–454].

critical grain density exists (the critical distance between granules b_c) when the critical temperature of the global transition $T_c \to 0$.

The dynamics of the phase $\varphi(\tau)$ of a single SC grain can be described by a simple imaginary-time action [455],

$$S_A[\varphi] = -\frac{G_A}{8\pi} \int\int_0^\beta d\tau d\tau' \, \frac{\cos[\varphi(\tau) - \varphi(\tau')]}{(\tau - \tau')^2}. \tag{12.62}$$

Here G_A is the Andreev subgap, normalized to $e^2/\hbar$, conductance (below will be given its definition).

At low values of $G_A \ll 1$ normal-superconducting transport across the interface is suppressed by the usual Coulomb blockade effect governed by the junction's charging energy $E_C = 2e^2/C$. This case was considered in the previous subsection.

For large G_A one can start from the Gaussian approximation for $S_A[\varphi(\tau)]$, that means to expand the $\cos[\varphi(\tau) - \varphi(\tau\prime)]$ in (12.62) up to the second order. Then

$$S_A + S_C = \frac{4}{\pi} \int_{-\infty}^{\infty} \left(|\omega| G_A + \omega^2/E_C \right) \varphi_\omega^2 d\omega. \tag{12.63}$$

For the correlator (12.60) introduced in the previous subsection one can find:

$$C(\tau) = e^{-\langle(\varphi(\tau)-\varphi(0))^2\rangle/2}/2. \tag{12.64}$$

Phase correlator can be calculated within the logarithmic accuracy and for $\tau > 1/E_C G_A$ gives

$$\langle |\varphi(0) - \varphi(\tau)|^2 \rangle = \frac{4}{\pi} \int_{-\infty}^{\infty} \frac{(1 - \cos \omega\tau)}{|\omega| G_A + \omega^2/E_C} d\omega \simeq \frac{8}{\pi G_A} \ln(\tau G_A E_C). \tag{12.65}$$

We have seen (Eq. (12.60)) that the critical concentration of grains is determined by the behavior of

$$\mathcal{C}(\beta) = \int_0^\beta C(\tau) d\tau \propto \beta^{1-4/\pi G_A} \tag{12.66}$$

Indeed, if $G_A > 4/\pi$, $\mathcal{C}(\beta \to \infty)$ diverges and for any small E_J at some temperature Eq. (12.60) is satisfied and the transition superconductor–normal metal takes place. It seems to indicate that at large G_A superconductivity is always stable at $T = 0$, in agreement with [456, 457]. Nevertheless all this consideration is wrong and is nothing else as the artifact of the Gaussian approximation (12.63).

Now we will demonstrate that in the case of a small density of granule the quantum fluctuations destroy the global superconductivity even at $T = 0$. Our goal will be to find such critical concentration b_c^{-D} when it happens. The crucial point is to note that the employed Gaussian approximation breaks down at a finite time scale t^*, due to renormalization of G_A. This renormalization is caused by the periodicity of the action $S_A[\varphi]$ as a functional of $\varphi(\tau)$, that is, in physical terms, by the charge quantization. This problem is analogous to the one studied by Kosterlitz [458]. Translating his results to the present case, one gets the renormalization group (RG) equation

$$dG_A(\zeta)/d\zeta = -4/\pi, \tag{12.67}$$

with $\zeta = \ln \omega_d t$. This equation is to be solved with the initial condition $G_A(0) = G_A$. As a result, at the time scale $\tau^* \sim E_C^{-1} e^{\pi G_A/4}$ the renormalized Andreev conductance $G_A(t^*)$ decays down to the value of order unity [452]. At longer time scales $C_0(\tau)$ decays approximately as τ^{-2}, so the integral $\mathcal{C}(0) \sim \tau^* \sim E_C^{-1} e^{\pi G_A/4}$. This means that the effective charge energy

$$\widetilde{E}_C = E_C e^{-\pi G_A/4} \tag{12.68}$$

Taking into account that according to Eq. (12.41) $E_J \sim b^{-D}$, where b is the average distance between grains, and using Eq. (12.60), one can compare $\widetilde{E}_C = E_J$ and obtain the critical distance between islands at the point of the superconductor-normal metal transition [452]:

$$b_c \sim d e^{\pi G_A/4D}, \tag{12.69}$$

where d is diameter of the grain.

Let us comment on the notion of Andreev's conductance G_A introduced above. It depends on the system properties. For the system of superconducting grains of diameter d immersed into a $3D$ metal with bulk resistivity ρ and the tunneling resistance $R_T \gg R_N = \rho/4\pi d$ (for $d > l_{\mathrm{tr}}$) $G_A = R_N/R_T^2$.

The $2D$ case turns out to be more sophisticated. Let us consider an array of small superconducting islands (of diameter d each) in contact with a thin film of dirty normal conductor with the dimensionless conductance (2.88) $G_\square \gg 1$. The average distance between neighboring islands is $b \gg d$, b^{-2} is the concentration of islands. The resistance R_T of the interface between each island and the film is low: $G_T = \hbar/e^2 R_T \gg 1$. The islands are thick enough, to prevent suppression of superconductivity inside them. The corresponding condition for the superconducting gap reads $\Delta \gg G_T/\nu V_i$, where V_i is the island's volume and ν is the DOS. In a simplified model [452] with sufficiently strong Cooper-channel repulsion in the film: $g_n \gg G_T/4\pi\nu G_\square$. The conductance $G_A = G_T^2/4\pi\nu G_\square g_n$ is the Andreev subgap conductance in the limit of weak proximity effect, valid under the condition $1 \ll G_T \ll 4\pi\nu g_n G_\square$ [452, 453].

Nevertheless as the most challenging it seems the case of high conductance $G_T \gg 4\pi\lambda_n G_\square$. In this case instead of the unique renormalization-group equation (12.67) one has to write down the system of large number of such equations for each grain [454]. The numerical solution of this system gives

$$\widetilde{E}_C = E_C e^{-2,5\pi\sqrt{G_\square}} \tag{12.70}$$

and in this case Eq. (12.69) gives for the critical grain concentration

$$b_c \sim d e^{1,25\pi\sqrt{G_\square}}. \tag{12.71}$$

12.6.3 *Phase transition in disordered superconducting film in strong magnetic field*[87]

In sections 12.4 and 12.5 the phase transitions in the systems of artificial granules were studied. Now we will consider how due to the quenched disorder the superconducting drops are formed in fields $H > \overline{H}_{c2}(0)$. We will see that their appearance changes the character of the traditional Abrikosov phase transition.

In homogeneous superconductor the transition temperature $T_c \to 0$ and the phase transition becomes quantum in the field $H = \overline{H}_{c2}(0)$ (see Eq. (A.5)) [77, 367, 459]. As it was demonstrated in section 12.2 for any field less than the Clogston limit in nonhomogeneous superconductor the domains appear with the value of local $H_{c2}(0) > \overline{H}_{c2}(0)$, where superconducting drops are formed. Without the accounting for quantum fluctuations at $T = 0$ the inter-drop proximity type interaction would result in the appearance of the global superconductivity in any field below the paramagnetic limit. Taking quantum fluctuations into account one obtains (as was done for the case of the artificial superconducting grains), that T_c goes to 0 at some finite density of the superconducting drops. As we already know this density exponentially decreases with the growth of magnetic field (see Eq. (12.28)). That is why the quantum phase transition at $T = 0$ takes place in the finite field which nevertheless can considerably exceed the value of $\overline{H}_{c2}(0)$.

To find the transition point, we use the action (12.19) which describes the dynamics of the superconducting order parameter. Let us present the superconducting order parameter in the following form

$$\Delta(\mathbf{r}, t) = \sum_i |\Delta_{0\,i}|\, \psi_i(\mathbf{r})\, e^{i\phi_i(t)}, \tag{12.72}$$

where ψ_i is defined in (12.22). We consider the islands in which the modulus of the order parameter is fixed by the static mean-field equations (12.27) and only the phase is allowed to fluctuate. In order to get the effective action describing the system of local superconducting islands and to express it in terms of their phases one can substitute the expression (12.72) to the (12.19). We know that the frequency enters in propagator $\mathcal{L}^{-1}$ as $|\omega|$, but here we need it in the real time representation:

$$|\omega|_t = \int |\omega|\, e^{i\omega(t-t')} d\omega = -\frac{1}{|t-t'|^2}. \tag{12.73}$$

[87] In this section we follow the paper [435].

Taking this into account one can obtain the desired action as:

$$\mathcal{S} = -\int dt \sum_i \left\{ \frac{G_A^{(i)}}{8\pi} \int dt' \frac{\cos\ [\phi_i(t) - \phi_i(t')]}{(t-t')^2} - \frac{1}{E_C}\left(\frac{\partial \phi_i}{\partial t}\right)^2 \right.$$
$$\left. + \sum_{j \neq i} \int dt' J_{ij}(t-t') \cos\ [\phi_i(t) - \phi_j(t') + A_{ij}] \right\}. \qquad (12.74)$$

(compare with Eq. (12.62)). The coefficient of the dissipative term $G_A^{(i)}$ is determined by the coefficient of $|\omega|$ in $\mathcal{L}^{-1}$ and, being related to the modulus of the order parameter (12.27), turns to be a random function:

$$G_A^{(i)} = \nu \left|\Delta_{0i}\right|^2 / e\overline{\mathcal{D}}H = G_\square |\Delta_{0i}|^2 \left(\frac{L_H^2}{\overline{\mathcal{D}}}\right)^2, \qquad (12.75)$$

where $G_\square = \nu\overline{\mathcal{D}} = \frac{e^2}{\hbar R_\square}$, and $R_\square$ is the average resistance per square of the film. Note that $G_A^{(i)}$, like $|\Delta_{0i}|$, is distributed according to Eq. (12.28), so its typical value is large: $G_A^{(i)} \sim G_\square \tilde{h} \gg 1$.

We keep the ω^2 term in the action. As we will see below, the effective charging energy appears only as a high-frequency cut-off. With logarithmic accuracy, the exact value of E_C is not important in our problem.

Let us integrate out the high-frequency degrees of freedom in the action. First, we consider strong enough magnetic fields so that the network of the superconducting islands is very dilute. This means that the distances between islands is exponentially large (see Eqs.(12.28) and (12.41)) and the average Josephson energy is exponentially small:

$$E_J \propto \frac{1}{R^2} \propto \rho(\Delta_0) \propto \exp\left\{ -\frac{1}{2I}\left[\tilde{h} + \frac{|\Delta_0|^2}{4\pi}\left(\frac{L_H^2}{\overline{\mathcal{D}}}\right)^2\right]^2 \right\}. \qquad (12.76)$$

In the domain $\omega \gg J$, only the first two terms in action (12.74) are important. In this case, the action can be written as a sum of single-island actions. The phases in different islands fluctuate independently. In this single-island action one can integrate out the high-frequency phase fluctuations using the renormalization group developed by Kosterlitz for a spin system with long-range interactions [458]. Since the first term in Eq. (12.74) is not Gaussian, coefficient η gets renormalized when integrating out fast variables. The corresponding renormalization group equation for $G_A\,(\ln E_C t)$ is identical to (12.67). The characteristic time t^* necessary for G_A to reach the value of the order of 1 determines the effective charge energy $\widetilde{E}_C$ (12.68).

In drops where G_A is not large fast fluctuations take place and these drops do not contribute to global superconductivity. At the same time the density of drops with very large G_A (i.e. with large Δ_{0i}) is exponentially small and the

Josephson interaction between them is very weak. The quantum phase transition to the glassy state in such system happens at zero temperature when the drops with the Josephson interaction $E_J \sim \widetilde{E}_C$ first time form the infinite cluster. Both E_J and $\widetilde{E}_C$ depend exponentially on Δ_0. Comparing the exponents in Eqs. (12.76) and (12.68), with the value of G_A defined by Eq. (12.75) one can find with the logarithmic accuracy:

$$\frac{1}{2I}\left[\tilde{h} + \frac{|\Delta_0|^2}{4\pi}\left(\frac{L_H^2}{\overline{\mathcal{D}}}\right)^2\right]^2 = \frac{\pi}{4} G_{\square} |\Delta_0|^2 \left(\frac{L_H^2}{\overline{\mathcal{D}}}\right)^2. \tag{12.77}$$

For large $\tilde{h}$ this equation does not have solution for any Δ_0,and the infinite cluster of the correlated drops just does not appear. The maximal $\tilde{h}$ when (12.77) still has solution is

$$\tilde{h}_{c2} = \frac{\pi^2}{2} G_{\square} I. \tag{12.78}$$

In the case of a weak mesoscopic disorder the fluctuations of the diffusion coefficient Eq. (12.17) are determined by the random impurities allocation and $I = I_1 \sim G_{\square}^{-2} \sim Gi^2$. At that according to Eqs. (12.24)–(12.26) the shift of the upper critical field is small: $\delta\tilde{h}_{c2} \sim G_{\square}^{-1} \sim Gi$. Let us note that this result may acquire some logarithmic corrections of the order of $\left(G_{\square}^{-1} \ln G_{\square}\right)$ which are, however, beyond the scope of our investigation. Within logarithmic accuracy, we cannot distinguish the phase fluctuation mesoscopic effects under consideration from the usual superconducting fluctuations of the order parameter modulus inside the Ginzburg region. The mesoscopic drops of such type were studied in [368,369], where it was demonstrated that their concentration is exponentially small. In spite of this fact without accounting for quantum fluctuations their presence results in the establishment of global superconductivity at $T = 0$. As we already demonstrated above the quantum fluctuations reduce the significance of such drops to the case when $\delta\tilde{h}_{c2} \sim Gi_{(2)}$. In this case their contribution cannot be separated from the contributions of other fluctuations.

The situation can be opposite in a system with quenched disorder, for example in a granular superconductor. In this case (strong disorder) the shift of the critical region can be large and it is described by Eqs. (12.78) with $I = I_2$ (see Eq. (12.26)). The phenomenological constant d measures the strength of disorder which can be connected with dislocation clusters, grain boundaries in polycrystal samples etc. Let us emphasize, that the pinning parameters in a superconductor are determined by d. For example, the critical current of a superconducting film [460] in the collective pinning regime is $j_c/j_{c0} \approx [H_{c2}(0)/H]\, d^2/L_{H_{c2}}^2$, where j_{c0} is the depairing current in zero field. Let us also note that a possible randomness in the BCS interaction constant g would lead to the same effects on the upper critical field.

Let us mention that result (12.78) can be obtained more formally by calculating correlator $C = \int_0^\infty \langle \exp\left[i\phi_j(t) - i\phi_j(0)\right]\rangle_S \, dt$. At the transition point, the correlator diverges. One can perform the virial expansion with respect to the density of islands in C. As in the theory of liquids and gases and in the theory of spin-glasses with RKKY interactions [461], the virial expansion cannot prove the very existence of the transition. However, it determines the transition point if there is one. The transition is defined as a point at which all terms of the virial expansion become of the same order. Comparing the contribution in correlator C from independent islands and the one from pairs, we find Eq. (12.78).

At finite temperatures, there is no phase transition in a strict sense. $2D$ superconductor placed in an arbitrary magnetic field always posses a finite, though exponentially small, resistance. One can define the upper critical field as a field at which a sharp fall in the resistance takes place. At very low temperatures, $\tilde{h}_{c2}$ is determined by Eq. (12.78). As the temperature increases, $\tilde{h}_{c2}(T)$ decreases very rapidly. First of all, at a finite temperature, the Josephson coupling decays exponentially at the distances larger than $\sqrt{\mathcal{D}/T}$. Second, the thermal fluctuations destroy the Josephson coupling at $J_{ij} \sim T$. The both effects lead the following estimate of the transition temperature $T_c(\tilde{h}) \sim T_{c0} \exp\left[-\tilde{h}^2/4I\right]$. Thus, in a relatively wide region $\sqrt{I} < \tilde{h} < \delta\tilde{h}_{c2}(0)$, the critical temperature depends on the external magnetic field exponentially.

The increase of H_{c2} at low temperatures has been observed in a number of experiments [462]. The discussed mechanism can give a possible explanation for the effect. The critical field H_{c2} of the granular superconductor is determined by the formation of the superconducting drops in result of the optimal fluctuations. The density of such drops is exponentially small. It is why the exponentially small temperature destroys Josephson connections between them and at higher temperatures the dependence $H_{c2}(T)$ is defined by the classical formula [34, 77] with small fluctuation corrections (see section 2.5). Let us mention again that the universal fluctuations shift the critical field H_{c2} by the value proportional Gi. In this case the effect of the drops formation is indistinguishable with the effect of other quantum fluctuations. The behavior of $H_{c2}(T)$ depends in this case on the way of the definition of H_{c2} in the case of the smeared phase transition. Consequently the growth of $H_{c2}(T)$ should be observed only in those samples, where the disorder is strong and the critical current is relatively large.

13

FLUCTUATIONS IN JOSEPHSON JUNCTIONS

The Josephson effect, like superconductivity itself, represents a macroscopic quantum phenomenon, which appears due to the fact that a macroscopically large number of Cooper pairs stay at the same quantum state. The study of weak superconductivity, which has been generated by the prominent discovery of Josephson [463], today turned into the powerful industry of the fundamental research and applied science [375]. Amongst other investigations, the latter requires detailed knowledge of the effect of fluctuations on the properties of Josephson junctions. Adepts of the former found here the intriguing interplay between the fluctuations of different nature: thermal and quantum, fluctuations of phase and modulus of the order parameter. Their investigations required to develop the ideas of quantum mechanics for the systems with friction, etc.

Actually in 1962 Josephson theoretically predicted two different new phenomena which today are called by his name. The first consists in possibility of the direct current to path through the tunneling barrier between two superconductors without any voltage applied and it is called the stationary Josephson effect. The second one, nonstationary Josephson effect, consists in the appearance of the alternate current in such system when voltage is applied. It is accompanied by the emission of the electromagnetic radiation. Fluctuations destroy both of them: they lead to the decay of stationary Josephson current and smear out the monochromatic line of the Josephson junction emission. Below we will present the variety of fluctuation phenomena in Josephson systems.

13.1 General properties of a Josephson junction

13.1.1 *Stationary Josephson effect*

The Josephson effect takes place in superconducting circuit with a weak link. The role of such weak link can play a constriction, tunnel or normal metal barrier. Let us demonstrate the manifestation of the stationary Josephson effect on the simple example of constriction in superconducting film. We will follow here the method formulated by Aslamazov and Larkin [464], based on the GL description of the Josephson junction and applicable to the constriction of an arbitrary shape. Let us suppose that two superconducting banks (massive electrodes) are separated by the small bridge of length $L \ll \xi$ and cross-section area $S \ll \xi^2$ (see Fig. 13.1). Moreover, let us assume that the same inequalities are valid for the magnetic field penetration depth: $L \ll \lambda, S \ll \lambda^2$.

The reduced order parameter $\psi(\mathbf{r}) = \Psi(\mathbf{r})/\Psi_{\text{bulk}}$ in this bridge has to obey to the GL equation (2.22). Due to the condition $L \ll \lambda$ the influence of the

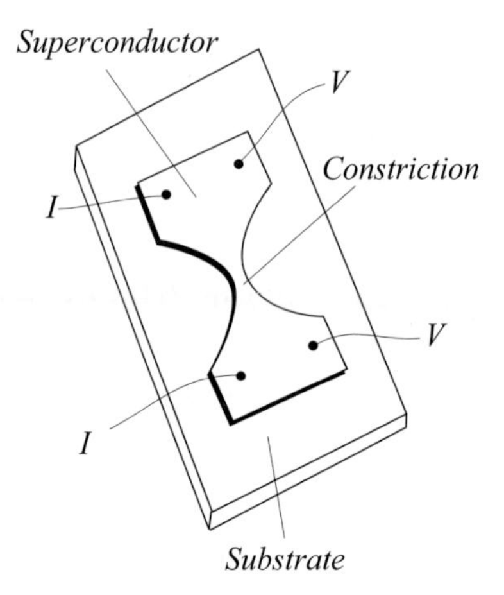

FIG. 13.1. The sketch of the Josephson junction.

intrinsic magnetic field on the current through the contact can be disregarded. Near to the constriction the order parameter $\psi(x)$ varies strongly within the length scale L, that is much smaller than ξ. Hence the main term in Eq. (2.22) is the first one ($\xi^2\psi'' \sim \psi\left(\xi^2/L^2\right) \gg \psi, \psi^3$) and in the first approximation the GL equation (2.22) corresponds to:

$$\nabla^2\psi(\mathbf{r}) = 0. \tag{13.1}$$

One can assume the junction banks to be in equilibrium, so $|\psi| = 1$ in both of them, while the order parameter phases χ_1 and χ_2 can be different. On the surface of the superconductor the boundary condition gives $\partial\psi(\mathbf{r})/\partial\mathbf{n} = 0$.

When $|\mathbf{r}| \gtrsim \xi$ the nonlinear term of the GL equation is already important. We assume that in this region the current density is small compared with the critical one, and therefore here $|\psi(\mathbf{r})| = 1$.

The general solution of Eq. (13.1) can be written in the form of a sum of two terms

$$\psi(\mathbf{r}) = f(\mathbf{r})\exp(i\chi_1) + (1 - f(\mathbf{r}))\exp(i\chi_2), \tag{13.2}$$

where $f(\mathbf{r})$ is the solution of the Laplace equation (13.1) which tends asymptotically to unity when the distance from the contact increases towards one of superconductors, and to zero with increasing distance towards the other. The phases χ_1 and χ_2 do not depend on coordinates, but can depend on the time.

Assuming that the mean free path of the electrons at the contact is much smaller than the contact dimensions, and that the temperature is close to critical, we get from Eq. (2.17) for the current density the expression

$$j = -\sigma\nabla\varphi - 2ieC\left[\Psi(\mathbf{r})\nabla\Psi^*(\mathbf{r}) - \Psi^*(\mathbf{r})\nabla\Psi(\mathbf{r})\right], \tag{13.3}$$

where the first term is the density of the normal current (σ is conductivity of the metal in the normal state; φ is scalar potential). The constant C can be expressed in terms of the coherence length:

$$C = \frac{1}{4m} = \alpha T_c \xi^2. \tag{13.4}$$

Applying the electroneutrality condition

$$\nabla \cdot \mathbf{j} = 0 \tag{13.5}$$

to Eq. (13.3) one can see that the scalar potential φ satisfies Eq. (13.1) and the same boundary condition on the surface of the superconductor as $\psi(\mathbf{r})$. It therefore can be expressed in terms of the function $f(\mathbf{r})$ namely $\varphi(\mathbf{r}) = f(\mathbf{r})V + \varphi_{-\infty}$, where $V = \varphi_{\infty} - \varphi_{-\infty}$ is the voltage on the contact. Substituting this expression and expression (13.2) to formula (13.3), we get

$$j = -\nabla f(\mathbf{r}) \left[\sigma V - C|\Psi_{\text{bulk}}|^2 \sin(\chi_1 - \chi_2)\right]. \tag{13.6}$$

In a superconductor far from the contact $\Psi(\mathbf{r})$ and ϕ do not depend on the coordinates yet can be dependent on time. Taking into account this fact in the TDGL equation (3.1)[88] one can see that $d\chi_{1,2}/dt = (1/2e)\,\phi_{\pm\infty}$. Integrating expression (13.6) for the current density over the contact cross-section, we obtain

$$IR = V + I_c R \sin\left(2e \int V dt\right), \tag{13.7}$$

where I is the fixed total current passing through the contact, R is the contact resistance being in the normal state, and I_c is the critical current of the contact:

$$I_c = \frac{|\Psi_{\text{bulk}}|^2}{4m\sigma R}. \tag{13.8}$$

In the case of constriction in the form of a wire

$$I_c^{(w)} = \frac{eS}{4mL}|\Psi_{\text{bulk}}|^2. \tag{13.9}$$

When the contact has the form of single void hyperboloid of revolution with a neck dimension a and the aperture angle 2θ, the current is

$$I_c^{(hb)} = \frac{\pi a}{8m}|\Psi_{\text{bulk}}|^2 \tan\frac{\theta}{2}. \tag{13.10}$$

The important consequence of Eqs. (13.7)–(13.6) is that the current can be different from zero when $\chi_1 - \chi_2 \neq 0$ even when there is no voltage applied to junction ($V = 0$). In this case

[88] Here we do not consider the effect of fluctuations and the Langevin forces in Eq. (3.1) are omitted.

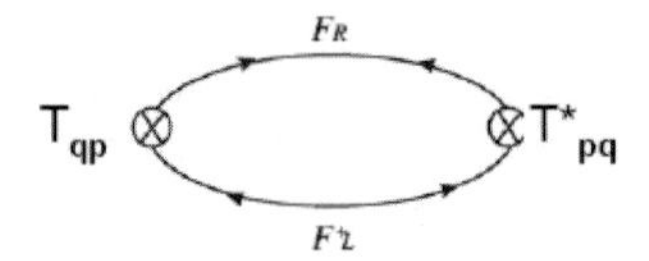

FIG. 13.2. The diagram for the calculation of the Josephson current.

$$I_J = I_c \sin(\chi_1 - \chi_2). \tag{13.11}$$

Equation (13.11) expresses the essence of the so-called stationary Josephson effect: constant current can flows through the superconducting circuit with a weak link[89] when the phase difference between its banks is created.

The GL energy of such a constriction can be calculated by the direct integration in Eq. (2.4) with the order parameter (13.2) that gives:

$$\Delta F(\phi) = E_J(1 - \cos\phi), \tag{13.12}$$

where $\phi = \chi_1 - \chi_2$ and

$$E_J = \frac{I_c}{2e}. \tag{13.13}$$

Kulik and Omelyanchuk [371] demonstrated that far below T_c the phase dependence of the Josephson current of constriction $I_J(\phi)$ contains many harmonics. With respect to the Josephson current of tunnel junction it was found [388]to appear in the sinusoidal form following from Eqs. (13.11)–(13.12) for all temperatures below T_c. Ambegaokar and Baratoff [388] also calculated the temperature dependence of the Josephson critical current I_c in the framework of the microscopic theory of superconductivity.

The Josephson current can be expressed in terms of the correlator of the matrix Green functions (7.10) [372, 399]. The stationary Josephson current is represented diagrammatically as the loop of two F-functions with the matrix elements of the tunneling Hamiltonian in the vertices (see Fig. 13.2).

We will need below the explicit expression for the value of the critical current, which in the approximation of the tunneling Hamiltonian can be expressed in the form:

$$I_c = 4e\sum_{\mathbf{p},\mathbf{q}} |T_{\mathbf{pq}}|^2\, T\sum_{\varepsilon_n} F_L^+(p,\varepsilon_n)\, F_R(q,-\varepsilon_n), \tag{13.14}$$

or

$$I_c = \frac{\pi}{eR_n} T\sum_{\varepsilon_n} \frac{\Delta_L}{\sqrt{\Delta_L^2+\varepsilon_n^2}}\frac{\Delta_R}{\sqrt{\Delta_R^2+\varepsilon_n^2}}, \tag{13.15}$$

where R_n is the normal state resistance of the tunnel junction per unit area:

[89]In our case this is constriction, but this could be the thin tunnel barrier, point contact, etc.

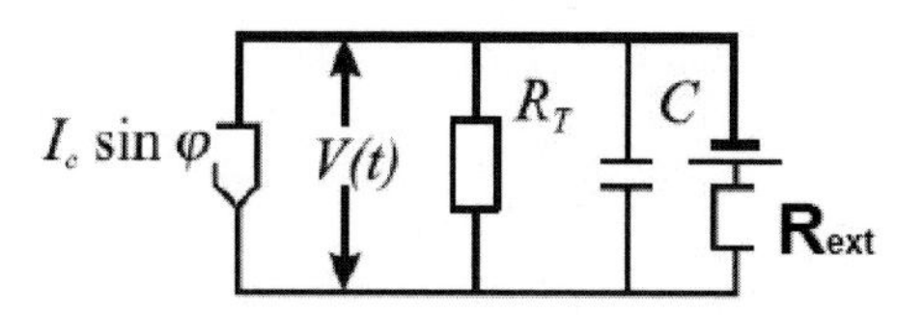

FIG. 13.3. Equivalent circuit.

$$\frac{1}{R_n} = 4\pi e^2 \sum_{\mathbf{p},\mathbf{q}} |T_{\mathbf{pq}}|^2 \, \delta\left(E_{\mathbf{p}} - \mu_1\right) \delta\left(E_{\mathbf{q}} - \mu_2\right) = 4\pi e^2 \nu_1 \nu_2 \left\langle |T_{\mathbf{pq}}|^2 \right\rangle. \quad (13.16)$$

In the case of a symmetric junction of two superconductors separated by thin insulator barrier the expression (13.15) is reduced to

$$I_{\mathrm{c}} = \frac{\pi \Delta(T)}{2eR_n} \tanh \frac{\Delta(T)}{2T}. \quad (13.17)$$

From this result and (13.8) it is seen that close to T_{c} the critical current growth linearly with the decrease of temperature.

13.1.2 *Nonstationary Josephson effect*

Non-stationary Josephson effect manifests itself when a voltage V is applied to a weak link of a superconducting circuit. A current biased Josephson tunnel junction can be represented by the electronic equivalent circuit of Fig. 13.3. Here R_{ext} is the resistance of the external circuit, C is the intrinsic junction capacitance and R_T is the junction resistance assumed as a linear ohmic element. The last assumption is the basis of the resistively and capacitively shunted junction (RCSJ) model. In this widely employed scheme, the actual nonlinearity of the quasiparticle branch in the I–V characteristics stemming from the superconducting gap is neglected. Depending on the relation between R_{ext} and R_T the Josephson junction can be in the situation with the preset voltage ($R_{\mathrm{ext}} \ll R_T$) or preset current ($R_T \ll R_{\mathrm{ext}}$).

Using the Josephson constitutive relations, the current balance in the circuit can be accounted by the following equation [375]

$$I = I_{\mathrm{c}} \sin\phi + \frac{V}{R_T} + C\frac{dV}{dt}, \quad (13.18)$$

where

$$\phi = 2e \int V dt \quad (13.19)$$

is the relative phase between the two superconductors ($\phi = \chi_L - \chi_R$).

The appearance in the Josephson junction of the alternate current when the constant voltage is applied leads to emission of the electromagnetic radiation with the frequency

$$\omega_J = 2e\overline{V}, \tag{13.20}$$

where $\overline{V}$ is the voltage at the barrier averaged over time. Such emission indeed was observed in experiment (see [375]), although in practice is more convenient to observe the related phenomenon, the so-called Shapiro steps.[90] Since both frequency and voltage can be measured with high accuracy this relation allowed one to define more precisely the value of electron charge. The modern standard of voltage is based on the Josephson effect.

13.2 Fluctuation broadening of the emission line

13.2.1 *Thermal fluctuations of the voltage*[91]

The important consequence of the non-stationary Josephson effect is the monochromaticity of the electromagnetic field emitted by junction. The natural mechanism of the corresponding line broadening is related to the thermal fluctuations of the voltage in the circuit containing the Josephson junction. Due to the fundamental relation (13.20) the voltage fluctuations are transformed in the fluctuations of the frequency, that results in the particular frequency modulation of the emitted signal.

Let us present the voltage applied to the Josephson junction as the sum of the constant part V_0 and the fluctuating one $V_1(t)$. We restrict our consideration here to the limit of high temperatures: $T \gg \omega, eV$. More general description of the problem is presented, for instance, in [466]. The Fourier component of the random function correlator $\overline{V_1(t)\,V_1(t+\tau)}$, calculated at various instants, but which differs by the same interval τ, can be related to the circuit impedance by means of the Nyquist theorem [467]:

$$N(\omega) = \frac{1}{2\pi}\int_{-\infty}^{\infty} dt e^{-i\omega t}\overline{V_1(t)\,V_1(t+\tau)} = \frac{T^*}{\pi}\operatorname{Re} Z(\omega), \tag{13.21}$$

where $Z(\omega)$ is the impedance of the Josephson junction equivalent circuit (see Fig. 13.3)

$$Z^{-1}(\omega) = -i\omega C + R_T^{-1} + R_{\text{ext}}^{-1} \tag{13.22}$$

while T^* is some characteristic temperature. When both the Josephson junction and the external part of the circuit are at the same temperature T then it has to be used in Eq. (13.21). The problem is that in practice the temperatures of a junction T and an external circuit T_{ext} can be substantially different.

[90] The steps appearing at the voltage–current characteristic of the Josephson junction under the effect of its irradiation by the electromagnetic wave of frequency (13.20).

[91] In this section we base on the results of [465].

The question which temperature one has to use in Eq. (13.21) requires special consideration.

In order to find the appropriate temperature let us use the equivalent scheme shown at Fig. 13.3. The first Kirchoff rule gives the balance of currents

$$-i\omega CV + I_c \sin\phi + \frac{V}{R_T} + \zeta_{\text{int}} = \frac{\mathcal{E} - V}{R_{\text{ext}}} + \zeta_{\text{ext}}, \tag{13.23}$$

where ζ_{int} and ζ_{ext} are the Langevin currents modeling the noises in Josephson junction and in the external part of the circuit.

According to the fluctuation–dissipation theorem

$$\langle \zeta_{\text{int}}(t)\, \zeta^*_{\text{int}}(t') \rangle = \frac{T}{R_T} \delta(t - t'), \tag{13.24}$$

and

$$\langle \zeta_{\text{ext}}(t)\, \zeta^*_{\text{ext}}(t') \rangle = \frac{T_{\text{ext}}}{R_{\text{ext}}} \delta(t - t'). \tag{13.25}$$

Averaging Eq. (13.23) over time one can find

$$\overline{I_c \sin 2e \int \left(\overline{V} + V_1(t)\right) dt} + \frac{\overline{V}}{R_T} = \frac{\mathcal{E} - \overline{V}}{R_{\text{ext}}}. \tag{13.26}$$

In the first approximation,[92] from (13.23) one can find:

$$V_1(t) = \frac{\zeta_{\text{ext}}(t) - \zeta_{\text{int}}(t)}{Z^{-1}(\omega)}. \tag{13.27}$$

Let us define

$$\operatorname{Re} Z^{-1}(\omega) = \left(\frac{1}{R_T} + \frac{1}{R_{\text{ext}}} \right) = \frac{1}{R^*}. \tag{13.28}$$

Now we insert (13.27) into the expression (13.21) and determine the effective temperature T^*:

$$N(\omega) = \frac{1}{2\pi} |Z(\omega)|^2 \left(\frac{T}{R_T} + \frac{T_{\text{ext}}}{R_{\text{ext}}} \right) = \frac{1}{2\pi} \frac{T/R_T + T_{\text{ext}}/R_{\text{ext}}}{(R^*)^{-2} + (\omega C)^2} = \frac{T^*}{\pi} \operatorname{Re} Z(\omega), \tag{13.29}$$

where

$$T^* = R^* \left(\frac{T}{R_T} + \frac{T_{\text{ext}}}{R_{\text{ext}}} \right). \tag{13.30}$$

One can see that T^* can be much higher than the cryostat temperature T.

[92]In this approximation we neglect the nonzero contribution of the oscillating Josephson current. The accounting for it in (13.23) would result in the appearance of the higher harmonics in the emission of the junction.

The spectrum, namely the Fourier transform of the time dependent current flowing through the junction is:

$$I(t) = I_c \sin\phi(t) = I_c \sin\left(2eV_0 t + 2e\int V_1(t')\,dt'\right). \tag{13.31}$$

It is centered at ω_J. In accordance with the relation (13.20) the voltage fluctuations manifest themselves in the smearing of the emission line with some width $\Delta\omega = \Gamma$. Our goal is to find the value of Γ. For that we consider the function $K(\omega)$

$$K(\omega) = \frac{2}{\pi}\int_{-\infty}^{\infty} e^{-i\omega\tau}\overline{I(t)\,I(t+\tau)}d\tau. \tag{13.32}$$

Substitution of the current $I(t)$ from Eq. (13.31) one can find

$$K(\omega) \sim \int_{-\infty}^{\infty} d\tau e^{-i(\omega-\omega_J)\tau}\overline{\exp\left[2ie\int_0^{\tau} V_1(t+t')\,dt'\right]}, \tag{13.33}$$

where $\omega_J = 2eV$. Equation (13.33) is proportional to the intensity of the emission in the frequency interval $(\omega, \omega + d\omega)$.

When $V_1(t) = 0$ Eq. (13.33) gives $(\omega > 0)$:

$$K(\omega) = \delta(\omega - \omega_J), \tag{13.34}$$

i.e. the emission line width is zero. As demonstrated in [465] the presence of the deterministic part in $V(t)$ does not result in the broadening of the line but leads only to appearance of the higher harmonics in the signal.

Let us return to the study of the emission line shape. After averaging the exponent in Eq. (13.33) one notice that for large time intervals τ the thermal noise $V_1(t)$ in the approximation (13.27) is of the Gaussian type. This means that the average product of any number of even multipliers V can be divided into the sum of the products of all possible pair averages. For such Gaussian averaging one can use the formula

$$\overline{e^{if(t)}} = \exp\left[-\overline{[f(t)]^2}/2\right] \tag{13.35}$$

and obtain

$$\begin{aligned} K(\omega) &\sim \int_{-\infty}^{\infty} d\tau e^{-i(\omega-\omega_J)\tau}\exp\left[-2e^2\int_0^{\tau}\int_0^{\tau} dt_1 dt_2 \overline{V_1(t+t_1)\,V_1(t+t_2)}\right] \\ &= \int_{-\infty}^{\infty} d\tau e^{-i(\omega-\omega_J)\tau}\exp\left[-8e^2\int_0^{\tau} d\omega N(\omega)\frac{\sin^2\omega\tau/2}{\omega^2}\right]. \end{aligned} \tag{13.36}$$

Substituting here Eq. (13.29) for $N(\omega)$ one can find

$$K(\omega) \sim \int_{-\infty}^{\infty} d\tau e^{-i(\omega-\omega_J)\tau} \exp\left\{-4e^2 T^* R^* \left[|\tau| + R^* C \left(e^{-|\tau|/R^* C} - 1\right)\right]\right\}$$
$$= \Phi\left(1, x+1+i(\omega-\omega_J) R^* C, x\right) \mathrm{Re} \frac{2R^* C}{x + i(\omega-\omega_J) R^* C}, \qquad (13.37)$$

where $x = 4e^2 R^{*2} T^* C$ and $\Phi(1, y, x)$ is the degenerated hypergeometric function.

In order to analyze this expression let us notice that the characteristic time τ in the integral (13.37) is of the order of the inverse width of the line. Let us introduce the "quality factor" of the Josephson junction for the RCSJ model

$$Q = \Gamma R^* C. \qquad (13.38)$$

It is identical to $\beta_c^{1/2}$, where β_c is the damping parameter introduced by Stewart and McCumber [468, 469] frequently used in the Josephson junctions physics. When the quality factor is low, $Q \ll 1$, Eq. (13.37) is reduced to

$$K(\omega) \sim \frac{1}{\pi} \frac{\Gamma}{(\omega-\omega_J)^2 + \Gamma^2}. \qquad (13.39)$$

Hence, in this case, the shape of the line turns out to be a Lorentzian with the width

$$\Gamma = 4e^2 R^* T^*. \qquad (13.40)$$

For fluctuations of the low frequencies $\omega \sim \Gamma$ the "capacity resistance" $(\Gamma C)^{-1}$ is large and the value of capacity does not enter in the answer.

In the opposite limit of a high quality factor, $Q \gg 1$, Eq. (13.37) gives

$$K(\omega) \sim \frac{1}{\sqrt{\pi}\widetilde{\Gamma}} \exp\left[-\left(\frac{\omega-\omega_J}{\widetilde{\Gamma}}\right)^2\right], \widetilde{\Gamma} = 2e\sqrt{\frac{2T^*}{C}}. \qquad (13.41)$$

In this case the line shape is of the Gauss type. Its width is determined by the junction capacity and depends on the effective resistance only by means of T^*:

$$K(V_1) \sim \frac{1}{\sqrt{\pi}\widetilde{\Gamma}} \exp\left(-\frac{CV_1^2}{2T^*}\right). \qquad (13.42)$$

Let us mention that Eq. (13.41) could be written down by simple statistical averaging if one notices that in the considered limit the fluctuations of V_1 are distributed following the Boltzmann law.

The important conclusion of this section consists in the formulation of the theoretical background for the fabrication of the Josephson junctions in the quality of voltage standard, where the line must be as narrow as possible. One can see that the junction must possess low resistance, high capacity and the measurements have to be done at as low temperature as possible.

13.2.2 *Thermal fluctuations of the order parameter*

The Josephson effect can be understood as the manifestation of the phenomenon of superconducting coherence, appearing in superconducting circuit with a weak link at temperatures below T_c. In the "Mr. Pickwick's sense", due to fluctuation pairings the superconducting coherence takes place also above T_c, but in contrast to the situation below critical temperature this coherence is limited in time and space. Such consideration makes obvious the absence of the fluctuation stationary Josephson effect above T_c: the Josephson supercurrent is proportional to the odd function of the phase, hence the order parameter phase fluctuations simply average out this coherent effect. Nevertheless it turns out that for the fluctuation emission of electromagnetic radiation by Josephson junction there is still a room [470]. Let us discuss fluctuation effects related to this possibility.

13.2.2.1 *Asymmetric Josephson junction* Let us start from the discussion of the case of asymmetric Josephson junction between two superconducting films, one of which has the thickness d less than corresponding $\xi_{GL}(T)$. Let us suppose that the critical temperatures T_{c1} and T_{c2} of the electrodes are substantially different and that the temperature of the system T is slightly above the lowest critical temperature T_{c1} ($T - T_{c1} \ll T_{c1}$), so the first electrode remains in the normal state where order parameter fluctuations take place. The second electrode is well below its critical temperature T_{c2} and has developed a superconducting order parameter Δ_2. The voltage V is applied to the junction.

The microscopic order parameter $\Delta_1^{(\mathrm{fl})}(\mathbf{r},t)$ in the first electrode has small amplitude, fluctuating in space and time. It corresponds to the phenomenological order parameter of the GL theory Ψ. As it was already mentioned above, due to the order parameter fluctuations, the coherence between two electrodes, limited in time and space, appears in spite of the normal state of the first electrode. The corresponding fluctuating current through the junction, in the assumption of the slow variation of $\Delta_1^{(\mathrm{fl})}(\mathbf{r},t)$ can be written in the spirit of Eq. (13.15)

$$I_c(t) = \frac{\pi}{eR}T\sum_n \int d^2\mathbf{r}\frac{\Delta_1^{(\mathrm{fl})}(\mathbf{r},t)}{|\varepsilon_n|}\frac{\Delta_2}{\sqrt{\Delta_2^2+\varepsilon_n^2}} = F_c\left(\frac{\Delta_2}{2\pi T_c}\right)\int \Delta_1^{(\mathrm{fl})}(\mathbf{r},t)\,d^2\mathbf{r}, \tag{13.43}$$

where

$$F_c(x) = \frac{1}{eR}\sum_{n=0}\frac{1}{n+1/2}\frac{1}{\sqrt{(n+1/2)^2+x^2}}. \tag{13.44}$$

The correlator (13.32) then can be calculated as

$$K(\omega) = \frac{2}{\pi}F_c^2\left(\frac{\Delta_2(T_{c1})}{2\pi T_{c1}}\right)\int d^2\mathbf{r}d^2\mathbf{r}' \times \int_{-\infty}^{\infty} e^{-i(\omega-\omega_J)\tau}\overline{\left\langle \Delta^{(\mathrm{fl})}(\mathbf{r},t)\,\Delta^{*(fl)}(\mathbf{r}',t+\tau)\right\rangle}d\tau$$

$$= \frac{16TS}{\pi^2 \alpha m \nu \xi^2} F_{\mathrm{c}}^2 \left(\frac{\Delta_2 (T_{c1})}{2\pi T_{c1}} \right) \frac{1}{\left[\tau_{\mathrm{GL}}^{-1} (\epsilon) - i (\omega - \omega_J) \right]}$$

(we used Eq. (3.5) for the correlator of order parameters and the relation (A.10)). Moreover it was taken into account that due to the applied voltage V the chemical potential is shifted and the order parameter $\Delta^{(\mathrm{fl})} (\mathbf{r}, t)$ acquires the constant phase shift $\omega_J t = 2eVt$ with respect to the order parameter of the first electrode. The correlator $\mathrm{Re}\, K (\omega)$ is responsible for the emission line shape. When the second electrode is well cooled

$$F_{\mathrm{c}} (x \to \infty) = \frac{1}{eR} \ln \frac{\Delta_2 (T_{c1})}{\pi T_{c1}} \tag{13.45}$$

and for the line shape one can write:

$$\mathrm{Re}\, K (\omega) = \frac{64 T_{c1}^2 S}{\pi^2 \nu e^2 R^2} \left(\ln^2 \frac{\Delta_2 (T_{c1})}{\pi T_{c1}} \right) \frac{\Gamma (\epsilon)}{\Gamma^2 (\epsilon) + (\omega - \omega_J)^2}, \tag{13.46}$$

where $\Gamma (\epsilon) = \tau_{\mathrm{GL}}^{-1} (\epsilon) = 8\epsilon T / \pi$ and $\epsilon = (T - T_{c1}) / T_{c1}$. In accordance with the results of the previous section it is clear, that the accounting for the voltage fluctuations would lead to the substitution of $\Gamma (\epsilon)$ by the value

$$\Gamma^* (\epsilon) = \tau_{\mathrm{GL}}^{-1} + 4e^2 R T_{\mathrm{c}} \tag{13.47}$$

describing the complete line widening. Nevertheless, the first item in this expression usually noticeably exceeds the second one.

13.2.2.2 *Symmetric Josephson junction* For the symmetric Josephson junction at temperatures slightly above T_{c} one has to accounting for fluctuations of the order parameters in both electrodes. Taking into account that $\Delta^{(\mathrm{fl})} (\mathbf{r}, t) \ll T$ the instant Josephson current can be written as:

$$I_{\mathrm{c}} (t) - \frac{\pi}{eR} T \sum_n \int d^2\mathbf{r} \frac{\Delta_1^{(\mathrm{fl})} (\mathbf{r}, t)}{\sqrt{\Delta_1^{(\mathrm{fl})2} + \varepsilon_n^2}} \frac{\Delta_2^{(\mathrm{fl})} (\mathbf{r}, t)}{\sqrt{\Delta_2^{(\mathrm{fl})2} + \varepsilon_n^2}}$$

$$= \frac{\pi}{2eRT} \int \Delta_1^{(\mathrm{fl})} (\mathbf{r}, t) \Delta_2^{(\mathrm{fl})} (\mathbf{r}, t) \, d^2\mathbf{r}. \tag{13.48}$$

The corresponding correlator $K (\omega)$ can be calculated in the same way as above. Carrying out the averaging over fluctuations in each electrode separately, passing to the Fourier components $\langle \Psi (0, 0) \Psi^* (r, \tau) \rangle_{\Omega, q \to 0}$, and calculating the appropriate convolution one can find

$$K (\omega) = \frac{\pi S}{2e^2 R^2 T^2} \left[\frac{1}{4 m \nu \xi^2} \right]^2 \int_{-\infty}^{\infty} e^{-i(\omega - \omega_J)\tau} \int d^2\mathbf{q} \langle \Psi \Psi^* \rangle_{t, q \to 0}^2 \, d\tau$$

$$= \frac{16\pi^3 TS}{7\zeta(3)\, e^2 R^2 \nu} Gi_{(2)} \ln \frac{T}{2\tau_{\rm GL}^{-1}(\epsilon) + i\delta\Omega}.$$

Finally for the line shape of the fluctuation emission of the symmetric junction we obtain:

$$\operatorname{Re} K(\omega) = \frac{16\pi^3 TS}{7\zeta(3)\, e^2 R^2 \nu} Gi_{(2)} \ln \frac{32T\Gamma(\epsilon)/\pi}{4\Gamma^2(\epsilon) + (\omega - \omega_J)^2}. \tag{13.49}$$

Let us compare the expressions (13.49) and (13.46). Both gives the Lorentz type broadening of the emission line. Nevertheless one can see that for the symmetric junction, when both electrodes are in the fluctuation regime, the line is smeared twice more. Moreover, the intensity of the line in the case of the asymmetric junction is proportional to the total area of the junction. This is possible to understand: one electrode is in the coherent state so the system works as the unique resonator. In the case of symmetric junction the total emission is formed as the sum of emissions of a large number of small independent junctions of the area $\sim \xi^2(\epsilon)$.

13.3 Fluctuation suppression of the Josephson current below $T_{\rm c}$[93]

The Ambegaokar–Baratoff formula (13.17) shows that just below critical temperature the nonzero critical current appears in the Josephson junction.[94] The critical current of a symmetric Josephson junction is proportional to the square of the order parameter modulus:

$$I_{\rm c} = \frac{\pi\Delta^2(T)}{4eTR_n}. \tag{13.50}$$

In the case of an asymmetric junction, close to the lower critical temperature, when the Josephson effect just appears, critical current is proportional to the first power of the modulus of the smallest order parameter.

In derivation of Eq. (13.15) the value of the order parameter $\Delta(T)$ was supposed to be determined by the BCS theory:

$$\Delta^2(T) \to \Delta_0^2(T) = \frac{8\pi^2 T_{c0}^2}{7\zeta(3)} |\epsilon_0|, \tag{13.51}$$

where $|\epsilon_0| = (T_{c0} - T)/T_{c0}$ is the reduced temperature counted from the MFA value of critical temperature T_{c0}. Hence the Ambegaokar–Baratoff theory predicts the linear (for strongly asymmetric junction this is the square root) increase

[93] In this section we base on the results of [399].

[94] In the next chapter will be demonstrated that thermal and quantum fluctuations result in the occur of the phase slip events and correspondingly in appearance of nonzero voltage in small size junctions. In this chapter we will consider junctions of a relatively large area, where the current saturates already at very small voltages. Namely this current $I_{\rm c}$ we will call as the critical.

of the critical current with the decrease of temperature in the region close to the BCS critical temperature T_{c0}.

Having in mind to accounting for the effect of order parameter fluctuations on the value of critical current below critical temperature let us average this formula over fluctuations. Important, that the order parameter fluctuates in each electrode independently, so for the symmetric junction the formula (13.50) has to be rewritten with the substitution $\Delta^2(T) \to \langle\Delta(T)\rangle^2$. Let us express the microscopic order parameter Δ averaged over fluctuations in terms of the phenomenological order parameter Ψ. Using the relation (A.10) and (2.139)–(2.140) one can write:

$$\begin{aligned}\langle\Delta(T)\rangle^2_{(\mathrm{fl})} &= \Delta_0^2 - \frac{1}{4mC_{(D)}}\left[3\langle\psi_r^2\rangle + \langle\psi_i^2\rangle\right] \\ &= \Delta_0^2(T) - \frac{8\pi^2T_{\mathrm{c}}^2}{7\zeta(3)\vartheta_D}Gi_{(D)}^{(4-D)/2}\alpha\xi^D\left(3\langle\psi_r^2\rangle + \langle\psi_i^2\rangle\right) \end{aligned} \quad (13.52)$$

(we used here also Eq. (A.9) for constant $C_{(D)}$ and definition of the Gi number Eq. (2.87)). The average values $\langle\psi_r^2\rangle$ and $\langle\psi_i^2\rangle$ were already calculated in chapter 3 and they are defined by Eqs. (2.139)–(2.140).

In accordance with Eq. (13.52) the total Josephson current amplitude $I_{\mathrm{c}}(T)$ can be presented as the sum of the BCS

$$I_{\mathrm{c}}^{(0)}(|\epsilon_0|) = \frac{2\pi^3T_{c0}}{7\zeta(3)eR_n}|\epsilon_0| \quad (13.53)$$

and fluctuating $\delta I_{\mathrm{c}}^{(\mathrm{fl})}(|\epsilon_0|)$ parts:

$$I_{\mathrm{c}}(T) = \frac{\pi\langle\Delta(T)\rangle^2_{(\mathrm{fl})}}{4eTR_n} = I_{\mathrm{c}}^{(0)}(|\epsilon_0|) + \delta I_{\mathrm{c}}^{(\mathrm{fl})}(|\epsilon_0|). \quad (13.54)$$

The further results are qualitatively different for $3D$ and $2D$ cases.

In case of the Josephson junction with thick electrodes ($d \gg \xi(|\epsilon_0|)$) fluctuations have $3D$ character, the main contribution to the integrals in Eqs. (2.139)–(2.140) originates from the large momenta $|\mathbf{q}| \sim \xi^{-1} \gg \xi^{-1}(|\epsilon|)$ and it leads to the shift of critical temperature Eq. (2.145). Renormalizing properly $T_{c0} \to T_{\mathrm{c}}$ and carrying out the remaining integral one finds fluctuation corrections both to $\langle\Delta\rangle^2$ and I_{c}. They have the same form as fluctuation correction to the superfluid density n_s (2.144) but with the different numerical factor:

$$\begin{aligned}\langle\Delta_{(3)}(T)\rangle^2_{(\mathrm{fl})} &= \frac{8\pi^2T_{c0}^2}{7\zeta(3)}\left[\frac{T_{c0}-T}{T_{c0}} - \frac{8}{\pi}\sqrt{Gi_{(3)}} + 3\sqrt{2Gi_{(3)}|\epsilon_0|}\right] \\ &\approx \frac{8\pi^2T_{\mathrm{c}}^2}{7\zeta(3)}\frac{T_{\mathrm{c}}-T}{T_{\mathrm{c}}}\left[1 + 3\sqrt{2\frac{Gi_{(3)}}{|\epsilon|}}\right],\end{aligned}$$

$$\delta I_{c(3)}\left(|\epsilon|\right)=\frac{6\pi^{2}T_{c0}}{7\zeta\left(3\right)eR_{n}}\sqrt{2Gi_{(3)}|\epsilon|}. \tag{13.55}$$

One can see that the inclusion in consideration of the order parameter fluctuations results in the change of the temperature dependence of I_c. Let us note that the fluctuation correction to the critical current became positive after the accounting for the fluctuation shift of the critical temperature in the Ambegaokar–Baratoff formula.

In case of the Josephson junction between two thin film electrodes of thickness $d \ll \xi_{GL}\left(|\epsilon_0|\right)$) ($2D$ case) the analysis of Eq. (13.52) turns out more sophisticated. The ultraviolet divergence of both $\langle\psi_r^2\rangle$ and $\langle\psi_i^2\rangle$expressions was already discussed above: it is related to the breakdown of the GL theory for short wave length fluctuations and corresponding logarithmic divergence has to be cut off at $q\sim\xi^{-1}$. Yet this is not all. The momentum integration in Eq. (2.140) for $\langle\psi_i^2\rangle$ is also divergent at lower limit of integration. This infrared divergence is known in theory of phase transitions and, being specifical for $2D$ case, will be discussed in chapter 15. We will cut it off at some large length scale $L\gg\xi$. Renormalization of the critical temperature in accordance with Eq. (2.148) can be done as above.[95] Substituting Eqs. (13.56) to (13.50) one can find

$$\begin{aligned}\left\langle\Delta_{(2)}\left(T\right)\right\rangle_{\text{fl}}^{2}&=\frac{8\pi^{2}T}{7\zeta\left(3\right)}\left[T_{c0}-T-2TGi_{(2)}\ln\frac{1}{|\epsilon_{0}|}-TGi_{(2)}\ln\left(\frac{L^{2}}{\xi^{2}\left(T\right)}\right)\right]\\&=\frac{8\pi^{2}T_{\mathrm{c}}^{2}}{7\zeta\left(3\right)}\left[|\epsilon|-2Gi_{(2)}\ln\frac{Gi_{(2)}}{|\epsilon|}-Gi_{(2)}\ln\frac{L^{2}}{\xi^{2}\left(T\right)}\right],\end{aligned} \tag{13.56}$$

what gives for the renormalized by fluctuations amplitude of Josephson current

$$\delta I_{c(2)}\left(|\epsilon|\right)=\frac{4\pi^{3}T_{\mathrm{c}}}{7\zeta\left(3\right)eR_{n}}Gi_{(2)}\left[\ln\frac{|\epsilon|}{Gi_{(2)}}+\ln\left(\frac{\xi\left(T\right)}{L_{J}}\right)\right]. \tag{13.57}$$

As the cut-off parameter L we have used here the, characteristic for the physics of large Josephson junction, new length scale L_J (Josephson length) which will be derived later.

As far as we consider the effect of superconducting fluctuations on Josephson current with respect to the MFA Ambegaokar–Baratoff curve (see Eq. (13.50), which turns zero at T_{c0} evidently turns down the curve $I_{\mathrm{c}}^{(0)}(|\epsilon_0|)$. Nevertheless it is necessary to mention that in practice T_{c0} is a fictitious, unobservable value: superconductivity in the electrodes appears at the renormalized by fluctuations critical temperature T_{c}. It is why the Ambegaokar–Baratoff curve has to be sketched in starting from T_{c} instead of T_{c0}. Correspondingly the renormalized

[95] Let us recall that in a $2D$ superconductor there is an additional shift of the critical temperature determined by the Coulomb potential fluctuations (see (2.150)). The same shift of critical temperature would appear in the expression for order parameter in Eq. (13.52) if one were include in consideration such fluctuations (see [399]). Evidently they do not effect on the shape of the functional dependence of $I_{\mathrm{c}}(T)$ but shift critical temperature. The microscopic consideration performed in [399] confirms Eq. (13.54).

by fluctuations curve Eq. (13.54) (with $\delta I_c(|\epsilon|)$ determined by Eqs. (13.55) and (13.57)) passes above $I_c^{(0)}(|\epsilon|)$.

Let us mention that for large $L_J^2/\xi^2(T)$ the fluctuation correction $\delta I_{c(2)}(|\epsilon|)$ can become of the order $I_c^{(0)}(T)$, which is small enough near T_c. This happens in the range of temperatures

$$|\epsilon| \sim Gi_{(2)} \ln\left(\frac{L_J^2}{\xi^2(T)}\right) \gg Gi_{(2)}. \tag{13.58}$$

The corresponding expression for the Josephson current in this region of strong fluctuations will be derived in the chapter 15.

13.4 Josephson current decay due to the thermal phase fluctuations

We have seen above how the thermal noise washes out the monochromaticity of the Josephson junction radiation. Now we will consider the other phenomenon related to the thermal fluctuations: the decay of the stationary Josephson current as a result of the temperature voltage fluctuations [471–473].

Let us return to a current biased Josephson tunnel junction represented by its equivalent circuit of Fig. 13.3. Equation (13.18) can be rewritten in the form:

$$M_C\frac{\partial^2\phi}{\partial t^2} + \eta\frac{\partial\phi}{\partial t} + \frac{\partial U(\phi)}{\partial\phi} = 0, \tag{13.59}$$

where $M_C = C/4e^2$, $\eta = \left(4e^2R_T\right)^{-1}$ and

$$U(\phi) = -(\frac{I}{2e}\phi + E_J\cos\phi). \tag{13.60}$$

Equation (13.59) can be treated as the equation of motion with friction (η is its viscosity) of a particle of mass M_C in the washboard potential[96] (13.60) (see Fig. 13.4). Accordingly, in such a mechanical analogy, space and velocity coordinates are $x = \phi$ and

$$v = 2eV = \frac{\partial\phi}{\partial t}. \tag{13.61}$$

The values of current I and amplitude I_c determine the slope of the potential $U(\phi)$ and the depth of valleys.

Above we referred to a junction with preset current but the following discussion also can be extended to a junction in a superconducting loop with the inductance L_i. In that case an equation for the total magnetic flux

[96]The term "washboard potential" may be unclear for a modern readers who wash their clothes in automatic laundry washes. Washboards, or chatter bumps, were in use before the latter found such widespread use as now. A washboard is a corrugated metal plate on which surface the wet clothes were rubbed.

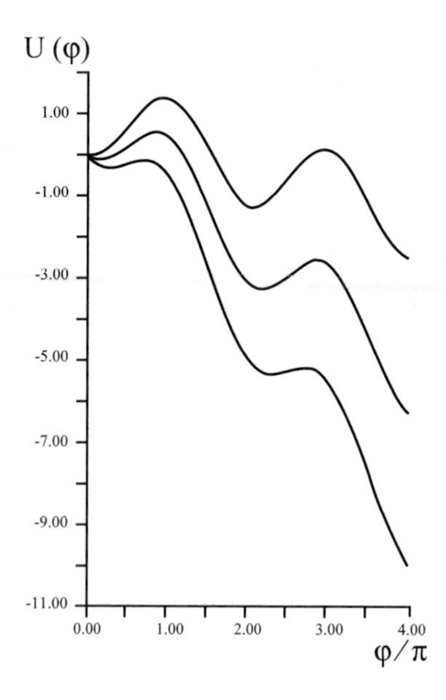

FIG. 13.4. Washboard potential.

$$\Phi = \Phi_{\text{ext}} + L_i I_s$$

in the superconducting loop can be written. Here Φ_{ext} and I_s are the applied magnetic flux and the superconducting current in the loop, the total flux is related to the phase as

$$\phi = 2\pi \frac{\Phi}{\Phi_0}.$$

In this case the mechanical analogy is with a motion of a particle in a sinusoidally modulated parabolic potential of the type:

$$U(\phi) = \frac{1}{2L_i} \left(\Phi_{\text{ext}} - \frac{\Phi_0 \phi}{2\pi} \right)^2 - I_{\text{c}} \frac{\Phi_0}{2\pi} \cos\phi. \tag{13.62}$$

Referring to the description in terms of the mechanical analogy let us first discuss the zero-noise case. Two situations can be considered: overdamped regime and underdamped regimes. In the overdamped regime (large dissipation), the $I - V$ curve of the junction is single valued. By increasing the current I we increase the average slope of the washboard potential while the particle remains in the valley. As soon as we exceed the critical current ($I > I_{\text{c}}$) the particle overcomes the barrier and starts to sweep down the slope. In this "running" state is thus $V > 0$ and, accordingly, $d\phi/dt \neq 0$, i.e. we are describing the finite voltage branch of the $I - V$ curve. Decreasing the current (the slope of $U(\phi)$)

as soon as we go down to the critical current, the friction will restore the $V = 0$ state.

In the underdamped case the capacitance effect (in terms of the mechanical analogy, the inertia of the particle) is large. The I–V curve is hysteretic. The lower branch of hysteresis corresponds to the circumstance that in this case, even when the current is reduced below I_c, the inertia tends to keep the running state of the particle or, equivalently, the capacitance maintains the voltage across the junction so that the back-switching in the I–V curve occurs at a current value smaller than I_c (hysteresis).

In the classical regime the fluctuations are accounted by adding a fluctuation term, $i(t)$, to the Kirchoff equation for the current balance in the junction (similarly Eq. (13.23)). Such a stochastic current is due to the Brownian motion of quasiparticles in the resistor R_T producing a usual Johnson noise. The resulting equation is therefore of Langevin type.

Let us study the case $I < I_c$. The minima of potential (13.60) correspond to the metastable states of the junction. The escape probability (activation rate) from a potential well is given by

$$\Gamma = A\frac{\omega_p}{2\pi} \exp\left[-\frac{\Delta U}{kT}\right], \tag{13.63}$$

where the characteristic plasma frequency ω_p is given by the interpolation formula

$$\omega_p = \left[\frac{1}{M_C}\frac{d^2U(\phi)}{d\phi^2}\right]^{\frac{1}{2}}_{\phi_{\min}} = \left(\frac{2\pi I_c}{\Phi_0 C}\right)^{1/2}\left[1-\left(\frac{I}{I_c}\right)^2\right]^{1/4} \tag{13.64}$$

and $\Delta U = U_{\max} - U_{\min}$ is defined by Eqs. (13.60) or (13.62).

In the limiting cases of the small current or, vice versa, current close to the critical one, the value can be expressed explicitly

$$\Delta U \simeq 2E_J \begin{cases} 1-\Phi_0 I/\left(4E_J\right), & I \ll I_c, \\ \frac{1}{3}\left(1-\left(\frac{I}{I_c}\right)^2\right)^{3/2}, & I - I_c \ll I_c. \end{cases} \tag{13.65}$$

The dimensionless pre-exponential factor A was calculated by Kramers in two limiting cases [474]. Let us follow his consideration. Instead the Langevin equation is convenient to use its equivalent, the Fokker–Planck equation for the distribution function $f\left(p_\phi, \phi\right)$, written in the space of momentum $p_\phi = M_C\frac{\partial\phi}{\partial t}$ and coordinate ϕ of the particle:

$$\frac{p_\phi}{M_C}\frac{\partial f}{\partial x} - \frac{\partial U}{\partial \phi}\frac{\partial f}{\partial p_\phi} = \eta\frac{\partial}{\partial p_\phi}\left(\frac{p_\phi}{M_C}f + T\frac{\partial f}{\partial p_\phi}\right).$$

This equation is written under the assumption that the particle lifetime in the well exceeds all the characteristics time scales.

In the limit of a low viscosity (large R_T) the particle initially oscillates in one of the minima with the plasma frequency (13.64).

In the absence of noise ($T = 0$) the particle energy in a well is conserved and it does not leave the well (the current does not decay). The nonzero noise leads to the establishment of the thermal equilibrium of the particle with the bath. As a result of the energy fluctuations the particle can possess the energy equal or larger than the barrier height (probability of such event is exponentially small) and leave the well. The flow of probability will be proportional to the viscosity (let us recall, that in the absence of viscosity there is no interaction with thermostat and the escapes from the well are impossible). Kramers found that in this conditions

$$A\left(\eta S_0 \ll M_C T\right) = \frac{\eta S_0}{M_C T}, \tag{13.66}$$

where

$$S_0 = 2\int_{\phi_1}^{0} \sqrt{2M_C\left[E - U(\phi)\right]}d\phi \tag{13.67}$$

is the action calculated along the trajectory of a particle with the energy E corresponding to the barrier height in absence of friction. The quantity S_0 appears in a natural fashion if, allowing for Eq. (13.63), the energy loss per particle oscillation in the well is calculated in an approximation linear in η:

$$\delta E = 2\int_{\phi_1}^{0} \eta\frac{\partial\phi}{\partial t}d\phi = \frac{2\eta}{M_C}\int_{\phi_1}^{0} p_\phi d\phi = \frac{\eta}{M_C}S_0.$$

Expression (13.66) is valid so long as $\eta S_0 \ll M_C T$. In the opposite limit of large viscosity $\eta S_0 \gg M_C T$ Kramers found

$$A = \left(1 + \frac{\eta^2}{4M_C^2\omega_{\mathrm{p}}^2}\right)^{1/2} - \frac{\eta}{2M_C\omega_{\mathrm{p}}}. \tag{13.68}$$

In this case the result depends on the ratio between the inverse decay time $\eta/M_C = \left(R_T C\right)^{-1}$ and the oscillation frequency ω_{p}. When $\omega_{\mathrm{p}} \gg \left(R_T C\right)^{-1}$ then $A = 1$. In the opposite case ($\omega_{\mathrm{p}} \ll \left(R_T C\right)^{-1}$)

$$A\left(\omega_{\mathrm{p}} \ll \left(R_T C\right)^{-1}\right) = \frac{M_C\omega_{\mathrm{p}}}{\eta}. \tag{13.69}$$

Melnikov [475] solved the Kramers problem completely and found the expression for the pre-exponential factor which valid for an arbitrary relation between ηS_0 and $M_C T$

$$A_{\mathrm{cl}} = \exp\left\{\int_{-\infty}^{\infty}\frac{dx}{2\pi}\ln\frac{1 - \exp\left[-\frac{\eta S_0}{M_C T}\left(x^2 + \frac{1}{4}\right)\right]}{x^2 + \frac{1}{4}}\right\}. \tag{13.70}$$

In the case of small viscosity this leads to the Kramers result (13.66). Allowing the next term we obtain in this limit

$$A_{\mathrm{cl}} = \frac{\eta S_0}{M_C T}\left[1 + \zeta\left(\frac{1}{2}\right)\left(\frac{\eta S_0}{\pi M_C T}\right)^{1/2}\right]. \tag{13.71}$$

The formula (13.63) gives the lifetime of the particle (junction) in the state with a fixed current. What will happen with a lapse of time depends on the strength of current and value of viscosity. At low viscosity and large current the particle can take a run up to a high speed, i.e. the voltage across the junction will be determined by its resistance R_T. When the viscosity is high the particle will be trapped in the neighboring well and will oscillate again trying to overcome the next barrier. The situation resembles downhill skiing on the nonmonotonous route in mountains. The average route slope corresponds to the value of current, viscosity corresponds to the friction.

In the case when particle is not able to overcome next barrier at full speed the diffusive motion with voltage pulse corresponding to each escape event takes place. The particle moves with an average velocity $< v >= 2eV = 2\pi\Gamma$. The I–V curve of the junction can be obtained through the dependence of Γ on the current I.

When the current is small the particle can return back and oscillate back and forth. Nevertheless, due to condition $I \neq 0$ the height of the back barrier is larger, than the height of the front one, so the probability of such return is less than to move forward. For $I \ll I_c$ the average drift velocity, i.e. the average voltage across the junction is equal to

$$\overline{V} = \frac{\pi}{e}\left(\Gamma_{\Rightarrow} - \Gamma_{\Leftarrow}\right) = \frac{\omega_{\mathrm{p}}}{2e} A \exp\left(-\frac{2E_j}{T}\right)\sinh\left(-\frac{\Phi_0 I}{2T}\right).$$

At low temperatures this value turns so small that the average voltage is determined by the direct *through* barrier quantum tunneling which we will discuss in the next section.

13.5 Macroscopic quantum tunneling

Above we have treated the Josephson junction in the framework of the electrical engineering style RCSJ model as the completely classical object. The only trace of its quantum mechanical nature was in the presence of the phase, but we completely ignored its superconducting origin. This approach works good for junctions with not too small capacity when its charge can be considered as the continues variable. Nevertheless, such carelessness is improper for small capacity junctions where the Cooper pair tunneling occurs apiece. The number of particles and phase in quantum mechanics are conjugated variables and the Heisenberg inequality must be respected: $\Delta\phi\Delta N \gtrsim 1$. Side by side with the number of Cooper pairs the phase becomes operator.

Let us remember that in our mechanical analogy the capacity plays the role of mass, so the quantization of the Josephson junction problem in the absence of viscosity is trivial: it is enough to substitute in its classical total energy

$$E = U(\phi) + \frac{CV^2}{2} = U(\phi) + \frac{p_\phi^2}{2M}$$

the momentum p_ϕ by its operator $-i\partial/\partial\phi$ and to obtain the Hamiltonian

$$\widehat{H} = U(\phi) + \frac{\widehat{p}_\phi^2}{2M} = -\frac{I}{2e}\phi - E_J\cos\phi - 4E_c\frac{\partial^2}{\partial\phi^2}, \qquad (13.72)$$

where $E_c = e^2/2C = E_C/4$ is the charging energy of the junction for a single electron charge. One can see that the quantitative criterion of the term "small capacity" consists in the comparison between the Josephson energy E_J and the charging energy of the junction E_c.

When the current is absent ($I = 0$) the Hamiltonian (13.72) is equivalent to that one of the electron moving in $1D$ periodic potential. The solution of the spectral problem in the latter is known: the electron spectrum becomes divided on "allowed bands" and "energy gaps".

The presence of the current ($I \neq 0$) makes the problem to be analogous to the charge transfer in the $1D$ crystal placed in electric field E. In the absence of viscosity (impurities in lattice), a particle first accelerates moving along the electric field direction, but then its motion is slowed down, when its energy reaches the band top it stops and changing direction of motion turns back. So one can right away conclude that in the absence of viscosity ($R_T \to \infty$)[97] the average voltage on the Josephson junction will be zero. Indeed, the conductivity of the ideal crystal placed in electric field equals zero too: electrons just perform the Bloch oscillations, no charge transfer occurs. Correspondingly applying the nonzero current through the junction in ballistic regime ($R_T \to \infty$) one can detect the voltage oscillations on it, but the average value $\overline{V}$ remains to be equal to zero.

The situation $E_c \sim E_J$ is realized in modern nanoscale systems. It presents the typical mesoscopic problem and stays beyond the problem of macroscopic quantum tunneling (MQT), which we are intended to discuss here. That is why below we will analyze the case $E_c \ll E_J$, which still corresponds to macroscopic junction.[98] Nevertheless let us stress again that we will study the quantum effects but in the quasiclassical approximation. This approach will allow us to demonstrate the essence of the MQT, to recognize the limits of its occurrence and possible competition with the thermal phase fluctuations.

[97]Let us note that tending $R_T \to \infty$ we still assume $R_T \ll R_{\text{ext}}$, in purpose to remain in the conditions of the preset current in the equivalent circuit.

[98]One has to remember that the expression for Josephson energy is written with the accuracy up to terms of the order $E_J\omega^2/\Delta^2$. This means that the term of the order e^2E_J/Δ^2 can appear in (13.59) as the corrections to the junction capacity C.

The value E_c characterizes kinetic energy of particle, so its smallness in comparison with the barriers heights means that in these conditions particle oscillates in vicinity of one of the potential $U(\phi)$ minima. The energy levels near the bottom of the well are

$$E_n = \omega_p \left(n + \frac{1}{2}\right), \tag{13.73}$$

where ω_p is determined by Eq. (13.64).

13.5.1 *Pair tunneling in the case of zero viscosity*

Let us discuss the properties of the system described by Hamiltonian (13.72) in the limit of infinite resistance of the Josephson junction, i.e. in the absence of viscosity ($\eta = 0$).

We start from the case of $I = 0$. The potential (13.60) has a series of minima, and according to the quantum mechanics a particle can tunnel between them. In general tunneling of particles from one minimum to another affects the spectrum dramatically. Let us write the corresponding Schroedinger equation

$$\left(-E_J \cos\phi - 4E_c \frac{\partial^2}{\partial\phi^2}\right) \psi(\phi) = E\psi(\phi)$$

which, by substitution $\phi = \pi + 2x$, takes the canonical form of Mathieu equation [476]:

$$\psi''(x) + (a - 2q\cos 2x)\psi(x) = 0, \tag{13.74}$$

with parameters $a = E/E_c$ and $q = E_J/(2E_c)$. First of all let us remember the Bloch theorem, according to which, due to the periodicity of the potential, the solutions of (13.74) satisfy the condition:

$$\psi(x + \pi) = e^{i\vartheta}\psi(x),$$

where ϑ is some phase.

The regime $E_J \gg E_c$ in the Hamiltonian (13.72) corresponds to $q \gg 1$. In this case, the eigenvalue a acquires the exponentially small part depending on ϑ [476]:

$$a(\vartheta) = -\frac{E_J}{E_c} + \frac{\omega_p}{2E_c} - 16\sqrt{\frac{2}{\pi}}q^{3/4}\exp(-4\sqrt{q})\cos\vartheta.$$

Let us compare this formula with the results of the tight-binding approximation for the electron spectrum in crystal. The energy band has the form

$$E(p) = 2t\cos\left(\frac{ps}{2\hbar}\right),$$

where t is the amplitude of the electron tunneling between two neighboring cites in crystal. Hence the tunneling amplitude for the transition between two minima of the washboard potential (13.60) in the absence of current is:

$$t = -\frac{4}{\pi^{1/2}} E_c \left(\frac{2E_J}{E_c}\right)^{3/4} \exp\left(-\sqrt{\frac{8E_J}{E_c}}\right). \qquad (13.75)$$

Due to tunneling the one-particle levels are split in tight bands of the width $2t$. In general case $I \neq 0$ and the matrix element t^2 can be obtained in the framework of the quasiclassical consideration:

$$t_n^2 = \Gamma_n = A_n \exp\left[-2\mathcal{S}(E_n)\right], \qquad (13.76)$$

$$\mathcal{S}(E_n) = \int \sqrt{2M\left[U(\phi) - E_n\right]} d\phi, \qquad (13.77)$$

where the integral is carried out in the domain of the underbarrier motion of the particle. The value of the pre-exponent factor is determined by the expressions [477]

$$A_n = \frac{\sqrt{2\pi}}{\Gamma(n+1)} \left(\frac{n+1/2}{e}\right)^{n+1/2} \approx \begin{cases} \sqrt{\frac{\pi}{e}} = 1.075, & n = 0. \\ 1, & n \gg 1. \end{cases}$$

It is necessary to remember, that in the Eq. (13.77) E_n is the level of the dimensional quantization, hence even for $n = 0$ $E_0 = \frac{\omega_p}{2} \neq 0$. This small difference of E_0 from zero results in the renormalization of the preexponential factor. In result when $I \ll I_c$ one can reproduce the expression (13.75).

When current almost reaches the critical value I_c

$$\Gamma_n = \frac{6\omega_p}{\sqrt{\pi}} \left(\frac{6\Delta U}{\omega_p}\right)^{1/2} \exp\left[-\frac{36\Delta U}{5\omega_p}\right], \qquad (13.78)$$

where ΔU is determined by the second asymptotic of Eq. (13.65).

The finite bias current $I \neq 0$ breaks the symmetry between the directions of motion and, as the electric field in the case of electron in a crystal, generates the particle Bloch oscillations. Its value determines the amplitude of such oscillations along the phase axis. In the absence of viscosity, at zero temperature, the particle stays in the coherent quantum stationary state and can perform such Bloch oscillations infinitely long.

It is clear that any finite viscosity will destroy the coherence of such a state of the particle. After this dephasing the particle will start its life from the very beginning. This means that at some value of viscosity the "distance" which the particle will be able to pass in the coherent state along the phase axis will reduce to the order of the potential period 2π, i.e. only tunneling to the neighboring well will be possible. Since the band width was found to be exponentially small it is clear that the same exponent will determine the critical viscosity and critical current restricting the long phase travels of the particle in the potential (13.60). This means that above such edge after the tunneling of the particle to the neighboring well the probability of its coherent return back will be negligible. Below we will discuss only such case of the viscosity higher than this exponentially small edge of the "nearest neighbor" coherence. This still will leave a room for the variety of different scenarios for tunneling probability (Γ) calculation. The concrete realization of either of them in practice, as we will see below, depends on the value of the quality factor Q.

13.5.2 *Case of low viscosity: $Q \gg 1$*

Let us assume the viscosity to be small enough ($Q = R_T C \omega_{\mathrm{p}} \gg 1$) not to affect noticeably the particle oscillations in the well and its under-barrier motion. At low temperatures ($T \ll \omega_{\mathrm{p}}$) the tunneling takes place from the lowest level of the energy spectrum (13.73) and its probability is determined by the formula (13.75)–(13.78). At high temperatures the results of the classical consideration (13.63)–(13.65) are valid. One can see that the principal difference between them consists only in the temperature dependence of the exponent: whereas in the classical case this is the activation Boltzmann exponential law with the exponent equal to the barrier height divided T, in the case of the quantum tunneling temperature in exponent has to be substituted by ω_{p}.

In the case of rectangular barrier of width d the height U the probability of tunneling at finite temperature is determined by the sum of both quantum and activation processes:

$$\Gamma \sim \exp\left(-\frac{U}{T}\right) + \exp\left(-2d\sqrt{2M_C U}\right). \tag{13.79}$$

When the barrier is smooth (like it the case of the Josephson junction) the combined tunneling occurs by the following scheme. First, the particle excites in the thermal activation manner and at some moment gets up to some energy E_n. The barrier transparency increases with increase of the number n and the tunneling probability Γ as the function of energy is determined by the product of the classical and quantum probabilities:

$$\Gamma \sim \exp\left(-\frac{E}{T}\right) \exp\left[-2\mathcal{S}\left(E\right)\right]. \tag{13.80}$$

The act of quantum tunneling happens when the particle reaches such level E_{tun} that the probability of the direct quantum tunneling through the barrier from it (see Eq. (13.76)) becomes larger than the probability of the activation jump on the higher energy levels with further tunneling through the barrier.[99]

[99] The described situation resembles the thermonuclear fusion when the fast nucleus belonging to the Maxwell distribution tail tunnels through the Coulomb barrier.

Quantitatively this can be formulated as the condition for the extremum of the exponent in Eq. (13.80)

$$\frac{\partial S(E)}{\partial E} = -\frac{1}{2T}. \tag{13.81}$$

Condition (13.81) implies that the period of the particle motion in the inverted potential is equal to $1/T$. With the growth of temperature the energy E_{tun} increases and when $T = \omega_{\text{p}}/2\pi$ it reaches the barrier height.[100] At higher temperatures the classical activation escape scheme is realized (see Eq. (13.63)). Corresponding numerical factor in the exponent of (13.78) decreases from 7.2 (at $T = 0$) to 6.28 ($T = \omega_{\text{p}}/2\pi$).

The escaped time as the function of the bias current was measured in [478] in the case of underdamped regime ($Q = 30$) and a good agreement with the presented above theory was found.

Another quantum effect is related not with the tunneling but with the energy quantization in the well. The existence of levels manifests itself brightly in the resonance reduction of the lifetime of a metastable state under the effect of an alternating current with frequency equal to the distance between some two levels. The effect consists in the equalization of the occupation numbers of the levels when the frequency of electromagnetic field coincides with their energy difference. Hence applying the a.c. field of the appropriate frequency and changing the bias current one can reach the peak in the barrier transparency and to observe the maximum in the escape time. Experimentally such resonance was observed in [479] and its microscopic theory was given in [480].

13.5.3 *Case of high viscosity: $Q \lesssim 1$*

Let us switch to discussion of the high viscosity case, i.e. of the properties of overdamped junction ($Q \lesssim 1$). The viscosity in this case changes dramatically both the character of its motion in the classically accessible domain as well as the character of its under-barrier tunneling. We will be interested in the calculation of the exponential dependence of the escape time. In this way we will study the peculiarities of the under-barrier particle motion. At that the discussion of the overdamping effect on the phase oscillations (which contributes in the pre-exponent factor) we will leave apart. It seems that the presence in the system of a high viscosity makes the Hamiltonian formalism to be unapplicable. Fortunately this is not completely correct and we will show below how the difficulty of the description of quantum system with friction can be overpassed. Friction can be understood as the interaction of the considered particle with the external bath. Introducing the thermostat in the Hamiltonian of the complex system we sacrifice the simplicity of an $1D$ problem (because of increase of the number of degrees of freedom) but return to the comfort of the Hamilton description. What concerns

[100] More precisely, ω_{p} is the oscillation frequency in the inverted potential. Nevertheless in the case of potential (13.60) it coincides with the plasma frequency (13.64) near the minimum.

the excess, corresponding to the bath degrees of freedom the statistical averaging has to be performed over them.

These were Caldeira and Legget [481] who first found solution of the MQT problem in general form and applied it to the case of low but finite viscosity. The opposite limit of high viscosity was investigated by Larkin and Ovchinnikov in [482–484]. They developed the methods of quantum field theory in order to take into account the viscous contribution in the under-barrier action $\mathcal{S}(E)$, which determines the escape time (see Eq. (13.76)). We will not stop here on the cumbersome technical details but will evaluate the viscous action qualitatively.

In the overdamped case one can omit the kinetic energy contribution and to say that the viscous action is determined by the work of the viscous friction force on the displacement $\Delta\phi \approx 1$ (when $I \sim I_c/2$) multiplied by the virtual tunneling time:

$$\mathcal{S} \sim \eta \left(\frac{d\phi}{dt}\right) \Delta\phi\Delta t \sim \eta \left(\Delta\phi\right)^2 \sim \eta. \tag{13.82}$$

Hence, according to Eq. (13.76),

$$\Gamma\left(Q \ll 1, I \sim I_c/2\right) \sim \exp\left[-2\eta\right]. \tag{13.83}$$

In the vicinity of critical current (when $I_c - I \ll I_c$): $\Delta\phi \approx 1 - I/I_c$ and

$$\Gamma\left(Q \ll 1, I_c - I \ll I_c\right) \sim \exp\left[-2\eta\left(1 - I/I_c\right)^2\right]. \tag{13.84}$$

The exact calculations of the decay probability into the neighboring minimum of the potential in the limit of high viscosity was calculated in [483, 485] and in the most general form can be written as:

$$\Gamma\left(T, I, Q \ll 1\right) = \frac{1}{8M^2}\left[\frac{2\eta^7}{\left(I/2e\right)^2 + \left(\pi T/2\eta\right)^2}\right]^{1/2} \times \exp\left\{-\pi\eta\left[1 - \ln\left(\frac{I}{I_c}\sqrt{1 + \pi^2T^2\eta^2/I^2}\right)\right] - \frac{I}{\pi T\eta}\arctan\frac{\pi T\eta}{I}\right\}. \tag{13.85}$$

In the limit of zero temperature it gives

$$\Gamma\left(T, I, Q \ll 1\right) \approx \frac{\sqrt{2}e\eta^{7/2}}{4M^2 I_c}\left(\frac{I}{I_c}\right)^{-\pi\eta}. \tag{13.86}$$

Therefore for currents $I < I_c$ both classical and quantum fluctuations lead to decay of the Josephson current, i.e. to appearance of the nonzero voltage drop on the junction. When $T = 0$ this voltage is proportional to Γ (see Eq. (13.86)) and strongly nonlinearly depends on the value of current.

14

PHASE SLIP EVENTS

14.1 Classical phase-slip events

In the case of a $1D$ wire Eq. (2.142) gives a small correction to n_s. At the same time it is clear that at $T \neq 0$ phenomenon of the superconductivity in $1D$ should not exist. This results from: (a) the general statement about the absence of the phase transitions in $1D$ systems; (b) the exponential decrease of the correlator $\langle \Psi^*(0)\Psi(\mathbf{r})\rangle$ for a $1D$ system below the mean-field transition point; (c) the classical partition function in $1D$ can be calculated exactly and it does not have any singularity.[101]

So, if the phase fluctuations are unable to destroy n_s, what is the mechanism for killing superconductivity in $1D$ systems? It turns out that it is phase-slip events, which also can be called instantons or topological excitations. At temperatures beyond the critical region ($|\epsilon| \gtrsim Gi_{(1)}$) the probability of such excitations is exponentially small and they cannot be found by the methods of perturbation theory.

As it was already mentioned there is a general statement on the absence of phase transitions in $1D$ systems. The superconducting phase must be unstable and supercurrents cannot persist for a long time, it must decay. Phase-slips are the mechanism of the supercurrent decay. Let us consider a closed loop made of a $1D$ superconducting wire. Without taking fluctuations into account the magnetic flux through this loop is a constant. If the phase of the wave function is a continuous function of the position, then it is not possible to change the supercurrent in the loop gradually. However, due to fluctuations, there is a finite probability to have the modulus of the order parameter vanish at some point. At that the phase is not defined at this point and a phase-slip event may occur (a change of the phase along the loop by 2π). Due to such phase-slip the supercurrent, which is determined by the gradient of the phase, decays. Let us note that this reasoning works for $1D$ systems only. In higher-dimensional systems, it is always possible to connect any two points by a path along which the phase is a continuous function.

With the aim of understanding the origin of phase-slip events let us start from a discussion of a small Josephson junction included in a superconducting

[101]It is possible to demonstrate [486] that the free energy of a D-dimensional classical field is equal to the ground state energy of the related $(D-1)$-dimensional quantum mechanical system. Indeed, in the $1D$ case the functional integral turns out to be the same as the Feynman integral for a particle moving in the potential $a|\Psi|^2 + \frac{b}{2}|\Psi|^4$. The ground state energy in this case is an analytic function of a and does not have a singularity at any temperature.

ring of induction L_i and in short will repeat the results of section 13.4. The potential energy related to the phase difference at the junction is the same as in a layered superconductor (2.66), so the energy of the ring can be expressed in terms of the magnetic flux Φ[102]

$$E_r(\Phi) = E_J \cos\frac{2\pi\Phi}{\Phi_0} + \frac{\Phi^2}{2L_i}. \tag{14.1}$$

Besides the ground state with $\Phi = 0$ there are many metastable states with Φ near to $n\Phi_0$ in such a system. The energies of such states are also the minima of function (14.1), but not absolute minima. In the limit of large E_J the energies of these minima are equal to $\Phi_0^2 n^2/2L_i$. In order to transit from one metastable state to another one with lower energy, the system has to pass through a potential barrier. The heights of these barriers are determined by the maxima of Eq. (14.1) and are equal to $2E_J$. The probability of such a "jump" (the so-called "phase-slip process") is proportional to $\exp\{-2E_J/T\}$. The flux flow decay, and respectively, the voltage appearing in the circuit are proportional to this probability:

$$V = \frac{d\Phi}{dt} \sim \exp\{-2E_J/T\}.$$

The positions of the maxima and minima of the function (14.1) are determined by the condition $\partial E_r/\partial\Phi = 0$.

Let us consider now the mechanism of the phase slip process in the circuit when the superconducting wire does not have a weak link and its free energy can be described by the GL functional (2.4). The minima of this expression are equal to $-a^2V/2b + \Phi_0^2 n^2/2L_i$ and they correspond to the magnetic flux values equal to multiple quanta $n\Phi_0$. The transition between the neighboring minima can only occur through the saddle points of the functional (2.4). These saddle points correspond to the unstable solutions of the GL equation

$$\Psi + \frac{b}{a}\Psi^3 + \xi^2(T)\Psi'' = 0$$

The solutions of this equation are continuous functions in the same way as those corresponding to the minima of the functional, but they can pass through zero at some points where the modulus of the order parameter vanishes while its phase changes by π [487]. Such a solution of the GL equation has form

$$\Psi(x) = \sqrt{-\frac{a}{b}}\tanh\left(\frac{x}{2\xi(T)}\right)$$

and is called instanton. Substituting it into Eq. (2.4) one can find that the energy of such an instanton in the wire of the cross-section S is [488]:

[102] The Josephson part of the GL energy, related to interlayer tunneling, was written as $\mathcal{J}|\Psi_n - \Psi_{n-1}|^2$ (see (2.66)). Supposing $\Psi_n = |\Psi|e^{i\varphi n}$ one can rewrite the corresponding energy as $2\mathcal{J}|\Psi|^2(1-\cos\varphi)$. Applying this result to the superconducting ring with a weak link one can express the phase difference by means of the magnetic field flux:
$\varphi = \oint \nabla\varphi dl = 2e\oint \mathbf{A}d\mathbf{l} = 2\pi\Phi/\Phi_0$.

$$\Delta F_0 = \frac{8\sqrt{2}}{3}\frac{a^2}{2b}S\xi(T) = \frac{\sqrt{2}}{3}\left(\frac{|\epsilon|}{Gi_{(wire)}}\right)^{3/2}T.$$

So the probability for the superconducting ring to change its magnetic flux by one flux quantum only implies that, being thermically activated, it passes over such a barrier. It turns out to be exponentially small, and the wire resistance is $R \sim \exp\{-\Delta F_0/T\}$.

In the case of a thin film the topological defects carrying the flux quanta are given by fluctuation vortices. The energy of such a vortex is equal to

$$E = \frac{n_s}{4m}\ln\frac{\lambda}{\xi}.$$

In a thin film λ is very large. In the framework of the Berezinskii–Kosterlitz–Thouless (BKT) theory (see section 15.3) it is assumed to be infinite. Thus the film resistivity below the critical temperature turns out to be zero.

14.2 Quantum phase slip events in nanorings[103]

As we have shown in chapter 12, the properties of superconducting grains change dramatically when their size shrinks to a few nanometers. The recent success in the manufacturing of superconducting nanowires [490] has raised similar questions regarding the superconductivity in $1D$ nanoobjects. It has been demonstrated experimentally [491] that as the nanowire becomes thinner, the superconducting transition in it disappears, and a finite resistance is observed at low temperatures. The suppression of the superconductivity in thin wires was attributed to the destruction of the phase coherence by quantum phase slips [492, 493].

To understand the specificity of the superconducting properties of thin wires one has to keep in mind that in an infinite $1D$ conductor all the electronic states are localized. The localization length L_c in a wire of cross-section S can be estimated as $L_c \sim p_F^2 Sl$. If the length L of the wire is shorter than L_c, it can be viewed as a metal grain which becomes a good superconductor if the gap Δ exceeds the quantum level spacing $E_F/p_F^3 SL$.[104] On the other hand, a wire of length $L \gg L_c$ cannot be viewed as a good conductor, and its superconducting properties are affected by localization.

If the attractive interaction between the electrons is weak, so that the bulk superconducting gap Δ is small compared to the level spacing $\delta_{L_c} \sim E_F/p_F^3 SL_c$ of a piece of the wire of length L_c, the superconductivity is suppressed everywhere in the wire. In the more interesting regime of $\Delta \gg \delta_{L_c}$, the superconducting gap is not affected by the localization. The important question in this regime is that of phase coherence between the different parts of the wire. Experimentally this

[103] In this section we base on the results of [489].

[104] One can check that this condition may be read as $Gi_{(1d)} \ll 1$ (see Table 2.1).

issue can be studied by measuring the persistent current in a nanowire ring as a function of the magnetic flux piercing it. The magnitude of the persistent current oscillations as a function of the flux is a direct measure of the superconducting phase coherence throughout the wire. Below, we demonstrate how the persistent current in a superconducting nanoring depends on the flux and the size of the ring.

If the wire is relatively thick, the electrons move freely between different parts of the wire, and therefore the superconducting phase φ is a well-defined classical variable. At low temperatures $T \ll \Delta$ there are no quasiparticle excitations in the nanoring. The allowed states of the system differ by the phase change $\phi = \varphi(L) - \varphi(0)$ accumulated over the circumference of the ring. At a given value, Φ, of the magnetic flux through the ring ϕ can assume the values $\phi_m = 2\pi\Phi/\Phi_0 + 2\pi m$, where m is an arbitrary integer, and Φ_0 is the flux quantum. The energies of these states are given by [105]

$$E_n = \frac{2\pi^2 N_s S}{m_e L}\left(\frac{\Phi}{\Phi_0} + m\right)^2. \tag{14.2}$$

Here N_s is the density of superconducting electrons, m_e is the electron mass.

The dependencies of the energy levels (14.2) on the flux are shown by the solid line in Fig. 14.1(a). The levels corresponding to different numbers m intersect at the points corresponding to the half-integer flux values. The persistent current can be found as a derivative of the ground state energy over flux $I = dE/d\Phi$ and it shows the characteristic sawtooth behavior, Fig. 14.1(b).

Classical instantons describe the lifetime of the excited states. Quantum instantons besides that show the change in the ground state energy and the character of its dependence on magnetic flux. The probability of the classical phase slip event is determined by the height of the energy barrier between two metastable states while the probability of the quantum phase-slip event is determined by the corresponding instanton action. In a complete analogy with the case of Josephson junction (section 13.5), there are two domains of superconducting nanowire which contribute to the instanton action: the instanton core ($x < \xi$) and its exterior ($x > \xi$). Let us start our consideration from the second domain. Here the order-parameter modulus fluctuates weakly and the action $\mathcal{S}_{\text{ext}}$

$$\mathcal{S}_{\text{ext}} = \int dx \int d\tau \mathcal{L}(x,\tau), \tag{14.3}$$

is determined by the phase Lagrangian [494, 495]:

$$\mathcal{L}(x,\tau) = \frac{g}{2\pi}\left[v(\partial_x\phi)^2 + v^{-1}(\partial_\tau\phi)^2\right]. \tag{14.4}$$

Here v is the velocity of the acoustic excitations in the system (plasmons in our case). The constant g can be calculated in terms of the microscopic parameters

[105] Below we will neglect the small energy of magnetic field related to circulating in the ring persistent current.

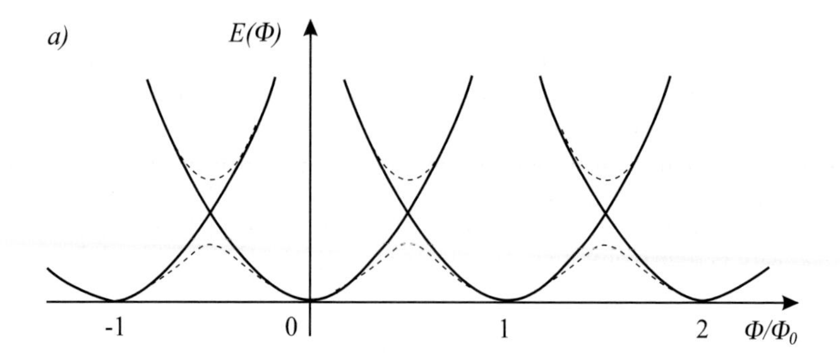

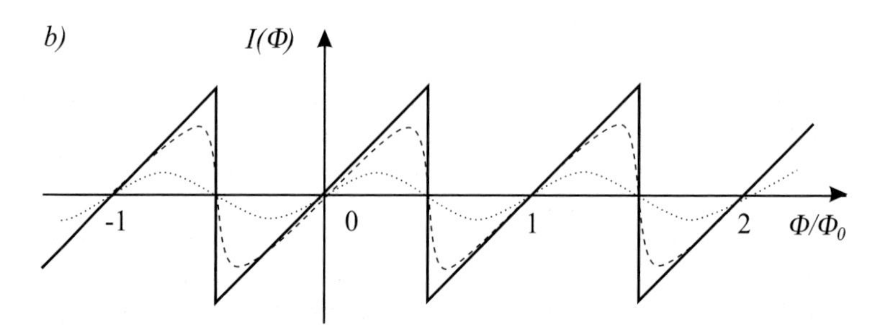

FIG. 14.1. (a) The energy of a nanoring as a function of the flux Φ through it. In the limit of relatively large cross-section S the energy demonstrates the classical behavior (14.2) shown by solid line. Quantum phase-slips result in level splitting shown in dashed line, and then lead to sinusoidal dependence of the ground state energy on the flux (dotted line). (b) The persistent current in the nanoring, $I(\Phi) = dE/d\Phi$. In the classical limit the current shows sawtooth behavior shown by solid line. As the wire becomes thinner, the sawtooth is rounded and eventually transforms to a sinusoidal oscillation.

and is related with the conductance. In the case of superconducting wire the first term corresponds to the kinetic energy of supercurrent hence

$$gv/2\pi = N_s S/m_e. \tag{14.5}$$

The second term in its corresponds to the charged energy and can be expressed in terms of the wire linear capacity:

$$g/2\pi v = \frac{1}{e^2} \ln \frac{L^2}{S}. \tag{14.6}$$

Combining these two expressions one can find

$$g = 2\pi \sqrt{\frac{N_s S}{m_e e^2} \ln \frac{L^2}{S}} \tag{14.7}$$

and to see that g for nanowires decreases with the decrease of its cross-section. The external action has the form of Eq. (14.4) not only for superconducting nanowire but also for other $1D$ systems, for example for charge density wave (CDW). Such an instanton action corresponding to the problem of the CDW tunneling through impurity was calculated in [496]. Both these problems formally are equivalent thus obtained in [496] result provides us with the necessary value of action (14.3) also for the case under consideration:

$$\mathcal{S}_{\text{ext}} = g \ln\left(\frac{L}{\xi}\right). \tag{14.8}$$

This result can be easily understood considering τ as the second space coordinate. Indeed, in this case the instanton action is nothing other than the energy of the vortex which maps the quantum $1D$ instanton in $2D$ classical problem. In the quantum $1D$ wire g^{-1} plays the role of temperature. We will see in the next chapter that in the $2D$ superconductor at some temperature $T = T_{BKT}$ the transition to the localized vortex–antivortex state (BKT transition) occurs. In the case of a quantum wire such transition corresponds to the point of the instanton formation while the temperature T_{BKT} corresponds to $g = 1$.[106] In the case of a thick film $g \gg 1$ (see Eq. (14.7)) and at large lengths the system exhibits superconducting properties. When $g \sim 1$ several different phases, depending on the gate voltage, can be realized [500]. In the case of a thin enough wire $g < 1$. We will demonstrate below that in this case superconducting correlations at large distances exponentially decrease, although the characteristic distances of their decay can still remain much larger than the localization length of such wire in normal state. We will find corresponding length scale L_c and will calculate the value of the persistent current in the ring of the diameter $\sim L_c$.

It turns out that when $g \ll 1$ the main contribution to the instanton action is given by the domain of its core ($x < \xi$). In this region not only the phase of the order parameter changes but also its modulus which at some point $x = x_c$ turns zero. In contrast to its classical analogue in the quantum slip event this happens only for one moment of time.

Let us evaluate the corresponding action. One needs to accounting for the fact that the quantum phase-slips must involve suppression of the superconducting gap Δ in the wire. The typical volume of space where the gap is suppressed is $V \sim S\xi$, where $\xi \sim \sqrt{v_F l/\Delta}$ is the coherence length. Given the condensation energy per unit volume $\varepsilon \sim m_e p_F \Delta^2$ and the typical correlation time $\tau \sim 1/\Delta$, one can estimate the phase-slip action in the wire $\mathcal{S}_0 \sim \varepsilon V \tau \sim \sqrt{\Delta/\delta_{L_c}} \sim L_c/\xi$. A more rigorous treatment [492, 493] of the quantum phase-slips gives rise to the same result, but unfortunately the numerical coefficient cannot be obtained analytically.

[106] The special role of the value $g = 1$ was pointed out for the problems of conductance of a Luttinger liquid [497] and of tunneling with dissipation [498, 499].

The above picture of levels (14.2) does not correspond to reality for thin wires, where the fluctuations of the superconducting order parameter cannot be neglected. The most important type of the fluctuations is the quantum phase-slip that changes the phase ϕ at a point x by $\pm 2\pi$. We will show that the effect of the rare phase-slips on the system reduces to quantum transitions between the levels (14.2). Such transitions are most important near the degeneracy points at half-integer values of Φ/Φ_0 and result in the small level splitting shown by dashed line in Fig. 14.1(a). We will also show that the multiple quantum phase-slips in very thin and/or long wires lead to sinusoidal behavior of the ground state energy and the persistent current, see dotted lines in Fig. 14.1(b).

The theory of quantum phase-slips in nanowires is rather complex. In particular, the most important quantitative parameter describing the phase-slip – its action – can be found only up to an unknown numerical coefficient. We will be interested below in the dependence of the oscillation amplitude (see Fig. 14.1) on the wire length. With the increase of the latter the problem becomes one of many instantons and for simplicity we first will study the useful model, where the one instanton action can be calculated exactly.

The effect of the phase-slips on the ground state of the nanoring can be studied using a simple model of a chain of coupled superconducting grains. Similarly to the nanowire, the chain of grains will have quantum phase-slips that affect the persistent current in the same way. We will discuss the relation between the two models in more detail below.

Our model of the chain of superconducting grains is defined by the following imaginary-time action [501]:

$$\mathcal{S}\left(\theta_1, \cdots, \theta_n\right) = \int_0^\beta d\tau \sum_{n=1}^{N} \left\{ \frac{\dot{\theta}_n^2}{16E_c} + E_J[1 - \cos\theta_n(\tau)] \right\}. \tag{14.9}$$

The grains are assumed to be connected by tunnel junctions with Josephson coupling energy E_J; the capacitance of each junction C gives rise to the charging energy $E_c = e^2/2C$; the variable θ_n is the phase difference across the n-th junction; $\beta = 1/T$. Each term of (14.9) is equivalent to the Hamiltonian (13.72) for the isolated junction. Nevertheless the junctions cannot be considered as independent since the sum of $\theta_n(\tau)$ is determined by the total flux Φ penetrating the ring. To model a closed chain pierced by flux $\Phi = (\phi/2\pi)\Phi_0$, one should impose an additional condition

$$\sum_{n=1}^{N} \theta_n(\tau) = \phi \tag{14.10}$$

on the phases θ_n. Note that the action (14.9) does not include the charging energy terms due to the self-capacitance of the grains. Our model is motivated by the fact that the electric field is very well screened in metal wires, and for the case $g \ll 1$ the electric field energy in the space beyond the wire can be neglected.

The ground state properties of the chain can be derived from the partition function

$$Z = \int \mathfrak{D}\theta_1 \cdots \mathfrak{D}\theta_N e^{-\mathcal{S}(\theta_1,\ldots,\theta_N)} \tag{14.11}$$

in the limit $\beta \to \infty$. Applying the Feynman method of functional integration in quantum mechanics one can transform the functional integral over the variables θ_n to the usual expression for the partition function in the form of sum over the eigenstates of Hamiltonian (13.72). We will primarily be interested in the case of strong coupling between the grains, $E_J \gg E_c$. In the limit $E_c \to 0$ the phases θ_n become classical variables, and the energy states of the chain can be found by minimizing the sum of the Josephson energy terms in the action (14.9) with the constraint (14.10). When number of contacts N is large all θ_n are close to minima points $2\pi m_n$. Hence the cosine in (14.9) can be expanded and minimization of the obtained quadratic form (14.9) with constraint (14.10) gives:

$$E_m = \frac{E_J}{2N}(\phi + 2\pi m)^2, \tag{14.12}$$

in complete analogy with Eq. (14.2) for a metal ring.

At finite, small E_c the fluctuations of $\theta_n(t)$ should be taken into consideration. Important fluctuations of the phases $\theta_n(\tau)$ involves the formation of an instanton (quantum phase-slip), i.e. a trajectory that begins near one of the minima (14.12) of the classical action (14.9) at $\tau = 0$ and ends near another minimum at $\tau = \beta$. For instance, a trajectory starting at $\theta_n(0) = \phi/N$ and ending at

$$\theta_n(\beta) = (\phi - 2\pi)/N + 2\pi\delta_{nk} \tag{14.13}$$

for arbitrary k in the interval $1 \leq k \leq N$ connects the minima (14.12) with $m = 0$ and $m = -1$. This means that the phase changes on each junction by the value $(\phi - 2\pi)/N$, but on the contact k phase-slip event by 2π occurs.

The shape of the instanton trajectory can be found by minimizing the classical action (14.9) with the above boundary conditions on $\theta_n(t)$. In the case of large number of junctions $N \gg 1$ the dominant contribution is due to the contact k where the phase-slip occurs, and the contributions of the other contacts can be neglected. Then one obtains the usual result [502]

$$\theta_k(\tau) = 4\arctan\left\{\exp\left[2\sqrt{2E_J E_c}\,(\tau - \tau')\right]\right\}, \tag{14.14}$$

where τ' is the arbitrary moment in imaginary time where the phase-slip is centered; we have assumed the limit of low temperature $T \ll \sqrt{E_J E_c}$. The action associated with this instanton trajectory is $\mathcal{S}_0 = 2\sqrt{2E_J/E_c}$.

The instantons account for the possibility of the system tunneling between the different minima (14.12) of the classical action (14.9). Let us stress that the corresponding exponent coincides with the instanton action $\mathcal{S}_0$, and the prefactor can be obtained by considering the problem (14.9) for the case of a single junction (without the constraint (14.10)). Corresponding amplitude of such tunneling process t was already calculated above and is given by Eq. (13.75); one can see that is exponentially small.

It is important to note that for any k the set of phases $\theta_n = (\phi - 2\pi)/N + 2\pi\delta_{nk}$ describes the same physical state of the chain. Thus to get the total amplitude $t_{0,-1}$ of the tunneling between the states with $m = 0$ and $m = -1$ due to the instantons (14.14) in all the N junctions one must to sum up the single amplitude (13.75), that results in the appearance of the additional factor of N in it: $t_{0,-1} = Nt$. The same tunneling amplitude describes any transition of system from the state m to $m \pm 1$: $t_{m,m-1} = t_{m,m+1} = Nt$.

It is clear that such transitions (instantons) change the ground state energy (14.12). Their effect can be taken into account by adding to the initial Hamiltonian the term corresponding to the nearest neighbor interaction calculated in the tight-binding approximation:

$$H\psi_m = E_m\psi_m - Nt(\psi_{m-1} + \psi_{m+1}). \tag{14.15}$$

Here ψ_m is the wave function of the system in the state m with energy E_m given by Eq. (14.12).

At $Nt \ll E_J/N$ the hopping term is small compared with the diagonal matrix elements in the Hamiltonian (14.15). Its effect is most significant when Φ/Φ_0 is half-integer, and the energy levels E_n are degenerate, Fig. 14.1(a). In this regime the instanton fluctuations give rise to the level repulsion shown by the dashed line in Fig. 14.1(a). In order to calculate the corresponding splitting one can use the perturbation theory for the degenerate level substituting Eq. (14.15) by the system of two equations. Their solution is reduced to the solution of the quadratic equation. As a result the shape of the current steps is given by

$$I = \frac{2eE_J}{N}\left[\chi - \frac{\pi\chi}{\sqrt{\chi^2 + (tN^2/\pi E_J)^2}}\right], \tag{14.16}$$

where $\chi = \phi - \pi = 2\pi(\Phi/\Phi_0 - 1/2) \ll 1$, and the respective rounding of the sawtooth in current is shown in Fig. 14.1(b).

In the opposite limit $Nt \gg E_J/N$ the instanton fluctuations affect the spectrum of the Hamiltonian (14.15) dramatically. To find the ground state energy $E(\phi)$ it is more convenient to apply the Hamiltonian (14.15) to the wave function ψ_m chosen in the "coordinate representation," $\psi(x) = \sum_m \psi_m e^{i(2m-\phi/\pi)x}$. The resulting Schroedinger equation then takes the form of the Mathieu equation (13.74), at that the phase ϕ enters the problem via a twisted boundary condition $\psi(x+\pi) = e^{-i\phi}\psi(x)$. It resembles the quasi-momentum of the particle moving in the periodic $1D$ potential, described by the same Mathieu equation (13.74). The same as was discussed above in the case of the single junction here the parameter a is proportional to the energy $E(a = 2NE/\pi^2 E_J)$, while q characterizes the effective hopping strength (instanton intensity): $q = N^2 t/\pi^2 E_J$. The important feature which differs the cases consists in the dependence of q on N^2, that can make this parameter to be larger than 1, in spite of the relation $t \ll E_J$.

The mentioned analogy between the discussed problem and $1D$ motion in a periodic potential immediately suggests an idea of the band motion. Indeed, the case $q \ll 1$ corresponds to the narrow gap and one arrives to the formula (14.16). The opposite case $q \gg 1$ corresponds to the narrow band of the allowable states. In this case the dependence of the eigenvalue a on the phase ϕ is exponentially small in amplitude: $a = -16\sqrt{2/\pi}q^{3/4}e^{-4\sqrt{q}}\cos\phi$ [476]. For the persistent current $I = 2e\,(dE/d\phi)$ we then find

$$I = 16\sqrt{2N}\,\frac{e}{\hbar}(E_J t^3)^{1/4}\exp\left(-\frac{4}{\pi}N\sqrt{\frac{t}{E_J}}\right)\sin\phi. \tag{14.17}$$

It is important to note that the Schroedinger equation for the Hamiltonian (14.15) coincides with the Mathieu equation (13.74) at arbitrary hopping strength q. Therefore by solving Eq. (13.74) with the appropriate twisted boundary condition, which can be easily done numerically, one can describe the magnetic flux dependence of the persistent current at arbitrary phase-slip strength q.

Despite the difficulties of the analytic treatment of the phase-slips in a nanowire, their effect on the persistent current can be understood in terms of the tight-binding Hamiltonian (14.15). However, in order to apply the above results to the nanoring, one should substitute the appropriate matrix elements. In particular, in calculating $t \propto e^{-\mathcal{S}_0}$ one should use $\mathcal{S}_0 \sim L_c/\xi$.

A more subtle effect of the random potential is its influence on the relative phases of instantons in different parts of the wire. It can be explored within our model by adding gates to the grains and applying random gate voltages. It has been shown recently [503] that the gate voltages give rise to Aharonov–Casher interference effects for the instantons in different junctions, and under certain conditions can suppress the effects of the phase-slips.

The most significant change in the properties of the persistent current caused by the random gate potential is that instead of $\ln I \propto -N$ a slower dependence $\ln I \propto -N^{3/4}$ on the length of the chain at large N was found. Respectively, in metal nanorings the dependence of the persistent current on the dimensions of the device is given by $\ln I \propto -L^{3/4}e^{-\mathcal{S}_0/2}$, where the instanton action $\mathcal{S}_0$ is proportional to the cross-section of the wire S.

Another type of systems to which the described theory should apply is the $1D$ arrays of Josephson junctions. Recently it became possible to tune the Josephson coupling energy in such devices by external magnetic field [504]. As a result, one should be able to study the whole crossover from sawtooth to sinusoidal dependence of $I(\Phi)$ in a single sample.

15

PHASE FLUCTUATIONS IN A 2D SUPERCONDUCTING SYSTEM

15.1 The crucial role of phase fluctuations in 2D systems

As we have seen above more than once, the phase fluctuations play a special role in 2D superconducting systems. For instance, one can notice that the fluctuation correction to the average value of order parameter $\langle \Psi \rangle$ below T_{c} is divergent (see (2.141)). This means that the appropriate corrections cannot be supposed to be weak even relatively far from the transition $|\epsilon| \gg Gi_{(2)}$, where the fluctuations of the value $\left\langle |\Psi(\mathbf{r})|^2 \right\rangle$ are still small. Let us discuss this problem in more details.

Neglecting the order parameter modulus fluctuations one can write the effective functional which describes the order parameter phase fluctuations only:

$$\mathcal{F}[\varphi] = \frac{n_s}{4m} \int d^2\mathbf{r} \left[\nabla \varphi \left(\mathbf{r} \right) \right]^2 = \frac{n_s}{4m} \sum_{\mathbf{k}} \mathbf{k}^2 \varphi_{\mathbf{k}}^2. \tag{15.1}$$

(see chapter 2). Let us calculate now the average value of the order parameter using this functional without the assumption of weak fluctuations:

$$\langle \Psi \rangle = |\Psi| \left\langle e^{i\varphi} \right\rangle = |\Psi| \left\langle \exp\left(-\sum_{\mathbf{k}} \frac{4mT}{n_s \mathbf{k}^2} \right) \right\rangle. \tag{15.2}$$

In the 2D case the sum in (15.2) diverges and $\langle \Psi \rangle = 0$ [505]. Nevertheless a phase transition in such system exists. In order to see this let us study the behavior of the correlation function $\langle \Psi^*(0)\Psi(\mathbf{r}) \rangle$ at large distances.

Above the mean-field critical temperature the Fourier component $\left\langle |\Psi_{\mathbf{k}}|^2 \right\rangle$ was already calculated (see Eq. (3.5) and the correlator $\langle \Psi^*(0)\Psi(\mathbf{r}) \rangle$ takes the form:

$$\langle \Psi^*(0)\Psi(\mathbf{r}) \rangle \left(T > T_{\mathrm{c}0} \right) = \sum_{\mathbf{k}} \left\langle |\Psi_k|^2 \right\rangle \exp\left(i\mathbf{k}\mathbf{r} \right) = \frac{mT}{\pi} \mathbf{K}_0 \left(\frac{r}{\xi(\epsilon)} \right), \tag{15.3}$$

where $\mathbf{K}_0(x)$ is the modified Bessel function. At large arguments $\mathbf{K}_0(x \gg 1) = \sqrt{\pi/(2x)} e^{-x}$ and we find that in the normal phase this correlator decreases exponentially at distances $r \gtrsim \xi(\epsilon)$.

Below the transition point

$$\begin{aligned}\langle \Psi^*(0)\Psi(\mathbf{r})\rangle &= |\Psi|^2 \langle \exp(i\varphi(0) - i\varphi(\mathbf{r}))\rangle \\ &= |\Psi|^2 \left\langle \exp\left(i\sum_{\mathbf{k}} [1 - \exp(i\mathbf{kr})]\,\varphi_{\mathbf{k}}\right)\right\rangle \\ &= |\Psi|^2 \exp\left(-\frac{1}{2}\sum_{\mathbf{k}} |1 - \exp(i\mathbf{kr})|^2 \langle \varphi_{\mathbf{k}}^2\rangle\right). \end{aligned} \tag{15.4}$$

The average value $\langle \varphi_{\mathbf{k}}^2\rangle$ has already appeared in our calculations (for instance, in Eq. (2.140)) as the phase fluctuation mode below T_c : $\langle \varphi_{\mathbf{k}}^2\rangle = 4mT\left(n_s\mathbf{k}^2\right)^{-1}$. The last integral (sum) in the exponent of (15.4) evidently converges at small $\mathbf{k}$. After the angular integration it can be expressed in terms of an integral of a Bessel function and it logarithmically diverges at the upper limit. This divergence must be cut-off at $\mathbf{k} \sim \xi^{-1}$, where the expression for $\langle \varphi_{\mathbf{k}}^2\rangle$ is not valid more. As a result below the transition point the correlator takes the form

$$\langle \Psi^*(0)\Psi(\mathbf{r})\rangle_{|\mathbf{r}|\gg\xi}(T) = |\Psi|^2 \exp\left(-\frac{mT}{\pi n_s}\ln\frac{r}{\xi}\right) = |\Psi|^2\left(\frac{r}{\xi}\right)^{-mT/(\pi n_s)}. \tag{15.5}$$

These two ((15.3) and (15.5)), quite different, asymptotics of the correlator for the low and high temperature limits were obtained for the regions $|\epsilon| \gtrsim Gi_{(2)}$. Nevertheless it is clear that there must be some point at which the high temperature exponential asymptotic changes to the low temperature power one. This temperature is reasonably called as the point of the superconducting transition. In the $2D$ case such a transition is nothing else as the BKT transition and corresponding temperature is called T_{BKT}. It is worth mentioning that, in spite of the fluctuation destruction of $\langle\Psi\rangle$, in the low temperature phase the observable physical quantity n_s is renormalized by fluctuations finitely and is different from zero. In the GL region ($|\epsilon| \gtrsim Gi_{(2)}$) this renormalization turns out to be even small (see (2.149)).

15.2 Exponential tail in the Josephson current close to T_c

Let us consider a thin superconducting film, being slightly below its critical temperature, which is separated by a tunnel barrier from a superconductor with higher critical temperature. The energy of such junction is defined by Eq. (13.12). Let us study the effect of fluctuations on the temperature dependence of $E_J = I_c/2e$. Three possible answers can be expected. Looking on Eq. (13.43) one can see that $I_c \sim \langle\Delta\rangle$. Hence, in accordance with Eq. (15.2), $\langle\Psi\rangle = \langle\Delta\rangle = 0$, and one could expect $I_c = 0$ or, in other words, the existence of a small but nonzero resistance at any temperature $T \neq 0$. The second answer could be the following. There are two points of different transitions: first one, T_{BKT}, when the superfluid density n_s along the film disappears and the second one, T_J, when the Josephson critical current vanishes. The latter temperature T_J one can estimate as the temperature at which the corresponding fluctuation correction reaches the value of already small Josephson current. Finally, the third option consists in the existence of the unique temperature T_{BKT}, while in the region $T_J < T < T_{\text{BKT}}$, the value of $I_{\text{c}}(T)$ is small but different from zero.

Let us demonstrate that this, third, possibility is realized in practice. At temperatures when the fluctuation correction to critical current of the Josephson junction is small, it is determined by Eq. (13.57). Let us find the value of the cut-off parameter for this formula assuming the area of the junction S being large enough. The fluctuation part of the junction energy is given by Eqs.(15.1) and (13.57). The latter for small $\varphi^{(\text{fl})}$ can be written as

$$\delta E_J^{(\text{fl})}(\epsilon) = \frac{E_J}{2S} \int d^2\mathbf{r} \left[\varphi^{(\text{fl})}(\mathbf{r})\right]^2. \tag{15.6}$$

Accomplishing the Fourier transform of fluctuating phase:

$$\varphi^{(\text{fl})}(\mathbf{r}) = \sum \varphi_{\mathbf{k}} \exp(i\mathbf{k}\cdot\mathbf{r}) \tag{15.7}$$

one obtains instead of Eq. (15.1):

$$F\left(\varphi^{(\text{fl})}\right) = \frac{n_s}{2m} \sum_{\mathbf{k}} \left(\mathbf{k}^2 + \frac{1}{L_J^2}\right) \varphi_{\mathbf{k}}^2, \tag{15.8}$$

where the Josephson length L_J is determined by relation

$$\frac{n_s}{2mL_J^2} = \frac{E_J}{2S}. \tag{15.9}$$

When I_{c} is small $L_J \gg \xi$ and the fluctuation correction (13.57) can become relatively large even for $\epsilon \gg Gi_{(2d)}$. In order to find the form of $I_{\text{c}}(\epsilon)$ in this nonperturbative region of temperatures let us calculate the average value

$$I_{\text{c}}(\varphi) = \frac{E_J}{2S} \left\langle \sin\left(\varphi + \varphi^{(\text{fl})}\right)\right\rangle \tag{15.10}$$

without expansion in series over $\varphi^{(\text{fl})}$. The averaging in Eq. (15.10) is carried out with the free energy (15.1) over all possible phase fluctuations. Carrying out the Gauss integral in Eq. (15.10) analogously this was done in Eqs. (15.2)–(15.5) one can find

$$\delta E_J^{(\text{fl})}(\epsilon) = E_{J0} \exp\left(-\frac{2Gi_{(2d)}}{|\epsilon|} \ln \frac{L_J}{\xi(T)}\right). \tag{15.11}$$

In the region $\epsilon \gg Gi_{(2d)}$ the superfluid density n_s can be taken in its unperturbed form (2.149) what gives

$$I_{\text{c}}(\epsilon) = I_{\text{c}}^{(AB)}(\epsilon) \exp\left(-\frac{\epsilon^*}{|\epsilon|}\right), \tag{15.12}$$

where

$$\epsilon^* = Gi_{(2d)} \ln \left[Gi_{(2d)} \left(\frac{L_J}{\xi} \right)^2 \right]. \tag{15.13}$$

When $|\epsilon| \gg \epsilon^*$ the correction is small and Eq. (15.12) reduces to Eq. (13.57), while for $|\epsilon| \lesssim \epsilon^*$ in the temperature dependence of the critical current the exponential tail has be observed.

15.3 Berezinskii–Kosterlitz–Thouless transition

Since the pioneering papers of Berezinskii [506] and Kosterlitz and Thouless [507] it is known that in $2D$ systems vortices can exist either as bound vortex–antivortex pairs or as dissociated free vortices and anitvortices (see for review [508, 509]). In the temperature axis these two phases are separated by the so-called Berezinskii–Kosterlitz–Thouless transition temperature T_{BKT}. As we already have seen, the above vortices in superconductors appear when a sufficiently strong magnetic field is applied. Nevertheless in the case of $2D$ superconductor vortices may appear As a result of thermal fluctuations. In absence of magnetic field the densities of vortices and antivortices are equal.

At $T > T_{\mathrm{BKT}}$ in superconductors there exist decoupled vortices and antivortices of some density. When the supercurrent flows they move in opposite directions and transfer magnetic flux. As a result an electric field, proportional to the density of decoupled vortices, is induced in the system, which means the appearance of the resistive state. Hence the decoupling of the vortex–antivortex pairs above T_{BKT} leads to the breakdown of superconductivity and namely the temperature T_{BKT} ($T_{\mathrm{BKT}} < T_{c0}$) has to be considered as the true transition temperature to superconducting state.

Below T_{BKT} all vortices and antivortices are bound and cannot transfer magnetic flux. It is remarkable that as temperature decreases at the transition point $T = T_{\mathrm{BKT}}$ the nonzero superfluid density appears in the system and this happens in a discontinuous way. The value of the superfluid density jump is universal [510]:

$$n_{s2}(T_{\mathrm{BKT}}) = \frac{4mT_{\mathrm{BKT}}}{\pi}. \tag{15.14}$$

Let us demonstrate how this jump appears from simple heuristic arguments due to Kosterlitz and Thouless [507]. The energy of a single vortex in a sample of linear dimension R evidently is

$$E_v = \frac{\pi n_{s2}(T)}{2m} \ln \frac{R}{\xi}. \tag{15.15}$$

The entropy is determined by the logarithm of the number of different states. The vortex core can be located anywhere in the sample,the area of its cross-section is of the order of $\pi\xi^2$. Therefore the number of different vortex states in the sample can be defined by the ratio of the sample area R^2 over the core area $\pi\xi^2$, while the free energy of the single vortex:

$$F = E - TS = \left[\frac{\pi n_{s2}(T)}{2m} - 2T\right] \ln \frac{R}{\xi}. \tag{15.16}$$

One can see that when $T < \pi n_{s2}(T)/4m$ the free energy is minimal for the state without vortices, while for $T > \pi n_{s2}(T)/4m$ the free energy is minimized when in system there are vortices. Vortices destroy stiffness so it is natural to identify the critical temperature of the transition with T_{BKT} determined by Eq. (15.14).

One can easily recognize that the value of the critical temperature T_{BKT}, separating the region where the vortices are bound in pairs from that where they are free, must be lower than the BCS (mean-field) critical temperature T_{c0}, the temperature at which the vortices disappear.

Using the ideas of the BKT theory we can clear up the behavior of the superfluid density in the transition region. Indeed, in chapter 2, in the framework of the GL theory we did not succeed in entering into the critical region, and the validity of the obtained result (2.149) was restricted to $|\epsilon| \gg Gi_{(2)}$ only. At the same time the Nelson–Kosterlitz jump Eq. (15.14) establishes the value of superfluid density $n_{s2}(T_{\mathrm{BKT}})$ at the point of the BKT transition. As a result an interpolation formula for the superfluid density of the 2D system can be written [74] by the unification of (2.149) and (15.14):

$$n_{s2}(T) = \frac{mT_{\mathrm{BKT}}}{\pi}\left[\frac{|\epsilon|}{Gi_{(2)}} - 2\ln\frac{Gi_{(2)}}{|\epsilon| + Gi_{(2)}} + 4\right], \tag{15.17}$$

where $\epsilon = (T - T_{\mathrm{BKT}})/T_{\mathrm{BKT}}$. When $\epsilon \gg Gi_{(2)}$ this formula reduces to Eq. (2.149), when $\epsilon = 0$ it reproduces the Nelson–Kosterlitz jump (15.14).

In absence of a magnetic field only thermally generated vortices are present in the system and part of them are free for $T > T_{\mathrm{BKT}}$. The characteristic size, the so-called vortex correlation length $\xi_+(T)$, physically represents the scale at which vortices begin to unbind at temperature $T > T_{\mathrm{BKT}}$. It evidently plays the role of GL coherence length $\xi(T)$ and diverges at $T \to T_{\mathrm{BKT}}$. At distances less than $\xi_+(T)$ vortices are still bounded in pairs, even though the temperature is above the nominal vortex unbinding temperature.

Since the discovery of the BKT state the dependence of ξ_+ on temperature had been the object of a great interest. By imposing that at the scale ξ_+ the vortex–antivortex interaction is completely reduced Kosterlitz obtained for it in the framework of the renormalization group analysis the unusual exponential form [458, 508]:

$$\xi_+(T) = \xi(T) \exp\left(b\sqrt{\frac{Gi_{(2)}}{\epsilon}}\right), \tag{15.18}$$

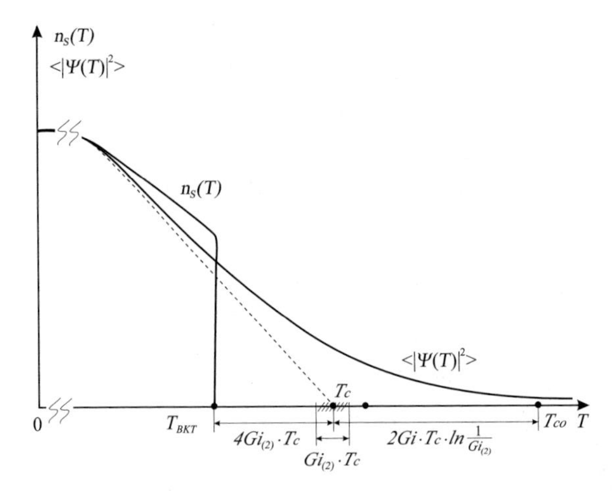

FIG. 15.1. Temperature dependencies of the superfluid density $n_{s2}(T)$ and the averaged over fluctuations square of the order parameter $\langle|\Psi(T)|^2\rangle$ in the vicinity of the transition. Dashed line corresponds to the BCS theory, although it is drawn starting not with the mean-field critical temperature T_{c0}, but with the renormalized by superconducting fluctuations critical temperature T_c. The latter turns out to be shifted with respect to T_{c0} by the value $T_{c0}Gi_{(2)}\ln Gi_{(2)}$. In its turn the temperature of BKT transition T_{BKT} turns out to be below T_c for the value $4T_{c0}Gi_{(2)}$, i.e. it is out of the critical region.

where b is a constant of the order of unity while $\xi(T)$ is the usual GL coherence length calculated for temperature T.[107] The appearance of the factor $Gi_{(2)}$ in the exponent is easy to understand: when $T - T_{\mathrm{BKT}} \geq T\, Gi_{(2)}$ the system is beyond the critical region and one cannot distinguish between fluctuation vortices and order parameter fluctuations, i.e. $\xi_+(T) \sim \xi(T_{\mathrm{BKT}})$.

One can find the density of free vortices at temperature slightly above T_{BKT}. Since all vortices within the clusters of size $\xi_+(T)$ are still paired, the free vortex density is

$$n_f = \frac{1}{\xi_+^2(T)}. \tag{15.19}$$

The temperature dependencies of the superfluid density and the average square of the order parameter in the vicinity of the transition and the relative position of different critical temperatures are presented in Fig. 15.1.

[107] This expression was specified by Minnhagen, taking into account the temperature dependence of the superfluid density [511]:

$$\xi_+(T) = \xi(T)\exp\left(b\sqrt{\frac{T_{c0} - T}{T - T_{\mathrm{BKT}}}}\right).$$

15.4 Manifestation of vortex fluctuations above BKT transition

We have now all essential to evaluate the effect of fluctuation unbinding of vortex–antivortex pairs on the normal state properties of $2D$ superconducting system slightly above T_{BKT}, which we already identified with the superconducting transition temperature. Since the difference between T_{BKT} and T_{c} lies in the limits of $Gi_{(2)}T_{\mathrm{c}}$ the matter of our main concern here will be the critical region $\epsilon \lesssim Gi_{(2)}$.

We started this book from the study of the fluctuation contribution to heat capacity of the superconducting grain. In order to do this the partition function with the GL free energy, including fourth order term was calculated exactly. As a result we found that in $0D$ case there is no any singularity at the mean field transition point $T_{\mathrm{c}0}$, fluctuations completely smear out phase transition. In spite of the mean-field theory prediction the same conclusion was done for the phase transition in $1D$ wire (see section 14.1). Renormalization group analysis of the $3D$ case (section 2.6) demonstrated the presence of a weak ($\sim \epsilon^{-1/10}$) singularity in heat capacity in the confines of the critical region. The question concerning the singularities of thermodynamic functions at the transition point of $2D$ system remained open and in this section we will discuss this problem.

The possibility of fluctuation dissociation of the vortex–antivortex pairs proves itself in the heat capacity of the BKT phase. The relative contribution to the heat capacity can be evaluated basing on relations (15.15)-(15.19). For temperatures $T - T_{\mathrm{BKT}} \ll Gi_{(2)}T_{\mathrm{BKT}}$ the main temperature dependence of energy originates from the exponential coherence length (15.18), so the superfluid density can be taken at the transition point (see (15.14)–(15.17)). As a result for the vortex-antivortex part of heat capacity with the logarithmic accuracy one obtains the expression

$$\frac{C_v\left(\epsilon \ll Gi_{(2)}\right)}{\Delta C} \sim -\exp\left(-2b\sqrt{\frac{Gi_{(2)}}{\epsilon}}\right). \tag{15.20}$$

This consideration demonstrates that the contribution to heat capacity due to fluctuation dissociation of the vortex–antivortex pairs, being exponentially small at the transition point, noticeably growths with the temperature increase in the confines of the critical region. Let us emphasize the negative sign of the vortex–antivortex pairs dissociation contribution, opposite to that one due to the long wavelength order parameter fluctuations. This fact means that the mean-field result for the heat capacity jump overestimates the true value and the vortex-antivortex pairs contribution smears it out.

The d.c. resistivity of the vortex–antivortex system at temperatures above T_{BKT} is related to the presence of unpaired vortices able to move in the electric field and hence to dissipate energy. They can be both free vortices, induced by an external magnetic field, or thermally induced ones appearing in the process of the vortex–antivortex pair dissociation. In $2D$ case, the pinning energy of a vortex is small in comparison with T hence the vortex system can be considered as a liquid.

The voltage associated with the vortex motion in a superconductor can be obtained from the Josephson relation (13.19):

$$V = \frac{1}{2e}\frac{d\phi}{dt},$$

where V is the voltage drop and ϕ is the phase difference between the two ends of the sample. Since a phase-slip of 2π occurs when the vortex crosses the sample one has:

$$|\frac{d\phi}{dt}| = 2\pi L n_f |v_d|,$$

where L is the length of the sample, n_f is the vortex concentration and $\mathbf{v}_d$ is the drift velocity of the vortices across the film (the direction of $\mathbf{v}_d$ is opposite for vortices of the opposite helicity). Neglecting the pinning effects one can say that this velocity is simply proportional to the Lorentz (Magnus) force: $\mathbf{v}_d = \mathbf{f}_i/\eta_v$, where η_v is the vortex viscosity and $\mathbf{f}_i = \pm\Phi_0 \widehat{z} \times \mathbf{j}_v$. Let us note that the vortex viscosity η_v is not singular near T_{BKT}. One can say that each fluctuation vortex carries the magnetic flux quantum $\Phi_0 = \pi/e$. Therefore, the electric field $E = V/L$ is determined by the rate of magnetic flux transfer:

$$E = \Phi_0^2 \frac{n_f}{\eta_v}\mathbf{j}_v.$$

In absence of applied magnetic field only thermally generated vortices present in the system and they are free for $T > T_{\mathrm{BKT}}$. Their concentration is determined by the formula (15.19) and the resistivity corresponding to this vortex–antivortex flow for $T > T_{\mathrm{BKT}}$ is

$$\rho_v\left(\epsilon \ll Gi_{(2)}\right) = \Phi_0^2 \frac{n_f}{\eta_v} = \frac{\Phi_0^2}{\eta_v \xi^2\left(T_{\mathrm{BKT}}\right)} \exp\left(-2b\sqrt{\frac{Gi_{(2)}}{\epsilon}}\right). \tag{15.21}$$

Substituting the explicit expression for the vortex viscosity (5.11) one finally finds

$$\rho_v\left(\epsilon \ll Gi_{(2)}\right) \sim \rho_n \sqrt{Gi_{(2)}}\left[1 + \frac{2}{T\tau_\varepsilon Gi_{(2)}}\right] \exp\left(-2b\sqrt{\frac{Gi_{(2)}}{\epsilon}}\right).$$

This expression gives an Ohmic resistance of the $2D$ system slightly above the BKT transition ($\epsilon \ll Gi_{(2)}$) and it describes rather well typical experimental data on $R(T)$ in this region of temperatures.

16

FLUCTUATIONS NEAR SUPERCONDUCTOR-INSULATOR TRANSITION

Quantum phase transitions occur at zero temperature when another parameter is varied (magnetic field, carriers density, etc). In the absence of static disorder, the classical transition with changing the crystalline structure in a sample of dimensionality D leads to the same singularity in thermodynamic characteristics as the quantum transition at space dimension $D - 1$ [4]. The static disorder changes the character of singularities even at a classical transition [445–447]. Strong disorder can suppress the transition temperature up to zero and gives rise to a quantum phase transition.

Disorder enhances the effect of Coulomb repulsion. As a result, the critical temperature T_c of $2D$ superconductors decreases [92, 93, 512] and can even go to zero [95]. At this quantum critical point the conductivity takes a universal value [513].

16.1 Quantum phase transition

It is usually supposed that the temperature of the superconducting transition does not depend on the concentration of nonmagnetic impurities (Anderson's theorem [235, 236]). Nevertheless when the degree of disorder is very high Anderson localization takes place in the normal phase of superconductor. At that it would be difficult to expect that under conditions of strong electron localization superconductivity can exist, even if there is inter-electron attraction. This means that at $T = 0$ a phase transition should takes place driven by the disorder strength or carrier concentration. Such a transition is called a quantum phase transition since at zero temperature the classical fluctuations are absent. Indeed, one can see from (1.5) that in the limit $T \to 0$ the thermal fluctuation Cooper pairs vanish.

In the metallic phase of a disordered system the conductivity is mostly determined by the weakly decaying fermionic excitations. Their dynamics results in the familiar Drude formula (the method which accounts for the fermionic excitations will be referred to as the Fermi approach). Inside the critical region the charge transfer due to fluctuation Cooper pairs turns out to be more important. We demonstrated above that up to some extension the fluctuation Cooper pairs may be considered as Bose particles. Therefore, the following approach dealing with the fluctuation pairs as the charge carriers will be called the Bose approach.

Let us suppose that at temperature $T = 0$ the superconducting state occurs in a weakly disordered system. In principle two scenarios of the development of the situation are possible with an increase of the disorder strength: the system at some critical disorder strength can go from the superconducting state to the metallic state or to the insulating state. The first scenario is natural and takes place in the following cases: if the effective constant of the inter-electron interaction changes its sign with the growth of the disorder; if the effective concentration of magnetic impurities increases together with the disorder growth; if the pairing symmetry of superconducting state is nontrivial it can be destroyed even by the weak disorder level. We will study here the scenario where the superconductor becomes an insulator with disorder increase. This means that at some disorder degree range, higher than the localization edge when the normal phase does not exist any more at finite temperatures, superconductivity can still survive. From the first glance this statement seems strange: what does superconductivity mean if the electrons are already localized? And if it really can take place beyond the metallic phase, at what value of disorder strength and in which way does the superconductivity finally disappears?

One has to have in mind that localization is a quantum phenomenon in its nature and with the approach to the localization edge the coherence length of localization L_l grows. From the insulator side of the transition vicinity this means the existence of large scale regions where delocalized electrons exist. If the energy level spacing in such regions does not exceed the value of superconducting gap Cooper pairs still can be formed by the delocalized electrons of this region.

The problem can be reformulated in another, already familiar, way: how does the critical temperature of the superconducting transition decrease with the increase of the disorder strength? In the previous sections we have already tried to solve it by discussing the critical temperature fluctuation shift. We have seen that the fluctuation shift of the critical temperature is proportional to $\sqrt{Gi_{(3)}}$ for a $3D$ superconductor and to $Gi_{(2)} \ln\{1/Gi_{(2)}\}$ for $2D$. This means that the critical temperature is not changed noticeably as long as the Gi number remains small. So one can expect the complete suppression of superconductivity when $Gi \sim 1$ only. For further consideration it is convenient to separate the $3D$ and $2D$ cases because the physical pictures of the superconductor–insulator transition for them are rather different.

16.2 $3D$ case

As one can see from Eq. (2.87) in the $3D$ case the Gi number remains small at $p_F l \sim 1$: $Gi_{(3d)} \approx (T_c/E_F) \ll 1$. Nevertheless, approaching the edge of localization, the width of the fluctuation region increases [514]. In the framework of the self-consistent theory of localization [515] such a growth of the width of the fluctuation region was found in paper [516].

Instead of the cited self-consistent theory let us make some more general assumptions concerning the character of the metal–insulator (M–I) transition in the absence of superconductivity [74]. We suppose that in the case of very strong disorder and not very strong Coulomb interaction the M–I transition is of second order. The role of "temperature" for this transition is played by the "disorder strength" which is characterized by the dimensionless value $g = p_F l/2\pi$. With its decrease the conductivity of the metallic phase decreases and at some critical value g_c tends to zero as

$$\sigma = e^2 p_F (g - g_c)^{\varkappa}. \tag{16.1}$$

This is the critical point of the Anderson (M–I) transition. We assume that the thermodynamic DOS remains constant at the transition point.

The electron motion in metallic phase far enough from the M–I transition has a diffusion character and the conductivity can be related to the diffusion coefficient $\mathcal{D} = p_F l/3m$ by the Einstein relation: $\sigma = \nu e^2 \mathcal{D}$. One can say that diffusion like "excitations" with the spectrum $\omega(q) = iDq^2$ propagate in the system. At the point of the M–I transition normal diffusion terminates and conductivity, together with $\mathcal{D}$, turns zero. In accordance with scaling ideas, the diffusion coefficient can be assumed here to be a power function of q: $\mathcal{D}(q) \sim q^{z-2}$, with the dynamical critical exponent $z > 2$. The anomalous diffusion excitation spectrum in this case would take the form $\omega \sim q^z$. Since hitherto the value of z is unknown we will investigate below different options.

In the insulating phase ($g < g_c$) some local, anomalous diffusion, confined to regions of the scale L_l, is still possible. It cannot provide charge transfer through out all the system, so $\mathcal{D}(q = 0) = 0$, but for small distances ($q \gtrsim L_l^{-1}$) anomalous diffusion takes place. Analogously, in the metallic phase ($g > g_c$) the diffusion coefficient in the vicinity of the transition has an anomalous dependence on q for $q \gtrsim L_l^{-1}$ ($\mathcal{D}(q) \sim q^{z-2}$) and weakly depends on it for $q \lesssim L_l^{-1}$. So one can conclude that the diffusion coefficient for $q \gtrsim L_l^{-1}$ from both sides of the transition has the same q-dependence as for all q in the transition point. It can be written in the form

$$\mathcal{D}(q) \sim \frac{g}{3m} \left[\frac{\varphi(qL_l)}{p_F L_l} \right]^{z-2}, \tag{16.2}$$

where

$$\varphi(x) = \begin{cases} x, \; x \gg 1, \\ 1, \; x \ll 1, g > g_c, \\ 0, \; x \ll 1, g < g_c. \end{cases} \tag{16.3}$$

The localization length L_l, characterizing the spatial scale near the transition, grows with the approach to the transition point like

$$L_l(g) \sim \frac{1}{p_F} (g - g_c)^{-\frac{\varkappa}{z-2}}. \tag{16.4}$$

The critical exponent in this formula is found from the Einstein relation in the vicinity of the M–I transition.

At finite temperatures, instead of the critical point g_c, a crossover from metallic to insulating behavior of $\sigma(g)$ takes place. The width of the crossover region is $\widetilde{g} - g_c$,where $\widetilde{g}$ is determined from the relation $\mathcal{D} L_l^{-2}(\widetilde{g}) \sim E_F [p_F L_l(\widetilde{g})]^{-z} \sim T$

(we have used the second asymptotic of (16.2)). In this region the diffusion coefficient is

$$\mathcal{D}(T) \sim TL_l^2 \sim \frac{T}{p_F^2}\left(\frac{E_F}{T}\right)^{2/z} \tag{16.5}$$

and it depends weakly on the $g - g_c$. Beyond this region the picture of the transition remains the same as at $T = 0$.

Let us consider now what happens to superconductivity in the vicinity of the localization transition. In the mean-field approximation (BCS) the thermodynamic properties of a superconductor do not depend on the character of the diffusion of excitations. This should be contrasted with the fluctuation theory, where such a dependence clearly exists. We will show that the type of superconducting transition depends on the dynamical exponent z. If $z > 3$, the transition to superconductivity occurs on the metallic side of the localization transition (we will refer to such a transition as S–N transition). If $z < 3$, the transition to superconductivity occurs from the insulating state directly (S–I transition).

Let us study how the superconducting fluctuations affect the transition under discussion. In spirit of the GL approach fluctuation phenomena in the vicinity of the transition can be described in the framework of the GL functional (2.30). The coherence length in the metallic region, far enough from the Anderson transition, was reported in Introduction to be equal to $\xi^2 = \xi_c l = 0.42\mathcal{D}/T$. In the vicinity of the M–I transition we still believe in the diffusive character of the electron motion resulting in the pair formation. The only difference from the previous consideration is the anomalous character of the quasiparticle diffusion. So in order to describe the superconducting fluctuations simultaneously near superconducting (in temperature) and Anderson (in g) transitions let us use the GL functional (2.30) with the k-dependent diffusion coefficient (16.2).

The value of Gi can be estimated from the expression for the fluctuation contribution to heat capacity (2.34) taken at $\epsilon \sim Gi$, where the fluctuation correction reaches the value of the heat capacity jump:

$$1 \sim \frac{T}{\nu}\int \frac{d^3q}{\left(TGi + \mathcal{D}(q)q^2\right)^2}, \tag{16.6}$$

with $T \simeq T_c$. Let us approach the M–I transition from the metallic side. If we are far enough from transition, Gi is small and the integral in (16.6) is determined by the region of small momenta $\mathcal{D}(q)q^2 \lesssim TGi$:

$$Gi \sim \frac{T}{\nu^2\mathcal{D}^3(q=0)}. \tag{16.7}$$

Two scenarios are possible: Gi becomes of the order of 1 in the metallic phase, or it remains small up to the crossover region, where finally it reaches its saturation value. In the first case we can use the second asymptotic of (16.2) for $D(q)$ and find:

$$Gi \sim \frac{T_{\rm c}}{E_F}(p_F L_l)^{3z-6}. \qquad (16.8)$$

One can see that Gi becomes of the order of 1 at $p_F L_M \sim (E_F/T)^{1/(3z-6)}$. Comparing this value with $p_F L_l(\widetilde{g}) \sim (E_F/T)^{1/z}$ at the limit of the crossover region we see that for $z > 3$ the first scenario is realized. Concluding the first scenario discussion we see that the superconducting critical temperature goes to zero at $L_l = L_M$, still in the metallic phase, so at $T = 0$ a S–N type quantum phase transition takes place.

The second scenario takes place for $z < 3$ when Gi remains small even at the edge of crossover region, reaching there the value

$$Gi \sim \left(\frac{T_{\rm c}}{E_F}\right)^{2(3-z)/z} \ll 1. \qquad (16.9)$$

In the crossover region the diffusion coefficient, and hence Gi, almost do not vary. That is why the temperature of superconducting transition remains almost frozen with further increase of disorder driving the system through the Anderson transition. The abrupt growth of Gi and decrease of $T_{\rm c}$ take place when the system finally goes from the crossover to the insulating region. In the insulator phase the diffusion coefficient $D(q \lesssim l^{-1}) = 0$ and from (16.6) one can find for Gi:

$$Gi \sim \sqrt{\frac{E_F}{T_{\rm c}}}\frac{1}{(p_F l)^{3/2}}. \qquad (16.10)$$

Comparing this result with the values in Table 2.1 one can see that it coincides with the Gi number for a $0D$ granule of size $L_l \ll \xi(T)$. Hence we see that in the second scenario the Ginzburg number reaches 1 and, respectively, $T_{\rm c} \to 0$ at $p_F L_I \sim (E_F/T_{\rm c})^{1/3}$, which is far enough from the M–I transition point. That is why in this case one can speak about the realization at $T = 0$ of a S–I type quantum phase transition. The scale L_I determines the size of the "conducting" domains in the insulating phase, where the level spacing reaches the order of the superconducting gap. It is evident that in the domain of scale $L_l \lesssim L_I$ superconductivity cannot be realized.

In the vicinity of a quantum phase transition one can expect the appearance of nonmonotonic dependencies of the resistance on temperature and magnetic field. Indeed, starting from the zero resistance superconducting phase and increasing temperature from $T = 0$, the system passes through the localization region, where the resistance is high, to high temperatures where some hopping charge transfer will decrease the resistance again. The analogous considerations are applicable to the magnetic field effect: first the magnetic field "kills" superconductivity and increases the resistance, then it destroys localization and decreases it.

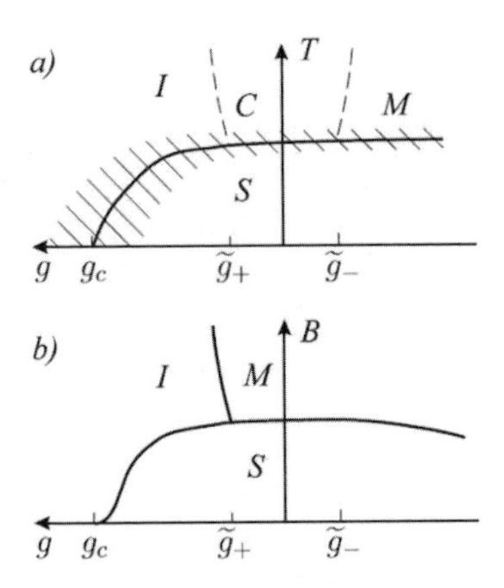

FIG. 16.1. Phase diagram in the temperature–disorder plane for a $3D$ superconductor.

The phase diagram in the (T, g) plane has the form sketched in Fig. 16.1. For $g_I = g_c - (T_c/E_F)^{3(z-2)/\varkappa}$, an S–I transition takes place at $T = 0$. Increasing the temperature from $T = 0$ in the region $0 \lesssim g \lesssim g_I$ we remain in the insulating phase with exponential dependence of resistance on temperature. For $g_I \lesssim g \lesssim \tilde{g}_-$, at low temperatures $0 \leq T < T_c(g)$, the system stays in the superconducting state which goes to the insulating phase at higher temperatures. In the vicinity of the Anderson transition ($\tilde{g}_- \lesssim g \lesssim \tilde{g}_+$) the superconducting state goes with growth of the temperature to some crossover metal–insulator state which is characterized by a power decrease of the resistivity with the increase of temperature. Finally at $g_c \lesssim g$ the superconducting phase becomes of the BCS type and at $T = T_c$ it goes to a metallic phase.

The phase diagram in the magnetic field–disorder plane is similar to that in the (T, g) plane with the only difference that at $T = 0$ there is no crossover region, instead a phase transition takes place.

16.3 $2D$ superconductors

16.3.1 *Preliminaries*

As was demonstrated in section 3.7, according to the conventional theory of paraconductivity, the sheet conductivity in the vicinity of the superconducting transition is given by a sum of the electron residual conductivity ge^2 (Fermi part) and the conductivity of the Cooper pair fluctuations (Bose part) (see (3.80)). This expression is valid in the GL region when the second term is a small correction to the first one. The width of the critical region can be determined from the requirement of equality of both contributions in (3.80):

$$\epsilon_{\rm cr} = \frac{1}{16g} = 1.3Gi_{(2d)}. \tag{16.11}$$

In accordance with general scaling ideas one can believe that inside the fluctuation region the conductivity should obey the form:

$$\sigma(T) = ge^2 f\left(\frac{\epsilon}{Gi_{(2d)}}\right). \tag{16.12}$$

Concerning the scaling function $f(x)$, we know its asymptotes in the mean field region ($x \gg 1$) and just above the BKT transition [506, 507]:

$$f(x) = \begin{cases} 1 + x^{-1}, & x \gg 1, \\ \exp\left[-b(x - x_{\rm BKT})^{-1/2})\right], & x \to x_{\rm BKT} = -4. \end{cases} \tag{16.13}$$

The BKT transition temperature $T_{\rm BKT}$ is determined by Eq. (15.14) and one can find its value by comparing the superfluid density n_s from (15.14) with that found in the Ginzburg–Landau scheme (see (2.149)):

$$T_{\rm BKT} = T_c(1 - 4Gi). \tag{16.14}$$

Here we assumed that the parameter Gi is small, so that the BKT transition temperature does not deviate much from that one calculated in the frameworks of the BCS scheme shifted by longwavelength fluctuations.

16.3.2 *Boson mechanism of $T_{\rm c}$ suppression*

The classical and quantum fluctuations reduce n_s and therefore, suppress $T_{\rm BKT}$. At some $g = g_{\rm c} \sim 1$, the superfluid density n_s, and simultaneously $T_{\rm BKT}$, go to zero. In the vicinity of this critical concentration of impurities $T_{\rm BKT} \ll T_c$. Thus a wide new window of intermediate temperatures $T_{\rm BKT} \ll T \ll T_c$ opens up. In this window, according the dynamical quantum scaling conjecture [513], one finds

$$\sigma = e^2\varphi\left(\frac{T}{T_{\rm BKT}}\right). \tag{16.15}$$

At $T - T_{\rm BKT} \ll T_{\rm BKT}$ the BKT law (16.12)–(16.13) should hold, so $\varphi(x) = f(x)$ and is exponentially small. In the intermediate region $T_{\rm BKT} \ll T \ll T_{\rm c}$ the duality hypothesis [517] gives $\varphi(x) = \pi/2$. Let us derive this relation.

We will start from the assumption that in the region $T_{\rm BKT} \ll T \ll T_c$ the conductivity is a universal function of temperature which does not depend on the pair interaction type. Being in the framework of the classical approach, let us suppose that in a weak electric field pairs move with the velocity $\mathbf{v} = \mathbf{F}/\eta$, where $\mathbf{F} = 2e\mathbf{E}$ is the force acting on the pairs. The current density $\mathbf{j} = 2en\mathbf{v} = \sigma\mathbf{E}$, (here n is the pair density), so one can relate the conductivity with the effective viscosity η: $\sigma = 4e^2n/\eta$.

Let us recall that we are dealing with a quantum fluid, so another view, taking superconductivity into account, on the problem of its motion near the quantum phase transition exists. One can say that with the increase of Gi the role of quantum fluctuations grows too and fluctuation vortices carrying the magnetic flux quantum $\Phi_0 = \pi/e$ are generated. With electric current flow in the system the Lorentz (Magnus) force acts on a vortex: $F = j\Phi_0$. The electric field is equal to the rate of magnetic flux transfer, i.e. to the density of the vortex current: $\mathbf{E} = \Phi_0 n_v \mathbf{v}_v = \Phi_0 n_v \mathbf{F}/\eta_\mathbf{v}$, where n_v is the density and η_v is the viscosity of the vortex liquid. As a result $\mathbf{E} = \Phi_0^2 n_v \mathbf{j}/\eta_\mathbf{v} = \mathbf{j}/\sigma$. So one can conclude that for vortices the velocity is proportional to the voltage, and the force is proportional to the current. For Cooper pairs (bosons) the situation is exactly the opposite.

The duality hypothesis consists in the assumption that at the critical point the pair and the vortex liquid density flows are equal: $n_v \mathbf{v}_v = n\mathbf{v}$. Comparing these quantities, expressed in terms of the conductivity from the above relations, one can find a universal value for the conductivity at the critical point

$$\sigma_c = \frac{2e}{\Phi_0} = \frac{2e^2}{\pi}. \tag{16.16}$$

One can restrict oneself to a less strong duality hypothesis, supposing the product $n\eta = CT^\delta$ with a universal δ exponent both for the pair and the vortex liquids, while the constant C for them is different. In this case, based on duality, is possible to demonstrate that $\delta = 0$ and the conductivity is temperature independent up to T_{c0} but its value is not universal any more and can vary from one sample to another.

To conclude, let us emphasize that in the framework of the boson scenario of superconductivity suppression, the BCS critical temperature is changed insignificantly, while the "real" superconducting transition temperature $T_{\mathrm{BKT}} \to 0$.

16.3.3 *Fermion mechanism of T_c suppression*

Apart from the above fluctuation (boson) mechanism of the suppression of the critical temperature in the 2D case, there exists another, fermionic mechanism. The suppressed electron diffusion results in a poor dynamical screening of the Coulomb repulsion which, in turn, leads to the renormalization of the inter-electron interaction in the Cooper channel and hence to the dependence of the critical temperature on the value of the high-temperature sheet resistivity of the film. As long as the correction to the non-renormalized BCS transition temperature T_{c0} is still small, one finds [92, 93, 512, 399]:

$$T_\mathrm{c} = T_{c0}\left(1 - \frac{1}{12\pi^2 g}\ln^3\frac{1}{T_{c0}\tau}\right). \tag{16.17}$$

At small enough T_{c0} this mechanism of critical temperature suppression turns out to be the principal one. The suppression of T_c down to zero in this case may happen in principle even at $g \gg 1$. A renormalization group analysis gives [95] the corresponding critical value of conductance

$$g_c = \left(\frac{1}{2\pi} \ln \frac{1}{T_{c0}\tau} \right)^2 . \tag{16.18}$$

Here we should recall that the typical experimental [518] values of g_c are in the region $g_c \sim 1 - 2$, and do not differ dramatically from the predictions of the boson duality assumption $g_c = 2/\pi$. If one attempts to explain the suppression of T_c within the fermion mechanism, one should assume that $\ln(1/T_{c0}\tau) > 5$. Then, according to Eq. (16.18), $g_c > 2/\pi$ and the boson mechanism is not important. On the contrary, if $\ln(1/T_{c0}\tau) < 4$, then Eq. (16.17) gives a small correction for T_c even for $g_c = 2/\pi$ and the fermion mechanism becomes unimportant. The smallness of the critical temperature T_c compared to the Fermi energy is the cornerstone of the BCS theory of superconductivity and it is apparently satisfied even in HTS materials. Nevertheless it is necessary to use the theoretically large logarithmic parameter with care, if one needs $\ln(1/T_{c0}\tau)$ to be as large as 4.

17

ROLE OF FLUCTUATIONS IN HIGH TEMPERATURE SUPERCONDUCTIVITY

The renaissance of interest in fluctuation phenomena became the consequence of the discovery of high temperature superconductivity. There are two principal reasons for this revival. The first consists in the fact that the traditional fluctuation effects, which we have discussed all along this book, turn out to be, in overdoped and optimally doped HTS, relatively large in magnitude and important for understanding their anomalous properties. The second reason is the dramatic growth of the normal state anomalies in the underdoped phases of HTS, and the inability of the traditional fluctuation theory to explain them. Actually these huge anomalies appear in the underdoped phase as the natural continuation of the corresponding fluctuation manifestations of the overdoped phases, hence the generalization of the fluctuation theory on strong fluctuation is required. Moreover, many of investigators suspect in fluctuations of this or that type the origin of intriguing peculiarities of the cuprate superconductivity itself.

As is recognized now the optimally or overdoped phases of HTS compounds present an example of a "bad" Fermi liquid. The accounting for superconducting fluctuations here is identical to including of the electron–electron interaction beyond the mean field approximation. Superconducting fluctuations turn out to be strong here by the following reasons: high values of critical temperature, low charge carriers concentration and effective two-dimensionality of oxide superconductors. All these factors can make $Gi_{(2)} \sim T_c/E_F$ high up to 0.1. As a result, the theory of fluctuation phenomena presented here allows us to explain a number of normal state anomalies of the optimally or overdoped phases of HTS.

Nevertheless, analyzing the rapid growth of the normal state anomalies with the decrease of the oxygen content below the optimal doping concentration one notices that it strongly overcomes the theoretical predictions. In the framework of the naive fluctuation theory it turns out to be impossible to explain huge pseudogap-like phenomena typical of the underdoped phase. This is due to the fact, that the quasiparticle approach, underlying in the basis of the traditional theory of superconductivity (see Part II), fails in the underdoped phase. Probably the reason of dramatic increase of the role of fluctuations in the underdoped region of the phase diagram is the same as that one of high values of the critical temperature: this is the vicinity of the system to the antiferromagnetic transition. Strong quantum antiferromagnetic fluctuations (paramagnons) change here the estimation for Gi number, the value of the effective Fermi energy and put

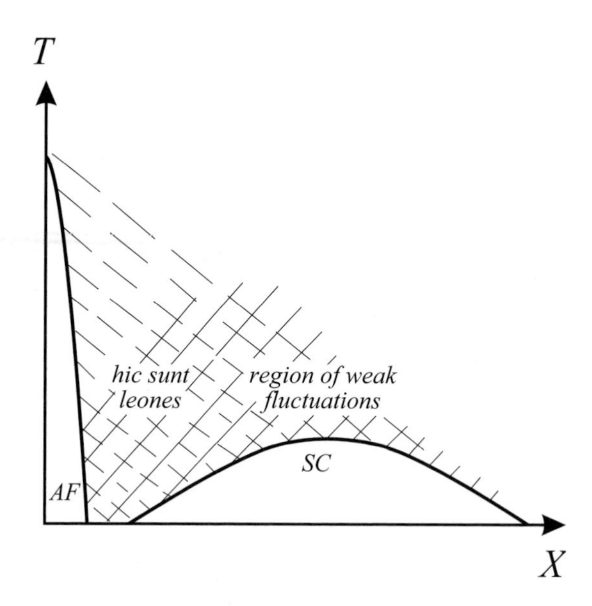

FIG. 17.1. Typical phase diagram of HTS. The land "hic sunt leones" corresponds to the domain of strong fluctuations.

under the question the applicability of the Fermi liquid scenario at all [519]. Fluctuations are so strong here that it is senseless to develop their theory basing on one or another version of the perturbation theory. New concepts are required in this matter and below we present some short review of already existing ideas.

There is no any quantitative microscopic theory of HTS at the moment. The variety of different semi-phenomenological models exists, which partially overlap, partially contradict each other. In this chapter we will give their comparative description, emphasizing the role of fluctuation in every model.

17.1 Phase diagram

The closeness to the antiferromagnetic transition one can adjust changing the degree of the oxygen doping x. For example, the typical phase diagram of HTS in $x - -T$ plane is presented in Fig. 17.1.

Overdoped compounds are far enough from the AF transition and the role of paramagnons here is reduced to that one of the phonons in the conventional superconductivity. The peculiarities of paramagnon exchange can be responsible for d-wave symmetry of the superconducting paring in HTS [519], but in other aspects superconductivity is of the BCS type. The region of superconducting fluctuations here is relatively narrow and it is well described by the theory presented in precedent chapters. At higher temperatures the normal state properties of the overdoped superconductors are well described by the Landau theory of Fermi liquid.

In optimally doped oxides ($x \approx 0.18$), where the critical temperature reaches its maximal value, some normal state anomalies (like in magnetoresistance) are still described by the theory of weak fluctuations. Other anomalies (like in Nernst effect [142]) can be explained in the assumption that superconducting fluctuations are still present at temperatures considerably higher than critical one.

In underdoped compounds the superconducting critical temperature approaching AF transition decreases, while the domain of the pseudogap[108] manifestation is extended, i.e. superconducting fluctuations grow.[109] Evidently, strong superconducting fluctuations have to be studied together with strong AF fluctuations. Nevertheless, such a unique theory still does not exist.

17.2 Resonating valence bond theory

The general picture describing the state with such strong fluctuations was proposed by Anderson [521–523] and is called the theory of the resonating valence bond (RVB). In accordance with RVB scenario, the electrons which posses the opposite spins and located in the neighboring cites of the crystal lattice form pairs. At fixed temperature with decrease of the doping degree these pairs are ordered in lattice and the state becomes antiferromagnetic. Vice versa, when the doping is fixed and temperature decreases, the pairs, being Bose particles, are condensed and the state becomes superconducting. The pseudogap manifests itself in the processes where the pair-breaking takes place: for instance in c-axis conductivity, in photoemission, in Knight shift etc.

17.2.1 *Strongly correlated Fermi systems*

The efforts to construct the microscopic theory of RVB state got the name "Theory of strongly correlated Fermi systems." The simplest models, like Hubbard one

$$\mathcal{H}_H = \sum_{ij} t_{ij} \left(c_{i\uparrow}^{+} c_{j\uparrow} + c_{i\downarrow}^{+} c_{j\downarrow} \right) + U \sum_{i} c_{i\uparrow}^{+} c_{i\downarrow}^{+} c_{i\uparrow} c_{i\downarrow}, \tag{17.1}$$

or t–J model

$$\mathcal{H}_{t-J} = \mathcal{P} \sum_{ij} t_{ij} \left(c_{i\uparrow}^{+} c_{j\uparrow} + c_{i\downarrow}^{+} c_{j\downarrow} \right) \mathcal{P} + \sum_{ij} J_{ij} \left(2 S_i S_j - \frac{1}{2} n_i n_j \right) \tag{17.2}$$

[108] The notion of the pseudogap, widely used in HTS literature, requires some special comments. Usually it is related to the opening of some smeared gap in the one-particle excitation spectrum, observed in the photoemission, tunneling, optical, out-of-plane conductivity and other measurements (for a recent review see e.g. [520]). As we already saw above, the accounting for superconducting fluctuations in the Fermi-liquid (overdoped) part of the phase diagram also results in opening of the pseudogap. As far as one moves to underdoped region superconducting fluctuations become strong and their effect oversteps the limits of the used perturbation theory. Hence one could attribute the large pseudogap observed there to the superconducting fluctuations. At the same time the closeness of the underdoped region to the antiferromagnetic state gives grounds to think about the spin origin of the pseudogap. The true theory of the pseudogap in HTS has to accounting for both magnetic and superconducting pseudogap origin.

[109] In Fig. 17.1, the corresponding domain is called "hic sunt leones": ancient Romans called in such a way on their maps unknown areas of the African continent.

are usually studied in such approach ($\mathcal{P}$ is the projection operator which removes double occupancy of any cite). Nevertheless, in spite of the simple form of the Hamiltonian of such models, the quantitative results allowing the experimental verification still have not been obtained in their frameworks. For example, the $t-J$ model, even in the case $J_{ij} = 0\,(U = \infty)$ and at zero temperature can be studied analytically only in the limits of the weak or strong doping. Indeed, in the case of strong doping $x \to 1$ the holes density is small and the gas approximation can be used [524]. The small parameter of such theory for $2D$ case is $-1/\ln(1-x)$. In the opposite limit of the completely filled band ($x = 0$), when each site is occupied by one particle, the system turn out to be strongly degenerated by spin variable and removing even one particle (adding of one hole) switches it to the ferromagnetic state (Nagaoka theorem [525]). Yet this ferromagnetic region of the phase diagram turns out to be narrow by numerical reasons (in the model where the Hamiltonian does not contain any parameter the ferromagnetism disappears at $x_c \ll 1$). The presence of such numerical smallness results in the rich physical picture. What are the properties of the state when $x_c < x \ll 1$ and the ferromagnetic state is already broken? Ioffe and Larkin argued [526] that for $J = 0$ this state is ferrimagnetic (i.e. there are two ferromagnetic sublattices with almost opposite spin directions) and superconducting.

Now let us consider the effects of interaction ($J > 0$). For half-filling, where each site is occupied by one electron, no hopping is possible, and the Hamiltonian reduces to the quantum Heisenberg Hamiltonian with the antiferromagnetic spin exchange. Although the $2D$ Heisenberg model cannot be solved analytically, there is strong numerical evidence, that at $T = 0$ the system stays in antiferromagnetic state.[110]

The presence of holes makes the picture more complicated. Each hole hopping from site to site creates changes in the spin configuration, unless the spin polarization is uniform. The resulting excitations in the system inhibit the hopping, since the hopping probability is reduced by a factor proportional to the overlap between the original and final spin wavefunctions. This effect results in narrowing of the one-particle excitations band, thus decreasing the kinetic contribution to the energy. The band width is maximized in the fully polarized state for which the spin configuration is unaffected by permutations of different spins. Thus while the J term in the Hamiltonian favors an antiferromagnetic ordering, the kinetic t term favors a ferromagnetic state. The result of competition between these terms depends on the hole density and interaction strength J/t, determining the physical consequences of the t–J model [529].

Probably the t–J model or some of its modifications can describe the complete phase diagram presented in Fig. 17.1. But in spite of the existence of the small parameters $x \ll 1$ and $J \ll t$ the analytic solution of this model still does not exist and the characteristic values of J and x, separating different domains of

[110] The exact diagonalization of the Hamiltonian for the system of the finite number of spins shows that the long range antiferromagnetic correlations take place in the ground state [527, 528].

the phase diagram, can be found only numerically. For instance, when $J = 0.3t$ the numerical calculations [530] restrict the antiferromagnetic region to $x < 0.1$. For $x = 0.1$ the Neel temperature $T_N = 0$ but the Nagaoka ferromagnetic state is already broken.

Some properties of the t–J model or Hubbard Hamiltonian ground state can be found by means of the variational procedure. In this way the *d*-wave symmetry of the superconducting pairing was successfully obtained side by side with the quasiparticle spectral weight, Drude weight in the range $0.1 < x < 0.3$, and even antiferromagnetic instability at very low doping [530–535]. Calculated in this way order parameter remains finite even at zero doping while the superfluid density n_s depends on the doping degree linearly.[111] In accordance with the BKT theory $T_c \sim n_s$ and this can explain the decrease of critical temperature of the transition to superconducting state as one goes deeper into the underdoped region. The variation procedure provides the ground state wave function. Nevertheless even the knowledge of the ground state wave function is not sufficient to find the excitations spectrum and to explain the thermal properties of the system in the domain "hic sunt leones". That is why semi-phenomenological theories are required.

17.2.2 *Spin–charge separation*

One of these semi-phenomenological theories of strongly correlated Fermi system is based on the analogy with the $1D$ Luttinger–Tomanaga model [536]. In this model there exist two types of the low-energy excitations: spinons – excitations with spin 1/2, and holons – spinless charged excitations [537]. In contents of HTS the idea of spin-charge separation was introduced by Anderson and coauthors [538] (see also [522, 539–544]). In the $2D$ case one can get the spin–charge separation in certain models with large number of fields [545, 546]. The latter allows to use the MFA. Fluctuations of this mean-field are described by the gauge field **a** which appears side by side with spinons and holons. The Hamiltonian describing the interaction of spinons and holons with this gauge field is

$$\mathcal{H}_{\text{int}} = (\mathbf{j}_s + \mathbf{j}_h)\,\mathbf{a} + \mathbf{j}_s \mathbf{A}, \tag{17.3}$$

where **A** is the vector-potential of the external electric field, $\mathbf{j}_s, \mathbf{j}_h$ are spinon and holon current operators. In (17.3) the external field influences only with spinons, but changing variables $\mathbf{a} \to \mathbf{a} - \mathbf{A}$ one can make the electric field to interact only with holons.

The gauge field **a** entering $2D$ RVB theory is similar to gluon field in QCD. Its role consists in joining of the particles of both type. The presence of such field results in the additivity of resistances instead of conductivities. As the

[111] In the framework of the Fermi liquid theory one can understand the decrease of n_s on doping assuming that the superfluid density depends on the Fermi liquid constant (In the Galilean system n_s does not depend on this parameter, but in the crystalline lattice such dependence may turn out to be strong).

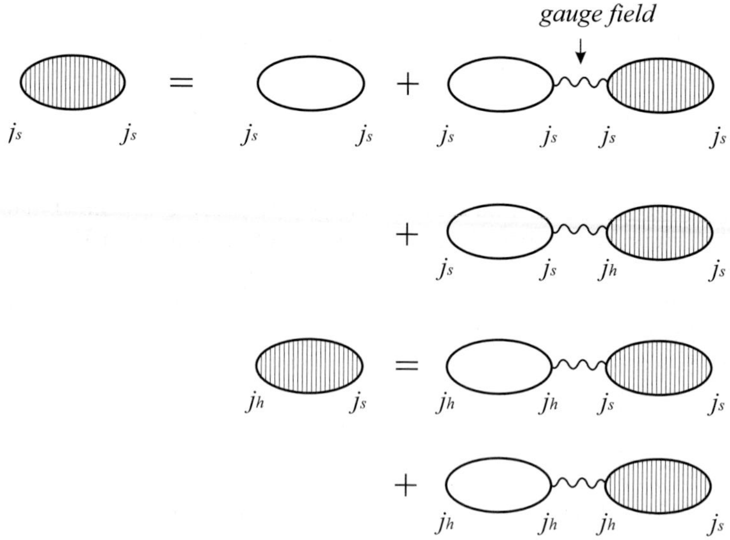

FIG. 17.2. Operator of electromagnetic response for the spinon-holon liquid with the gauge invariant field

consequence superconductivity sets in when both holon and spinon subsystems become superfluid.

In order to understand this rule of resistance summation let us study the electromagnetic response of the spinon–holon liquid. Since only spinons transfer charge, we will be interested in the current $\mathbf{j}_s(\omega) = -Q_{ss}^R(\omega) A_\omega$. It is important that a strong gauge field $\mathbf{a}$ acts between both spinons and holons, not discriminating between them. One can drawn the graphic equations for $Q_{ss}^R(\omega)$ renormalized by the gauge interaction (see Fig. 17.2). The corresponding algebraic equations have the form (we suppose the simplest version of interaction independent on momenta etc.):

$$\begin{aligned} Q_{ss} &= \langle j_s j_s \rangle + \langle j_s j_s \rangle \mathcal{G}\left[Q_{ss} + Q_{hs}\right], \\ Q_{hs} &= \langle j_h j_h \rangle \mathcal{G}\left[Q_{ss} + Q_{hs}\right], \end{aligned} \tag{17.4}$$

where Q_{hs} is the auxiliary holon–spinon response operator which appeared to make complete our system and $\mathcal{G}$ is the gauge field propagator. Correlators $\langle j_s j_s \rangle, \langle j_s j_s \rangle$ are nonrenormalized by gauge field response operators of the corresponding spinon and holon subsystems.

One can resolve system (17.4) and in the limit of infinitely strong interaction $\mathcal{G}$ to get

$$\lim_{\mathcal{G}\to\infty} Q_{ss} = \lim_{\mathcal{G}\to\infty} \frac{\langle j_s j_s \rangle \left(1 - \langle j_h j_h \rangle \mathcal{G}\right)}{1 - \left(\langle j_s j_s \rangle + \langle j_h j_h \rangle\right)\mathcal{G}} = \frac{1}{\langle j_s j_s \rangle^{-1} + \langle j_h j_h \rangle^{-1}}. \tag{17.5}$$

In accordance with the above argumentation the conductivity of system is proportional to Q_{ss}, hence

$$\frac{1}{\sigma_{tot}} = \frac{1}{\sigma_s} + \frac{1}{\sigma_h}. \tag{17.6}$$

hence resistances of spinon and holon subsystems must be summed [546].

Let us discuss the expected from the point of view of the RVB theory variation of the critical temperature of superconducting transition along the phase diagram. We start from the region of high dopings (overdoped phase) where the holon critical temperature $T_c^{(h)}$ is high and superconducting transition is determined by the lower spinon critical temperature $T_c^{(s)}$, like in BCS. Hence here, in overdoped phase, $T_c^{(h)}$ does not significate any real transition between some qualitatively different states of normal metal but just crossover between them, for example between non-Fermi and Fermi liquid behaviors.

In overdoped part of the phase diagram the fluctuation pseudogap is observed only in the narrow temperature region above T_c (see chapters 8-11). With the decrease of doping degree the gap in the spinon spectrum increases and gradually transforms from the BCS gap to the Hubbard one. Simultaneously, due to the approaching to the antiferromagnetic transition and corresponding growth of the role of paramagnon interactions, $T_c^{(s)}$ increases and together with it the superconducting critical temperature T_c grows. Nevertheless this growth of T_c continues only up to the value of the order of $T_c^{(h)}$. Indeed, the matter of fact that with the decrease of doping the carries concentration diminishes and $T_c^{(h)}$ falls with both of them. Hence in the region of small doping namely $T_c^{(h)}$ determines the true superconducting critical temperature. In this scenario the "hic sunt leones" phase is nothing else as the holons Bose gas above its condensation point.

The spinon-holon scenario of HTS is attractive in its ability to give the qualitative description of the phase diagram. Unfortunately the idyllic picture where holons subsystem is a weak nonideal Bose gas does correspond neither theory nor experiment. For example, it does not explain the complete destruction of superconductivity at non-zero doping. In paper [547] it was argued that strong interaction with gauge field can reduce the superfluid density and temperature of the Bose–Einstein condensation and even bring them to zero. As a result the Bose liquid appears which particles does not condense more, which is often called Bose metal. Its properties we will discuss below.

Let us mention that in some gauge field models the excitation with the fractional statistics (anyons) appear [548], or the bosons can be transformed in marginal fermions [549].

17.2.3 *Hidden order phase*

Another pretender for the description of the "hic sunt leones" phase is the so-called "hidden order" model. The authors of [550] develop the idea of the hidden broken symmetry of $d_{x^2-y^2}$-type in underdoped phases of HTS. They argue that the pseudogap observed in underdoped cuprates is an actual gap in the one particle excitation spectrum at the wave vector $(\pi, 0)$ and symmetry related points of the Brillouin zone associated with the development of this new order. On phenomenological grounds the authors propose the orbital antiferromagnetism with the d-wave symmetry of the wave functions [551–553] as the potential candidate on this order.

Other candidate for hidden order is the stripe phase [554, 555]. "Stripes" is a term that is used to denote unidirectional density wave states, which can involve unidirectional charge modulations ("charge stripes") or coexisting charge and spin density order ("spin stripes"). The corresponding scenario of phase fluctuations in the underdoped phase assumes the existence of such density waves of fluctuation nature. Authors of [554] believe that the physics of doped insulator including antiferromagnetism and superconductivity, is driven by the lowering of the zero-point kinetic energy. The motion of a single hole in an antiferromagnet is frustrated because it stirs up the spins and creates strings of ferromagnetic bonds. Consequently, a finite density of holes forms self organized structures, designed to lower the zero-point kinetic energy. This is accomplished in three stages: (a) formation of charge inhomogeneity (stripes), (b) the creation of local spin pairs, and (c) the establishment of a phase-coherent high-temperature superconducting state.

In the recent revisited RVB version [556] Anderson argues that the high temperature underdoped region of the phase diagram belongs to the non-Fermi liquid state of the charge carriers but it is characterized by some order parameter. It is why there is no sharp phase transition between the Fermi liquid and non-Fermi liquid domains but some crossover region takes place. This part of the phase diagram corresponds to the destruction of spinon bound states by the fluctuation motion of holes. Anderson substantiates some linear decrease of the upper crossover line $T^*(x)$ as the function of doping. The pseudogap phase, corresponding to the charge-spin separation, is disposed below the crossover domain.

17.3 Bose-metal

In this picture the presence of fermionic excitations is ignored, since in the underdoped phase they have a large gap in spectrum. The only ones important are the Bose fields: Cooper pairs, vortices, paramagnons. That is why the part of the phase diagrams with $T > T_c$, corresponding to low doping oxygen concentrations, got the name of Bose-metal. This scenario resembles strongly the picture of the quantum phase transition between superconducting and insulating phases (see section 16.3 and [513]), but here the role of impurities play paramagnons. Indeed, in the case of S-I quantum phase transition the localization edge is reached by means of the of the impurity concentration growth, which increases superconducting fluctuations and suppresses T_c. In the case of the HTS the tuning parameter is nothing other than the doping degree x: decreasing it we move along the phase diagram toward the antiferromagnetic state, hence grow antiferromagnetic fluctuations. These fluctuations are strong and their theory is still to be created. Nevertheless there are several qualitative scenarios what could happen here.

17.3.1 *Staggered flux state*

Some authors argue that the superconducting transition in HTS is driven by thermal or quantum fluctuations of the superconducting order parameter [554, 557] implying that in the wide region above the resistivity determined T_c one has local superconducting order without global phase coherence. In such a situation one would expect strong superconducting fluctuations, i.e. a large paraconductivity, magnetoresistance and fluctuation diamagnetism.

There is a weak point in the phase fluctuation scenario proposed above which can be seen from the following arguments [558, 559]. Above T_c the density of fluctuation vortices ("pancakes" in $2D$ case) should be high. The vortex density at $T > T_c$ can be high when their energy is of the order of T_c and together with it tends to zero when $x \to 0$. Some part of the vortex energy originates from the domain $r > \xi$, where the order parameter modulus changes weakly. The contribution of this region to the vortex energy is determined by the phase gradient square with the coefficient proportional to the superfluid density. As was already mentioned above the variational method of numerical analysis allows us to believe that $n_s \sim x \to 0$ and it seems that the formulated above requirement is satisfied. Nevertheless, it is necessary to remember that also the core domain contributes to the vortex energy. Keeping in mind that in the scenario discussed the vortices are present also in the normal phase, it would be difficult to justify dependence of the vortex energy on x. In ordinary superconductors, the BKT temperature is close to the mean field value, and the core energy rapidly becomes small ($\sim E_F\epsilon$). However, in the case of HTS it is postulated that the mean-field critical temperature is high, so that a large core energy is expected. Indeed, in the region of a core of conventional vortex the order parameter (and also gap) vanishes since its energy cost is too high: $\nu\Delta_0^2$ per unit area. Using as a core radius the value $\xi = v_F/\Delta_0$, it energy can be estimated as $\sim E_F$. In our case we can replace E_F by the exchange energy J of the t–J model. If this were the case, the proliferation of vortices would not happen until a high temperature $\sim J$, independent of the doping level x is reached. Namely this problem is the trouble of the fluctuation scenario under consideration. Thus, for the phase fluctuation scenario to work, it is essential to have "cheap" vortices with energy cost of order T_c. Then the essential problem is to understand what the vortex core is made of. One has to identify the state living inside the vortex core which could decrease its with respect to the Abrikosov's solution. The candidate for this state according to the authors of [558, 559] is the staggered flux state with an orbital currents.

The authors based their work on an $SU(2)$ formulation of the t–J model. In this formulation the staggered flux plays a central role being the progenitor of the Néel state at half-filling on the one hand and being the close competitor to the

d-wave superconductor with small doping on the other hand. Here the dynamics is determined by the virtual pancakes, which in $2D$ case conduct themselves as particles but transfer a flux instead of charge. The authors argue that the low-energy vortices, where the staggered flux state is stabilized in the core, are needed to explain consistently the pseudogap phenomena in the underdoped phase.

In the paper [531], the ground state energy of the staggered flux phase was calculated by means of the variation procedure. It turns out to be higher than that of the superconducting phase, but their difference tends to zero with decrease of the doping degree ($x \to 0$). This means that the energy of the vortex core tends to zero, too, if it resides in staggered flux state. In the same paper the Cooper pais size is evaluated. It turns out to be large for the over- or optimally-doped compounds, while it decreases down to the lattice spacing for underdoped phase. This means that the Ginzburg–Levanyuk number $Gi \sim 1$.

17.3.2 *Vortices localization*

One can believe that when $Gi_{(2)}^{\text{strong}} \sim 1$ the wide fluctuation region, analogous to the BKT transition, exists. Physical properties of the system are determined in this case by vortices (pancakes in $2D$ case). Indeed, we already have seen that vortices can participate in different transport phenomena. In many respects their behavior is analogous to that of particles. Therefore one has not be surprised that it was demonstrated recently [560] that the quantum corrections to vortex conductivity exist, similar to conductivity of disordered metal.

Let us discuss this phenomenon qualitatively. How do the two localizations differs? First of all, the Einstein relation of diffusion coefficient $\mathcal{D}$ and conductivity σ in the case of vortex motion takes another form [513,547]. In contrast to charge carriers vortices transfer magnetic flux. As a result their diffusion coefficient turns out to be proportional to resistivity instead of conductivity. Let us denote the vortex “conductance” by the same symbol $g = \nu\mathcal{D} = n/\eta$, where n is the vortex density and η is their viscosity. The Lorentz (Magnus) force $F_L = j\Phi_0$ and in conditions of a steady flow it is counterpoised by the viscous force $\eta\mathbf{v}$. The intensity of electric field turns out to be proportional to vortex flow velocity: $\mathbf{E} = \Phi_0 n\mathbf{v}$. As a result the resistance per square of the $2D$ layer $\rho = E/j$ is

$$\rho = \frac{\Phi_0^2 n}{\eta} = \frac{\pi^2}{e^2} g. \tag{17.7}$$

In the quantum region vortices behave themselves correspondingly as quantum particles (for instance they can tunnel through potential barrier). Hence, analogously to the situation in electron localization theory one can expect that quantum coherent scattering and vortex interaction renormalize the diffusion coefficient. It turns out that the effect of simple impurities on vortices is analogous to the effect of the random magnetic field on electrons and hence, the WL correction is absent. The main contribution to the diffusion coefficient occurs due to exchange interaction, which for the boson-vortices has the opposite sign with respect to fermion–electrons (due to the rules of diagrammatic technique each fermion loop brings in the appropriate expression −1). Das and Doniach [560] calculated this quantum correction and obtained

$$\rho = \rho_0 \left[1 + \frac{(2 - 2\ln 2)}{\pi p_F l} \ln \frac{1}{T\tau} \right]. \tag{17.8}$$

Plausibly this result is related to the experiment [364], where superconductivity in HTS compounds was killed applying very high pulse magnetic fields and the logarithmic growth of the resistance was observed down to very low temperatures.

17.3.3 *Bipolarons*

Another scenario of Bose metal is the bipolaron one. In this model electrons are coupled strongly in bipolarons [561] due to the electron–phonon interaction which is supposed to be with the characteristic energy considerably exceeding the Fermi energy. As a result the mobile bipolarons are formed in normal state with their further Bose–Einstein condensation. Therefore superconductivity here is the superfluidity of the almost ideal Bose gas of bipolarons. The author extends the BCS theory towards an intermediate and strong-coupling regime [562] and demonstrates the existence of two energy scales in it, a temperature independent incoherent gap and a temperature dependent coherent gap combining into one temperature dependent global gap.

17.4 BCS scenarios

As it was repeatedly mentioned above, the BCS picture adequately describes the properties of overdoped and optimally doped HTS. The paramagnon exchange between electrons results in their attraction. In contrast to the electron–phonon interaction in original BCS scheme, the electron(hole)–paramagnon interaction may be not weak (at least at the part of Fermi surface). That is why the BCS picture has to be modified. Some versions of such modification are reviewed below.

17.4.1 *Strong coupling BCS-like schemes*

The effect of soft paramagnons, with the characteristic energies small with respect to pseudogap, on superconducting properties of HTS became the subject of studies in [563, 519]. The main conclusion of such consideration reduces to the statement that the soft phonons or paramagnons influence on superconductivity is analogous to the effect of elastic impurities. These models are analogous to the electron-phonon superconductivity with strong coupling. The authors compare in details the spin fluctuation (paramagnon) approach with the Migdal–Eliashberg approach [229] to phonon superconductors and showed that despite the absence of the small electron to ionic mass ratio justifying Eliashberg theory for phonons, an Eliashberg type approach to the spin–fermion model is still allowed, but only at strong coupling $g \gtrsim 1$.

Let us remember that superconducting fluctuations in the Eliashberg model of superconductivity with electron–phonon strong coupling were studied in [230].

As a result with the growth of effective interaction constant decreases the diffusion constant $\mathcal{D}$ and increases the Gi number:

$$Gi_{(2)}^{strong} \sim g^3 Gi_{(2)} \sim g^3 \frac{T_c}{E_F}. \tag{17.9}$$

Since $Gi_{(2)} \lesssim 0.1$ one can see that even the interaction with $g \sim 2.5$ makes $Gi_{(2)}^{strong}$ to be equal to 1.

As was already discussed in chapter 2 such large Gi number indicates not only on the fluctuation broadening of the critical region but also on suppression of the critical temperature itself. This consideration sheds light on the enigmatic properties of the underdoped phase. Indeed, with the decrease of carriers concentration (doping) the system more and more approaches the antiferromagnetic state, the role of paramagnons grows, they increase $Gi_{(2)}^{\mathrm{strong}}$ number and suppress critical temperature. The dependence $T_c(x)$ goes down while all domain between it and its mean field value T_{c0} is placed in the disposal of fluctuations. In this way such scenario explains the region "hic sunt leones" of the Fig. 17.1.

17.4.2 *BEC–BCS scenario*

In the picture of this group of scenarios part of electrons interacts so strongly that they form bounded Bose particles with the binding energy larger that the Fermi one. It is important that another electron subsystem exists, which interacts weakly or can be considered as noninteracting at all.

The BCS superconducting instability of a Fermi system and the Bose–Einstein Condensation (BEC) of bosons below a critical temperature can be unified by following the continuous evolution between these two limits as the strength of the fermion attraction increases. Within this scheme, the phase diagram of cuprate superconductors can be interpreted in terms of a crossover from Bose-Einstein condensation of preformed pairs to BCS superconductivity, as doping is varied. A large amount of papers is devoted to the development of this idea. We enumerate here only small part of them [564–570] and proceed to the brief discussion of the two models.

Any Bose-type scenario would lead to the conclusion that superconducting transition is similar to Bose condensation and has a wide fluctuation region near T_c, i.e. that the Gi number is of the order of 1. At the same time the experiments show that the smearing of the resistive transition is relatively narrow. Plausibly this circumstance is related to the fact that the large gap in the electron spectrum develops not for all directions of the momentum space and as a result the soft fermionic modes reduce Gi.

Let us remember the temperature dependence of the superfluid density $n_{s2}(T)$ of $2D$ conventional superconductor. In accordance with the BCS formula it decreases continuously with temperature increase and becomes small ($\sim Gi_{(2)} n_{s2}(0)$) at the edge of critical region ($\sim Gi_{(2)} T_c$). With the further increase of temperature it jumps to zero (see (15.14). Similar behavior of the superfluid density was observed in HTS too and this fact is very troublesome both for Bose-metal and

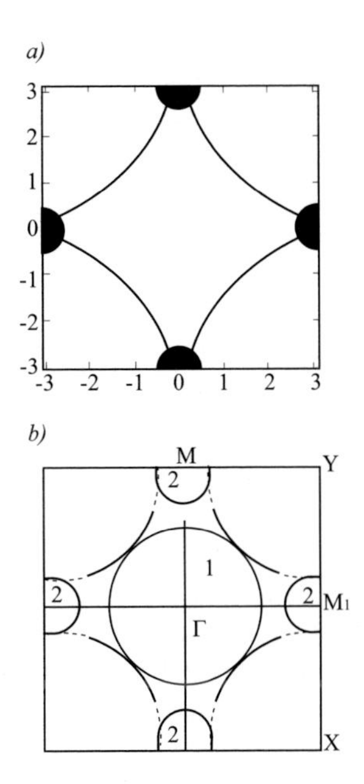

FIG. 17.3. Sketch of the Fermi surface of underdoped cuprates. (a) Tight band model calculated Fermi line. The shaded circles represent the part of momentum space where a pseudogap was observed in the experiment. The fermions in these regions are assumed to be paired in the bosons. (b) The same Fermi line with denoted quasiparticles arcs (thick solid line) and patches of quasi-localized states (thick dotted line). The two band model Fermi line is shown by the thin solid lines.

holon scenarios, where $Gi_{(2)} \sim 1$ and it is difficult to explain the decrease of superfluid density near T_c 10 times with respect to $n_{s2}(0)$. That is why the authors of [571] supposed that the strongly interacting electrons, forming Bose pairs, occupy only the small part of Fermi surface, while weakly interacting electrons occupy the major part of Fermi surface and namely this large group provides the necessary smallness of the $Gi_{(2)}$.

In the paper [571] was demonstrated that the ideas of the BEC description and the mean-field essence of superconducting transition of the BCS model can be reconciled if the former occurs against the background of the Fermi liquid and processes that convert bosons into the fermions are allowed.

It is reasonable to assume that fermions near the corners of the Fermi line (which lie inside the circles shown in Fig.17.3 (a) are paired into boson, which charge $2e$ and no dispersion; this is the key assumption of the model. So one-particle fermionic excitations acquire a gap; the soft modes appearing instead of these fermionic excitation are spinless bosons. Note that in this model Bose condensation does not occur because bosons have no dispersion (i.e. are infinitely heavy). Another assumption of the model is that interaction, transferring electrons from the "circles", where they are paired to the other parts of the Fermi line (where Fermi velocity is large), is weak.

In spite of the fact that the presence of bosons can affect on the value of the critical temperature the fluctuations are determined by the fermion subsystem and the Gi number is

$$Gi \sim \frac{T_c}{t}, \tag{17.10}$$

where t is the band width.

The authors of [572] explore fluctuations in the underdoped cuprate superconductor focusing on the consequences of a strongly anisotropic inter-electron interaction. The model is implemented by a two-band system with different intraband and interband pairing interactions (see Fig. 17.3(b)). The electrons of the first band have a large Fermi velocity v_{F1} and a small attraction g_{11} giving rise to largely overlapped Cooper pairs; superconducting fluctuations are weak here. On the contrary, the electrons of the other band have a small v_{F2} and a large attraction g_{22} resulting into tightly bound pairs having strong fluctuations. The interaction between electrons of these two bands is supposed to be weak $g_{12} \ll g_{22}$ and even with this restriction its role is found to be crucial (see below).

At variance from the models of mixed fermions and bosons, the fermionic origin of both the weakly and strongly bound Cooper pairs is kept. This way of thinking develops ideas of [571], where the strong-coupling limit of one component was considered as the composite bosons. The expression for the effective Gi number is found in the form:

$$Gi^{\text{eff}} = \left[1 + \frac{g_{12}^2}{g_{11}^2 g_{22}^2} \ln^2 \frac{T_c^{(0)}}{T_{c1}^{(0)}}\right] \frac{Gi^{(1)} Gi^{(2)}}{Gi^{(1)} + \left(\frac{g_{12}^2}{g_{11}^2 g_{22}^2} \ln^2 \frac{T_c^{(0)}}{T_{c1}^{(0)}}\right) Gi^{(2)}}, \tag{17.11}$$

where $T_{c1}^{(0)}$ and $T_{c2}^{(0)}$ are mean field critical temperatures of the corresponding band in the absence of interaction between them ($g_{12} = 0$) and

$$T_c^{(0)} = \sqrt{T_{c1}^{(0)} T_{c2}^{(0)}} \exp\left[\frac{1}{2}\sqrt{\ln^2 \frac{T_{c2}^{(0)}}{T_{c1}^{(0)}} + 4\frac{g_{12}^2}{g_{11}^2 g_{22}^2}}\right] \tag{17.12}$$

is the mean-field pairing temperature accounting for the interband interaction.

In the framework of models discussed above it was supposed that strong interaction between holes (excitations of the Fermi type) takes place only at small patches of the Fermi surface, so called "hot points". This model describes well such superconducting properties of the system as small Gi and fast decrease of n_s with the growth of temperature. In order to describe the normal properties of the system at temperatures above T_c or in the region of the vortex core in superconducting state the other model turns out to be more convenient. It is called "the cold points" model [573]. In this model the quasiparticle interaction is supposed to be strong almost everywhere at Fermi surface and only in narrow apertures along the diagonals $p_x = \pm p_y$ some long-living Fermi excitations exist. Such model allows to explain the most enigmatic property of the HTS normal state, its linear temperature dependence of resistivity. Plausibly the "the cold points" model is the one-particle vision of the staggered phase or hidden order pictures.

All these models are semi-phenomenological and it is unclear could they be derived from the microscopic t–J model or not. The same question can be brought up in more physical way: is it possible to get the superconducting state with small Gi and strong temperature dependence of n_s starting from the t–J Hamiltonian? If the answer is positive (at least experimentally) for optimally doped or slightly underdoped compounds so how the Gi number depends on doping level x? The variational method demonstrates that with the decrease of x the Gi increases although still do exist neither experimental nor theoretical answers on the question: how small are x_{cr} corresponding to the region of strong fluctuations, where $Gi \sim 1$.

In all models above we assumed the homogeneity of samples, i.e. space independence of the doping level. This assumption appears reasonable and justified for optimally doped superconductors where the oxygen contents is close to the stoichiometrical value. The situation can be quite different in the case of strongly underdoped compounds, where the drops with higher oxygen concentration may posses the higher critical temperature. Such a drop model was already discussed in chapter 12. In the context of HTS analogous models were discussed in [450, 574].

Since today all these questions do not have answers we call the corresponding part of the phase diagram as "terra incognita" or "hic sunt leones".

17.5 Phenomenology

Besides the mentioned semi-phenomenological models there are purely phenomenological ones. In some of them it is supposed that the phenomenological description of the fermionic subsystem is possible after integration over all Bose fields. In other theories the integration over fermionic fields reduces the problem to the GL type phenomenology. Let us review shortly both types of such theories.

17.5.1 *Marginal Fermi liquid*

We will start from the description of the marginal Fermi liquid model [575]. In this model the one-electron Green function has the usual form

$$G(\mathbf{k}, \omega) = \frac{1}{\omega - \varepsilon(\mathbf{k}) - \sum(\mathbf{k}, \omega)} \tag{17.13}$$

but unusual form has the self-energy part $\sum(\mathbf{k}, \omega)$. In their calculation of $\sum(\mathbf{k}, \omega)$ the authors postulate the single hypothesis: over a wide range of momentum $\mathbf{q}$

there exist excitations in the system which make the following contribution to both charge and spin polarizability $P(\mathbf{q},\omega)$:

$$\operatorname{Im} P(\mathbf{q},\omega) \sim -\nu(0) \begin{cases} \omega/T, & |\omega| < T, \\ \operatorname{sign}\omega, & |\omega| > T. \end{cases} \tag{17.14}$$

This hypothesis grows from the idea of renormalization of the electron Green function due to electron interaction with low energy (soft) excitations. This means that the paramagnon states with small ω are highly occupied. The microscopic form of the corresponding propagator in view of the strength of interaction is still unknown and the authors simply assume its influence on $\operatorname{Im} P(\mathbf{q},\omega)$ in the form (17.14). The consequence of this assumption is the appearance of very specific excitations in the Fermi system, spread up to the limits of the quasiparticle definition:

$$\sum(\mathbf{k},\omega) \sim g^2\nu^2(0)\left[\omega \ln \frac{\max(|\omega|,T)}{\omega_c} - i\frac{\pi}{2}\max(|\omega|,T)\right], \tag{17.15}$$

where ω_c is the ultraviolet cutoff parameter. Equation (17.15) is quite different from the one-particle self-energy in a conventional Fermi liquid where, at $T = 0, \operatorname{Re}\sum \sim \omega$ and $\operatorname{Im}\sum \sim \omega^2$ and a quasiparticle representation for the one-particle Green function is valid:

$$G(\mathbf{k},\omega) = \frac{z_k}{\omega - E(\mathbf{k}) + i\Gamma_{\mathbf{k}}} + G_{\text{incoh}}, \tag{17.16}$$

with finite z_k and $\Gamma_{\mathbf{k}} \ll E(\mathbf{k})$. Substitution of Eqs. (17.13) to (17.15) results in

$$z_k^{-1} = \left[1 - \frac{\partial}{\partial\omega}\operatorname{Re}\sum(\mathbf{k},\omega)\right]_{\omega=E(\mathbf{k})} \sim \ln\frac{\omega_c}{E(\mathbf{k})}. \tag{17.17}$$

Near the Fermi surface $E(\mathbf{k}) = 0$ thus the quasiparticle weight vanishes logarithmically and Green function becomes entirely incoherent. Marginal Fermi liquid theory allows to explain the series of HTS normal state anomalies, including the linear increase of resistivity, peculiarities of the specific heat, a.c. conductivity, etc.

17.5.2 *Bose phenomenology*

One of the efforts to construct the GL scheme, including both superconducting and antiferromagnetic fluctuations was undertaken in [576], where the so-called SO(5) model was proposed. In this phenomenological model antiferromagnetic and superconducting transitions are described by the GL equation with $5D$ order parameter (three its components correspond to the magnetization vector and other two components are related to the complex scalar of the superconducting order parameter). Such unification of superconductivity and antiferromagnetism seems unjustified to us, nevertheless some generalized phenomenology of both types of fluctuations could be very useful.

Speaking about phenomenological theories it is necessary to underline that the GL theory of superconductivity turned out to be very effective still before formulation of the BCS theory. Even today, half a century after the creation of the microscopic theory of superconductivity, it is convenient to describe many properties of conventional superconductors in terms of this theory (see Part I). Being phenomenological GL theory does not assume any specific mechanism of superconductivity and one can believe that it can describe some phenomena even in underdoped phases of HTS.

Moreover, in order to understand the enigmatic properties of underdoped cuprate systems it may turn out to be useful to look for more general phenomenological theories, taking into account their specifics.

APPENDIX A

RELATIONS BETWEEN PARAMETERS OF MICRO- AND MACROSCOPIC THEORIES

A.1 Coherence length

As the characteristic size of the fluctuation Cooper pair we use the GL coherence length

$$\xi(T) = \frac{\xi}{\sqrt{\epsilon}}, \quad \epsilon = \frac{T - T_c}{T_c}. \tag{A.1}$$

The microscopic theory (see Eq. (6.33)) in the case of an isotropic Fermi surface gives for the coherence length ξ the precise expression:

$$\xi^2 = -\frac{v_F^2 \tau^2}{D}\left\{\psi\left(\frac{1}{2} + \frac{\hbar}{4\pi k_B T \tau}\right) - \psi\left(\frac{1}{2}\right) - \frac{\hbar}{4\pi k_B T \tau}\psi'\left(\frac{1}{2}\right)\right\}, \tag{A.2}$$

where $\psi(x)$ is the digamma function and $D = 3, 2, 1$ is the space dimensionality. In the clean (c) and dirty (d) limits:

$$\xi_c = 0.133\frac{\hbar v_F}{k_B T_c}\sqrt{\frac{3}{D}} = 0.74\xi_0\sqrt{\frac{3}{D}}, \tag{A.3}$$

$$\xi_d = 0.36\sqrt{\frac{\hbar v_F l}{k_B T_c}\frac{3}{D}} = 0.85\sqrt{\xi_0 l}\sqrt{\frac{3}{D}}. \tag{A.4}$$

Here $l = v_F \tau$ is the electron mean free path and $\xi_0 = \hbar v_F / \pi\Delta(0)$ is the conventional BCS definition of the coherence length of a clean superconductor. One can see that (A.3) and (A.4) coincide with the above estimations.

Let us stress some small numerical difference between our Eq. (A.2) and the usual definition of the coherence length. We are dealing near the critical temperature, so the definition (A.2) is natural and allows us to avoid many numerical coefficients in further calculations. The BCS coherence length $\xi_0 = \hbar v_F / \pi\Delta(0) = 0.18\hbar v_F / k_B T_c$ was introduced for zero temperature and an isotropic $3D$ superconductor.

It is convenient to determine the coherence length also from the formula for the upper critical field:

$$H_{c2}(T) = A(T)\frac{\Phi_0}{2\pi\xi^2(T)}. \tag{A.5}$$

The function $A(T_c) = 1$, while its value at $T = 0$ depends on the impurities concentration. For the dirty case the appropriate value was found by Maki [577] $A_d(0) = 0.69$, for the clean case by Gor'kov [34] $A_c^{2D}(0) = 0.59$, $A_c^{3D}(0) = 0.72$.

The value $A(0)$ and the small difference mentioned above between the BCS and (A.2) definitions of the coherence length make the difference between the linear interpolation of the second critical filed $\widetilde{H_{c2}}(0)$ from the critical temperature (see Eq. (2.54)) and its exact value (A.5).

A.2 Other GL functional parameters

As was demonstrated in Part II, close to T_c the microscopic theory of superconductivity allows to write down the free energy of superconductor in the form of GL type series over the order parameter $\Delta(\mathbf{r})$:

$$\mathcal{F}[\Delta(\mathbf{r})] = \int \left[A\Delta^2 + \frac{B}{2}\Delta^4 + C_{(D)} (\nabla\Delta)^2 \right] dV \tag{A.6}$$

with

$$A = \nu \ln \frac{T}{T_c}, \tag{A.7}$$

$$B = \frac{7\zeta(3)}{8\pi^2 T^2}\nu. \tag{A.8}$$

The coefficient $C_{(D)}$ is related to the square of coherence length (6.33):

$$C_{(D)} = \nu \xi_{(D)}^2 (T\tau). \tag{A.9}$$

The value of Δ in the simplest version of the BCS theory for homogeneous superconductor in the absence of magnetic field and paramagnetic impurities coincides with the gap in the energy spectrum. One could identify the phenomenological order parameter Ψ with the microscopic parameter Δ:

$$\Psi = \sqrt{4mC_{(D)}}\Delta. \tag{A.10}$$

In this case the precise values for the coefficients α and b can be carried out from the microscopic theory:

$$4m\alpha T_c = \xi^{-2}; \alpha^2/b = \frac{8\pi^2}{7\zeta(3)}\nu, \tag{A.11}$$

where $\zeta(x)$ is the Riemann zeta function, $\zeta(3) = 1.202$.

One can notice that the arbitrariness in the normalization of the order parameter amplitude leads to the ambiguity in the choice of the Cooper pair mass. Indeed, this value enters in (2.6) as the product with coefficient α therefore one of these parameters has to be fixed. In the phenomenological theory m usually is accepted as the electron mass that results in the value of Cooper pair mass equal to $2m$ (see in (2.4)). In such choice the parameter α for a clean D-dimensional superconductor is defined as

$$\alpha_{(D)} = \frac{2D\pi^2}{7\zeta(3)} \frac{T_c}{E_F}. \tag{A.12}$$

As the consequence of the voluntarist identification of the GL parameter m with the electron mass, due to the impurity concentration dependence of ξ, both GL parameters α and b turn out to be impurity concentration dependent too. Vice versa, the microscopic parameters A and B do not depend on impurities concentration, only $C_{(D)}$ depends on it.

APPENDIX B

PROPERTIES OF THE EULER GAMMA FUNCTION AND ITS LOGARITHMIC DERIVATIVES

B.1 Euler gamma function

The gamma function $\Gamma(z)$ was defined by Euler in the form

$$\Gamma(z) = \int_0^\infty t^{z-1} e^{-t} dt, \tag{B.1}$$

where $\mathrm{Re}\, z > 0$ is supposed. Its other, useful for our purposes, definition can be given in the form of an infinite product [578, 579]:

$$\Gamma(z) = \lim_{n_\mathrm{c} \to \infty} \frac{n_\mathrm{c}!\, n_\mathrm{c}^{z-1}}{z(z+1)(z+2)\dots(z+n_\mathrm{c}-1)}. \tag{B.2}$$

The factorial here can be substituted by its asymptotic expression according to Stirling's formula

$$n_\mathrm{c}! \approx \left(\frac{n_\mathrm{c}}{\mathrm{e}}\right)^{n_\mathrm{c}} \sqrt{2\pi n_\mathrm{c}}, \qquad \ln(n_\mathrm{c}!) \approx \left(n_\mathrm{c} + \frac{1}{2}\right) \ln n_\mathrm{c} - n_\mathrm{c} + \ln\sqrt{2\pi}, \tag{B.3}$$

that gives the asymptotic expression for the gamma function for large argument ($z \gg 1$)

$$\ln\Gamma(z) \approx \left(z - \frac{1}{2}\right) \ln z - z + \frac{1}{2}\ln(2\pi) + \frac{1}{12z}. \tag{B.4}$$

Finally, let us report several useful partial values of the gamma function:

$$\Gamma(n+1) - n!, \tag{B.5}$$

$$\Gamma\left(\frac{1}{2}\right) = \sqrt{\pi}, \tag{B.6}$$

and mention the functional relation

$$\Gamma(z)\Gamma(-z) = -\frac{\pi}{z \sin \pi z}. \tag{B.7}$$

B.2 Digamma function and its derivatives

In the theory of fluctuations the fundamental role plays the logarithmic derivative of the Euler gamma function or, the so-called digamma function. Let us report some of their properties (see e.g. [578, 579]).

The definition of the digamma function $\psi(z) \equiv \psi^{(0)}(z)$ is

$$\psi(z) \equiv \frac{\mathrm{d}}{\mathrm{d}z} \ln \Gamma(z) = \lim_{n_c \to \infty} \left\{ - \sum_{n=0}^{n_c - 1} \frac{1}{n+z} + \ln n_c \right\}. \tag{B.8}$$

It is analytic everywhere except the points $z_m = 0, -1, -2, \cdots$ where, following the gamma function, it has the simple poles. High order derivatives of the digamma function can be found directly from (B.8). They are expressed by the convergent series

$$\psi^{(N)}(z) = (-1)^{N+1} N! \sum_{n=0}^{\infty} \frac{1}{(n+z)^{N+1}}. \tag{B.9}$$

and can be related to the Hurwitz zeta functions [578, 579] with integer first argument:

$$\psi^{(N)}(z) = \frac{\mathrm{d}^N}{\mathrm{d}z^N} \psi(z) = (-1)^N N! \, \zeta(N+1, z). \tag{B.10}$$

where

$$\zeta(q, z) = \sum_{n=0}^{\infty} \frac{1}{(n+z)^q} \tag{B.11}$$

is Hurwitz, or generalized Riemann, zeta function. For the case, important for further discussion $z = 1/2$ this relation can be simplified and expressed in terms of the Riemann zeta function $\zeta(z)$:

$$\psi^{(N)}(\frac{1}{2}) = (-1)^{N+1} N! \left(2^{N+1} - 1\right) \, \zeta(N+1). \tag{B.12}$$

For example

$$\psi^{(1)}\left(\frac{1}{2}\right) = 3\zeta\left(2\right) = \frac{\pi^2}{2}, \tag{B.13}$$

$$\psi^{(2)}\left(\frac{1}{2}\right) = -14\zeta(3), \tag{B.14}$$

$$\zeta\left(3\right) = \frac{7\pi^3}{180} - 2\sum_{k=1}^{\infty} \frac{1}{k^3 \left(e^{2\pi k} - 1\right)} = 1.20206... \tag{B.15}$$

One can introduce the function $\psi^{(-1)}(z)$ generating digamma function and its derivatives, which with the accuracy of the constant coincides with the $\ln \Gamma(z)$:

$$\psi^{(-1)}(z) \equiv \lim_{n_c \to \infty} \left\{ -\sum_{n=0}^{n_c-1} \ln(n+z) + \left(n_c - \frac{1}{2} + z\right) \ln(n_c) - n_c \right\} = \ln \frac{\Gamma(z)}{\sqrt{2\pi}}.$$

In conclusion let us present several asymptotic expressions and useful relations. For large argument $z \gg 1$,

$$\psi(z) \approx \ln z - \frac{1}{2z} - \frac{1}{12z^2}, \tag{B.16}$$

$$\psi^{(1)}(z) \equiv \zeta(2, z) \approx \frac{1}{z} + \frac{1}{2z^2} + \frac{1}{6z^3}, \tag{B.17}$$

$$\psi(\frac{1}{2} + z) \approx \ln z + \frac{1}{24z^2}, \tag{B.18}$$

$$\psi^{(-1)}(z \to \infty) \approx \left(z - \frac{1}{2}\right) \ln z - z + \frac{1}{12z}, \tag{B.19}$$

$$\psi(z \to 0) = -\frac{1}{z}. \tag{B.20}$$

Functional relations

$$\psi(1+z) - \psi(z) = \frac{1}{z}, \tag{B.21}$$

$$\psi\left(\frac{1}{2} + iz\right) - \psi\left(\frac{1}{2} - iz\right) = \pi i \tanh \pi z, \tag{B.22}$$

$$-\psi(1) = C_{\text{Euler}} = \lim_{n_c \to \infty} \left\{ \sum_{n=1}^{n_c-1} \frac{1}{n} - \ln(n_c) \right\} = 0.577216\ldots, \tag{B.23}$$

$$\psi\left(\frac{1}{2}\right) = -\ln(4\gamma_E) = -2\ln 2 - C_{\text{Euler}}. \tag{B.24}$$

We used the definition $\gamma_E = e^{C_{\text{Euler}}} = 1.78..$

APPENDIX C

INTEGRALS OF THE LAWRENCE–DONIACH THEORY

Below we present some integrals which often appear in the theory of fluctuations in layered superconductors:

$$\int_0^{2\pi} \frac{d\theta}{2\pi} \ln\left[\epsilon + \frac{r}{2}(1-\cos\theta)\right] = 2\ln\frac{\sqrt{\epsilon}+\sqrt{\epsilon+r}}{2}, \tag{C.1}$$

$$\int_0^{2\pi} \frac{d\theta}{2\pi} \cos\theta \ln\left[\epsilon + \frac{r}{2}(1-\cos\theta)\right] = \frac{2}{r}\left[\sqrt{\epsilon(\epsilon+r)} - (\epsilon + r/2)\right], \tag{C.2}$$

$$\int_0^{2\pi} \frac{d\theta}{2\pi} \frac{1}{\epsilon + \frac{r}{2}(1-\cos\theta)} = \frac{1}{\sqrt{\epsilon(\epsilon+r)}}, \tag{C.3}$$

$$\int_0^{2\pi} \frac{d\theta}{2\pi} \frac{1}{\left[\epsilon + \frac{r}{2}(1-\cos\theta)\right]^2} = \frac{\epsilon + \frac{r}{2}}{\left[\epsilon(\epsilon+r)\right]^{3/2}}, \tag{C.4}$$

$$\int_0^{2\pi} \frac{d\theta}{2\pi} \frac{1}{\left[\epsilon + \frac{r}{2}(1-\cos\theta)\right]^3} = \frac{(\epsilon+r)\,\epsilon + \frac{3}{8}r^2}{\left[\epsilon(\epsilon+r)\right]^{5/2}}. \tag{C.5}$$

For any complex parameter $z \neq 1$ is valid identity:

$$\int_0^{2\pi} \frac{dx}{2\pi} \frac{1}{\cos x - z} = -\frac{1}{\sqrt{z^2-1}} \tag{C.6}$$

with the corresponding choice of the square root branch.

One more useful integral

$$\int_0^{\infty} \frac{x^2\,dx}{\sinh^2 x} = \pi^2/6. \tag{C.7}$$

REFERENCES

[1] Landau, L.D. (1937). *Zhurnal Eksperimentalnoi i Teoreticheskoi Fisiki*, **7**, 19.
[2] Onsager, L. (1944). *Physical Review*, **65**, 117.
[3] Bornholdt, S. and Wagner, F. (2002). *Physica*, **A316**, 453-468.
[4] Vaks, V. and Larkin, A.I. (1965). *Zhurnal Eksperimentalnoi i Teoreticheskoi Fisiki*, **49**, 975 [*Soviet Physics - JETP*, **22**, 678 (1966)].
[5] Kadanoff, L.P. (1966). *Physics*, **2**, 263.
[6] Patashinskii, A.Z. and Pokrovsky, V.L. (1966). *Zhurnal Eksperimentalnoi i Teoreticheskoi Fisiki*, **50**, 439 [*Soviet Physics - JETP*, **23**, 1292 (1966)].
[7] Gribov, V.N. and Migdal, A.A. (1968). *Zhurnal Eksperimentalnoi i Teoreticheskoi Fisiki*, **55**, 1498 [*Soviet Physics - JETP*, **28**, 784 (1969)].
[8] Polyakov, A.M. (1968). *Zhurnal Eksperimentalnoi i Teoreticheskoi Fisiki*, **55**, 1026 [*Soviet Physics - JETP*, **28**, 533 (1969)].
[9] Halperin, B.I. and Hohenberg, P.C. (1969). *Physical Review*, **168**, 898.
[10] Lee, T.D. and Yang, C.N. (1957). *Physical Review*, **105**, 1119.
[11] Levanyuk, A.P. (1959). *Zhurnal Eksperimentalnoi i Teoreticheskoi Fisiki*, **36**, 810 [*Soviet Physics–JETP*, **9**, 571 (1959)].
[12] Larkin, A.I. and Khmelnitski, D.E. (1969). *Zhurnal Eksperimentalnoi i Teoreticheskoi Fisiki*, **56**, 2087-2098 [*Soviet Physics–JETP*, **29**, 1123 (1969)].
[13] Di Castro, C. and Jona-Lasinio, G. (1969). *Physics Letters*, **A29**, 322.
[14] Wilson, K.G. and Fisher, M. (1972). *Physical Review Letters* , **28**, 240.
[15] Wilson, K.G. (1972). *Physical Review Letters* , **28**, 548.
[16] Tzuneto, T. and Abrahams, E. (1973). *Physical Review Letters*, **30**, 217.
[17] Wilson, K.G. (1971). *Physical Review*, **B4**, 3174; ibid. 3184.
[18] Kamerling Onnes, H. (1911). *Leiden Communications*, **122b**, 124.
[19] Bardeen, J., Cooper, L.N. and Schrieffer, J.R. (1957). *Physical Review*, **106**, 162; ibid. **108**, 1175.
[20] London, F. and London, H. (1935). *Proceedings of the Royal Society*, **A149,** 71.
[21] Meissner, W. and Ochsenfeld, R. (1933). *Naturwissenschaften*, **21,** 787.
[22] Peierls, R. (1936). *Proceedings of the Royal Society*, **A155**, 613.
[23] Ginzburg, V.L. and Landau, L.D. (1950). *Zhurnal Eksperimentalnoi i Teoreticheskoi Fiziki,* **20**, 1064.
[24] De Gennes, P.G. (1966). *Superconductivity of Metals and Alloys,* W.L. Benjamin, Inc, New York–Amsterdam.
[25] Abrikosov, A.A. (1957). *Zhurnal Eksperimentalnoi i Teoreticheskoi Fisiki*, **32**, 1442 [*Soviet Physics–JETP* **5**, 1174 (1957)].

[26] Shubnikov, L.V., Khotkevich, V.I., Shepelev, Yu.D. and Riabinin, Yu.N. (1937). *Zhurnal Eksperimentalnoi i Teoreticheskoi Fisiki* , **7**, 221.
[27] Maxwell, E. (1950) *Physical Review*, **78**, 477; ibid. **79**, 173.
[28] Reynolds, C.A., Serin, B., Wright, W.H. and Nesbitt, L.B. (1950). *Physical Review*, **78**, 487.
[29] Froelich, H. (1950). *Physical Review* **79**, 845.
[30] Migdal, A.B. (1958). *Zhurnal Eksperimentalnoi i Teoreticheskoi Fisiki* , **34**, 1438 [*Soviet Physics–JETP*, **7**, 996 (1958)].
[31] Cooper, L.N. (1956). *Physical Review*, **104**, 1189.
[32] Bogolyubov, N.N. (1958). *Zhurnal Eksperimentalnoi i Teoreticheskoi Fisiki* **34**, 58, ibid. 73 [*Soviet Physics–JETP*, **7**, 41, ibid. 51 (1958)].
[33] Gor'kov, L.P. (1958). *Zhurnal Eksperimentalnoi i Teoreticheskoi Fisiki***34**, 735 [*Soviet Physics–JETP*, **6**, 505 (1958)].
[34] Gor'kov, L.P. (1959). *Zhurnal Eksperimentalnoi i Teoreticheskoi Fisiki***37**, 1407 [*Soviet Physics–JETP*, **9**, 1364 (1959)].
[35] Ginzburg, V.L. (1960). *Soviet Solid State Physics* **2**, 61.
[36] Aslamazov, L.G. and Larkin, A.I. (1968). *Soviet Solid State Physics*, **10**, 875; *Physics Letters*, **26A**, 238 (1968).
[37] Glover, R.E. (1967). *Physical Review Letters*, **A25**, 542.
[38] Ausloos, M. and Varlamov, A.A. (eds.) (1997). *Fluctuation phenomena in high temperature superconductors*, NATO-ASI Series, Kluwer, Dordrecht.
[39] Bok, J., Deutcher, Pavuna, G.D, and Wolf, S.A. (eds.) (1998). *Gap symmetry and fluctuations in high temperature superconductors* . Plenum, London.
[40] Maki, K. (1968). *Progress in Theoretical Physics*, **39**, 897; ibid. **40**, 193.
[41] Thompson, R.S. (1970). *Physical Review* **B1**, 327.
[42] Schmidt, V.V. (1966). *Pisma v Zhurnal Eksperimentalnoi i Teoreticheskoi Fiziki*, **3**, 141; In *Proceedings of the 10th International Conference on Low Temperature Physics*, **C2**, p. 205, VINITI, Moscow.
[43] Tsuboi, T. and Suzuki, T. (1977). *Journal of Physical Society of Japan*, **42**, 654.
[44] Varlamov, A.A. and Yu, L. (1991). *Physical Review*, **B44**, 7078.
[45] Tsuzuki, T. (1972). *Journal of Low Temperature Physics*, **9**, 525.
[46] Prober, D.E., Beasley, M.R. and Schwall, R.E. (1977). *Physical Review*, **B15**, 5245.
[47] Landau, L.D. and Lifshitz, E.M. (1978). *Quantum Mechanics. Course of Theoretical Physics, Vol.3,* Pergamon Press, Oxford.
[48] Mishonov, T.M. and Penev, E.S. (2000). *International Journal of Modern Physics*, **14,** 3831.
[49] Klemm, R.A., Beasley, M.R. and Luther, A. (1973). *Physical Review*, **B8**, 5072.
[50] Lawrence, W.E. and Doniach,S. (1971). In *Proceedings of the 12th International Conference on Low Temperature Physics*, (ed. E. Kanda), p.361. Academic Press of Japan, Kyoto.

[51] Yamaji, K. (1972). *Physics Letters*, **A38**, 43.
[52] Aslamazov, L.G. and Larkin, A.I. (1973). *Zhurnal Eksperimentalnoi i Teoreticheskoi Fisiki* **67**, 647 [*Soviet Physics–JETP*, **40**, 321 (1974)].
[53] Schmid, A. (1969). *Physical Review* **180**, 527.
[54] Schmidt, H. (1968). *Zeitschrift für Physik*, **B216**, 336.
[55] Prange, R.E. (1970). *Physical Review* **B1**, 2349.
[56] Patton, B.R.,Ambegaokar, V. and Wilkins, J.W. (1969). *Solid State Communications*, **7**, 1287.
[57] Kurkijarvi, J. Ambegaokar, V. and Eilenberger, G. (1972). *Physical Review*, **B5**, 868.
[58] Gollub, J.P., Beasley, M.R., Newbower, R.S. and Tinkham, M. (1969). *Physical Review Letters* , **22**, 1288.
[59] Gollub, J.P., Beasley, M.R. and Tinkham, M. (1970). *Physical Review Letters* , **25**, 1646.
[60] Gollub, J.P., Beasley, M.R., Callarotti, R. and Tinkham, M. (1973). *Physical Review*, **B7**, 3039.
[61] Lee P.A. and Shenoy S.R. (1972). *Physical Review Letters* , **28**, 1025.
[62] Farrant, S.P. and Gough, C.E. (1975). *Physical Review Letters*, **34**, 943.
[63] Skocpol, W.J., and Tinkham, M. (1975). *Reports on Progress in Physics*, **38**, 1094.
[64] Thompson,R.S. and Kresin, V.Z. (1988). *Modern Physics Letters*, **B2**, 1159.
[65] Quader, K.F. and Abrahams, E. (1988). *Physical Review*, **B38**, 11977.
[66] Lee, W.C., Klemm, R.A. and Johnson, D.C. (1989). *Physical Review Letters* , **63**, 1012.
[67] Carretta, P., Lascialfari, A., Rigamonti, A., Rosso, A. and Varlamov, A.A. (2000). *Physical Review*, **B61**, 12420.
[68] Buzdin, A.I. and Dorin, V.V. (1997). In *Fluctuation phenomena in high temperature superconductors*, (ed. M. Ausloos and A.A. Varlamov), NATO-ASI Series, Kluwer, Dordrecht.
[69] Hikami, S., Fujita, A. and Larkin, A.I. (1991). *Physical Review*, **B44**, 10400.
[70] Baraduc, C., Buzdin, A.I., Henry, J-Y., Brison, J.P. and Puech, L. (1995). *Physica* **C248**, 138.
[71] Junod, A., Genoud, J-Y. and Triscone,G. (1998). *Physica* **C294**, 115.
[72] Lascialfari, A. Mishonov, T., Rigamonti, A., Tedesco, P. and Varlamov A.A. (2002). *Physical Review*, **B65**, 144523.
[73] Blatter, G., Feigel'man, M.V., Geshkenbein,V.B., Larkin, A.I. and Vinokur, V.M. (1994). *Reviews of Modern Physics*, **66**, 1180.
[74] Larkin, A.I. (1999). *Annals of Physics*, (Leipzig), **8**, 785.
[75] Larkin, A.I. and Ovchinnikov, Yu. N. (2001). *Zhurnal Eksperimentalnoi i Teoreticheskoi Fisiki*, **119**, 595 [*Soviet Physics–JETP*, **92**, 519 (2001)].
[76] Galitski, V. M. and Larkin, A. I. (2001). *Physical Review*, **B63**, 174506.
[77] Maki, K. (1964). *Physics* , **1**, 127.

[78] Caroli, C., Cyrot, M. and De Gennes, P.G. (1966). *Solid State Communications*, **4**, 17.
[79] Saint-James, D., Sarma, G. and Thomas, E. (1969). *Type II superconductors.* Pergamon Press, Oxford.
[80] Maki, K. (1964). *Physics* , **1**, 21; ibid. 127; ibid. 201.
[81] Gor'kov, L.P. (1959). *Zhurnal Eksperimentalnoi i Teoreticheskoi Fisiki*, **36**, 1918; ibid. **37**, 833, 1407 [*Soviet Physics–JETP* **9**, 1364 (1959); ibid. **10**, 593 (1960)].
[82] Landau, L.D., Abrikosov, and A.A, Khalatnikov, I.M. (1954). *Doklady Akademii Nauk of the USSR*, **95**, 437, ibid. 773, ibid. 1177.
[83] Diatlov, I.T., Sudakov, V.V. and Ter-Martirosyan, K.A. (1957). *Zhurnal Eksperimentalnoi i Teoreticheskoi Fisiki* **32**, 767 [*Soviet Physics–JETP*, **5**, 631 (1957)].
[84] Sudakov, V.V. (1956). *Doklady Akademii Nauk of the USSR* **111**, 338.
[85] Ahlers, G. (1971). *Physical Review* **A3**, 696; Ahlers, G. (1973). *Physical Review* **A8**, 530; Mueller, K.H., Pobell, F. and Ahlers, G. (1975). *Physical Review Letters*, **34**, 513.
[86] Gell-Mann, M. and Low, F.E. (1954). *Physical Review*, **95**, 1300.
[87] Stueckelberg, E.S. and Peterman, A. (1951). *Helvetia Physics Acta* **24**, 153.
[88] Patashinskii, A.Z. and Pokrovsky, V.L. (1982). *The Fluctuation Theory of Phase Transitions*. Nauka, Moscow.
[89] Efetov, K.B. and Larkin, A.I. (1977). *Zhurnal Eksperimentalnoi i Teoreticheskoi Fisiki*, **72**, 2350 [*Soviet Physics–JETP*, **45**, 1236 (1977)].
[90] Buzdin, A.I. and Vuyichich, B. (1990). *Modern Physics Letters* **B4**, 485.
[91] Glazman, L.I. and Koshelev, A.E. (1990). *Zhurnal Eksperimentalnoi i Teoreticheskoi Fisiki*, **97**, 1371 [*Soviet Physics–JETP*, **70**, 774 (1990)].
[92] Ovchinnikov, Yu.N. (1973). *Zhurnal Eksperimentalnoi i Teoreticheskoi Fisiki* **64**, 719 [*Soviet Physics–JETP* **37**, 366 (1973)].
[93] Maekawa, S. and Fukuyama, H. (1982). *Journal of Physical Society of Japan*, **51**,1380.
[94] Takagi, H. and Kuroda, Y. (1982). *Solid State Communications*, **41**, 643.
[95] Finkelstein, A.M. (1987). *Pisma v Zhurnal Eksperimentalnoi i Teoreticheskoi Fiziki*, **45**, 46 [*JETP Letters*, **45**, 37 (1987)].
[96] Kos, S., Millis, A., Larkin, A.I. (2004). *Physical Review*, **B70**, 214531.
[97] Halperin, B.I., Lubenski, T.C. and Ma, S. (1974). *Physical Review Letters*, **32**, 292.
[98] Schmid, A. (1966). *Physik Kondensierter Materie*, **5**, 302.
[99] Caroli, C. and Maki, K. (1967). *Physical Review*, **159**, 306, ibid. 316.
[100] Abrahams, E. and Tsuneto, T. (1966). *Physical Review*, **152**, 416.
[101] Woo, J.W.F. and Abrahams, E. (1968). *Physical Review*, **169**, 407.
[102] Di Castro, C. and Young, W. (1969). *Il Nuovo Cimento*, **B62**, 273.
[103] Ullah, S. and Dorsey, A.T. (1991). *Physical Review*, **B44**, 262; (1990). *Physical Review Letters*, **65**, 2066.

[104] Gor'kov, L.P. and Eliashberg, G.M. (1968). *Zhurnal Eksperimentalnoi i Teoreticheskoi Fiziki,* **54**, 612 [*Soviet Physics–JETP* **27**, 328(1968)].
[105] Landau, L.D. and Khalatnikov, I.M. (1954). *Doklady Akademii Nauk of the USSR*, **96**, 469,
[106] Buzdin, A.I. and Varlamov, A.A. (1998). *Physical Review*, **B58**, 14195.
[107] Varlamov, A.A., Balestrino, G., Livanov, D.V. and Milani, E. (1999). *Advances in Physics*, **48**, 655.
[108] Aslamazov, L.G. and Varlamov, A.A. (1980). *Journal of Low Temperature Physics*, **38**, 223.
[109] Aronov, A.G., Hikami, S. and Larkin, A.I., (1995). *Physical Review*, **B51**, 3880.
[110] Efetov, K.B. (1979). *Zhurnal Eksperimentalnoi i Teoreticheskoi Fisiki* **76**, 1781 [*Soviet Physics–JETP*, **49**, 905 (1979)].
[111] Blatter, G., Geshkenbein, V.B. and Larkin, A.I. (1992). *Physical Review Letters*, **68**, 875.
[112] Dekker, C. (1999). *Physics Today* **52**, 22.
[113] Bockrath, M., Cobden, D. H., Lu, J., Rinzler, A.G., Smalley, R. E., Balents, L. and McEuen, P.L. (1999). *Nature* (London), **397**, 598.
[114] Egger, R., and Gogolin, A. O. (1997). *Physical Review Letters*, **79**, 5082.
[115] Kane, C.L., Balents, L. and Fisher, M. P. A. (1997). *Physical Review Letters*, **79**, 5068.
[116] Egger, R. (1999). *Physical Review Letters*, **83**, 5547.
[117] Bachtold, A., Strunk, C., Salvetat, J. P., Bonard, J. M.,Forro, L., Nussbaumer, T. and Schoennberger, C. (1999). *Nature* (London), **397**, 673.
[118] Reulet, B., Kasumov, A.Yu., Kosiak, M., Deblock, R., Khodos, I. I., Gorbatov, Yu.B.et al. (2000). *Physical Review Letters*, **85**, 2829.
[119] Tang, Z.K., Zhang, L., Wang, N., Zhang, X. X., Wen, G. H., Li, G. D. et al. (2001). *Science*, **292**, 2462.
[120] Kosiak, M., Kasumov, Gueron, S., A.Yu., Reulet, B., Khodos, I. I., Gorbatov, Yu.B. et al. (2001). *Physical Review Letters*, **86**, 2416.
[121] Livanov, D.V. and Varlamov, A.A. (2002). *Physical Review* **B66**, 104522.
[122] Egger, R. and Gogolin, A.O. (2001). *Physical Review Letters*, **87**, 066401.
[123] Little, W. A. and Parks, R. D. (1962). *Physical Review Letters*, **9**, 9.
[124] Buzdin, A.I. and Varlamov, A.A. (2002). *Physical Review Letters*, **89**, 076601.
[125] Larkin, A.I. and Matveev, K.A. (1997). *Physical Review Letters*, **78**, 3749.
[126] Aslamazov, L.G. and Larkin, A.I. (1971). In *Proceedings of the first European conference on condesed matter*. European Physical.Society, Geneva, Switzerland.
[127] Mishonov, T., Pozheninnikova, A. and Indekeu, J. (2002). *Physica*, **B65**, 064519.
[128] Luttinger, J. M. (1964). *Physical Review,* **A135**, 1505.
[129] Balesku, R. (1978). *Equilibrium and nonequilibrium statistical mechanics. Vol.2*, Mir, Moscow.

[130] Caroli, C. and Maki, K. (1967). *Physical Review*, **164**, 591.
[131] Abrikosov, A.A. (1988). *Fundamentals of the theory of metals*, North-Holland, Elsevier, Groningen.
[132] Schmidt, V.V. (1997). *The physics of superconductors*, (ed. P.H. Muller and A.V. Ustinov), Springer-Verlag, Berlin-Heidelberg.
[133] Ono, Y. (1971). *Progress in Theoretical Physics*, **46**, 757.
[134] Hu, C.-R. (1976). *Physical Review*, **B13**, 4780.
[135] Cooper, N. R., Halperin, B. I. and Ruzin, I. M. (1997). *Physical Review*, **B55**, 2344.
[136] Maki, K. (1974). *Journal of Low Temperature Physics*, **14**, 419.
[137] Zeh, M., Ri, H.-C., Kober, F., Huebner, R.P., Ustinov A.V., Mannhart, J., Gross, R. et al. (1990). *Physical Review Letters*, **64**, 3195.
[138] Haggen, S.J., Lobb, C.J., Greene, R.L., Forrester, M.G. and Talvacchio, J. (1990). *Physical Review*, **B42**, 6777.
[139] Clayhold J.A., Linnen (Jr), A.W., Chen, F. and Chu, C.W. (1994). *Physical Review*, **B50**, 4252.
[140] Holn, C., Galffy, M. and Freimuth, A. (1994). *Physical Review*, **B50**, 15875;
[141] Palstra, T.T.M., Battlog, B., Schneemeyer, L.F. and Waszczak, J.V. (1990). *Physical Review Letters*, **64**, 3090.
[142] Xu, Z.A., Ong, N.P., Wang, Y., Kakeshita, T. and Uchida, S. (2000). *Nature*, **406**, 486.
[143] Wang, Y., Xu, Z.A., Kakeshita,T., Uchida, S., Ono, S. Ando, Y. et al. (2001). *Physical Review*, **B64**, 224519.
[144] Wang, Y., Ong, N. P., Xu, Z. A., Kakeshita, T., Uchida, S.D., Bonn, A., Liang, R.and W. N. Hardy (2002). *Physical Review Letters*, **88**, 257003.
[145] Capan, C., Behnia, K., Hinderer, J., Jansen, A.G.M., Lang, W., Marcenat, C. et al. (2002). *Physical Review Letters*, **88**, 056601.
[146] Capan, C., Behnia, K., Li, Z.Z., Raffy, H., Marin, C. (2003). *Physical Review*, **B67**, 100507.
[147] Wen, H.H., Liu, Z.Y., Xu, Z.A., Weng, Z.Y., Zhou, F., and Zhao, Z.X. (2003). *Europhys. Lett.*, **B63**, 583.
[148] Xu, Z.A., Shen, J.Q., Zhao, S.R., Zhang, Y.J., and Ong, N.P.(2005). *Physical Review*, **B72**, 144527.
[149] Wang, Y., Li, L., and Ong, N.P. (2006). *Physical Review*, **B73**, 024510.
[150] Li, P., and Greene, R.L. (2007). *Physical Review*, **B76**, 174512.
[151] Hartnoll, S.A., Kovtun, P.K., Mueller, M., and Sachdev. (2007) *Physical Review*, **B76**, 144502.
[152] Podolsky, D., Raghu, S., and Vishwanath, A. (2007). *Physical Review Letters*, **99**, 117004.
[153] Raghu, S., Podolsky, D., Vishwanath, A., and Huse, D.A. (2008). arXiv: 0801.2925v2.
[154] Ussishkin, I., Sondhi, S.L. and Husse, D.A. (2002). *Physical Review Letters*, **89**, 287001.
[155] Ussishkin, I. (2003). *Physical Review*, **B68**, 024517.

[156] Ussishkin, I., and Sondhi, S.L. (2004) *Int. J. Mod. Phys.*, **B18**, 3315.
[157] Pourret, A., Aubin, H., Lesueur, J., Marrache-Kikuchi, C.A., Berge, L., Dumoulin, L., and Behnia, K. (2006). *Nature Phys.*, **2**, 683.
[158] Pourret, A., Aubin, H., Lesueur, J., Marrache-Kikuchi, C.A., Berge, L., Dumoulin, L., and Behnia, K. (2007). *Physical Review Letters*, **76**, 214504.
[159] Reizer, M.Yu. and Sergeev, A.V. (1994). *Physical Review*, **B50**, 9344.
[160] Obraztsov, Yu.N. (1964). *Soviet Phys. Solid State*, **6**, 331 .
[161] Sondheimer, E.H. (1948). *Proc. R. Soc.*, **A193**, 484.
[162] Varlamov, A.A., Livanov, D.V. and Federici, F. (1997). *Pisma v Zhurnal Eksperimentalnoi i Teoreticheskoi Fiziki*, **65**, 196 [*JETP Letters*, **65**, 182 (1997)].
[163] Varlamov, A.A., Livanov, D.V. and Reggiani, L. (1992) *Physics Letters*, **A165**, 369.
[164] Varlamov, A.A. and Livanov, D.V. (1991) *Zhurnal Eksperimentalnoi i Teoreticheskoi Fisiki* **99**, 1816 [*Soviet Physics–JETP*, **72**, 1016 (1991)].
[165] Howson, M.A., Salamon, M.B., Friedmann, T.A., Rice, J.P. and Ginsberg, D. (1990). *Physical Review*, **B41**, 300.
[166] Zavaritski, N.V., Samoilov, A.V. and Yurgens, A.A. (1992). *Pisma v Zhurnal Eksperimentalnoi i Teoreticheskoi Fiziki*, **55**, 133 [*JETP Letters*, **55**, 127 (1992)].
[167] Keshri, S. and Barhai, P.K. (1997). *Chech Journal of Physics*, **47**, 249.
[168] Mosqueira, J., Viera, J.A. and Vidal, F. (1994). *Physica*, **C229**, 301; Mosqueira, J., Viera, J.A, Masa, J., Cabeza, O. and Vidal, F. (1995). *Physica*, **C253**, 1.
[169] Abrahams, E., Redi, M. and Woo, C. (1970). *Physical Review*, **B1**, 208.
[170] Niven, D.R. and Smith, R.A. (2002). *Physical Review*, **B66,** 214505.
[171] Vishveshwara, S. and Fisher, M. P. A. (2001). *Physical Review*, **B64**, 134507.
[172] Tinkham, M. (1996). *Introduction to superconductivity*, (2nd edition). McGraw Hill, New York.
[173] Blatter, G. and Geshkenbein, V.B. (2002). In: *The physics of superconductors*, (ed. K.H. Bennemann and J.B. Ketterson), Vol.1, chapter 10, Springer-Verlag, Berlin.
[174] Larkin, A.I. and Ovchinnikov, Yu.N. (1986). In *Nonequilibrium superconductivity*, (ed. D.N. Langenberg and A.I. Larkin), Elsevier Science Publishers, New York-Amsterdam.
[175] Bardeen, J. and Stephen, M.J. (1965). *Physical Review* **140**, A1197.
[176] Gor'kov, L.P. and Kopnin, N.B. (1973). *Zhurnal Eksperimentalnoi i Teoreticheskoi Fisiki* **65**, 396 [*Sov.Phys.JETP,* **38**, 195 (1973)].
[177] Bardeen, J. and Sherman, R.D. (1975) *Physical Review*, **B12**, 2634.
[178] Larkin, A.I. and Ovchinnikov, Yu.N. (1976). *Pisma v Zhurnal Eksperimentalnoi i Teoreticheskoi Fiziki,* **23**, 210 [*JETP Letters*, **23**, 187 (1976)].
[179] Larkin, A.I. and Ovchinnikov, Yu.N. (1973). *Zhurnal Eksperimentalnoi i Teoreticheskoi Fiziki,* **65**, 1704 [*Soviet Physics–JETP*, **38**, 854 (1974)].

[180] Larkin, A.I. and Ovchinnikov, Yu.N. (1979). *Journal of Low Temperature Physics*, **34**, 409.

[181] Labusch, R. (1969). *Crystal Lattice Defects,* **1**, 1.

[182] Larkin, A.I., (1970). *Zhurnal Eksperimentalnoi i Teoreticheskoi Fisiki* , **58**, 1466 [*Soviet Physics–JETP*, **31**, 784 (1970)].

[183] Nattermann, T. (1990). *Physical Review Letters*, **64,** 2454.

[184] Korshunov, S.E. (1993). *Europhysics Letters* , **11**, 757.

[185] Giamarchi, T. and Le Doussal, P. (1994). *Physical Review Letters*, **72,** 1530; *Physical Review*, **B52**, 1242 (1995).

[186] Larkin, A.I. and Ovchinnikov, Yu.N. (1978). *Pisma v Zhurnal Eksperimentalnoi i Teoreticheskoi Fiziki*, **27**, 301 [*JETP Letters*, **27**, 280 (1978)].

[187] Fischer, D.S. (1986). *Physical Review Letters*, **56,** 1964.

[188] Bucheli, H., Wagner, O.S., Geshkenbein, V.B., Larkin, and A.I., Blatter, G. (1998) *Physical Review*, **B57**, 7642; ibid. (1999) **B59,** 11551.

[189] Lee, P.A. and Fukuyama, H. (1978). *Physical Review,* **B17,** 535.

[190] Gor'kov, L.P. and Dolgov, E.N. (1979). *Zhurnal Eksperimentalnoi i Teoreticheskoi Fisiki,* **77**, 396 [*Soviet Physics–JETP,* **50**, 203 (1979)].

[191] Lee, P.A., and Rice, T.M. (1979). *Physical Review*, **B19,** 3970.

[192] Larkin, A.I. and Khmelnitski, D.E. (1979). *Microscopic theory of friction*, Preprint of the Landau Institute, Russian Academy of Sciences, Moscow.

[193] Imry, Y. and Ma, S.K. (1975). *Physical Review Letters*, **35,** 1399.

[194] Anderson, P.W. (1962). *Physical Review Letters*, **9**, 309.

[195] Anderson, P.W. and Kim,Y.B. (1964). *Reviews of Modern Physics*, **36**, 39.

[196] Geshkenbein, V.B. and Larkin, A.I. (1989). *Zhurnal Eksperimentalnoi i Teoreticheskoi Fisiki,* **95**, 1108 [*Soviet Physics–JETP*, **68**, 639(1989)].

[197] Feigel'man, M.V., Geshkenbein, V.B., Larkin, A.I. and Vinokur, V.M. (1989). *Physical Review Letters*, **63** , 2303.

[198] Yeshurun, Y. and Malozemoff, A.P. (1988). *Physical Review Letters*, **60**, 2202.

[199] Yeshurun, Y., Malozemoff, A.P., Holtzberg, F.H. and Dinger, T.R. (1988). *Physical Review*, **B38**, 11828.

[200] Yeshurun, Y., Malozemov, A.P., Worthington, T.K., Yandrofski, R.M., Krusin-Elbaum, L. and Holtzberg, F.H. (1989). *Cryogenics,* **29**, 258.

[201] Hagen, C.W. and Griessen, R. (1989). *Physical Review Letters*, **62**, 2857.

[202] Campbell, I.A., Fruchter, L. and Cabanel, R. (1990). *Physical Review*, **B42**, 1561.

[203] Lairson, B.M., Sun, J.Z., Geballe, T.H., Beasley, M.R. and Bravman, J.C. (1991). *Physical Review*, **B43**, 10405.

[204] Malozemoff, A.P. (1991). *Physica,* **C185-189**, 264.

[205] Kim, Y.B., Hempstead, C.F. and Strnad, A.R. (1962). *Physical Review Letters*, **9**, 306.

[206] Beasley, M.R., Labusch, R. and Webb, W.W. (1969). *Physical Review*, **181**, 682.

[207] Fischer, M.P.A. (1989). *Physical Review Letters*, **62,** 1415.
[208] Glazman, L.I. and Fogel, N.Ya. (1984). *Fizika NizkihTemperatur (USSR)*, **10**, 95.
[209] Larkin, A.I., Ovchinnikov, Yu.N. and Schmid, A. (1988). *Physica*, **B152**, 266.
[210] Fisher, M.P.A., Tokuyasu, T.A. and Young, A.P. (1991). *Physical Review Letters*, **66,** 2931.
[211] Blatter, G., Geshkenbein, V.B. and Vinokur, V.M. (1991). *Physical Review Letters*, **66,** 3297.
[212] Ivlev, B.I., Ovchinnikov, Yu.N. and Thompson, R.S. (1991). *Physical Review*, **44,** 7023.
[213] Liu, Y., Havilland, D.B., Glazman, L.I. and Goldman, A.M. (1992). *Physical Review Letters*, **68,** 2224.
[214] Nelson, D.R. (1988). *Physical Review Letters*, **60**, 1973.
[215] Bray, A.J. (1974). *Physical Review*, **B9**, 4752.
[216] Thouless, D.J. (1975). *Physical Review Letters*, **34,** 949.
[217] Ruggeri, G.J. and Thouless, D.J. (1976). *Journal of Physics*, **F 6,** 2063.
[218] Hikami, S. and Fujita, A. (1990). *Physical Review*, **B41,** 6379; *Progress in Theoretical Physics*, **83,** 443 (1990).
[219] Ikeda, R., Ohmi, T. and Tsuneto, T. (1989). *Journal of Physical Society of Japan Letters*, **58**, 1377; ibid. **60**, 1051 (1991).
[220] Bresin, E., Fujita, A. and Hikami, S. (1990). *Physical Review Letters*, **65,** 1949; ibid. **65,** 2921 (1990).
[221] Houghton, A., Pelcovits, R.A. and Sudbo, A. (1989). *Physical Review*, **B40,** 6763.
[222] Nordborg, H. and Blatter, G. (1997). *Physical Review Letters*, **79,** 1925; *Physical Review*, **B58,** 14556 (1998).
[223] Koshelev, A.E. and Nordborg, H. (1999). *Physical Review*, **B59,** 4358.
[224] Nelson, D.R., Halperin, B. I., (1979), *Physical Review*, **B9**, 2457.
[225] Gammel, P.L., Schneemeyer, L.F. and Bishop, D.J. (1991). *Physical Review Letters*, **66**, 953.
[226] Fischer, D.S., Fischer, M.P.A. and Huse, D.A. (1991). *Physical Review*, **B43,** 130.
[227] Geshkenbein, V.B., Ioffe, L.B. and Larkin, A.I. (1997). *Physical Review*, **B55**, 3173.
[228] Abrikosov, A.A., Gor'kov, L.P., Dzyaloshinski, I.E. (1963). *Methods of the quantum field theory in statistical physics.* Dover Publications, New York.
[229] Eliashberg, G.M. (1960). *Zhurnal Eksperimentalnoi i Teoreticheskoi Fisiki*, **38**, 966 [*Soviet Physics–JETP* **11**, 696 (1960)].
[230] Narozhny, B.N. (1993). *Zhurnal Eksperimentalnoi i Teoreticheskoi Fisiki* **104**, 2825 [*Soviet Physics–JETP* **77**, 301 (1993)]; (1994). *Physical Review* **B49**, 6375.
[231] Klemm, R.A. and Liu, S. H. (1994). *Physical Review*, **B49**, 6375.

[232] Liu, S. H. and Klemm, R.A. (1993). *Physical Review*, **B48**, 4080; ibid. 10650.

[233] Svidzinski, A.V. (1982). *Inhomogeneous problems of the theory of superconductivity*. Nauka, Moscow.

[234] Larkin, A.I. (1965). *Zhurnal Eksperimentalnoi i Teoreticheskoi Fisiki* **48**, 232 [*Soviet Physics–JETP*, **21**, 153 (1965)].

[235] Abrikosov, A.A. and Gor'kov, L.P. (1958). *Zhurnal Eksperimentalnoi i Teoreticheskoi Fisiki*, **35**, 1558; ibid. **36**, 319 (1959) [*Soviet Physics–JETP* **8**, 1090; ibid. **9**, 220 (1959)].

[236] Anderson, P.W. (1959 *Journal of Physical Chemistry of Solids*, **11**, 26).

[237] Di Castro, C., Castellani, C., Raimondi, R. and Varlamov, A.A. (1990). *Physical Review*, **B42**, 10211.

[238] Ioffe, L.B., Larkin, A.I., Varlamov, A.A. and Yu, L. (1993). *Physical Review*, **B47**, 8936; *Physica*, **C235-240**, 1963 (1994).

[239] Gray, K.E. and Kim, D.H. (1993). *Physical Review Letters*, **70**, 1693.

[240] Dorin, V.V., Klemm, R.A., Varlamov, A.A. Buzdin, A.I., and Livanov, D.V. (1993). *Physical Review*, **B48**, 12951.

[241] Altshuler, B.L., Reizer, M.Yu. and Varlamov, A.A. (1983). *Zhurnal Eksperimentalnoi i Teoreticheskoi Fisiki* **84**, 2280 [*Soviet Physics–JETP* **57**, 1329 (1983)].

[242] Larkin, A.I. and Khmelnitski, D.E. (1982). *Soviet Physics Uspekhi*, **136,** 536.

[243] Altshuler, B.L. and Aronov, A.G. (1985). In: *Electron-electron interaction in disordered systems*, (eds. A.L.Efros, and M.Pollak), North Holland, Amsterdam.

[244] Eliashberg, G.M. (1961). *Zhurnal Eksperimentalnoi i Teoreticheskoi Fisiki* **41**, 1241 [*Soviet Physics–JETP*, **14**, 856 (1961)].

[245] Klemm, R.A. (1974). *Journal of Low Temperature Physics*, **16**, 381.

[246] Baraduc, C., Pagnon, V., Buzdin, A.I., Henry, J. and Ayache, C. (1992). *Physics Letters*, **A166,** 267.

[247] Ami, S. and Maki, K. (1978). *Physical Review* **B18**, 4714.

[248] Yip, S. (1990). *Physical Review*, **B41**, 2612.

[249] Carretta, P., Rigamonti, A., Varlamov, A.A. and Livanov, D.V. (1996). *Physical Review* **B54**, R9682.

[250] Varlamov, A.A. (1994). *Europhysics Letters*, **28**, 347.

[251] Patton, B.R. (1971). *Physical Review Letters*, **27**, 1273.

[252] Keller, J. and Korenman, V. (1971). *Physical Review Letters* **27,** 1270; *Physical Review*, **B5,** 4367 (1972).

[253] Altshuler, B.L., Aronov, A.G., Khmelnitski, D.E. and Larkin, A.I. (1982). In: *Quantum theory of solids,* (ed. I.M.Lifhsitz), Mir, Moscow.

[254] Vavilov, M., Glazman, L. and Larkin, A.I. (2003). *Physical Review*, **B68**, 075119.

[255] Gorkov, L.P. (2002). In: *The Physics of superconductors,* (ed. K.H. Bennemann and J.B. Ketterson), Vol.1, chapter 4, Springer-Verlag, Berlin.

[256] Altshuler, B.L., Aronov, A.G., Khmelnitski, D.E. (1982). *Journal of Physics*, **C15**, 7367
[257] Breing, W., Chang, M.C., Abrahams, E., and Wolfle,P. (1985). *Physical Review*, **B 31**, 7001.
[258] Gordon, J.M., Lobb. C.J. and Tinkham, M. (1984). *Physical Review* **B29**, 5232.
[259] Gordon, J.M. and Goldman, A.M. (1986). *Physical Review* **B34**, 1500.
[260] Reizer, M.Yu. (1992). *Physical Review* , **B45**, 12949.
[261] Skvortsov, M.A., Larkin, A.I. and Feygel'man, M.V. (2004). *Physical Review*, **B**, in press
[262] Ausloos, M. and Laurent, Ch. (1988). *Physical Review,* **B37**, 611.
[263] Frietas, P.P., Tsuei, C.C. and Plaskett, T.S. (1987). *Physical Review*, **B36**, 833.
[264] Hikita, M. and Suzuki, M. (1990). *Physical Review*, **B41**, 834.
[265] Akinaga, M., Abukay, D. and Rinderer, L. (1988). *Modern Physics Letters*, **2**, 891.
[266] Poddar, A., Mandal, P.,Das, A.N., Ghosh, B. and Choudhury, P. (1989). *Physica*, **C159**, 231.
[267] Kim, D.H., Goldman, A.M., Kang, J.H., Gray, and K.E., Kampwirth, R.T. (1989). *Physical Review*, **B39**, 12275.
[268] Balestrino, G., Nigro, A. and Vaglio, R. (1989). *Physical Review* **B39**, 12264.
[269] Kumm, G. and Winzer, K. (1990). *Physica*, **B165-166** , 1361.
[270] Cimberle, M.R., Ferdeghini, C., Marrè, D., Putti, M., Siri, S., Federici, F. et al. (1997). *Physical Review*, **B55**, R14745.
[271] Reggiani, L. Vaglio, R. and Varlamov, A.A. *Physical Review*, **B44**, 9541 (1991).
[272] Balestrino, G., Milani, E. and Varlamov, A.A. *Physica*, **C210**, 386 (1993).
[273] Federici, F. and Varlamov, A.A. (1996). *Pisma v Zhurnal Eksperimentalnoi i Teoreticheskoi Fiziki*, **64**, 397 [*JETP Letters* **64**, 497 (1996)]; *Physical Review*, **B57**, 6071 (1997).
[274] Bieri, J.B. and Maki, K. (1990). *Physical Review*, **B42**, 4854.
[275] Livanov, D.V., Savona, G. and Varlamov, A.A. (2000), *Physical Review*, **B62,** 8675.
[276] Larkin, A.I. and Ovchinnikov, Yu.A. (1973). *Journal of Low Temperature Physics*, **10**, 407.
[277] Varlamov, A.A. and Dorin, V.V. (1983). *Zhurnal Eksperimentalnoi i Teoreticheskoi Fisiki*, **84**, 1868 [*Soviet Physics–JETP* **57**, 1089 (1983)].
[278] Randeria, M. and Varlamov, A.A. (1994). *Physical Review* **B50**, 10401.
[279] Eschrig, M., Rainer, D. and Souls, J. (1999). *Physical Review* **B59**, 12095.
[280] Axnaes, J. (1999). *Journal of Low Temperature Physics*, **117**, 259.
[281] Bieri, J.B., Maki, K. and Thompson, R.S. (1991). *Physical Review*, **B44**, 4709.

[282] Aleiner, I., Glazman, L.I. and Rudin, A. (1997). *Physical Review* **B55**, 322.

[283] Larkin, A.I. (1980). *Pisma v Zhurnal Eksperimentalnoi i Teoreticheskoi Fiziki,* **31**, 239 [*JETP Letters*, **31**, 219 (1980)].

[284] Abrahams, E. Anderson, P.W., Lichardello, D.C. and Ramakrishnan,T.V. (1979). *Physical Review Letters*, **42**, 673.

[285] Abrahams, E. (1985). *Journal of Statistical Physics*, **38**, 89; Lopes dos Santos, J.M.B., Abrahams, E. (1985). *Physical Review*, **B31**, 172.

[286] Gershenzon, M.E., Gubankov, V.N. and Zhuravlev, Yu.E. (1983). *Solid State Communications*, **45**, 87.

[287] Rosenbaum, R. (1985). *Physical Review*, **B32**, 2190.

[288] Mori, N. (1987). *Japan Journal of Applied Physics (Supplement)*, **26**, 1339.

[289] Belevtsev, B.I., Komnik, Yu.F. and Fomin, A.V. (1985). *Journal of Physics*, **B58**, 111; *Fizika Nizkih Temperatur (USSR)*, **10,** 850 (1984).

[290] Altshuler, B.L. and Aronov, A.G. (1979). *Zhurnal Eksperimentalnoi i Teoreticheskoi Fisiki*, **77**, 2028 [*Soviet Physics–JETP*, **50**, 968 (1979)].

[291] Altshuler, B.L., Aronov, A.G., and Lee, P.A. (1980). *Physical Review Letters*, **44**, 1288.

[292] Altshuler, B.L., Khmelnitski, D.E., Larkin, A.I. and Lee, P.A. (1980). *Physical Review*, **B22**, 5142.

[293] Altshuler, B.L., Aronov, A.G., Larkin, A.I. and Khmelnitski, D.E. (1981). *Zhurnal Eksperimentalnoi i Teoreticheskoi Fisiki*, **81**, 768, [*Soviet Physics–JETP* **54**, 411 (1981)].

[294] Fukuyama, H. (1981). *Journal of Physical Society of Japan*, **50**, 3407.

[295] Finkel'shtein, A.M. (1983). *Zhurnal Eksperimentalnoi i Teoreticheskoi Fisiki*, **84**, 168 [*Soviet Physics–JETP*, **57**, 97 (1983)].

[296] Altshuler, B.L. and Aronov, A.G. (1981). *Solid State Communications*, **38**, 11.

[297] Lee, P.A., Ramakrishnan, T.V. (1982). *Physical Review*, **B26**, 4009.

[298] Abrahams, E. Anderson, P.W. and Ramakrishnan,T.V. (1979). *Physical Review Letters*, **43**, 718.

[299] Maki, K. and Thompson, R.S. (1989). *Physica* **C162-164,** 1441.

[300] Larkin, A.I. and Ovchinnikov, Yu.N. (1971). *Zhurnal Eksperimentalnoi i Teoreticheskoi Fiziki* **61**, 2147 [*Soviet Physics–JETP* **34**, 1144 (1972)].

[301] Varlamov, A.A. Dorin, V.V. and Smolyarenko, I.E. (1988). *Zhurnal Eksperimentalnoi i Teoreticheskoi Fisiki*, **94**, 257 [*Soviet Physics–JETP*, **67**, 2536 (1988)].

[302] Schmid, A. and Schön, G. (1975). *Journal of Low Temperature Physics* **20**, 207.

[303] Larkin, A.I. and Ovchinnikov, Yu. N. (1975). *Zhurnal Eksperimentalnoi i Teoreticheskoi Fisiki* **68**, 1915 [*Soviet Physics–JETP*, **41**, 960 (1976)].

[304] Larkin, A.I. and Ovchinnikov, Yu.N. (1977). *Zhurnal Eksperimentalnoi i Teoreticheskoi Fisiki* **73**, 299 [*Soviet Physics–JETP*, **46**, 155 (1977)].

[305] Suzuki, M. and Hikita, M. (1989). *Physical Review*, **B39**, 4756; (1991). *Physical Review,* **B44**, 249.

[306] Matsuda, A., Ishii, T., Kinishita, K. and Yamada, T. (1989). *Physica*, **C162-164**, 371;Matsuda, A., Hirai, T. and Komiyama, S. (1988). *Solid State Communications*, **68**, 103.

[307] Sugawara, J., Iwasaki, H., Kobayashi, N., Yamane, H. and Hirai, T. (1992). *Physical Review*, **B46**, 14818.

[308] Holm, W. Andersson, M., Rapp, O., Kulikov, M.A. and Makarenko, I.N. (1993). *Physical Review*, **B48**, 4227.

[309] Lang, W., Heine, G. Kula, W. and Sobolewski, R. (1995). *Physical Review* **B51**, 9180.

[310] Holm, W., Rapp, O., Johnson, C.N.L. and Helmersson, U. (1995). *Physical Review*, **B52**, 3748.

[311] Latyshev, Yu.I., Laborde, O. and Monceau, P. (1995). *Europhysics Letters* **29**, 495.

[312] Lang, W. (1995). *Zeitschrift für Physik*, **B97**, 583.

[313] Watanabe, T. and Matsuda, A. (1996). *Physica* **C263**, 313.

[314] Lang, W. (1995). *Physica*, **C245**, 69.

[315] Volz, W., Razavi, F.S., Quiron, G., Habermeier, H.-U. and Solovjov, A.L. (1997). *Physical Review*, **B55** 6631.

[316] Sekirnjak, C., Lang, W., Proyer, S. and Schwab, P. (1995). *Physica*, **C243**, 60.

[317] Hikami, S. and Larkin, A.I. (1988). *Modern Physics Letters* **B2**, 693.

[318] Aronov, A.G., Hikami, S. and Larkin, A.I. (1989). *Physical Review Letters* **62**, 965.

[319] Balestrino, G., Milani, E. and Varlamov, A.A. (1995). *Pisma v Zhurnal Eksperimentalnoi i Teoreticheskoi Fiziki* **61**, 814 [*JETP Letters* **61**, 833 (1995)].

[320] Yan, Y.F., Matl, P., Harris, J.M. and Ong, N.P. (1995). *Physical Review*, **B52**, R751; Ong, N.P., Yan, Y.F. and Harris, J.M. (1994). In *Procedings of CCAST symposium on high Tc superconductivity and the C60 family*, Beijing.

[321] Lang, W. (1997). *Physica*, **C282-287**, 233.

[322] Nakao, K., Takamaku, K., Hashimoto, K., Koshizuka, N. and Tanaka, S. (1994). *Physica*, **B201**, 262.

[323] Hashimoto, K. Nakao, K., Kado, H. and Koshizuka, N. (1996). *Physical Review*, **B53**, 892.

[324] Heine, G., Lang, W., Wang, X.L. and Wang, X.Z. (1996). *Journal of Low Temperature Physics*, **105,** 945.

[325] Kimura, T., Miyasaka, S., Takagi, H., Tamasaku,K., Eisaki, H., Uchida, S. et al. (1996). *Physical Review*, **B53**, 8733.

[326] Axnas, J., Holm, W. Eltsev, Yu. and Rapp, O. (1996). *Physical Review Letters*, **77**, 2280.

[327] Wahl, A., Thopart, D.,Villard, G., Maignan, A., Hardy, V., Soret, J.C. et al. (1999). *Physical Review*, **B59**, 7216.
[328] Nygmatulin, A.S., Varlamov, A.A., Livanov, D.V., Balestrino, G. and Milani, E. (1996). *Physical Review* **B53**.
[329] Wahl, A., Thopart, D., Villard, G., Maignan, A., Simon, Ch., Soret, J.C. et al. (1999). *Physical Review*, **B60**, 12495.
[330] Thopart, D., Wahl, A., Simon, Ch., Soret, J.C., Ammor, L., Ruyter, A., et al. (2000). *Physical Review*, **B62**, 9721.
[331] Abrikosov, A.A. and Gor'kov, L.P. (1960). *Zhurnal Eksperimentalnoi i Teoreticheskoi Fisiki*, **39**, 1781 [*Soviet Physics–JETP* **12**, 1243 (1961)].
[332] Livanov, D.V., Milani, E., Balestrino, G. and Aruta, C. (1997). *Physical Review*, **B55,** R8701.
[333] Maki, K. (1973). *Physical Review Letters*, **30**, 648.
[334] Bruynseraede, Y., Gijn, M., van Haesendonck, C. and Deutcher, G. (1983). *Physical Review Letters* **50**, 277.
[335] Gordon, M., Lobb, C.G. and Tinkham, M. (1983). *Physical Review*, **B28**, 4046.
[336] Raffy, H., Labowitz, R.B., Chaudari, P. and Maekava,S. (1983). *Physical Review*, **B28**, 6607.
[337] Santanham, P. and Prober, D.E. (1984). *Physical Review*, **B29**, 3733.
[338] Begmann, G. (1984). *Physical Review*, **B29**, 6114.
[339] Fukuyama, H., Ebisawa, H. and Tuzuki, T. (1971) . *Progress in Theoretical Physics*, **46**,1028.
[340] Abrahams, E., Prange, R.E. and Stephen, M.E. (1971). *Physica*, **55,** 230.
[341] Inoue, T., Miwa, S., Okamoto, K., and Awano, M. (1979). *Journal of Physical Society of Japan*, **46**, 418.
[342] Varlamov, A.A. and Livanov, D.V. (1990). *Zhurnal Eksperimentalnoi i Teoreticheskoi Fisiki*, **98**, 584 [*Soviet Physics–JETP*, **71**, 325 (1990)].
[343] Aronov, A.G. and Rapoport, A.B. (1992). *Modern Physics Letters*, **B6**, 1093.
[344] Angilella, G.G.N. Pucci, R., Varlamov, A.A. and Onufrieva, F. (2003). *Physical Review*, **B67**, 134525.
[345] Dorsey, A.T. (1992). *Physical Review*, **B46**, 8376.
[346] Nagaoka, T., Matsuda, Y., Obara, H., Sawa, A., Terashima, T., Chong, I. et al. (1998). *Physical Review Letters*, **80**, 3594.
[347] Lifshitz, I.M. (1960). *Zhurnal Eksperimentalnoi i Teoreticheskoi Fisiki*, **38**, 1569 [*Soviet Physics–JETP*, **11**, 1130 (1960)].
[348] Varlamov, A.A. Egorov, V.S. and Pantsulaya, A.V. (1989). *Advances in Physics*, **38**, 469.
[349] Blanter, Ya.M., Kaganov, M.I., Pantsulaya, A.V. and Varlamov, A.A. (1994). *Physics Reports*, **245**, 159.
[350] Blanter, Ya.M. Pantsulaya, A.V. and Varlamov, A.A. (1992). *Physical Review*, **B45**, 6267.
[351] Livanov, D.V. (1999). *Physical Review* , **B 60,** 13439.

[352] Lang, W. Heine, G., Schwab, P. and Wang, X.Z. (1994). *Physical Review*, **B49**, 4209.
[353] Wang, L.M., Yang, H. and Horn, H.E. (1999). *Physical Review*, **B59**, 14031.
[354] Onufrieva, F, Pfeuty, P. and Kisilev, M. (1999). *Physical Review Letters*, **82**, 2370.
[355] Angilella, G.G.N., Piegari, E.,and Varlamov, A.A. (2002). *Physical Review*, **B66**, 014501.
[356] Pavarini, E., Dasgupta, I., Saha-Dasgupta, T., Jepsen, O. and Andersen, O.K. (2001). *Physical Review Letters*, **87**, 047003.
[357] Onufrieva, F. Petit, S. and Sidis, Y. (1985). *Physical Review*, **B54**, 12464 (1996).
[358] Bulaevski, L.N. (1974). *Zhurnal Eksperimentalnoi i Teoreticheskoi Fiziki*, **66**, 2212 [*Soviet Physics–JETP* **39**, 1090 (1974)].
[359] Beloborodov, I.S., Efetov, K.B., (1999). *Physical Review Letters* **82**, 3332.
[360] Beloborodov, I.S., Efetov, K.B. and Larkin, A.I. (2000). *Physical Review*, **B61**, 9145.
[361] Gantmakher, V.F. (1998). *International Journal of Modern Physics* **B12,** 3151.
[362] Gantmakher, V.F., Ermolov, S.N., Tsydynzhapov, G.E., Zhukov, A.A. and Baturina, T.I. (2003). *Pisma v Zhurnal Eksperimentalnoi i Teoreticheskoi Fiziki*, **77**, 424 [*JETP Letters*, **77**, 498 (2003)].
[363] Gantmakher, V.F., Golubkov, M.V., Dolgopolov, V.T., Tsydynzhapov, G.E. and Shashkin, A.A. (1998). *Pisma v Zhurnal Eksperimentalnoi i Teoreticheskoi Fiziki*, **68**, 320 [*JETP Letters*, **68**, 344 (1998)].
[364] Tyler, A.W. Ando, Y., Balakirev, F.F., Passner, A., Boebinger, G.S., Schofield, A.J. et al. (1998). *Physical Review*, **B57**, R728.
[365] Maki, K. (1966). *Physical Review*, **148**, 362.
[366] Caroli, C. and De Gennes, P.G. (1966). *Solid State Communications* **4**, 17.
[367] Helfand, E. and Werthamer, N. R. (1966). *Physical Review*, **147**, 288.
[368] Spivak, B. and Zhou, F. (1995). *Physical Review Letters* **74**, 2800.
[369] Zhou, F. and Spivak, B. (1998). *Physical Review Letters*, **80**, 5647.
[370] Cohen, R.W. and Abels, B. (1968). *Physical Review*, **168**, 444.
[371] Kulik, I.O., Omelyuanchuk, A.N. (1975). *Pisma v Zhurnal Eksperimentalnoi i Teoreticheskoi Fiziki*, **21**, 216 [*JETP Letters* **21**, 96 (1975)].
[372] Larkin, A.I. and Ovchinnikov, Yu.N., (1966). *Zhurnal Eksperimentalnoi i Teoreticheskoi Fisiki*, **51**, 1535 [*Soviet Physics–JETP* **24**, 1035 (1967)].
[373] Scalapino, D.J. (1970). *Physical Review Letters*, **24**, 1052.
[374] Kulik, I. O. and Yanson, I.K. (1972). *The Josephson Effect in Superconducting Tunneling Structures*. Halsted, Jerusalem.
[375] Barone, A. and Paterno, G.F. (1982). *Physics and applications of Josephson effect*. Willey-Interscience, New York.

[376] Belogolovski, M., Chernyak, O. and Khachaturov, A. (1986). *Fizika Nizkih Temperatur (USSR)*, **12**, 630.
[377] Park, M., Isaacson, M.S. and Parpia, J.M. (1995). *Physical Review Letters,* **75**, 3740.
[378] Tao, H.J., Lu, F. and Wolf, E.L. (1997). *Physica*, **C282-287,** 563.
[379] Watanabe, T. et al. (1997). *Physical Review Letters,* **79,** 2113; Matsuda, A., Sugita,S. and Watanabe, T. (1999). *Physical Review* , **B60,** 1377; Watanabe, T., Fujii, T. and Matsuda, A. (2000). *Physical Review Letters,* **84,** 5848.
[380] Suzuki, M., Karimoto, S. and Namekawa, K. (1998). *Journal of Physical Society of Japan* **67,** 34.
[381] Renner, Ch., Revaz, B., Genoud, J-Y., Kadowaki, K. and Fischer, O. (1998). *Physical Review Letters*, **80,** 149.
[382] Cucolo, A.M., Cuoco, M. and Varlamov, A.A. (1999). *Physical Review*, **59,** R11675.
[383] Reizer, M.Yu. (1993). *Physical Review,* **B48**, 13703.
[384] Ferrel, R.A. (1969). *Journal of Low Temperature Physics,* **1**, 423.
[385] Giaever, I. (1969). In *Tunneling phenomena in solids* (Ed. E.Burstein, S.Lundquist), Plenum Publishing Corporation, New York.
[386] Takayama, H. (1971). *Progress in Theoretical Physics,* **46**, 1.
[387] Andreev, A.F. (1964). *Zhurnal Eksperimentalnoi i Teoreticheskoi Fisiki,* **46**, 185 [*Soviet Physics–JETP* **19**, 117(1964)].
[388] Ambegaokar, V. and Baratoff, A. (1963). *Physical Review Letters* , **11**, 486; ibid. **11**, 104.
[389] Cini, M. (1980). *Physical Review* , **B22**, 5887.
[390] Volkov, A.F., Zaitsev, A.V. and Klapwijk, T.M. (1993). *Physica*, **C 210**, 21.
[391] Feigel'man, M.V., Larkin, A.I. and Skvortsov, M.A. (2000). *Physical Review,* **B63**, 134507.
[392] Anderson, J.T., Carlson, R.V., Goldman, A.M. and Tan, H.-T. (1973). In *Proceedings of the 13th International Conference on Low Temperature Physics*, (ed. K.D. Timmerhaus, W.J. O'Sullivan, and E.F. Hammel), p.709, Plenum Publishing Corporation, New York.
[393] Carlson, R.V., Goldman, A.M. (1975). *Physical Review Letters,* **34**, 11.
[394] Anderson, J.T., Goldman, A.M. (1970). *Physical Review Letters*, **25,** 743.
[395] Goldman, A.M. (1981). In *Nonequilibrium superconductivity, phonons, and Kapitza boundaries*, (ed. K.E. Gray), Plenum Publishing Corporation, New York.
[396] Anderson, J.T., Carlson, R.V., Goldman, A.M. (1972). *Journal of Low Temperature Physics,* **8**, 29.
[397] Carlson, R.V., Goldman, A.M. (1973). *Physical Review Letters*, **31**, 880.
[398] Dinter, M. (1977). *Journal of Low Temperature Physics* **26**, 39; *Journal of Low Temperature Physics* **32**, 529 (1978).

[399] Varlamov, A.A. and Dorin, V.V. (1986). *Zhurnal Eksperimentalnoi i Teoreticheskoi Fisiki*, **91**, 1955 [*Soviet Physics–JETP* **64**, 1159 (1986)].

[400] Brieskorn, G., Dinter, M. and Schmidt, H. (1974). *Solid State Communications*, **15**, 757.

[401] Maki, K. and Sato, H. (1974). *Journal of Low Temperature Physics,* **16**, 557.

[402] Aleiner, I.L. and Altshuler, B.L. (1997). *Physical Review Letters*, **79**, 4242.

[403] Kee, H.-Y., Aleiner, I.L. and Altshuler, B.L. (1998). *Physical Review*, **B58**, 5757.

[404] Black, C.T., Ralph, D.C. and Tinkham, M. (1996). *Physical Review Letters*, **76**, 688.

[405] Braun, F., von Delft, J., Ralph, D.C. and Tinkham, M. (1997). *Physical Review Letters*, **79**, 921.

[406] Gerber, A., Milner, A., Deutscher, G. , Karpovsky, M. and Gladkikh, A. (1997). *Physical Review Letters*, **78**, 4277.

[407] Chui, T., Lindenfeld, P., McLean,W. L. and Mui, K. (1981). *Physical Review Letters*, **47**, 1617.

[408] Gantmakher, V. F., Golubkov, M., Lok, J. G. S. and Geim, A. K. (1996). *Zhurnal Eksperimentalnoi i Teoreticheskoi Fisiki*, **109**, 1264 [*Soviet Physics–JETP*, **82**, 951(1996)].

[409] Efetov, K.B. (1999). *Supersymmetry in Disorder and Chaos*, Cambridge University Press, Cambridge.

[410] Maniv, T. and Alexander, S. (1977). *Journal of Physics*, **C10**, 2419.

[411] Maniv, T. and Alexander, S. (1976). *Solid State Communications*, **18**, 1197.

[412] Kuboki, and K., Fukuyama, H. (1989). *Journal of Physical Society of Japan,* **58,** 376.

[413] Heym, J. (1992). *Journal of Low Temperature Physics,* **89**, 859.

[414] Mosconi, P., Rigamonti, and A. Varlamov, A.A. (2000). *Applied Magnetic Resonance,* **19,** 345.

[415] Mitrovic,V., Bachman, H.N., Halperin, W.B., Eschrig, and M, Sauls, J.A. (1999) *Physical Review Letters* **82**, 2784-87.

[416] Zimmermann, H., Mali, M., Bankey, M. and Brinkmann, D. (1991). *Physica*, **C185–189**, 1145; Brinkmann, D. (1995). *Applied Magnetic Resonance*, **8**, 67.

[417] Gorny, K., Vyasilev, O.M., Martindale, J.A., Nandor, V.A., Pennington, C.H., Hammel, P.C. et al. (1999). *Physical Review Letters*, **82**, 177-180.

[418] Zheng.G., Clark, W.G., Kitaoka, Y., Asayama, K., Odama, Y., Kuhns, et al. (1999). *Physical Review* **B60**, R9947-50; Zheng.G., Osaki, H., Clark, W.G., Kitaoka, Y., Kuhns, P., Reyes, A.P., et al. (2000). *Physical Review Letters* **85**, 405-08.

[419] Fay, D., Appel, J. Timm, C. and Zabel, A. (2001). *Physical Review* **B63**, 064509.

[420] Ralph, D.C., Black, C.T. and Tinkham, M. (1995). *Physical Review Letters*, **74,** 3241.

[421] Black, C.T., Ralph, D.C. and Tinkham, M. (1996). *Physical Review Letters*, **76,** 688.

[422] von Delft, J. and Ralph, D.C. (2001). *Physics Reports*, **345**, 61.

[423] Migdal, A.B. (1965). *The theory of finite Fermi-systems and properties of the atomic nuclea,* Nauka, Moscow.

[424] Richardson, R.W. (1963). *Physics Letters*, **3,** 277.

[425] Janko, B., Smith, A. and Ambegaokar, V. (1994). *Physical Review*, **B50**, 1152.

[426] Golubev, D.S. and Zaikin, A.D. (1994). *Physics Letters*, **A195**, 380.

[427] Lafarge, P., Joyez, P., Esteve, D., Urbina, C. and Devoret, M.H. (1993). *Physical Review Letters,* **70**, 994.

[428] Tuominen, M.T., Hergenrother, J.M., Tighe, T.S. and Tinkham, M. (1992). *Physical Review Letters,* **69**, 1997.

[429] Averin, D.V., and Likharev, K.K. (1991). In *Mesoscopic Phenomena in Solids,* (ed. B.L. Altshuler, P.A. Lee, and R.A. Webb). Elsevier, Amsterdam.

[430] Matveev, K.A. and Larkin, A.I. (1997). *Physical Review Letters.* **78,** 3749.

[431] von Delft, J.,Golubev, D.S., Tichy, W. and Zaikin, A.D. (1996). *Physical Review Letters,* **77,** 4962.

[432] Mastellone, A., Falci, G. and Fazio, R. (1998). *Physical Review Letters,* **80,** 4542.

[433] Berger, S.D. and Halperin, B.I. (1998). *Physical Review*, **B58**, 5213.

[434] Ioffe, L.B. and Larkin, A.I. (1981). *Zhurnal Eksperimentalnoi i Teoreticheskoi Fisiki,* **81**, 707 [*Soviet Physics–JETP* **54**, 378 (1981)].

[435] Galitski, V. M. and Larkin, A. I. (2001). *Physical Review Letters,* **87**, 087001; ibid. **89**, 109704 (2002).

[436] Halperin, B. I. and Lax, M. (1966) *Physical Review*, **148**, 722; see also: Narozhny B. N., Aleiner, I.L., Larkin, A.I. (2000). *Physical Review*, **B62**, 14898.

[437] Zittarz, J. and Langer, J.S. (1966). *Physical Review*, **148**, 741.

[438] Usadel, K. (1970). *Physical Review Letters ,* **25,** 507.

[439] Vavilov, M.G., Brower, P.W., Ambegaokar, and V., Beenakker, C.W.J. (2001). *Physical Review Letters.* **86,** 874.

[440] Ostrovsky, P.M., Skvortsov, M.A. and Feigel'man, M.V. (2001). *Physical Review Letters*, **87,** 027002; Ostrovsky, P.M., Skvortsov, M.A. and Feigel'man, M.V. (2002). *Pisma v Zhurnal Eksperimentalnoi i Teoreticheskoi Fiziki,* **75**, 284 [*JETP Letters*, **75**, 336 (2002)].

[441] Lamacraft, A. and Simons, B.D. (2001). *Physical Review Letters*, **85,** 4783.

[442] Lamacraft, A. and Simons, B.D. (2001) *Physical Review*, **B64,** 014514.

[443] Aslamazov, L.G. Larkin, A.I. and Ovchinnikov, Yu.N. (1968). *Zhurnal Eksperimentalnoi i Teoreticheskoi Fisiki,* **55,** 323 [*Soviet Physics- JETP,* **28**, 171 (1969)].

[444] Larkin, A.I. and Ovchinnikov, Yu. N. (1971). *Zhurnal Eksperimentalnoi i Teoreticheskoi Fisiki* **61**, 1221 [*Soviet Physics–JETP*, **34**, 651 (1972)].
[445] Harris, A.B. (1974). *Journal of Physics,* **C7**, 1671.
[446] Khmelnitski, D.E. (1980). *Zhurnal Eksperimentalnoi i Teoreticheskoi Fisiki,* **68**, 1961 [*Soviet Physics–JETP*, **41**, 981 (1975)].
[447] Lubensky, T.S. (1975). *Physical Review*, **B11**, 3573; A. B. Harris and T. C. Lubensky *Physical Review Letters*, **33**, 1540 (1974).
[448] Shander, E.F. (1978). *Journale de Physique* (Paris), **L423,** 11.
[449] Entin-Wolman, O., Kapitulnik, A. and Shapira, Y. (1981). *Physical Review*, **B24**, 6464.
[450] Geshkenbein, V.B., Ioffe, L.B. and Millis, A.J. (1998). *Physical Review Letters* , **80**, 5078.
[451] Efetov, K.B. (1980). *Zhurnal Eksperimentalnoi i Teoreticheskoi Fisiki,* **78**, 2017 [*Soviet Physics–JETP,* **51**, 1015 (1980)].
[452] Feigel'man, M.V., and Larkin, A. I. (1998). *Chemical Physics* **235**, 107.
[453] Feigel'man, M.V., Larkin, A.I. and Skvortsov, M.A. (2000). *Physical Review*, **B 61**, 12361.
[454] Feigel'man, M.V., Larkin, A.I. and Skvortsov, M.A., (2001). *Uspekhi Fizicheskih Nauk (Supplement)* **171**, 76.
[455] Ambegaokar, V., Ekkern, U. and Shön, G. (1982). *Physical Review Letters,* **48**, 1745.
[456] Schmid, A. (1983). *Physical Review Letters*. **51**, 1506.
[457] Bulgadaev, S. A. (1984). *Pisma v Zhurnal Eksperimentalnoi i Teoreticheskoi Fiziki,* **39**, 264 [*JETP Letters* **39**, 314 (1984)].
[458] Kosterlitz, J.M. (1976). *Physical Review Letters*, **37**, 1577.
[459] Gor'kov, L.P. (1959). *Zhurnal Eksperimentalnoi i Teoreticheskoi Fisiki* , **37**, 833 [*Soviet Physics–JETP* **14**, 628 (1962)].
[460] Ovchinnikov, Yu.N. (1980). *Zhurnal Eksperimentalnoi i Teoreticheskoi Fisiki,* **79**, 1825 [*Soviet Physics–JETP* **52**, 923 (1980)].
[461] Khmel'nitskii, D.E., Larkin, and A.I., Melnikov, V.I. (1971). *Zhurnal Eksperimentalnoi i Teoreticheskoi Fisiki,* **60**, 846 [*Soviet Physics–JETP,* **33**, 458 (1971)].
[462] Okuma, S. (1983). *Journal of Physical Society of Japan*, **52**, 3269; Hebrad, A. F. and Paalanen, M. A.. (1984) *Physical Review*, **B30**, 4063.
[463] Josephson, B.D. (1962). *Physics Letters*, **1**, 251.
[464] Aslamazov, L.G. and Larkin, A.I., (1968). *Pisma v Zhurnal Eksperimentalnoi i Teoreticheskoi Fiziki,* **9**, 150-154 [*JETP Letters* **9**, 87 (1969)].
[465] Larkin, A.I. and Ovchinnikov, Yu.N. (1967). *Zhurnal Eksperimentalnoi i Teoreticheskoi Fisiki,* **53**, 2159 [*Soviet Physics–JETP* **26**, 1219 (1968)].
[466] Likharev, K.K. (1985). *Introduction to the Josephson junctions dynamics.* Moscow, Nauka.
[467] Landau, L.D., Lifshitz, E.M. and Pitaevski, L.P. (1985). *Electrodynamics of continuous media. Course of Theoretical Physics, Vol.8.* Elsevier Science Ltd.

[468] Stewart, C. and McCumber, D.E. (1968). *Applied Physics Letters*, **12**, 277.
[469] McCumber, D.E. (1968). *Journal of.Applied Physics*, **39**, 3313.
[470] Kulik, I.O. (1969). *Pisma v Zhurnal Eksperimentalnoi i Teoreticheskoi Fiziki,* **10**, 488 [*JETP Letters* **10**, 313 (1969)].
[471] Ivanchenko, Yu.M. and Zilberman, L.A. (1969). *Zhurnal Eksperimentalnoi i Teoreticheskoi Fisiki,* **55**, 2395 [*Soviet Physics–JETP* **28**, 1272 (1969)].
[472] Ambegaokar, V. and Halperin, B.I. (1969). *Physical Review Letters* , **22**, 1364.
[473] Anderson, J.T. and Goldman, A.M. (1969). *Physical Review Letters*, **23**, 128.
[474] Kramers, H.A. (1940). *Physica (Utrecht)* **7**, 284.
[475] Melnikov, V.I. (1984). *Zhurnal Eksperimentalnoi i Teoreticheskoi Fisiki,* **60**, 380 [*Soviet Physics–JETP* **60**, 380 (1984)].
[476] Abramowitz, M. and Stegun, I.A. (1974). *Handbook of Mathematical Functions*. Dover, New York.
[477] Surry, W,H., (1947). *Physical Review*, **71**, 360.
[478] Devoret, M.H., Martinis, J.M. and Clarke, J. (1985). *Physical Review Letters* , **55**, 1908.
[479] Martinis, J.M., Devoret, M.H. and Clarke, (1985). *Physical Review Letters,* **55**, 1543.
[480] Larkin, A.I. and Ovchinnikov, Yu.N. (1986). *Zhurnal Eksperimentalnoi i Teoreticheskoi Fisiki* **91**, 318 [*Soviet Physics–JETP* **64**, 185(1986)].
[481] Caldeira, A. and Legget, A.J. (1981). *Physical Review Letters*, **46**, 211.
[482] Larkin, A.I., and Ovchinnikov, Yu.N. (1983). *Pisma v Zhurnal Eksperimentalnoi i Teoreticheskoi Fiziki,* **37**, 322 [*JETP Letters* **37**, 287].
[483] Larkin, A.I., and Ovchinnikov, Yu.N. (1984). *Zhurnal Eksperimentalnoi i Teoreticheskoi Fisiki* **86**, 719 [*Soviet Physics–JETP* **59**, 420 (1984)].
[484] Larkin, A.I., and Ovchinnikov, Yu.N.(1992). In *Quantum tunneling in condensed media*, (ed. Yu. Kagan and A.J. Legget), Elsevier Science Publishers.
[485] Korshunov, S.E. (1987). *Zhurnal Eksperimentalnoi i Teoreticheskoi Fisiki* **92**, 1828 [*Soviet Physics–JETP* **65**, 1025 (1987)].
[486] Vaks, V.G. and Larkin, A.I. (1965). *Zhurnal Eksperimentalnoi i Teoreticheskoi Fisiki,* **49**, 975 [*Soviet Physics–JETP*, **22**, 678 (1966)].
[487] Little, W.A. (1967). *Physical Review*, **166**, 398.
[488] Langer, J.S. and Ambegaokar, V. (1967). *Physical Review*, **164**, 498.
[489] Matveev, K.A., Larkin, A.I. and Glazman, L.I. (2002). *Physical Review Letters*, **89**, 96802.
[490] Bezryadin, A. Lau, C.N. and Tinkham, M. (2000). *Nature*, **404**, 971.
[491] Lau, C.N., Markovic, N., Bockrath, M., Bezryadin, A. and Tinkham, M. (2001). *Physical Review Letters* , **87**, 217003.
[492] Zaikin, A.D., Golubev,D.S., van Otterlo, A. and Zimanyi, G.T. (1997). *Physical Review Letters* , **78** , 1552.

[493] Golubev,D.S. and Zaikin, A.D. (2001). *Physical Review* , **B64**, 014504.
[494] Efetov, K.B. and Larkin A.I., (1975). *Zhurnal Eksperimentalnoi i Teoreticheskoi Fisiki*, **69,** 764 [*Soviet Physics- JETP*, **42**, 390 (1975)].
[495] Haldane, F.D.M., (1981). *Journal of Physics*, **C 14,** 2585.
[496] Larkin, A.I. and Lee, P.A., (1978). *Physical Review*, **B17**, 1596.
[497] Kane, C.L. and Fischer, M.P.A., (1990) *Physical Review*, **B46**, 15233.
[498] Leggett, A.J, Chakravarty, S., Dorsey, A., Fisher, M.P.A. Garg Anupam, Zwerger, W. (1987) *Review of Modern Physics* . **59,** 1.
[499] Schon, G. and Zaikin, A.D. (1990). *Physics Reports*, **B54**, 237.
[500] Glazman, L.I. and Larkin, A.I., (1997). *Physical Review Letters,* **79**, 3736.
[501] Bradlay, R.M. and Doniach, S. (1984). *Physical Review*, **B30**, 1138.
[502] Brézin, E. and Zinn-Justin, J. (1990). *Fields, strings and critical phenomena.* North Holland, Amsterdam.
[503] Ivanov, D.A., Ioffe, L.B., Geshkenbein, V.B. and Blatter, G. (2002). *Physical Review*, **B65**, 024509.
[504] Chow, E., Delsing, P. and Haviland, D.B. (1998). *Physical Review Letters*, **81**, 204.
[505] Mermin, N.D., Wagner, H. (1966). *Physical Review Letters*, **17**, 1133; Hohenberg, P.C. (1967). *Physical Review*, **158**, 383.
[506] Berezinskii, V.S. (1971). *Zhurnal Eksperimentalnoi i Teoreticheskoi Fisiki*, **61**, 1144 [*Soviet Physics–JETP* **34**, 610 (1972)].
[507] Kosterlitz, J.M. and Thouless, P.J. (1973). *Journal of Physics,* **C6**, 1181.
[508] Minnhagen, P. (1987). *Reviews of Modern Physics*, **59**, 1001.
[509] Minnhagen, P. (1997). In *Fluctuation phenomena in high temperature superconductors*, (ed. M. Ausloos and A.A. Varlamov), NATO-ASI Series, Kluwer, Dordrecht.
[510] Nelson, D.R. and Kosterlitz, J.M. (1977). *Physical Review Letters*, **39**, 1201.
[511] Minnhagen, P. (1981). *Physical Review* , **B24**, 6758.
[512] Takagi, H. and Kuroda, Y. (1982). *Solid State Communications*, **41**, 643.
[513] Fisher, M.P.A. (1990). *Physical Review Letters* , **65**, 923.
[514] Kapitulnik, A. and Kotliar, G. (1985) *Physical Review Letters,* **54**, 474.
[515] Wollhardt, P. and Wolfle, D. (1982). In *Anderson localization,* (ed. Y. Nagaoka and H. Fukuyama), Springer Verlag, Berlin.
[516] Bulaevski, L.N., Varlamov, and A.A. Sadovski, M.V. (1986). *Soviet Solid State Physics*, **28**, 997.
[517] Das, D. and Doniach, S. (1981). *Physical Review*, **B24**, 5063.
[518] Goldman, A.M. and Markovic, N. (1998). *Physics Today*, N 11 and refs. inside.
[519] Chubukov, A.V., Pines, D. and Schmalian, J. (2002). In: *The Physics of superconductors,* (ed. K.H. Bennemann and J.B. Ketterson), Vol.1, chapter 7, Springer-Verlag, Berlin.
[520] Timusk, T. and Statt, B. (1999). *Reports on Progress in Physics*, **62**, 61.
[521] Anderson, P.W. (1987). *Science*, **235**, 1196.

[522] Anderson, P.W., Baskaran, G., Zou, Z. and Hsu, T. (1987). *Physical Review Letters*, **58**, 2790.

[523] Baskaran, and G. Anderson, P.W. (1988). *Physical Review*, **B37**, 580.

[524] Galitski, V. (1958). *Zhurnal Eksperimentalnoi i Teoreticheskoi Fisiki* , **34**, 151 [*Soviet Physics–JETP* **7**, 104 (1958)].

[525] Nagaoka, Y. (1966). *Physical Review*, **147**, 392.

[526] Ioffe, L.B. and Larkin, A.I. (1988). *Physical Review* **B37**, 5730.

[527] Manousakis, E. (1991). *Review of Modern Physics* **63**, 1.

[528] Dagotto, E. (1994). *Review of Modern Physics* **66**, 763.

[529] Eisenberg, E., Berkovits, R., Huse, D.A. and Altshuler, B.L. cond-mat/0108523, 17 January 2002.

[530] Himeda, A. and Ogata, M. (1999). *Physical Review*, **B60**, 9935.

[531] Ivanov, D.A. and Lee, P.A. (2003). Cond-mat /0305143.

[532] Paramekanti, A., Randeria, M. and Triverdi, N. (2001). *Physical Review Letters*, **87**, 217002.

[533] Yokoyama, H. and Shiba, H. (1988). *Journal of Physical Society of Japan* **57**, 2482; *Journal of Physical Society of Japan*, **65**, 3615 (1996).

[534] Gros,C. (1988). *Physical Review*, **B38**, 931; *Annals of Physics*, **189**, 53 (1989).

[535] Lee, P.A. and Wen, X.-G. (1997). *Physical Review Letters*, **78**, 4111.

[536] Gogolin, A.O., Nersesyan, A. A. and Tsvelik,A. M. (1998). *Bosonization in Strongly Correlated Systems*, Cambridge University Press.

[537] Dzyaloshinski, I.E. and Larkin, A.I. (1973). *Zhurnal Eksperimentalnoi i Teoreticheskoi Fisiki*, **65**, 411 [*Soviet Physics–JETP* **38**, 202 (1974)].

[538] Anderson, P.W., Lee, P.A., Randeria, M., Rice, T.M., Triverdi, N. and Zhang, F.C. (2003). *ArXiv: cond-mat/* 0311467.

[539] Baskaran, G. and Anderson, P.W., (1988). *Physical Review*, **B37**, 580.

[540] Nagaosa, N. and Lee, P.A., (1990). *Physical Review Letters*, **64**, 2450.

[541] Weng, Z.Y., Sheng, D.N. and Ting C.S. (1998). *Physical Review Letters*, **80**, 5401.

[542] Shraiman, B. and Siggia, E. (1989). *Physical Review Letters*, **62**, 1564.

[543] Wiegmann, P.B. (1988). *Physical Review Letters*, **60**, 821.

[544] Lee, P.A. (1989). *Physical Review Letters* , **63**, 680.

[545] Afleck, I. and Marston, J.B. (1988). *Physical Review*, **B37**, 3774.

[546] Ioffe, L.B. and Larkin, A.I. (1989). *Physical Review* **B39**, 8988.

[547] Feigel'man, M.V., Geshkenbein, V.B., Ioffe, L.B. and Larkin, A.I. (1993). *Physical Review*, **B48**, 16641.

[548] Wilchek, F. (1990). *Fractional statistics and anion superconductivity,* World Scientific, Singapore.

[549] Reizer, M.Yu. (1989). *Physical Review*, **B39**, 1602.

[550] Chakravarty, S., Laughlin, R.B., Morr, D.K. and Nayak, C. (2000). cond-mat/0005443.

[551] Halperin, B.I. and Rice, T.M. (1968). *Solid State Physics* **21**, 115.

[552] Volkov, B.A., Gorbatsevich, A.A., Kopaev, Yu.V. and Tugushev, V.V.

(1981). *Zhurnal Eksperimentalnoi i Teoreticheskoi Fisiki*, **81**, 729 [*Soviet Physics– JETP*, **54**, 391 (1981)].
[553] Varma, C.M. (2000). *Physical Review*, **B61**, R3804.
[554] Emery, V.J. and Kivelson, S.A. (1995). *Nature*, **374**, 434.
[555] Emery, V.J. and Kivelson, S.A. (1999). cond-mat/9902179.
[556] Anderson, P.W. (2001). cond-mat/0104332.
[557] Doniach, S. and Inui, M. (1990). *Physical Review*, **B41**, 6668.
[558] Wen, X.-G. and Lee, P.A. (1996). *Physical Review Letters*, **76**, 503; P.A. Lee, and X.-G. Wen, *Physical Review* **B63**, 224517 (2001);
[559] Lee, P.A. (2002). *Orbital currents and cheap vortices in underdoped cuprates*, cond-mat/0210113.
[560] Das, D. and Doniach, S. (1990). *Physical Review*, **B41**, 11697.
[561] Alexandrov, A. and Mott, N.F. (1994). *Reports on Progress in Physics*, **57**, 1197-1288.
[562] Alexandrov, A. (2003). *Theory of superconductivity: from weak to strong coupling*, IOP Publishing.
[563] Haslinger, R., Abanov, Ar. and Chubukov, A. (2002). *Europhysics Letters*, **58**, 271.
[564] Kulik, I.O. (1988). *International Journal of Modern Physics*, **B2**, 851.
[565] Friedberg, R. and Lee, T.D. (1989). *Physical Review*, **B40**, 6745.
[566] Ioffe, L.B., Larkin, A.I., Ovchinnikov, Yu.N. and Yu,L. (1989). *International Journal of Modern Physics* , **B3**, 2065.
[567] Sa de Melo, C.A.R., Randeria, M. and Engelbrecht, J.R. (1993). *Physical Review Letters*, **71**, 3202; Randeria, M. (1994) *Physica*, **B199-200**, 373; for review see Randeria, M. In *Bose-Einstein condensation*, (ed. A. Griffin, D.W. Snoke, and S. Stringari), Cambridge University Press, Cambridge, NY (1995).
[568] Ranninger, J. and Robin, J.M. (1995). *Physica* **C253**, 279.
[569] Maly,J., Levin, K. and Liu, D.Z. (1996). *Physical Review*, **B54**, R15168; for review see: Levin, K., Chen, Q, Janko, B. (2003). In *The physics of superconductors,* (ed. K.H. Bennemann and J.B. Ketterson), Vol.2, chapter 5, Springer-Verlag, Berlin.
[570] Pistolesi, F. and Strinati, G.C. (1994). *Physical Review*, **B49**, 6356, ibid. B53, 15168 (1996). for review see Pieri, P. and Strinati, G.C. (2000). *Physical Review*, **B62**, 15370.
[571] Geshkenbein, V.B., Ioffe, L.B. and Larkin, A.I. (1997). *Physical Review*, **B55**, 3173.
[572] Perali, A., Castellani, C., Di Castro, C., Grilli, M., Piegari, E. and Varlamov, A.A. (2000). *Physical Review*, **B62**, R9295.
[573] Ioffe, L.B. and Millis, A.J. (1998). *Physical Review*, **B58**, 11631.
[574] Ovchinnikov Yu.N., Wolf, S.A. and Kresin, V.Z. (1999). *Physical Review*, **B60**, 4329.
[575] Varma, C.M., Littlewood, P.B., Schmitt-Rink, S., Abrahams, E. and Ruckenstein, A.E. (1989). *Physical Review Letters* , **63**, 1996.

[576] Zhang, S.-C. (1997). *Science,* **275**, N2, 1089.
[577] Maki, K. (1969). In *Superconductivity* (ed. R.D. Parks), Vol.2, chapter 18, Marcel Dekker, New York.
[578] Beytman, G. and Erdeyi, A. (1953). *Higher Transcedental Functions, Vol.1,* Mc Graw-Hill Book Company, Inc., New York.
[579] Gradshtein, I.S. and Ryzhik, I.N. (1994). *Tables of Integrals, Series, and Products*, Alan Jeffrey Editor, Academic Press, San Diego.

GLOSSARY

$\mathbf{A}$ vector-potential

$\mathbf{A_k}$ Fourier component of the vector-potential

$A = \nu \ln \frac{T}{T_c}$ coefficient of the Δ^2 term of the GL functional derived from the microscopic theory

$\mathfrak{a}$ interatomic distance

a_{Λ_n} renormalized variable of the renormalization group theory

$\widehat{a}^+, \widehat{a}$ creation and annihilation operators

$a_\triangle = \left(\frac{2}{\sqrt{3}}\right)^{1/2} \left(\frac{\Phi_0}{B}\right)^{1/2}$ period of the triangular Abrikosov's vortex lattice

$a = \alpha T_c \epsilon$ coefficient of the GL functional

α coefficient of the GL theory. Its value depends on the chosen normalization of the order parameter. When the GL parameter m is assumed to be equal to the free electron mass $\alpha = \frac{2\pi^2 D}{7\zeta(3)} \frac{T_c}{E_F}$. When the modulus of the order parameter is identified with the value of the gap in the spectrum of quasiparticles $\alpha = \frac{1}{4mT_c}\xi^2$ (see(A.2))

$B = \frac{7\zeta(3)}{8\pi^2 T^2}\nu$ coefficient of the $\Delta^4/2$ term of the GL functional derived from the microscopic theory

$B_m(T)$ value of the melting field

B_{2m} Bernoulli numbers

$B_\rho = B\overline{\chi_\rho^4}$ coefficient of the GL functional for the case of ρ-symmetry pairing

$\mathbf{B} = \mathbf{H} + 4\pi\mathbf{M}$ magnetic field induction

$\mathbf{B}_{(h)}(\mathbf{q}, \Omega_k, \omega_\nu)$ integrated three Green functions block appearing in the heat transport

$B_\alpha(q, \omega_\mu, \omega_\nu)$ integrated three Green functions block appearing in the AL contribution to conductivity

$B_{n,m}(\Omega_k + \omega_\nu, \Omega_k)$ integrated three Green functions block in the Landau representation

$B_x^{RR}(\mathbf{q}, \Omega_k, \omega_\nu)$, $B_x^{RA}(\mathbf{q}, \Omega_k, \omega_\nu)$, $B_x^{RA}(\mathbf{q}, \Omega_k, \omega_\nu)$ functions of diverse analyticity corresponding to $B_\alpha(q, \omega_\mu, \omega_\nu)$

b coefficient of the fourth-order term of the GL functional and the bare vertex of the RG theory

$\widetilde{b}$ effective interaction of the RG theory

b_{Λ_n} renormalized variable of the renormalization group theory

$b_m(t) = B_m(T)/B_{c2}(T)$ dimensionless induction of magnetic field

$\beta = 1/T$ inverse temperature

$\beta_A = 1.18$ parameter of the Abrikosov's theory of type-II superconductors

$\beta^{\alpha\beta}$ thermoelectric tensor

$\beta_{(\text{film})}^{DOS}(\epsilon)$ DOS fluctuation contribution to the thermoelectric tensor

$\beta_{(\text{film})}^{AL}(\epsilon)$ AL fluctuation contribution to the thermoelectric tensor

$\beta(T,\tau_\varphi)$ "universal" function appearing in the magnetoconductivity of a dirty film

$C = \frac{1}{4m} = \alpha T_c \xi^2$ coefficient of the gradient term of the GL functional, derived from the microscopic theory by Gor'kov

C elastic modulus tensor, C_{ik} are its components: compression C_{11}, tilt, C_{44} and shear, C_{66}

$C(\mathbf{q},\varepsilon_1,\varepsilon_2) = \frac{1}{2\pi\nu\tau}\lambda(\mathbf{q},\varepsilon_1,\varepsilon_2)$ four-leg impurity vertex, the so-called "Cooperon"

C_{Euler} Euler constant

C_{ij} capacity matrix

$C_\rho = C\overline{\left(\chi_\rho^2(\mathbf{p})\,\mathbf{v}_\mathbf{p}^2\right)}/v_F^2$ coefficient of the gradient term of the GL functional in the case of ρ-symmetry pairing

$\Delta C = C_S - C_N$ heat capacity jump at the phase transition point

c_L empirical Lidemann parameter

$U(j)$ "pinning barrier" height which depends on the carrying current.

c_n coefficient of the RG perturbation theory

$\mathcal{D} = \begin{cases} \tau v_F^2/D, & T\tau \ll 1, \\ v_F^2/2DT, & T\tau \gg 1, \end{cases}$ generalized electron diffusion coefficient

$\mathcal{D}(q) \sim q^{z-2}$ diffusion coefficient in the region of anomalous diffusion

$\widehat{\mathcal{D}}q^2 = \tau\langle(\Delta\xi(\mathbf{q},\mathbf{p})|_{|\vec{p}|=p_F})^2\rangle_{F.S.}$. Generalized diffusion operator $\widehat{\mathcal{D}}$ is introduced in order to deal with an arbitrary anisotropic spectrum

$2D$, $3D$, $4D$ notation for two-, three- or four-dimensional space

$D = 4 - \varepsilon$ space dimensionality,

d film thickness or grain diameter

$\Delta(\mathbf{r},\tau)$ order parameter field fluctuating in space and imaginary time

Δ_{pg} pseudogap

$\delta = (\nu V)^{-1}$ spacing of the dimensional quantization between levels

$\partial_{(k)} = -i\nabla - 2e(-1)^k\mathbf{A}(\mathbf{r})$

$E_0(T)$ energy scale at which the DOS renormalization occurs: in the clean case $E_0^{(c)} \sim \sqrt{T_c(T-T_c)}$, while in the dirty case $E_0^{(d)} \sim T - T_c$

$E(\mathbf{p})$ electron(hole) energy spectrum of the normal metal

$\mathcal{E}(\mathbf{p})$ Cooper pair energy spectrum

$E_Z = g_L\mu_B H$ Zeeman splitting

$E_{i\uparrow(\downarrow)} = E_i \mp E_Z/2$, where E_i orbital energy of i-th state

E_T Thouless energy

E_p vortex energy in the field of a pinning center

$E(\phi_i - \phi_j) = E_J\cos(\phi_i - \phi_j)$ interaction energy of neighboring superconducting drops

$E_v = \frac{\pi n_{s2}(T)}{2m}\ln\frac{R}{\xi}$ energy per unit length of a single vortex in a sample of linear dimension R

$E_C = \frac{1}{2}\sum_{i,j} C_{ij}V_iV_j$ Coulomb energy

$\mathcal{E}_0(H)$ first harmonics of a single grain energy related to the orbital effect

$\widetilde{E}_C$ effective charge energy for Cooper pair

$E_c = e^2/2C$ charging energy of the junction for a single-electron charge

$E_J = \frac{|\Delta|}{4e^2R}\tanh\frac{|\Delta|}{2T} = \frac{I_c}{2e}$ Josephson energy

E_F Fermi energy

$Ei(x) = \int_x^\infty \frac{\exp(-y)}{y} dy$ integral exponent

$e_{\alpha\beta\gamma}$ unitary antisymmetric tensor

$\mathrm{erf}(x)$ probability integral

$\epsilon = \ln(\frac{T}{T_c}) \approx \frac{T-T_c}{T_c}$ reduced temperature

$\widetilde{\epsilon}_H = \epsilon + h$ reduced temperature shifted by magnetic field in the Hartree approximation

ϵ_{cr} size of the critical region

$\varepsilon_n = (2n+1)\pi T$ fermionic Matsubara frequency

$\widetilde{\varepsilon}_n = \varepsilon_n + \frac{1}{2\tau} sign(\varepsilon_n)$ Matsubara frequency renormalized by impurity scattering

$\varepsilon_{\mathbf{p}} = \alpha T_c(\epsilon + \widehat{\xi}^2 \mathbf{p}^2)$ eigenvalue of the GL Hamiltonian

$F_{(D)} = (\mathcal{F}(\Psi))_{\min} = \mathcal{F}(\widetilde{\Psi})$ GL free energy which is determined by the minimum of the GL functional $\mathcal{F}(\Psi(\mathbf{r}))$

F_N GL free energy of the normal state

F_{fr} force of dry friction

$F_\Delta(\mathbf{r}, \mathbf{r}', \tau, \tau')$ anomalous component of Gor'kov Green function

$F^{(1)}, F^{(2)}$ fluctuation corrections to free energy in the one- and two-loop approximations

$F_L = j_{tr}\Phi_0/c = \eta_v v_v$ Lorentz force

F_{n,Λ_n} renormalized variable of the renormalization group theory

$\mathcal{F}[\Psi(\mathbf{r})]$ GL functional

$F(\epsilon, H)$ fluctuation part of the free energy in a magnetic field

$F_H(x) = \psi(x) + x\psi'(x) - 1 - \psi(1/2 + x)$ the function describing the behavior of fluctuation correction to the $2D$ Hall paraconductivity

$\Delta F(\phi) = E_J(1 - \cos\phi)$ GL energy of the Josephson junction, where $\phi = \chi_L - \chi_R$ is the relative phase between the two superconducting electrodes

$\mathbf{f} = -\partial E_p/d\mathbf{r}$ attractive force due to a single pinning center

$f(p_\phi, \phi)$ Fokker–Planck distribution function

$\Phi_0 = \pi/e$ elementary magnetic flux

Φ_{ext} magnetic flux applied to the loop

$\Phi(1, y, x)$ degenerated hypergeometric function

$\phi_{nk_z}(\mathbf{r})$ fluctuation Cooper pair wave function

$\varphi(\mathbf{r})$ scalar potential of the electric field.

$\varphi(r, t)$ phase of the order parameter

$G(V)$ differential tunnel conductance

$G_n(0)$ background value per unit area of the Ohmic conductance supposed to be bias independent

$\delta G(V) = G(V) - G_n(0)$

$G(\mathbf{p}, \varepsilon_n) = \frac{1}{i\varepsilon_n - \xi(\mathbf{p})}$ one-electron normal state Green's function

G_A Andreev subgap, normalized to $e^2/\hbar$

$G_\square$ dimensionless conductance

$G_T = \hbar/e^2$ dimensionless tunnel conductance

$Gi_{(D)}$ Ginzburg–Levanyuk number

$\widehat{G}_\Delta$ electron Green function of superconducting state written in the matrix Nambu representation

$-g$ negative constant of electron–electron attraction

$g = \frac{p_F l.}{2\pi}$ dimensionless value characterizing the "disorder strength"

g_c corresponds to the critical point of the Anderson (M-I) transition

g_L Landé g-factor

$\Gamma(x)$ gamma function

$\widehat{\Gamma}_\alpha = \widehat{\Gamma}_{\alpha\gamma}\widehat{q}_\gamma$ vertex part

Γ escape probability (activation rate) from a potential well

$\Gamma = 4e^2R^*T^*$ line width in the low quality factor line

$\widetilde{\Gamma} = 2e\sqrt{\frac{2T^*}{C}}$ analogue of Γ in the case of a high quality factor

$\Gamma^*(\epsilon) = \tau_{\mathrm{GL}}^{-1} + 4e^2RT_{\mathrm{c}}$

$\gamma_\varphi = \frac{2\eta}{v_F^2\tau\tau_\phi} \to \frac{\pi}{8T\tau_\phi}$ phase-breaking rate related to τ_ϕ

$\widehat{\gamma}^i_{(e)} = e\frac{\mathbf{p}_i}{m_e}$ electric current vertex

$\widehat{\gamma}^i_{(h)} = \frac{i(\varepsilon_n+\varepsilon_{n+\nu})}{2}\frac{\mathbf{p}_i}{m_e}$ heat current vertex

$\gamma_E = e^{C_{\mathrm{Euler}}} = 1.78..$ Euler constant,

γ gyromagnetic ratio

γ_a mass anisotropy

$\gamma_{\mathrm{GL}} = \pi\alpha/8$ dimensionless coefficient of the TDGL

$\mathcal{J}$ phenomenological constant proportional to the Josephson coupling between adjacent planes. The value of $\mathcal{J}$ is related to the coherence length along the z-direction: $\mathcal{J} = 2\alpha T_{\mathrm{c}}\xi_z^2/s^2$

J effective energy of the electron motion in the perpendicular direction in a layered superconductor

$\mathbf{J}^i_{(E)} = \int \mathcal{T}^i_0 d\mathbf{x}$ energy flow operator

$\widehat{\mathbf{J}}^i_{(h)} = \widehat{\mathbf{J}}^i_{(E)} - \frac{\mu}{e}\widehat{\mathbf{J}}^i_{\mathrm{tr}}$ heat flow operator

$\widehat{\mathbf{J}}^i_{\mathrm{tr}} = \frac{ie}{2m_e}\left(\psi\nabla^i\psi^+ - \psi^+\nabla^i\psi\right)$ electric current operator

$\mathbf{j}_s$ supercurrent

j_{c} critical current

$j_{GL} \sim en_s^{(2D)}/4m$ GL depairing current

$\mathbf{j}_{\mathrm{tr}} = \mathbf{j}^{(e)}_{\mathrm{tr}}$ electric transport current

$\mathbf{j}^{(Q)} = \mathbf{j}^{(h)}_{\mathrm{tr}} + \mathbf{j}^{(h)}_{\mathrm{magn}}$ current related to the heat flow,

$\mathbf{j}^{(h)}_{\mathrm{tr}} = \mathbf{j}^{\mathcal{E}} - (\mu/e^*)\,\mathbf{j}^{(e)}_{\mathrm{tr}}$ heat transport current

$\mathbf{j}^{\mathcal{E}}$ full energy flow

$\mathbf{j}^{(h)}_{\mathrm{magn}} = \mathbf{M} \times \mathbf{E}$ equilibrium magnetization current

$\mathcal{H}$ Hamiltonian

$\mathcal{H}_{BCS}$ BCS Hamiltonian

$\widehat{H}_i$ electron Hamiltonian in the potential of impurities

$\widehat{\mathcal{H}}_T = \sum_{\mathbf{p},\mathbf{k}}\left(T_{\mathbf{p},\mathbf{k}}\widehat{a}^+_{\mathbf{p}}\widehat{b}_{\mathbf{k}} + T^*_{\mathbf{k},\mathbf{p}}\widehat{a}_{\mathbf{p}}\widehat{b}^+_{\mathbf{k}}\right)$ tunneling Hamiltonian

$\mathbf{H}$ external magnetic field

$\widetilde{H}_{\mathrm{c2}}(0) = 2T_{\mathrm{c}}\left|dH_{\mathrm{c2}}/dT\right|_{Tc} = 2m\alpha T_{\mathrm{c}}/e = \Phi_0/2\pi\xi^2$ the definition of the second critical field as the linear extrapolation to zero temperature of the GL formula. The exact definition of $H_{\mathrm{c2}}(0)$ contains in comparison with (2.54) the numerical coefficient $A(0)$

$\overline{H_{\mathrm{c2}}(0)}$ value of second critical magnetic field averaged over the granular system

H_0 critical field destroying superconducting gap in a single grain

H_{c}^{spin} Clogston field

$H_{\mathrm{c}}^{\mathrm{or}}$ critical orbital magnetic field destroying the superconducting gap

$h = \frac{\omega_{\mathrm{c}}}{2\alpha T_{\mathrm{c}}} = \frac{eH}{2m\alpha T_{\mathrm{c}}} = \frac{H}{\widetilde{H}_{\mathrm{c2}}(0)}$ reduced magnetic field

$\widetilde{h} = \frac{H-H_{\mathrm{c2}}(T)}{H_{\mathrm{c2}}(T)}$ reduced magnetic field used at low temperatures

$h_\pm$ the components of the magnetic field at the nuclear site transverse with respect to the c axis both for NQR as well as for NMR

$\chi_P = e^2 v_F/4\pi^2$. Pauli paramagnetic susceptibility

$\chi = -\frac{e^2}{c^2}\frac{N_s^{(D)}}{m^*}\langle R^2\rangle$ Langevin diamagnetic susceptibility

$\chi_{\mu\nu}^{(D)}(T) - \frac{\partial^2 F_{(D)}}{\partial H_\mu \partial H_\nu}$ fluctuation susceptibility

$\chi_{+-}^{(R)}(\mathbf{k},\omega) = \chi_{+-}(\mathbf{k}, i\omega_\nu \to \omega + i0^+)$ dynamic susceptibility

$\chi_s = \chi_{+-}^{(R)}(\mathbf{k} \to 0, \omega = 0)$ spin susceptibility

$\Delta\chi(0)$ diamagnetic susceptibility jump between normal and superconducting phases

$\int \frac{d\Omega_{\mathbf{p}}}{4\pi} = \langle\rangle_{F.S.}$ means the averaging over the Fermi surface

$\int \mathfrak{D}\Psi(\mathbf{r})\cdots$ functional integration

$\int d^D\mathbf{k}/(2\pi)^D = \mu_D \int k^{D-1}dk$

$\int (d\mathbf{q}) = \int d^D q/(2\pi)^D$

$\int d^3q \equiv \int d^2\mathbf{q} \int_{-\pi/s}^{\pi/s} dq_z$ momentum space integral transformation for a layered superconductor

$I_{\alpha\beta}(q,\omega_\mu,\omega_\nu)$ integrated four Green functions block entering in the MT contribution

$I_T(V)$ tunneling current

$I_{qp}(V)$ quasiparticle tunneling current

I_c critical current of the junction

$I_c^{(w)}$ critical current of the junction in the case of a constriction in the form of a wire

$I_c^{(hb)}$ critical current of the junction in the case of a constriction in the form of single void hyperboloid of revolution

I_s applied superconducting current in the loop

$K(\omega_\nu)$ correlator of Green functions of different electrodes

$k_{\max}$ cut-off parameter of the GL theory, $k_{\max}^2/4m \sim \alpha T$

$\kappa^{\alpha\beta}(\mathbf{H})$ heat conductivity tensor

$\kappa, \tilde{\kappa}$ the functions of the impurity concentration entering in the DOS contribution in the case of an arbitrary impurity concentration

$\hat{\kappa}(\omega, T, \tau^{-1})$ function determining the frequency and temperature dependence of the pseudogap in a.c. conductivity

$\varkappa_2(T)$ generalization of the GL parameter $\varkappa$ for the entire range of temperatures below T_c

$\varkappa_2(0) \approx 1.2\varkappa_d$ in the clean case

$\varkappa_d = \frac{3c}{2\pi^2 e v_F l}\left[\frac{7\zeta(3)}{\pi\nu}\right]^{1/2}$ in the dirty case

$\varkappa_D = \pi\sqrt{\pi D/7\zeta(3)}$

$L_{x,y,z}$ the sample dimensions in appropriate directions

L_c correlation length

L sample size

L_i inductance

$L(\mathbf{q},\Omega_k)$ fluctuation propagator, i.e. the vertex part of the electron–electron interaction in the Cooper channel

$\widehat{L}$ TDGL operator

$L(\mathbf{p},\Omega)$ fundamental solution of the TDGL operator

$L_{\{m\}}(\Omega)$ eigenvalue of the TDGL operator

$L_l(g) = \frac{1}{p_F}(g - g_c)^{-\frac{\varkappa}{z-2}}$ localization length, characterizing the spatial scale near the transition

$\mathcal{L}(p,p',q)$ two-particle Green function

$\mathcal{L}$ Lagrangian of the system of interacting electrons with impurities

l electron mean free path

$l_T = \sqrt{\frac{D}{T}}$ diffusion length

Λ cut-off parameter of the renormalization group theory

λ_J Josephson penetration length

$\lambda_F = 2\pi/p_F$ electron wavelength

$\lambda(\mathbf{q}, \varepsilon_1, \varepsilon_2)$ three leg impurity vertex part in the particle–particle channel

$\lambda_0^g(\varepsilon_n, -\varepsilon_n)$ impurity vertex in the superconducting grain placed in magnetic field

λ penetration depth of magnetic field in superconductor

$M^{(1)}$, $M^{(2)}$ fluctuation corrections to the magnetization in the one- and two-loop approximations

$M(\epsilon, H)$ fluctuation magnetization

$M = (4\mathcal{J}s^2)^{-1}$ effective mass along the z-direction of the LD functional

$M_C = C/4e^2$ mass of the Josephson junction mechanical analogy

$m^* = 2m = \left(2\alpha T_c \xi^2\right)^{-1}$ Cooper pair mass

μ chemical potential

μ dynamical critical exponent of the theory of collective creep

$\mu_D = \frac{D}{2^D \pi^{D/2} \Gamma(1+D/2)}$

μ_B Bohr magneton

$\mathcal{N}$ total number of layers

N_e electron concentration

N_s concentration of superconducting electrons in the London theory

$n_s = N_s/2$ concentration of Cooper pairs, superfluid density

$n_f = \xi_+^{-2}(T)$ density of free vortices at a temperature slightly above T_{BKT}

$n_{s2}(T_{\mathrm{BKT}}) = \frac{4mT_{\mathrm{BKT}}}{\pi}$ universal jump condition at the BKT transition point

$\mathbf{n}$ quantum number related to the degenerate Landau state

$n_c \sim T/\omega_c \sim 1/h$ cut-off parameter, the number of the last Landau level at which the summation is interrupted.

n_i impurity concentration

$n_{\mathbf{p}}(t)$ fluctuation Cooper pair distribution function

$n_{\mathbf{p}}^{(0)} = < |\Psi_{\mathbf{p}}|^2 > = T/\varepsilon_{\mathbf{p}}$

ν critical exponents of the RG theory

ν one electron DOS for one spin

$\delta\nu(E, \epsilon) = -\frac{1}{\pi} \operatorname{Im} \int \frac{d^D \mathbf{p}}{(2\pi)^D} \delta G^R(\mathbf{p}, E)$ electron DOS

$\widetilde{P}(\mathbf{r}) = \widetilde{\psi}_{\downarrow}(\mathbf{r}) \widetilde{\psi}_{\uparrow}(\mathbf{r})$ operator of the Cooper pair annihilation in the Heisenberg representation

$\widetilde{P}_\rho(\mathbf{r})$ operator of the Cooper pair annihilation for the case of ρ-symmetry in the Heisenberg representation.

$\mathcal{P}(\mathbf{q}, \varepsilon_1, \varepsilon_2)$ correlator of two one-electron Green functions

$\Pi(\mathbf{q}, \Omega_k) = \sum_{\varepsilon_n} \mathcal{P}(\mathbf{q}, \varepsilon_n + \Omega_k, -\varepsilon_n)$ polarization operator

$\Pi^{\alpha\beta} = \gamma^{\alpha\lambda} \left(\sigma^{-1}\right)^{\lambda\beta}$ Peltier coefficient

$p_0 \sim \hbar/\xi(T)$ maximum value of the fluctuation Cooper pair center of mass momentum

$Q_{\alpha\beta}(\omega_\nu)$ electromagnetic response operator

$Q = \Gamma R^* C$ "quality factor" of the Josephson junction in RCSJ model. It is identical with $\beta_c^{1/2}$, where β_c Stewart and McCumber damping parameter

$q_\alpha - 2ee_{\alpha\beta\gamma}\widehat{x}_\beta H_\gamma$ gauge invariant momentum

R_N junction resistance in the normal state, and

R_n Ohmic resistance for unit area

R_T equivalent resistance of the Josephson junction in RCSJ model

R_{ext} resistance of the external part of the circuit in the RCSJ model

$R_\square$ average resistance per square of the film

r_p recombination probability

$\widehat{\mathbf{r}}^\beta_{\{li\}} = i\frac{\widehat{\mathbf{v}}^\beta_{\{li\}}}{\varepsilon_{\{i\}} - \varepsilon_{\{l\}}}$ matrix element of the coordinate operator

$r = \frac{2\mathcal{J}}{\alpha T} = \frac{4\xi_z^2(0)}{s^2}$ Lawrence-Doniach anisotropy parameter

$\rho_{xy}(T)$ and $\rho_z(T)$ the components of the resistivity tensor in xy-plane and along z-axis

$\rho_f = E/j_{\text{tr}} = \frac{B\Phi_0}{\eta_v c^2}$ flux flow resistivity

$\rho^{\alpha\beta}$ resistivity tensor

$\mathcal{S}_{\text{eff}}[\Delta, \Delta^*]$ effective action functional

$\mathcal{S}[\Delta(\mathbf{r}, \tau)] = \mathcal{S}_0 + \mathcal{S}_\Delta$ action, $\mathcal{S}_\Delta$ fluctuation part of the effective action functional

$S^{\alpha\beta} = -\beta^{\alpha\lambda}\left(\sigma^{-1}\right)^{\lambda\beta}$ differential thermopower (so-called Seebeck coefficient)

$S(j)$ under-barrier action

s inter-layer spacing

σ^{AL} AL contribution

σ^{DOS} fluctuation DOS correction to conductivity

$\sigma^{MT} = \sigma^{\text{MT(reg)}} + \sigma^{\text{MT(an)}}$ MT contribution

$\sigma_n = \frac{N_s^{(D)} e^2 \tau}{m^*}$ Drude conductivity

$\sigma_n^{\alpha\beta}$ Drude conductivity tensor

$\sigma^{\alpha\beta}(\epsilon, H, \omega)$ paraconductivity tensor

σ_{wl} WL correction to conductivity

σ_s conductivity of the spinon subsystem

σ_h conductivity of the holon subsystem

$\mathcal{T}_\mu^\nu$ energy–momentum tensor

$T_0 = \frac{2\gamma_E E_F}{\pi}\exp\left(\frac{1}{\nu|g|}\right)$

T_{c} critical temperature

T_{c0} mean field critical temperature

T_{BKT} critical temperature of the Berezinskii–Kosterlitz–Thouless transition

$T_{\mathbf{p},\mathbf{k}}$ tunneling matrix element

T_τ time ordering operator

$T_{\alpha\beta}$ operator of the direct scaling transformation

T^* effective temperature of the Josephson junction in the Resistively and Capacitively Shunted Junction model

T_m melting temperature

$1/T_1$ NMR relaxation rate

∇T temperature gradient

δT_{c} fluctuation shift of the critical temperature

$t_\xi^{-1} = \mathbf{D}\xi^{-2} \sim \tau_{\text{GL}}^{-1} \sim T - T_c$ inverse of the time necessary for the electron to diffuse over a distance equal to the coherence length $\xi(T)$

$t_\xi^{-1} \sim v_F\xi^{-1} \sim \sqrt{T_c(T - T_c)}$ the same value for the ballistic motion

t transfer integral between the nearest-neighbor sites in the theory of d-pairing

t tunneling amplitude of the electron hopping between two neighboring cites in crystal

t_0 characteristic time of the collective pinning theory

$t = T/T_c$ dimensionless temperature

$\tau_{\mathbf{p}} = \gamma_{GL}/\varepsilon_{\mathbf{p}}$ momentum dependent fluctuation Cooper pair lifetime

$\tau_i (i = x, y, z)$ the Pauli matrices in the Gor'kov-Nambu space

$\tau_{GL} = \pi\hbar/8k_B(T - T_c)$ characteristic time of the TDGL theory; plays the role of a fluctuation Cooper pair lifetime in the vicinity of T_c

$\tau_{bal} \sim \hbar/k_B T$ time of electron ballistic motion

τ_φ phase-breaking time

τ_ε energy relaxation time

τ electron scattering time

$\theta(x)$ Heavyside step function

θ angle between the magnetic field and the perpendicular to the layers plane

$\vartheta_D = \frac{\Gamma(2-D/2)}{2^D \pi^{D/2}}$: $\vartheta_1 = 1/4, \vartheta_2 = 1/4\pi$ $\vartheta_3 = 1/8\pi$

$U(\phi) = -(\frac{I}{2e}\phi + E_J \cos\phi)$ "washboard" potential of the Josephson junction mechanical analogy

U_c activation energy

$U_i(\mathbf{r})$ impurity potential: $\langle U(r)\rangle = 0$, $\langle U(r)U(r')\rangle = \langle U^2\rangle \delta(r - r')$

u displacement of vortices caused by a force f

$\sqrt{\langle u_T^2\rangle}$ average thermal displacement of a vortex $\overline{V}$ voltage at the barrier averaged over time.

V specimen volume

$V(\mathbf{r})$ potential of a single impurity

$V_{ee}(\mathbf{p}, \mathbf{p}_1, \mathbf{q})$ most general form of the inter-electron interaction

$\widehat{\mathbf{v}}$ velocity operator

$v_\alpha(p) = \frac{\partial\xi(p)}{\partial p_\alpha}$ quasiparticle velocity

v_F Fermi velocity in ab-plane

$v_z(p) = \frac{\partial\xi(p)}{\partial p_z} = -Js\sin(p_z s)$ electron velocity along the c-axis direction

v_v velocity of the vortex motion Z_Δ fluctuation contribution to the partition function

$\xi(\mathbf{p}) = E(\mathbf{p}) - E_F$ quasiparticle spectrum of the normal metal

ξ_l electron energy in the basis of the eigenfunctions of the Hamiltonian with the exact impurity potential

$\xi(T) = \frac{\xi}{\sqrt{\epsilon}}$ fluctuation Cooper pair size, its coherence length

ξ_{xy} in-plane BCS coherence length of a layered superconductor

ξ_z out-of-plane BCS coherence length of a layered superconductor

$\xi^2_{(D)}(T\tau) = \frac{\pi}{8T}\widehat{\mathcal{D}} = (4m\alpha T)^{-1} = \eta_{(D)}$ coefficient of the gradient term of D-dimensional GL theory

ξ_0 zero temperature coherence length

$\widehat{\xi} = -\frac{\nabla^2}{2m} + U_i(\mathbf{r}) - \mu$

$\xi_+(T) = \xi(\epsilon)\exp\left(\widetilde{b}\sqrt{\frac{Gi_{(2)}T_{\mathrm{BKT}}}{T - T_{\mathrm{BKT}}}}\right)$ vortex–antivortex pair coherence length

$Z_{(D)}$ partition function of the D dimensional system

$Z(\omega)$ impedance of the Josephson junction equivalent circuit

$z > 2$ dynamical critical exponent

ζ_{int} and ζ_{ext} the Langevin currents modelling the noises in Josephson junction and in the external part of the circuit

$\zeta(x)$ Riemann zeta function

$\zeta(\mathbf{r}, t)$ Langevin force

$\eta = \left(4e^2R\right)^{-1}$ viscosity of the Josephson junction mechanical analogy

η_v viscosity coefficient for the vortex motion

Ψ_l order parameter of the l-th superconducting layer

Ψ_{n,k_z} the Fourier coefficients of order parameter $\Psi_l(\mathbf{r})$ on the basis of $\phi_{nk_z}(\mathbf{r})$ eigenfunctions:

$\Psi(\mathbf{r}) = \widetilde{\Psi} + \psi(\mathbf{r}) = \widetilde{\Psi} + \psi_r + i\psi_i$. fluctuation order parameter, which can be treated as "Cooper pair wave function"

$\Psi_{\mathbf{k}} = \frac{1}{\sqrt{V}} \int \Psi(\mathbf{r}) e^{-i\mathbf{kr}} dV$

$\widetilde{\Psi}$ equilibrium value of the order parameter. When α is defined according to (A.12) $|\widetilde{\Psi}|^2$ coincides with the superfluid density n_s

$\Psi(r,\theta)$ coordinate dependent order parameter of the isolated vortex ($\xi \ll a$)

$\psi_{|k|>\Lambda}$ "fast" fluctuation mode of the RG theory

$\psi_{|k|<\Lambda}$ "slow" fluctuation mode of the RG theory

$\psi(x)$ and $\psi^{(n)}(x)$ the digamma function and its derivatives respectively

$\widetilde{\psi}^{+}_{\mathbf{p},\sigma}$ and $\widetilde{\psi}_{\mathbf{p},\sigma}$ the creation and annihilation field operators in the Heisenberg representation

$\psi_{np_y}(\mathbf{r})$ eigenfunction for an electron in a magnetic field in the Landau gauge

$\psi(x,t)$ electron field operator

$\psi_{,i} = \partial\psi/\partial x_i$ $(i = 1,2,3)$

$\Omega_{k\pm n} = \Omega_k \pm \varepsilon_n$, sum (difference) of Matsubara frequencies

$\omega_J = 2e\overline{V}$ frequency of emission of electromagnetic radiation by the Josephson junction

$\omega_{\mathrm{p}} = \sqrt{\frac{4\pi e^2 N}{m}}$ plasma frequency

$\omega_{mk} = \xi_k - \xi_m$

$\tilde{\omega} = \frac{\pi\omega}{16(T-T_{\mathrm{c}})}$ dimensionless electromagnetic field frequency

ω_{c} cyclotron frequency

ω_D Debye frequency

INDEX